FORTSCHRITTE IM KONSTRUKTIVEN INGENIEURBAU

o. Professor Dr.-Ing. Dr.-Ing. E. h. Gallus Rehm
am 18. Oktober 1984 60 Jahre

GALLUS REHM ZUM 60. GEBURTSTAG

FORTSCHRITTE IM KONSTRUKTIVEN INGENIEURBAU

Mit Beiträgen von

A.W. Beeby D. Briesemann A.S.G. Bruggeling J. Defourny, A. Bragard J. Eibl, J. Kobarg R. Eligehausen W. Fastenau A. Fischer L. Franke F. Herkommer H.K. Hilsdorf, J. Kropp, M. Günter D. Jungwirth H.P. Killing G. König, M. Krips K. Kordina, U. Quast H. Kupfer, R. Mang H. Martin C. Menn W. Menz, J. Schlaich U. Nürnberger, B. Neubert H. Paschen, H. Malonn H.W. Reinhardt F.S. Rostásy D. Rußwurm H.D. Seghezzi M. Stocker B. Thürlimann, H.R. Ganz O. Wagner N.V. Waubke M. Weiser, L. Preis H. Weitzmann, W. Dening G. Wischers

Herausgegeben von

Rolf Eligehausen und Dieter Rußwurm

1984

Verlag für Architektur und technische Wissenschaften Berlin

CIP-Kurztitelaufnahme der Deutschen Bibliothek

Fortschritte im konstruktiven Ingenieurbau:
Gallus Rehm zum 60. Geburtstag gewidmet /
mit Beiträgen von A. W. Beeby ...
Hrsg. von Rolf Eligehausen und Dieter Rußwurm. –
Berlin: Ernst, Verlag für Architektur u. techn. Wiss., 1984.
ISBN 3-433-01010-2
NE: Beeby, A. W. [Mitverf.]; Eligehausen, Rolf [Hrsg.];
Rehm, Gallus: Festschrift

Gesamtherstellung: Passavia Druckerei GmbH, D-8390 Passau 2

Printed in the Federal Republic of Germany

Zum Geleit

Professor Dr.-Ing. Gallus Rehm vollendet am 18. Oktober 1984 sein 60. Lebensjahr. Ein Anlaß für Rückblick und Besinnung:
Gallus Rehm legte trotz einer fünf Jahre währenden kriegsbedingten Unterbrechung seiner Ausbildung mit 27 Jahren die Diplom-Hauptprüfung ab und begann 1951 seine wissenschaftliche Laufbahn beim Materialprüfungsamt München als Mitarbeiter bei Professor Dr.-Ing. Dr.-Ing. E. h. Rüsch. Seither, in mehr als drei Jahrzehnten hingebungsvoller Lehr- und Forschungstätigkeit an den Technischen Universitäten von München, Braunschweig und Stuttgart, hat er sich große Verdienste um die Entwicklung der Massivbauweise erworben.
Seine richtungsweisenden Arbeiten über die Grundlagen der Verbund- und Rißgesetze, über Korrosion und Korrosionsschutz, über Ermüdungsfragen, über Schweißbarkeit von Betonstählen, über die Anwendung von Kunstharzen und Kunststoffen im Betonbau, zur Befestigungstechnik, und nicht zuletzt über Fragen der Bewehrungsführung im Stahlbetonbau haben höchste Anerkennung im In- und Ausland gefunden. Wichtige Entwicklungen zur Verbesserung der Eigenschaften unserer Betonstähle wären ohne seine eigenen Leistungen und Initiativen kaum zustande gekommen. In zahlreichen Veröffentlichungen hat Rehm die Ergebnisse seiner Arbeit der Fachöffentlichkeit vorgestellt.
Ich hatte in den vergangenen Jahren oftmals Gelegenheit, mit Gallus Rehm in nationalen und internationalen Gremien und Fachausschüssen zusammenarbeiten zu können; es war stets ein Vergnügen, zu beobachten, in wie überlegener Weise er seine Auffassung vorzustellen und die Entscheidung durch überzeugende Argumentation zu beeinflussen vermochte. FIP, RILEM, CEB und der Deutsche Ausschuß für Stahlbeton verdanken ihm viele wertvolle Beiträge; wichtige Abschnitte in DIN 1045 tragen seine Handschrift. Gallus Rehm genießt heute auf seinen Arbeitsgebieten als die maßgebende Fachautorität im nationalen und internationalen Raum hohe Anerkennung.
Mit unermüdlichem persönlichen Einsatz und glänzendem Organisationstalent hat sich Gallus Rehm die seinem Engagement und seiner Tatkraft entsprechenden Voraussetzungen geschaffen, um neben seiner Lehr- und Forschungstätigkeit auch engen Kontakt mit Aufgaben der Praxis halten zu können.
Die Zugehörigkeit zum gleichen Institut der TU München führte uns seinerzeit zusammen, wenngleich wir mit sehr unterschiedlichen Aufgaben betraut waren. Gallus Rehm war mit der Durchführung von Versuchen zur Klärung der Verbund- und Rißgesetze beauftragt: Mit Erstaunen und Respekt beobachteten wir, mit welchem Geschick und durchgreifendem Organisationsvermögen er Versuchsdurchführung und -auswertung systematisch und termingerecht abzuwickeln verstand und so schon nach wenigen Jahren die am besten gegliederte und effizienteste Abteilung aufzubauen vermochte. Hierbei unterstützte ihn sein bewundernswertes Feingefühl und seine untrügliche Sicherheit bei der Auswahl von Mitarbeitern, die bereit waren, ihn bei der Verfolgung seiner Ziele zu unterstützen.
Es fügte sich so glücklich, daß Gallus Rehm als Ordinarius für Baustoffkunde und Stahlbetonbau für die Technische Universität Braunschweig gewonnen werden konnte und gemeinsam mit mir die Leitung des damals neu geschaffenen Gemeinschaftsinstituts übernahm. Diese enge und von gegenseitigem Verständnis für die Bestrebungen des anderen getragene Zusammenarbeit hat viele wertvolle fachliche Anregungen gebracht, dem Institut neue Arbeitsgebiete erschlossen und nicht zuletzt unsere schon in München entstandene Freundschaft gefestigt. Ich denke gerne an diese fruchtbaren Jahre des gemeinsamen Aufbaues zurück.
Leider schon nach fünf Jahren folgte Gallus Rehm einem ehrenvollen Ruf an die Universität Stuttgart; eine Reihe seiner Mitarbeiter begleiteten ihn, um diesen glänzenden Mentor nicht zu verlieren – ein Beweis

aber auch dafür, daß Gallus Rehm die Zuneigung und innere Verbundenheit zu seinen Mitarbeitern gewinnen und erhalten konnte. So ist es nur allzu verständlich, daß seine Freunde und Schüler ihm zu seinem Geburtstag mit dieser Festschrift Freude bereiten und Dank bezeugen wollen, indem aus diesem Anlaß der Fachöffentlichkeit neue Erkenntnisse zugänglich gemacht werden. Ich wünsche dem Buch eine freundliche Aufnahme in der Fachwelt und verbinde damit meine herzlichsten Glückwünsche an meinen Freund Gallus Rehm.

KARL KORDINA

Gallus Rehm 60 Jahre

In der mehr als 100 Jahre zurückreichenden Entwicklung des Stahlbetonbaues zeichnet sich die letzte, von 1950 bis in die heutige Zeit reichende Phase durch einen steilen Anstieg aus. Abgestützt auf Ideenreichtum hervorragender Ingenieure bei der Entwicklung neuer Techniken sowie auf Arbeiten von Forschern und Hochschullehrern zur Absicherung wissenschaftlicher Grundlagen für die Bemessung und Ausführung von Stahlbetonbauten, konnte den ständig gesteigerten Anforderungen hinsichtlich Gestaltung, Nutzung und Wirtschaftlichkeit entsprochen werden; gleichzeitig wurden aber auch völlig neue Anwendungsgebiete erschlossen. Doch sind stürmische Entwicklungen oft von Euphorie getragen, die neue Technologien für die baupraktische Anwendung nicht immer hinreichend ausreifen läßt; so blieben auch Rückschläge nicht aus, die zu einer kritischen Überprüfung veranlaßten.

In den wissenschaftlichen Arbeiten von Gallus Rehm schlagen sich diese, den Fortschritt im Stahlbetonbau übergeordnet bestimmenden Parameter deutlich nieder. So sind seine Arbeiten ebenso geprägt durch die Entwicklung neuer Techniken, die er wissenschaftlich betreute, wie auch durch seine Bemühungen, Gewonnenes zu konsolidieren, um Schäden vorzubeugen. Ein paar Beispiele aus seinem weitreichenden Tätigkeitsfeld mögen dies belegen:

Sein konsequentes Engagement für den *Betonstahl* beeinflußte nachhaltig die Entwicklung dieses heute weitgehend ausgereiften Grundbaustoffes des Stahlbetons. Die seinerzeit noch gemeinsam mit Hubert Rüsch durchgeführten Untersuchungen zur Ermittlung der für den Verbund Beton – Stahl maßgeblichen Gesetzmäßigkeiten und der Einflüsse der Oberflächengestaltung gerippter Betonstähle auf deren Dauerschwingfestigkeit waren wesentliche Grundlage für die diesen Entwicklungsarbeiten unmittelbar folgende gesteigerte Ausnutzbarkeit von Stahlbetonquerschnitten, namentlich auch für den Bereich dynamischer Beanspruchungen.

Seine Bemühungen, die *Bewehrungstechnik* durch Einführung industrieller Fertigungsmethoden weitgehend zu rationalisieren, sind ein weiteres Beispiel für Forschungsarbeiten, die darauf abzielten, die Bauart zu verbessern und gleichzeitig ausführungsfreundliche, wirtschaftliche Lösungen anzubieten. So wurden in umfangreichen zielstrebig durchgeführten Untersuchungen die für zweckmäßig erachteten Maßnahmen wissenschaftlich abgesichert und durch vergleichende Betrachtungen zur Wirtschaftlichkeit bei Bauausführungen erprobt; die gewonnenen Erkenntnisse wurden schließlich durch Neufassung des Abschnittes 18 von DIN 1045, die unter Leitung von Rehm erarbeitet wurde, für die allgemeine baupraktische Nutzanwendung aufbereitet.

Rehms Arbeiten für die moderne *Dübel-Befestigungstechnik* liefern gleichartige Beispiele.

Wirksame *Qualitätssicherung* und -steuerung sind wesentliche Grundlage für eine risikoarme Ausnutzung der sich mit neuen Technologien eröffnenden vielfältigen Möglichkeiten, insbesondere dann, wenn an die Sorgfalt der Ausführung hohe Ansprüche zu stellen sind. Nachhaltig widmete Rehm sich auch diesem Bereich; so gab er den Arbeiten für eine nach neuzeitlichen Erkenntnissen aufgebaute Güteüberwachung von Betonstahl – entsprechend DIN 488 Blatt 6 – wesentliche Impulse. Im Jahre 1974 als Vornorm – zur Sammlung von Erfahrungen mit der Anwendung – herausgegeben, erwies sich diese von ihm maßgeblich mitgestaltete Norm mit ihren damals sehr fortschrittlichen statistischen Betrachtungsweisen als so vorausschauend, daß die zur Überwachung und Prüfung getroffenen Festlegungen auch nach 10 Jahren Anwendung den heute gestellten Anforderungen durchaus gerecht werden.

Bedingt durch veränderte Umwelteinflüsse trat mehr und mehr das Problem der *Dauerhaftigkeit* des Stahlbeton- und des Spannbetonbaues in den Vordergrund. Schadensanalysen lieferten einen Anhalt für Maßnahmen zu ihrer Verbesserung, wobei Rehm dem Korrosionsschutz der Bewehrung seine besondere Aufmerksamkeit widmete. Durch grundlegende Arbeiten über die verschiedenen Korrosionsarten gewann er Erkenntnisse, die die Voraussetzungen für eine Bewältigung dieses Problemkreises lieferten, der den Stahlbeton- und Spannbetonbau stark belastete.

Auch in den Beiträgen zu dieser Festschrift, die Freunde und Fachkollegen für Gallus Rehm zu seinem 60. Geburtstag erarbeitet haben, spiegeln sich recht deutlich Geist und Wirken des hiermit Gewürdigten wider. Möge sie in der Fachwelt eine gute Aufnahme finden.

HANNO GOFFIN

Lebenslauf

18.10.1924	geb. in Nyomja/Ungarn
1930 – 1942	Schulbesuch
1942 – 1945	Wehrdienst
1945 – 1946	Kriegsgefangenschaft
1946 – 1947	Aufbaudienst TH München
1947 – 1951	Studium an der TH München, Fakultät für Bauwesen
1951	Abschlußprüfung Dipl.-Ingenieur an der TH München
1951–1968	Wissenschaftlicher Mitarbeiter an der MPA der TH München
1958	Promotion zum Dr.-Ing.
1965	Gründung einer anerkannten Prüfanstalt für Betonstähle
1.2.1968	Ordinarius für Baustoffkunde und Stahlbetonbau an der TU Braunschweig
7.9.1973 – jetzt	o. Professor für Werkstoffe im Bauwesen, Universität Stuttgart Direktor der Forschungs- und Materialprüfungsanstalt Baden-Württemberg (FMPA) – Otto-Graf-Institut –
1979	Gründung eines Ingenieurbüros
13.10.1984	Verleihung der Würde eines Doktoringenieurs E.h. durch die TU Braunschweig

Zusammenstellung von Veröffentlichungen

„The fundamental law of bond“ Abschlußbericht, Band II, RILEM-Symposium „Bond and Crack Formation in Reinforced Concrete“, Stockholm, 1957

„Stress distribution in reinforcing bars embedded in concrete“, Abschlußbericht, Band II, RILEM-Symposium „Bond and Crack Formation in Reinforced Concrete“, Stockholm, 1957

„Notes on relation between crack spacing and crack width in members subjected to bending.“ Abschlußbericht, Band II, RILEM-Symposium „Bond and Crack Formation in Reinforced Concrete“, Stockholm, 1957 – mit H. Rüsch

„Notes on crack spacing in members subjected to bending“, Abschlußbericht, Band II, RILEM-Symposium „Bond and Crack Formation in Reinforced Concrete“, Stockholm, 1957 – mit H. Rüsch

„Zur Frage der Korrosion von Stahl in Beton“, Zement–Kalk–Gips, Heft 5, 1960 – mit H. Moll

„Beitrag zur Frage der Ermüdungsfestigkeit von Bewehrungsstählen“, Vorbericht zum VI. Kongreß des IVBH, Stockholm, 1960

„Über die Grundlagen des Verbundes zwischen Stahl und Beton“, Schriftenreihe des Deutschen Ausschusses für Stahlbeton, Heft 138, Berlin, 1961

„Versuche mit Betonformstählen“, Schriftenreihe des Deutschen Ausschusses für Stahlbeton, Heft 140, Berlin, 1963 – mit H. Rüsch

„Versuche mit Betonformstählen“, Schriftenreihe des Deutschen Ausschusses für Stahlbeton, Heft 160, Berlin, 1963 – mit H. Rüsch

„Versuche zur Bestimmung der Übertragungslänge von Spannstählen“, Schriftenreihe des Deutschen Ausschusses für Stahlbeton, Heft 137, Berlin, 1963 – mit H. Rüsch

„Schäden an Spannbetonbauteilen, die mit Tonerdeschmelzzement hergestellt wurden.“ Beton-Zeitung, Heft 12, 1963

„Versuche mit Betonformstählen“, Schriftenreihe des Deutschen Ausschusses f. Stahlbeton, H. 165, Berlin, 1964 – mit H. Rüsch

„Versuche zum Studium des Einflusses der Rißbreite auf die Rostbildung an der Bewehrung von Stahlbetonbauteilen“, Schriftenreihe des Deutschen Ausschusses für Stahlbeton, Heft 169, Berlin, 1964 – mit H. Moll

„Beobachtungen an alten Stahlbetonbauteilen hinsichtlich Karbonatisierung des Betons und Rostbildung an der Bewehrung“, Schriftenreihe des Deutschen Ausschusses für Stahlbeton, Heft 170, Berlin, 1965 – mit H. Moll

„Korrosion von Stahl in Beton“, im Jahrbuch des Bauwesens 1966, Deutsche Verlagsanstalt Stuttgart – mit H. Rauen

„Betonrippenstähle“, beton, Heft 7, 1966

„Die Prüfung und das Verhalten von Spannbetonbauteilen, die durch Anbringen (Kleben, Verdübeln) von Spannbeton-, Stahl- und Glasfaserelementen verstärkt wurden“, Materialprüfung, Heft 11, 1967 – mit D. Briesemann

„Untersuchungen über den Verlauf der Karbonatisierungstiefe im Betonfuß von Fertigteilträgern“, beton, Heft 4, 1968 – mit D. Briesemann

„Korrosion von Stahl in Beton – Grundlagen, Schadensarten und Hinweise zur Vermeidung von Mängeln“, Betonstein-Zeitung Heft 5, 1968 – mit H. Rauen

„Statistische Methoden bei der Qualitätskontrolle von Bewehrungsstählen“, Betonstein-Zeitung, Hefte 5 + 6, 1968 – mit H. Rehm

„Möglichkeiten zur Wiederherstellung der Tragfähigkeit beschädigter Fertigteilträgerdecken“, beton, Heft 7, 1968, – mit D. Briesemann

„Zur Frage der Rißbegrenzung im Stahlbetonbau“, Beton- und Stahlbetonbau, Heft 8, 1968 – mit H. Martin

„Über die Anwendung des Schweißens im Stahlbetonbau, Untersuchungen über den derzeitigen Stand des Einsatzes des Schweißens im Stahlbetonbau, Beschreibung der Problematik des Schweißens und der damit zusammenhängenden Fragen“, Betonstein-Zeitung, Hefte 11 + 12, 1968 – mit D. Russwurm

„Baustoffprüfungen, Prüfregelungen zur Gütesicherung, Grundlagen und Zusammenhänge statistischer Methoden“, Werners Kleiner Baukalender 1969, Werner-Verlag GmbH, Düsseldorf

„Korrosionsschutz von Stahl in Beton“, beton, Heft 4, 1969

„Kriterien zur Beurteilung von Bewehrungsstählen mit hochwertigem Verbund“, in Berichte aus Forschung und Praxis (Festschrift Rüsch), Verlag W. Ernst und Sohn, München/Berlin, 1969

„Zur Frage der Prüfregelung bei der Qualitätssicherung von Bewehrungsstählen“, Betonstein-Zeitung, Heft 3, 1969 – mit H. Rehm

„Möglichkeiten der konstruktiven Gestaltung von Stößen in vorgefertigten Stahlbetonbauteilen“, Betonstein-Zeitung, Heft 2, 1970

„Untersuchungen über Reaktionen des Zinks unter Einwirkung von Alkalien im Hinblick auf das Verhalten verzinkter Stähle im Beton“, Betonstein-Zeitung, Heft 6, 1970 – mit A. Lämmke

„Biegefeste Verbindung von Stahlbetonfertigteilen“, Betonstein-Zeitung, Heft 7, 1970 – mit H. Martin

„Die Verwendung vorgefertigter, geschweißter Bewehrungskörbe“, Festschrift zur Eröffnung des Stachusbauwerkes München, November 1970

„Untersuchungen zur Frage der Versprödungs- und Spannungsrißkorrosion an Spannstählen bei hohen ph-Werten“, Mitteilungen des Inst. für Bautechnik, Berlin, H. 2, 1971 – mit G. Rieche

„Untersuchungen über das Verhalten kohlenstoffarmer Betonstähle gegenüber Spannungsrißkorrosion“, Stahl und Eisen 91, Heft 12, 1971 – mit U. Nürnberger

„Zur Spannungsrißkorrosion bei kohlenstoffarmen Betonstählen, Stahl und Eisen 91, Heft 12, 1971 – mit G. Rieche

„Spannungskorrosion hochfester Spannstähle“, Cement XXIV, Heft 3, 1972 – mit U. Nürnberger

„Rationalisierung der Bewehrung im Stahlbetonbau“, Betonwerk + Fertigteil-Technik, Heft 5, 1972 – mit R. Eligehausen

„Untersuchungen an Betonstählen aus Viehstalldecken im Hinblick auf Spannungsrißkorrosionserscheinungen“, Betonwerk + Fertigteil-Technik, Heft 9, 1973 – mit U. Nürnberger und N.V. Waubke

„GFK-Stäbe als Bewehrung", Betonwerk + Fertigteil-Technik, Heft 9, 1973

„Faserbewehrter Beton – Welche Probleme ergeben sich?", Betonwerk + Fertigteil-Technik, Heft 9, 1973

„Dauerschwingverhalten von widerstandspunktgeschweißten Baustahlmatten aus kaltgewalztem Betonrippenstahl", Schweißen + Schneiden, 5. Jahrgang, 1974 – mit U. Nürnberger

„Kunstharzgebundene Glasfaserstäbe als Bewehrung im Betonbau", Die Bautechnik, Heft 4, 1974 – mit L. Franke

„Untersuchungen über das Tragverhalten von Übergreifungsstößen geschweißter Betonstahlmatten aus Betonrippenstahl unter Schwellast", Betonwerk + Fertigteil-Technik, Heft 10, 1974 – mit R. Eligehausen

„Anmerkungen zur Qualitätskontrolle von Betonstahl nach DIN 448", Materialprüfung, Heft 1, 1975 – mit D. Rußwurm

„Rationalisierung der Zeichenarbeiten im Stahlbetonbau", Betonwerk + Fertigteil-Technik, Hefte 5 + 6, 1975 – mit R. Eligehausen und R. Mallée

„Untersuchungen über die Eigenschaften verzinkter hochfester Spannstähle", Schriftenreihe des Deutschen Ausschusses für Stahlbeton, Heft 242, Berlin 1974 – mit U. Nürnberger und G. Rieche

„Korrosionsverhalten verzinkter Stähle in Zementmörtel und Beton", Schriftenreihe des Deutschen Ausschusses für Stahlbeton, Heft 242, Berlin, 1974 – mit A. Lämmke

„Ultraschallimpulstechnik für Fertigteile", Schriftenreihe des Deutschen Ausschusses für Stahlbeton, Heft 243, Berlin, 1975 – mit J. Neisecke und N. V. Waubke

„Übergreifungsstöße geschweißter Betonstahlmatten", Beton- und Stahlbetonbau", Heft 4, 1976 – mit R. Eligehausen und R. Tewes

„Zur Frage der Dauerhaltbarkeit von Spannstählen", Betonwerk + Fertigteil-Technik, Heft 9, 1976 – mit U. Nürnberger

„Anmerkungen zur Güte von Betonstählen", Betonwerk + Fertigteil-Technik, Hefte 1 + 2, 1977 – mit D. Russwurm

„Korrosionsprobleme im Stahlbetonbau", Betonwerk + Fertigteil-Technik, Heft 2, 1977

„Verhalten von kunstharzgebundenen Glasfaserstäben bei unterschiedlichen Beanspruchungszuständen", Die Bautechnik, Heft 4, 1977 – mit L. Franke

„Metallgußverankerungen für Zugglieder aus hochfesten Drähten", Draht 28, Heft 4, 1977 – mit U. Nürnberger und M. Patzak

„Einfluß einer nichtruhenden Belastung auf das Verbundverhalten von Rippenstählen", Betonwerk + Fertigteil-Technik, Heft 6, 1977 – mit R. Eligehausen

„Beurteilung von Betonstählen hergestellt nach dem Tempcore-Verfahren", Betonwerk + Fertigteil-Technik, Heft 6, 1977 – mit D. Russwurm

„Übergreifungsstöße von Rippenstählen unter nicht ruhender Belastung", Beton- und Stahlbetonbau, Heft 7, 1977 – mit R. Eligehausen

„Keil- und Klemmverankerungen für dynamisch beanspruchte Zugglieder aus hochfesten Drähten", Bauingenieur 52, Heft 8, 1977 – mit U. Nürnberger und M. Patzak

„Neue Bewehrungsrichtlinien, Manuskript der Neufassung von DIN 1045, Abschnitt 18, Erläuterungen der Neubearbeitung", Beton-Verlag GmbH, 1977 – mit R. Eligehausen und B. Neubert

„Korrosion im Bauwesen – Theorie und Praxis", VDI-Bericht Nr. 283, Düsseldorf, 1977

„Technische Möglichkeiten zur Erhöhung der Zugfestigkeit von Beton", Schriftenreihe des Deutschen Ausschusses für Stahlbeton, Heft 283, Berlin, 1977 – mit P. Diem, R. Zimbelmann

„Übergreifungsstöße von Rippenstäben unter schwellender Belastung", Schriftenreihe des Deutschen Ausschusses für Stahlbeton, Heft 291, Berlin, 1977 – mit R. Eligehausen

„Übergreifungsstöße geschweißter Betonstahlmatten", Schriftenreihe des Deutschen Ausschusses für Stahlbeton, Heft 291, Berlin, 1977 – mit R. Eligehausen und R. Tewes

„Verankerungen von Betonrippenstählen in Kunstharzmörtel und Kunstharzbeton", Bauingenieur, Heft 1, 1978 – mit L. Franke

„Rationalisierung der Bewehrungstechnik im Stahlbetonbau, vereinfachte Schubbewehrung in Balken", Betonwerk + Fertigteil-Technik, Hefte 3 + 4, 1978 – mit R. Eligehausen und B. Neubert

„Durability of Prestressing Steel-Stress Corrosion", Proceedings of the Eighth Congress of the Fédération International de la Précontraine, Part II, London, 1978 – mit U. Nürnberger und R. Frey

„Eigenschaften von feuerverzinkten Überzügen auf kaltumgeformten Spannstählen", Stahl und Eisen 98, Heft 8, 1978 – mit U. Nürnberger

„Technologische Eigenschaften kaltgezogener und nachträglich feuerverzinkter Spannstähle", Stahl und Eisen 98, Heft 8, 1978 – mit U. Nürnberger

„Korrosion von Bewehrungen im Spannbetonbau", Deutsche Bauzeitung, Heft 11, 1978 – mit R. Frey und U. Nürnberger

„Untersuchungen über den Korrosionsschutz von Spannstählen unter Spritzbeton", Schriftenreihe des Deutschen Ausschusses für Stahlbeton, Heft 298, Berlin 1978 – mit U. Nürnberger und R. Zimbelmann

„A vasbeton szerkezetek korszerü vasalása" (Die zeitgemäße Bewehrung von Stahlbetonbauteilen), veröffentlicht als Sonderdruck XXVIII EVFOLYAM, 1978, 3.SZAM

„Assessment of Concrete Reinforcing Bars made by Tempcore Process", in SEAISI Quarterly, Juli 1978 – mit D. Russwurm

„Zur Frage der Prüfung und Bewertung des Verbundes zwischen Stahl und Beton von Betonrippenstäben", in Forschungsbeiträge für die Baupraxis (Festschrift Kordina), Verlag W. Ernst und Sohn, Berlin/München/Düsseldorf 1979

„Dauerschwingfestigkeit (Betriebsfestigkeit) von Betonstahlmatten", Betonwerk + Fertigteil-Technik, Heft 3, 1979 – mit D. Russwurm

„Schweißen von Tempcore-Betonstahl", Betonwerk + Fertigteil-Technik, Heft 4, 1979 – mit D. Russwurm und J. Defourny

„Untersuchungen zum Spannungsrißkorrosionsverhalten von Seildrähten", Vorberichte zum Kolloquium: 2. Internationales Symposium „Weitgespannte Flächentragwerke", Sonderforschungsbereich 64, Universität Stuttgart, 1979 – mit U. Nürnberger und R. Frey

„Hinweise zur DIN 1045, Ausgabe Dezember 1978, Erläuterung der Bewehrungsrichtlinien", Schriftenreihe des Deutschen Ausschusses für Stahlbeton, Heft 300, Berlin 1979 – mit R. Eligehausen und B. Neubert

„Bond of Ribbed Bars Under High Cycle Repeated Loads", Journal of the American Concrete Institute, Heft 2, 1979 – mit R. Eligehausen

„Kunstharzgebundene Glasfaserstäbe als Bewehrung im Betonbau", Schriftenreihe des Deutschen Ausschusses für Stahlbeton, Heft 304, Berlin 1979 – mit L. Franke und M. Patzak

„Zur Frage der Krafteinleitung in kunstharzgebundene Glasfaserstäbe", Schriftenreihe des Deutschen Ausschusses für Stahlbeton, Heft 304, Berlin, 1979 – mit L. Franke

„Versuche zur Ermittlung der Korrosionsempfindlichkeit von Spannstählen", Forschung, Straßenbau und Straßenverkehrstechnik, Heft 1, 1980 – mit U. Nürnberger und R. Frey

„Dauerschwingfestigkeit von Betonrippenstählen", Betonwerk + Fertigteil-Technik, Heft 1 und 2, 1980 – mit D. Russwurm

„Warmbiegefähigkeit von Betonstählen", Betonwerk + Fertigteil-Technik, Heft 7, 1980 – mit D. RUSSWURM

„Kunstharzmörtel und Kunstharzbetone unter Kurzzeit- und Dauerstandbelastung", Schriftenreihe des Deutschen Ausschusses für Stahlbeton, Heft 309, Berlin, 1980 – mit L. FRANKE und K. ZEUSS

„Untersuchungen über die Schwingfestigkeit geschweißter Betonstahlverbindungen, Teil 1 Schwingfestigkeitsversuche", Schriftenreihe des Deutschen Ausschusses für Stahlbeton, Heft 317, Berlin, 1981 – mit W. HARRE und D. RUSSWURM

„Untersuchungen über die Schwingfestigkeit geschweißter Betonstahlverbindungen, Teil 2, Werkstoffkundliche Untersuchungen", Schriftenreihe des Deutschen Ausschusses für Stahlbeton, Heft 317, Berlin, 1981 – mit U. NÜRNBERGER

„Erläuterungen zu DIN 4227, Abschnitt 10" „Rissebeschränkung bei Spannbetonbauteilen", Schriftenreihe des Deutschen Ausschusses für Stahlbeton, Heft 320, Berlin, 1980 – mit R. ELIGEHAUSEN

„Entwicklungstendenzen", Kunststoffe im Bau, Heft 2, 1981 – mit L. FRANKE

„Zur Korrosion und Rißkorrosion bei Spannstählen", Werkstoffe und Korrosion 32, 1981 – mit R. FREY und U. NÜRNBERGER

„Zur Korrosion und Spannungsrißkorrosion von Spannstählen bei Bauwerken mit nachträglichem Verbund", Bauingenieur 56, 1981 – mit R. FREY und U. NÜRNBERGER

„Actual Damage of Prestressing Steel Due to Stress Corrosion", Proceedings of Third Symposium FIP, ATEP: „Stress Corrosion of Prestressing Steel", Madrid, 1981 – mit U. NÜRNBERGER und R. FREY

„Transferability of Results of SCC-Laboratory Tests in View of the SCC-Behaviour Under Practical Conditions", Proceedings of Third Symposium FIP, ATEP: „Stress Corrosion of Prestressing Steel", Madrid, 1981 – mit U. NÜRNBERGER und R. FREY

„Study of Different Types of Stress Corrosion Tests", Proceedings of Third Symposium FIP/ATEP: „Stress Corrosion of Prestressing Steel", Madrid, 1981 – mit U. NÜRNBERGER und R. FREY

„SCC-Behaviour of Metal-Coated Prestressing Steels not Embedded in Concrete", Proceedings of Third Symposium FIP, ATEP: „Stress Corrosion of Prestressing Steel", Madrid, 1981 – mit U. NÜRNBERGER und R. FREY

„Zur Frage der Vorschädigung von Spannstählen infolge geringer abtragender Korrosion", Bauingenieur 57, 1982 – mit R. FREY

„Neue Methoden zur Beurteilung des Spannungsrißkorrosionsverhaltens von Spannstählen", Betonwerk + Fertigteil-Technik, Heft 5, 1982 – mit U. NÜRNBERGER

„Kleben im konstruktiven Betonbau", Schriftenreihe des Deutschen Ausschusses für Stahlbeton, Heft 331, Berlin 1982 – mit L. FRANKE

„Aufgaben und Ziele der Forschung sowie Umsetzung der Ergebnisse in die Praxis, aufgezeigt am Beispiel der Befestigungstechnik", Betonwerk + Fertigteil-Technik, Heft 9, 1982

„Rationalisierung der Bewehrungstechnik – Ein unerschöpfliches Forschungsthema oder eine Möglichkeit zur Kostensenkung und Qualitätssteigerung im Stahlbetonbau?", Bauingenieur, Heft 12, 1982 – mit R. ELIGEHAUSEN, R. LEHMANN und B. NEUBERT

„Zur Frage der Beschränkung der Schrägrißbreiten in Stahlbetonbauteilen", Betonwerk + Fertigteil-Technik, Heft 6, 1983 – mit R. ELIGEHAUSEN und R. MALLÉE

„Korrosionsschutz des Bewehrungsstahls in Beton unter praxisnahen Bedingungen", Tiefbau – Ingenieurbau – Straßenbau, Heft 5, 1983

„Auswirkungen der modernen Befestigungstechnik auf die konstruktive Gestaltung im Stahlbetonbau", Betonwerk + Fertigteil-Technik, Heft 6, 1984 – mit R. ELIGEHAUSEN

„Auswirkungen von Fehlstellen im Einpreßmörtel auf die Korrosion des Spannstahls", Schriftenreihe des Deutschen Ausschusses für Stahlbeton, zur Zeit im Druck, Berlin, 1984 – mit R. FREY und D. FUNK

Dissertationen und Habilitationen, über die Prof. G. Rehm Bericht erstattet hat

BÖDEKER, W.: „Die Stahlblech-Holz-Nagelverbindung und ihre Anwendung – Grundlagen und Bemessungsvorschläge", Universität Braunschweig, 7.4.1970

BRIESEMANN, D.: „Zur Frage der Korrosionsinhibitoren für den Stahl in Beton", Technische Universität München, 5.3.1971

RUSSWURM, D.: „Das Schweißen der Bewehrung im Stahlbetonbau; kritische Betrachtung der mit dem Schweißen der Bewehrung im Stahlbetonbau zusammenhängenden Probleme", Universität Braunschweig, 12.1.1972

MARTIN, H.: „Zusammenhang zwischen Oberflächenbeschaffenheit, Verbund und Sprengwirkung von Bewehrungsstählen unter Kurzzeitbelastung", Technische Universität München, 7.7.1972

NÜRNBERGER, U.: „Zur Frage des Spannungsrißkorrosionsverhaltens kohlenstoffarmer Betonstähle in Nitratlösungen unter Berücksichtigung praxisnaher Verhältnisse", Universität Braunschweig, 13.7.1972

WAUBKE, N. V.: „Über einen physikalischen Gesichtspunkt der Festigkeitsverluste von Portlandzementbetonen bei Temperaturen bis 1000 °C", Technische Universität Braunschweig, Januar 1973

RIECHE, G.: „Mechanismen der Spannungskorrosion von Spannstählen im Hinblick auf ihr Verhalten in Spannbetonkonstruktionen", Universität Braunschweig, 14.2.1973

EL-AROUSY, H. T.: „Über die Steinkohlenflugasche und ihre Wirkung auf die Eigenschaften von Leichtbeton mit geschlossenem Gefüge im frischen und festen Zustand", Universität Braunschweig, 15.2.1973

NEISECKE, J.: „Ein dreiparametriges, komplexes Ultraschall-Prüfverfahren für die zerstörungsfreie Materialprüfung im Bauwesen", Universität Braunschweig, 12.7.1974

SCHIESSL, P.: „Zur Frage der zulässigen Rißbreite und der erforderlichen Betondeckung im Stahlbetonbau unter besonderer Berücksichtigung der Karbonatisierung des Betons", Technische Universität München, 19.7.1974

ROSTÁSY, F. S.: „Zwang und Rissebeschränkung bei Außenwänden aus Stahlleichtbeton", Universität Stuttgart, 11.6.1975

FRANKE, L.: „Zur Frage des Verbundkriechens von Stahl in Beton", Universität Stuttgart, 10.7.1975

KOCH, R.: „Verformungsverhalten von Stahlbetonstäben unter Biegung und Längszug im Zustand II auch bei Mitwirkung des Betons zwischen des Rissen", Universität Stuttgart, 24.2.1976

Kottmann, A.: „Über die Ursache von Rohrbrüchen in Versorgungsleitungen“, Universität Stuttgart, 24. 10. 1978

Eligehausen, R.: „Übergreifungsstöße zugbeanspruchter Rippenstäbe mit geraden Stabenden, Universität Stuttgart, 11. 1. 1979

Patzak, M.: „Die Bedeutung der Reibkorrosion für nicht ruhend belastete Verankerungen und Verbindungen metallischer Bauteile des konstruktiven Ingenieurbaus“, Universität Stuttgart, 17. 7. 1979

Mallée, R.: „Zum Schubtragverhalten stabförmiger Stahlbetonelemente“, Universität Stuttgart, 30. 6. 1980

Seeger, H. F.: „Beitrag zur Ermittlung des horizontalen Bettungsmoduls von Böden durch Seitendruckversuche im Bohrloch“, Universität Stuttgart, 9. 7. 1980

Spies, K.: „Konstruktives Entwerfen – Von der Grundrißdisposition zum Tragwerk“, Universität Stuttgart, 12. 11. 1981

Hermann, G.: „Zum Bruchverhalten gerichteter Glasfaserverbunde“, Universität Stuttgart, 22. 7. 1982

Pusill-Wachtsmuth, P.: „Tragverhalten von Metallspreizdübeln in unbewehrtem Beton unter zentrischer Zugbeanspruchung“, Universität Stuttgart, 24. 2. 1983

Epple, A.: „Untersuchungen über Einflüsse auf die Spannungsverteilung in aufgeleimten Holzlaschen und hölzernen Knotenplatten“, Universität Hamburg, 25. 2. 1983

Wessolly, L.: „Vorgespannte Seilnetztragwerke – Zur Sicherheit gegen Weiterreißen“, Universität Stuttgart, 14. 6. 1983

Weischede, D.: „Untersuchungen zum methodischen Konstruieren im Stahlbetonbau“, Universität Stuttgart, 27. 6. 1983

Schober, H.: „Ein Modell zur Berechnung der Verformungen und Risse im Stahl- und Spannbetonbau, Universität Stuttgart, 17. 1. 1984

Hock, B.: „Über die Verformungen und Beanspruchungen von Stahlbetonskelettbauten infolge von Temperatur- und Feuchtigkeitsänderungen“, Universität Stuttgart, 13. 2. 1984

Inhalt

Werkstoff-Forschung, Beton und Stahl

Befestigungstechnik

Konstruktion, Bemessung und Bewehrungstechnik

The function of research in the construction industry

Andrew W. Beeby

1 Introduction

There can be a substantial communication gap between those involved in research and those involved in design and construction in practice. This gap might be removed, or at least reduced, if it was clearer to both sides just what function research serves. The object of this paper is to discuss some of the issues involved in defining the true function of research in the construction industry. In particular, the following questions will be considered.

- Why do we need research and theory?
- What is the nature of engineering research?
- How should research findings be introduced into practice?
- How complex or rigorous should design methods be?

Many people, particularly those involved in research, would immediately answer 'it's obvious' to the first of these questions. In fact, it is not so obvious why we need research in the construction industry as is the case in some other industries. Consider, for example, the electronics or pharmaceutical industry: these invest 10–20% of their turnover on research and development. The functions of research and development are clear here: next year's product results from this year's R & D. Figures have been produced which suggest that there is a strong relationship between R & D expenditure and market share in many areas of industry [1]. Research serves no such self-evident function in the construction industry. Indeed some designers have expressed the view that the influence of research is merely to add unnecessary complexity to design methods. A couple of quotations will illustrate this.

"Theory and research appear always to take precedence over experience and empirical solutions. This results in unnecessary complexity, unrealistic precision and additional cost of both design and construction." (From a paper circulated by the Campaign for Practical Codes of Practice.)

"It appears that sound experience takes second place to the latest laboratory research, which is untried or unproven in practice and of little value to the designer" [2].

These remarks place great emphasis on 'experience' as the appropriate source of inspiration for design methods. Is this unreasonable? Great buildings which we have inherited from the past, such as cathedrals, palaces, fortresses etc, were designed entirely by the application of experience. Little or no 'theory' as we understand it today was applied in their design yet they served their design function adequately and many remain as some of the great technological achievements of mankind. What can research and theory offer that cannot be provided by experience? What is the justification for the investment in university departments and research establishments in the construction field?

Those directly concerned with research have a responsibility for demonstrating this value to the supporting industry.

2 Experience and change

It is the author's contention that the value of research and theoretical studies lies in the assistance they can give in coping with change. When change is very slow, then experience is a very powerful tool. When change is rapid experience is less reliable and we need a clearer understanding of the behaviour of the materials we employ. This understanding arises from research. This dichotomy will be developed by discussion of changes that have taken place in reinforced concrete design but, first, it is necessary to develop some idea of what consitutes rapid change.

'Rate of change' is a relative concept: whether change is rapid or slow depends upon the timescale against which we are measuring change. The geological history of the earth appears as a period of incredibly rapid change when viewed against the timescale of the life of the universe but to us, geological change is so slow that, for practical purposes (earthquakes excluded) we can consider geology to be unchanging. Many designers perceive change in reinforced concrete as being relatively slow. This perception is, however, incorrect and would seem to derive from a confusion of time scales: the time scale by which the average designer judges change being considerably shorter than that appropriate to building performance. The general consensus is that the design life for buildings should be of the order of 50 years. This defines the timescale we are interested in: if little change has occurred over a 50 year period then change is slow; if substantial changes have occurred then change is rapid.

Reinforced concrete is a relatively new structural material; it started to be used at the end of the last century, however it really only became a common material during the second and third decades of this century. For example, the first reinforced concrete building in Manchester was built in 1911. In London, the by-laws were only amended to permit the use of reinforced concrete in buildings in 1916. It is probable that the great majority of reinforced concrete buildings now existing were built after 1950. From this it will be seen that few of these buildings have so far lasted a full design life. Our ability to judge the adequacy of design methods would thus be limited, even if no changes had taken place during the last 50 years. However, there have been very substantial changes over the period in question: some of these are illustrated below.

Figure 1 shows how the maximum permissible stress in reinforcement permitted by British design documents under service loads has changed during this century. It will be seen that, over the last 50 years, maximum design stresses have increased from just over 100 N/mm^2 to roughly 280 N/mm^2. This last figure is approximate since the Limit State CP110 does not explicitly give a value directly comparable with earlier Codes. This increase arises from two sources: increases in the strength of reinforcing steels which were permitted to be used and decreases in the safety factors.

Figure 2 shows how the 28-day works cube strength expected from a 1:2:4 volume batched mix and the permissible stress under service load for this mix have changed over the same period. The dashed portion of the service stress line covers the period when the Code had changed from using elastic section design to ultimate load design. The dashed line has been obtained by calculating the stress on an elastic, modular-ratio basis corresponding to a balanced section obtained using the ultimate load equations. The same mix can thus now be used at stress levels that are roughly three times what they were 50 years ago. As with steel, this change arises partly from changes in the materials and partly from reductions in safety factors. The changes in cement strength with time are shown in Figure 3, taken from [3].

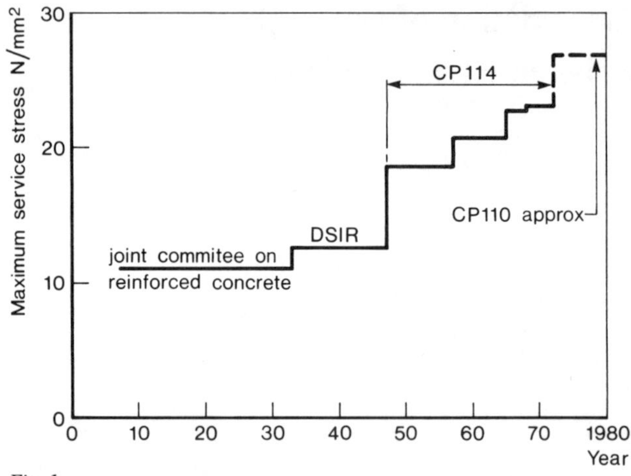

Fig. 1
Maximum permissible steel stress under service loads since 1907.

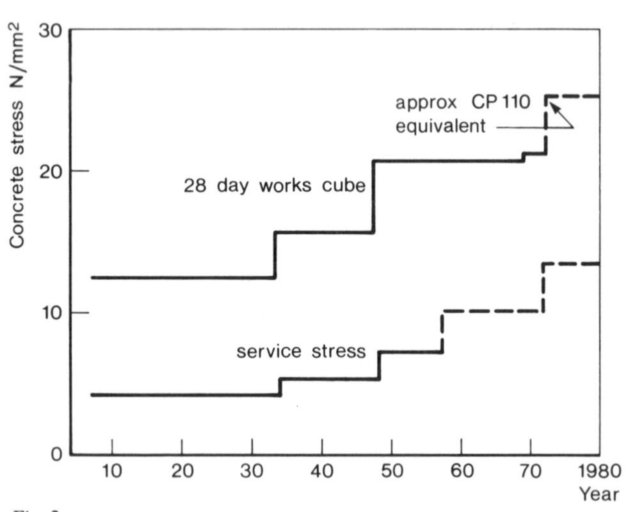

Fig. 2
Specified works cube strengths and service stress permitted for 1:2:4 concrete since 1907.

Two consequences of the changes should be noted.

Higher stress levels will result in the use of less material. Figure 4, reproduced from [4], shows how the weights of building structures in Sweden have changed over a roughly 50 year period. It seems reasonable to

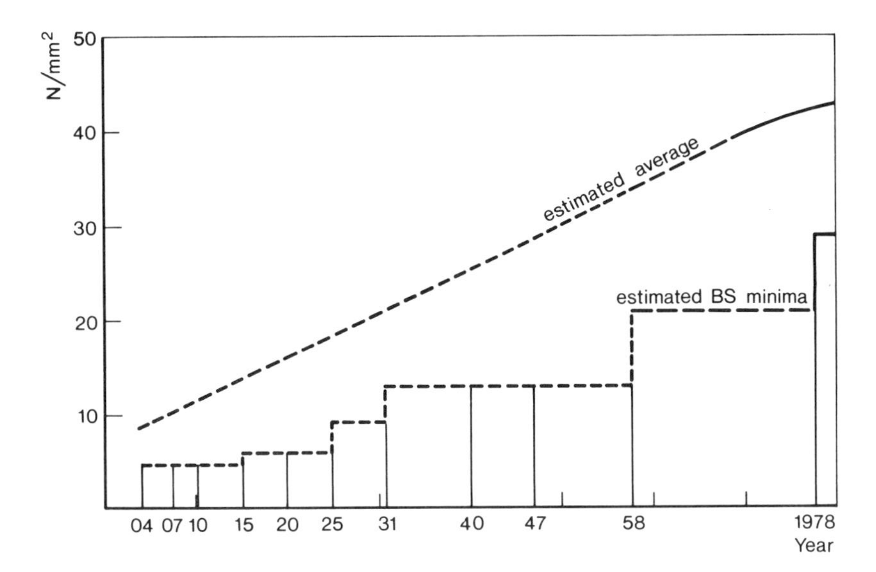

Fig. 3
Estimates of actual and British Standard minimum 28 day cement strengths in terms of concrete strength with 0.6 water/cement ratio (from Reference [3]).

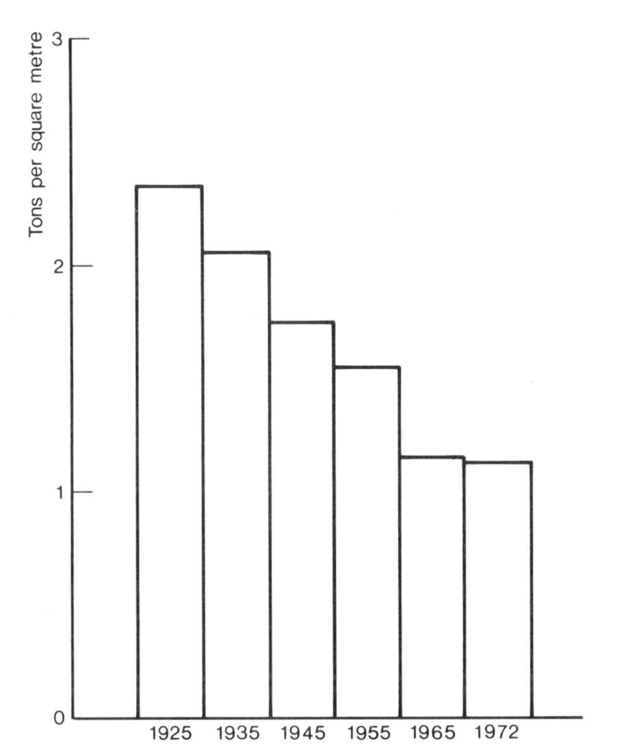

Fig. 4
The development of building weights in Sweden during the period 1925–1972 (from Reference [4]).

assume that the same order of change has also occurred elsewhere. Over this period, the weight of structures has apparently roughly halved. From this it seems not unreasonable to conclude that a doubling of stress levels and a halving of structural weight indicates an increase in flexibility of structures by a factor of about 4.

The increase in the strength of cement can have another, rather different consequence. If the dominant criterion used for the design of concrete mixes is strength, then modern cements can give mixes of the same strength as previously with much lower cement contents and higher water/cement ratios. As an example, a water/cement ratio of 0.45 would have been required to give a cube strength of 40 N/mm^2 in 1950 whereas the same strength could now be obtained using a water/cement ratio 0.65. While the strengths are the same, the durability of these two concretes would be very different. Roughly, if 20 mm of cover would have protected the steel from the ingress of carbonation for 50 years in 1950, 45 mm would be required to achieve the same in 1983.

Changes are not only technological, such as changes in the materials or in the design methods but can be entirely social, such as changes in the type of structures required, the influence of changing financial constraints or of political decisions. An example of this last type of change is the increasing emphasis placed by many countries on keeping the road network clear of snow and ice through the winter. This has resulted in very large increases in the use of de-icing salts during the last 25 years. This is illustrated very clearly in Figure 5 which shows the use of road salt in the United States since 1945 [5].

From the above examples it must be seen that change has been substantial and continuous over the last 50 years. This very seriously limits the usefulness of experience in the development of design methods. The effectiveness of a particular design method can really

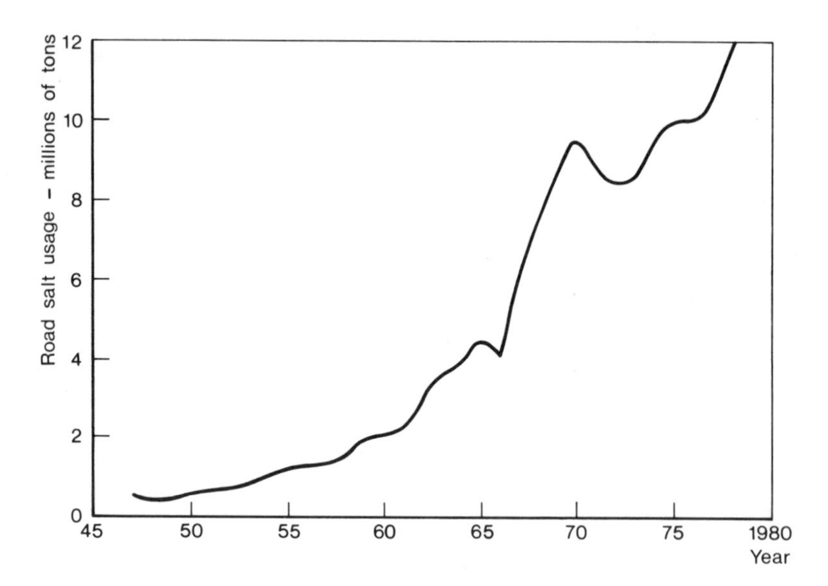

Fig. 5
Road salt usage in the United States (from Reference [5]).

only be judged where conditions have remained constant over a period similar to a full design life. Certainly, it would appear to take a period of 15–20 years for even highly unsatisfactory practices to be identified (consider, for example, the abuse of calcium chloride or the use of High Alumina Cement in the UK). Experience, therefore, may provide information about the adequacy of design methods as they were applied 15 or more years ago. In the interim, practices and materials have changed and the relevance of this experience to current conditions would need very careful consideration.

Another problem with experience as a test of design methods is that it is fundamentally a very inefficient test. This can be illustrated by the diagram below.

Quality of design procedure

3	wildly conservative	A
2	very conservative	A
1	on the conservative side	A
0	about right	A
−1	less than ideal	A
−2	unsatisfactory	B
−3	totally unacceptable	B

It is probably possible to judge whether a design method is 'unsatisfactory' or 'totally unacceptable' (zone B) within a relatively short time. However experience can give little guidance as to whether a method is 'less than ideal', 'about right', 'on the conservative side', 'very conservative' or even 'wildly conservative' (zone A). The best that can normally be said about a design method which is generally believed to be acceptable on the basis of experience is that it is probably within zone A in the diagram above.

The discussion of experience and change set out above shows that experience alone connot form a sufficient basis for the development of design methods for present or future conditions. A further essential ingredient is an understanding of the behaviour of structures and materials. This is developed from theoretical and experimental studies and provides the only means by which designers can anticipate the consequences of change. Experimental and theoretical studies alone are, however, also an insufficient basis for the development of design methods and the final proof is the construction of satisfactory structures. Only time will provide this proof.

3 'Scientific' and 'Engineering' research

Popper [6] describes the 'scientific method' as a process consisting of the following stages:

(i) construction of a hypothesis
(ii) testing the hypothesis
(iii) when the hypothesis fails a test, it is discarded and the process is repeated.

For the purposes of this paper, this will be described as 'Scientific Research'.

Engineers employ theory and experiment rather differently; they use theories as approximate models of reality. This can be illustrated by considering two basic theories used by engineers in treating structures: the theories of elasticity and plasticity. Both are highly mathematical theories developed to a state of great rigour over many years. They are used constantly in structural design despite the fact that the materials may violate the basic assumptions on which both theories are founded. Elasticity is the most commonly used theory even though it has been well known from earliest days of the use of reinforced concrete that concrete is not elastic and nor is reinforced concrete. Plastic theory is a more recent development and is used in section design and also in methods such as yield line analysis even though we know that concrete is not plastic. Philosophically, the situation becomes even more confused when we employ elastic methods for analysis and then what are effectively plastic methods for the design of critical sections.

It is important to understand that engineers are not primarily concerned with accurate prediction of behaviour: their objective is the economic provision of structures which will adequately serve their specified purpose. If a highly approximate and possibly even philosophically unsound method will achieve this, then further sophistication and rigour is unnecessary and counterproductive.

Engineers develop a repertoire of approximate models which are accepted as having adequate reliability within defined limits. These models are calibrated and adjusted and their limitations defined by experiment. Experiment can possibly best be described as 'structured experience' in this context. Theory forms the skeleton for these models and also, by providing an insight into the mechanisms involved, should help the engineer to assess the limitations of particular models. This process, which, for convenience will be called 'Engineering Research', may be explored further by considering a very simple example: the ultimate strength of singly reinforced sections subjected to pure flexure. The theoretical framework may be summarised as follows:

a) Plane sections remain plane under the action of flexure.
b) Moment is resisted by a tensile force in the reinforcement and a compressive force carried by the concrete.
c) The tensile and compressive forces are equal in magnitude and the moment of resistance is equal to either of these forces multiplied by the distance between the centroids of the two forces (the lever arm).

If the stress-strain curves of the materials are known, this provides sufficient information to define a relationship between applied moment and deformation. This would permit a maximum moment to be obtained.

Stress-strain curves for steel and concrete can be obtained experimentally. For steel, this is relatively straightforward but for concrete there are complications since the result is influenced by the shape of the specimen, the way load is applied and the rate of loading. Experimental work has therefore concentrated on testing compression zones and defining for these the maximum average stress developed and the corresponding position of the centre of compression and maximum strain. This avoids having to define an exact shape for the stress-strain response. The experiments show that, in general, it is reasonable to assume a maximum strain of around 0.0035, independent of strength, and this can therefore be used as a criterion of failure. Simplified stress-strain curves can be proposed which give the correct average stress and centroid positions at failure. The CEB parabolic-rectangular and rectangular diagrams are such simplified curves: they are purely empirical and owe little or nothing to theory.

The above example illustrates well the interaction between theory and experiment: the theory has provided the insight into the mechanisms involved and has defined the information required from the experimental work. The experimental work has been formulated to provide this information and hence design equations could be developed. It will be seen that this is a totally different relationship between theory and experiment from that envisaged in 'scientific research'. Experiment is not being used simply to test a theory but is an essential component of the process of developing a design method.

It may, at first sight, appear that the theory could be dispensed with and that design methods could be developed from experiment alone. For example, if enough beams were tested, design charts for flexural strength could be produced directly from the experimental results without reference to a theory. There are a number of reasons why, for reinforced concrete at least, this is impractical. Some of these will be discussed below.

Probably the major factor is that a theory reduces the amount of testing required to a practical level. It would not be possible to test all practical arrangements of reinforcement in all practical sized beams with all practical strengths of steel and concrete. It is essential to have some means of deciding upon the important parameters and an idea of how results from a limited testing program can be generalised and extrapolated to the much wider field of practical sizes and shapes of beam. A theory provides an answer to these questions. As an example, we would feel reasonable confident in applying the basic theoretical assumptions and parabolic-rectangular stress block to assess the flexural strength of, say, a circular section even though all experimental work had been carried out on rectangular sections. We recognise that the method will not be as accurate but our understanding of the theory suggests that any errors will be acceptable. This would not be true of fields where our theoretical understanding was less developed such as, for example, shear.

A final important feature of theory is that it may indicate areas where experimental results should be discarded. To those involved in the 'scientific method' this may appear as heresy but, as an example will show, it isn't. It has been agreed that in assessing ultimate flexural strengths, the tensile strength of the concrete should be ignored. One reason for doing so is that, in practice, tensile stresses may well have developed due to shrinkage, thermal movements or other effects not directly related to loading. In laboratory tests, however, these effects will not be present and the strength of lightly reinforced members can appear to depend dominantly on the tensile strength of the concrete. Results from such tests are clearly not relevant to the development of design procedures.

Hopefully, this discussion has demonstrated that theoretical and experimental studies are essential and complementary factors in the development of design methods.

There remain two practical points related to the development of design methods using experimental results which need to be mentioned. These are:

1. What level of safety should be incorporated into a new method and how should this be done?
2. What level of complexity or rigour should be accepted?

These two will be looked at in slightly more detail.

The basic question in (1) above can be illustrated by an entirely hypothetical example. Figure 6 illustrates the results of a research study to investigate the relationship between some parameter β and strength. The line A is the 'best fit' to the data and would be the line given by the researcher as defining the relationship

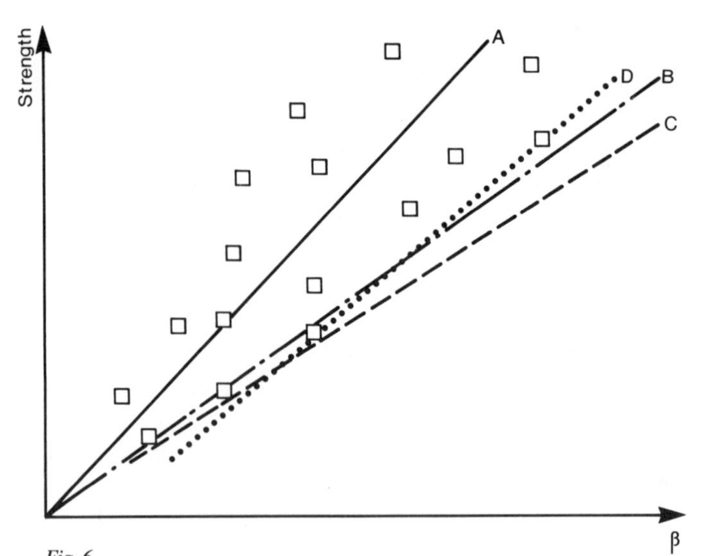

Fig. 6
Hypothetical relationship between strength and some parameter, β.

between strength and β. The question in deriving a 'design' relationship is:

a) Should line A be used directly with all the necessary safety built into the definition of β?
   or
b) Should we define a 'characteristic' relationship with, say a 5% probability that the strength will be below this combined with some lesser safety factor included in the definition of β?
   or
c) Should we use a 'lower bound' relationship such as line C or D?

From experience with a number of Codes of Practice, it is clear that no consistent approach is currently used. An alternative way of stating the problem is to pose the question "how should model uncertainty be taken into account in the derivation of a design equation?". Model uncertainty is the uncertainty introduced due to imperfections in the relationship. In the above example, errors may arise because strength is not directly proportional to β as is assumed or because β is not an ideal parameter; it should include factors which have been ignored. Should this uncertainty be allowed for by including it in γ_f or γ_m or in the definition of the design line or in some new γ factor or in some combination of these? Existing codes are silent on this question.

The second question posed at the start of this section is possibly amenable to a direct statistical answer once question (1) is answered satisfactorily. A more rigorous design method is one which will reduce the model uncertainty. This should permit a reduction in overall safety factor. Uncertainty arises from many sources such as, uncertainty about loading, about the properties of the actual material in structure, about analysis, about workmanship and model uncertainty. Only if a reduction in the model uncertainty is sufficient to make a significant reduction in the overall uncertainty and hence in the overall safety factor required can a more rigorous method offer any real economy. It should be possible to develop a statistical framework by which proposed design methods could be tested. It should be noted that such a test would work equally well as a tool in developing more simplified design approaches. The benefits to be gained from having established answers to these two questions are:

a) that the relative merits of different design methods can be compared on a rational basis.
b) that arguments about the required complexity or simplicity of design methods can be settled on a rational basis.
c) that researchers and code writers are both clear about how research information should be used to develop design methods.
d) that the consistency of design methods should be improved and that researchers should be clearer about what information they need to supply.

4 Theories

It remains to clarify how theories evolve. Theories like elasticity or plasticity are not real thories in the scientific sense as they do not attempt to offer an explanation of behaviour: they are simply exercises in mathematics. One cannot prove or disprove the theory of elasticity: one simply uses it or not depending upon whether it is appropriate for a particular situation. The type of theory that is important to understanding the behaviour of reinforced concrete is one which offers some conceptual picture of how a member or section behaves under the particular action considered. The theory outlined for a singly reinforced section in flexure is a very simple example of such a theory. For engineering purposes the theories used do not have to be absolutely correct but they should convey the correct general picture of behaviour, indicate the significant variables and the correct general form of the relationships between these. Clearly, such theories have to be tested to discover whether they achieve this. This is a completely different exercise from the use of experiment in the development of a design method and such an exercise is effectively one of pure classical scientific research.

5 Concluding discussion

In the first part of this paper it has been argued that concrete design and construction is in an era of rapid change. As a consequence, experience is of only limited value in defining and developing adequate design methods even though it is subsequent experience which, in the end, provides the final judgement as to whether a method is satisfactory or not.

Theoretical and experimental research provides a means of reacting to change and of generalising experience to handle a wider range of problems. Two basic classes of research have been recognised

a) Scientific Research. This is concerned with improving understanding of structural and material behaviour.
   The basic activity in this type of work is the development of theories that explain behaviour and the testing of these theories by experiment. The target is a theory that can be shown to provide a reasonably accurate understanding of known behaviour.
b) Engineering Research. This is concerned with the development of design methods. In this form of research, theory and experiment are complementary: the former providing a framework while the latter provides the numerical values. The objective is a reliable design method which will result in the achievement of acceptable performance in practice without excessive design effort.

Various aspects of Engineering Research have been discussed in some detail in the paper.

In practice, both researchers and practising engineers commonly confuse these two quite different forms of activity. Many research projects set out to achieve the objectives of both types of research and finish by achieving neither. Practising engineers and the general public constantly pressure research organizations to do ‘practical’ research. By this they seem to mean engineering research where the testing is limited to tests on specimens which accord with current practice. This very seriously limits the real value of much of the work that is done by limiting the range of applicability of any resulting method while adding little to our understanding or ability to react to change. The industry needs to clarify its ideas about what it really requires, what it can really expect from research and how this might best be obtained. For this reason, some further discussion of the issues involved seems justified.

A point which should be understood first about the relationship between research and design is that, in construction, the nature of this relationship has traditionally been very different from that in some other industries. It will help to distinguish two types of relationship here:

a) Research leading. This is typically the situation in electronics or other high technology industries where 'this year research produces next year's product'. Here, it is research which leads to changes.
b) Research reacting. In this case, research is a reaction to changes imposed from outside. This is almost always the case in the construction industry. For example, the changes in steel and concrete discussed in the previous section were not dominantly brought about by those within the industry but by materials suppliers. As far as designers were concerned, their problem was to react to these changes; they did not cause them. This is also obviously true in the case of the use of de-icing salts.

There are obvious weaknesses in using research in the 'reacting' manner. Designers have an exaggerated regard for experience and tend to extrapolate this to new situations unjustifiably. Research is only encouraged when this extrapolation has produced problems: the result can be very expensive in repair costs and in loss of confidence in the abilities of the industry. Research is then started urgently with solutions to the problem being required in far too short a timescale for these solutions to be adequately tested. The current international interest in durability is a classic example of this process in action.

Research, then, is seen by the construction industry (but not necessarily by research workers) as a means of solving problems, not as a means of developing new methods except where these are necessary to cope with problems.

Our main hope of avoiding troubles resulting from changes in practice such as those discussed in the first part of this paper lies in having sufficient understanding of behaviour to be able to foresee the consequences of change. Such unterstanding is developed from 'scientific' research rather than 'engineering' research. Work done to increase understanding is of permanent value whereas work done to develop formulae using 'practical' tests is ephemeral and has to be repeated every few years as conditions change. Indeed it occasionally happens that the results are out of date before they are even published. This can be particularly true of such work as exposure testing for durability where a test programme may easily take 10–15 years to complete: a period during which substantial changes in materials and stress levels may have occurred.

It is the author's opinion that research aimed at increasing understanding is currently much undervalued and, for the sake of the future, should be increased.

Another question which should be considered is whether the industry is operating in its own best long-term interests by viewing research as dominantly a 'trouble shooting' activity with research being initiated following the identification of a problem rather than its use in a more leading role. This will not be discussed further here but it is hoped that the paper has pointed out some of the weaknesses of the present situation.

6 References

[1] Pavitt, K.: Technical Innovation and British Economic Performance. London, Macmillan Press 1980

[2] Curtin, W. G.: Less code more practice. Architects Journal, 16 March 1983.

[3] Corish, A. T. and Jackson, P. J.: Portland Cement Properties – past and present. Concrete. July 1982.

[4] Oden, K.: 'Trends and development of building materials; Proceedings of IABSE Symposium on the selection of structural form'. London 1981.

[5] Baboian, R. (Editor): 'Automotive corrosion by deicing salts.' National Association of Corrosion Engineers, Houston, Texas 1981.

[6] Popper, K.: The logic of scientific discovery. Hutchinson, London 1980.

Bedeutet Vorspannen nicht auch Vorbelasten?

Antonius S. G. Bruggeling

1 Einleitung

1941 hat EUGÈNE FREYSSINET seine Gedanken über Vorspannen von Betontragwerke an die Öffentlichkeit gebracht und dabei erklärt „Vorspannen sei eine Revolution für die Kunst des Bauens“ [1].
Seit 1948 hat der Autor sich mit der Entwicklung der Vorspannung befaßt. Damals hat man in den Niederlanden mit der Planung einer Dachkonstruktion aus Spannbetongurten und Spannbetonträgern für eine Industriehalle begonnen. Im Oktober desselben Jahres fand an einem kleinen Versuchsträger mit einer Länge von 8,5 m die erste Forschung auf dem Gebiet des Spannbetons statt. Seit dieser Zeit hat die Vorspanntechnik in den Niederlanden eine wirklich revolutionäre Entwicklung durchgemacht. 32 Jahre nach diesen ersten Versuchen werden dreidimensional vorgespannte Pfeiler für das Oosterschelde-Wehr gefertigt. Der Bau dieses Wehres ist nur möglich, weil heute solche gewaltigen Pfeiler in Spannbeton hergestellt werden können. Die 66 Pfeiler, jeder mit einem Eigengewicht von 18000 Tonnen, wurden in einem Trockendock vorgefertigt. Ende 1983 sind schon eine große Anzahl dieser Pfeiler mit einem speziellen Hebeschiff „Ostrea“ versetzt worden (Bild 1). Mit diesem Beispiel und mit vielen anderen Beispielen, z. B. Bauten für die Gewinnung von Öl und Erdgas aus dem Nordmeer, ist klar gemacht, daß die Voraussage von FREYSSINET richtig war. Vorspannen hat dem Massivbau einen erstaunlichen Aufschwung gegeben. Es ist heute eine unbestrittene Tatsache, daß durch die Vorspanntechnik die Stahlbetonbauweise seit dem Zweiten Weltkrieg revolutionär erweitert wurde.
Der in 1953 in London gegründete internationale Spannbetonverein F.I.P. hat auf seinen bis heute organisierten neun Kongressen gezeigt, wie weltweit diese Revolution sich durchgesetzt hat. Viele große Brükken, Viadukte, Industriehallen, Gebäude, Behälter, Meeresbauten usw. sind seither in Spannbeton gebaut worden. Das breite Anwendungsgebiet der Vorspannung hat dazu geführt, daß in Nachfolge der Pioniere der vierziger Jahre sich eine Gruppe Spezialisten gebildet hat; die „Spannbetoningenieure“. Spannbeton war ein neues Anwendungsgebiet des Massivbaues geworden; ein Gebiet mit eigener Betrachtungsweise, eigenen Vorschriften, eigenen Verbänden.
Es waren die Pioniere unter diesen Spannbetoningenieuren, welche bei ihren Entwürfen u. a. den Nachweis der Bruchsicherheit schon 1950 eingeführt haben.
Auch die Entwicklung von neuen, sicheren Spannstählen, gefertigt mit gut überwachten Herstellungsverfahren, ist seitdem in Gang gesetzt worden. Man hat dabei auch einige Rückschläge gehabt; aber immer mehr ist die Qualität der Spannstähle verbessert worden. Gleichzeitig wurden neue Spannverfahren und Verankerungen entwickelt, welche sich in der Praxis bewähren.
Nicht nur im Massivbau hat das Vorspannen eine Rolle gespielt. Der Stahlbau (Spannbolzen), der Holzbau (vorgespannte Holzfachwerke) und der Grundbau

Bild 1
Hebeschiff „Ostrea“ mit einem Pfeiler der Oosterschelde-Wehr.

(Erdanker, Felsanker) haben auch das Vorspannprinzip benutzt und dadurch teilweise diese Revolution mitgemacht.
Vorspannen war wirklich eine Revolution!
Die Frage ist jetzt, ob die Gedanken von FREYSSINET noch Anlaß dazu geben, die „Revolution" auf anderen Gebieten weiterzuführen. Dazu muß man die Definition von Vorspannen, die von FREYSSINET gegeben wurde, genauer betrachten.

2 Die Definition von E. FREYSSINET

1941 hat FREYSSINET die folgende Definition des Spannbetons formuliert (siehe Bild 2).
Die deutsche Übersetzung der Definition von FREYSSINET sieht wie folgt aus:
„... Dies führt uns zur Definition der vorgespannten Konstruktionen, über welche ich nachfolgend berichten werde:
Darunter verstehe ich Konstruktionen, welche – bevor sie Lasten zu tragen haben oder gleichzeitig mit den ständigen Lasten – einem künstlich erzeugten, dauerhaft wirkenden Kräftesystem unterworfen werden, mit dem Ziel, zusätzliche, vorzugsweise den Lasten entgegengesetzte Spannungen zu bewirken. Dabei sollen die resultierenden Spannungen infolge der Lasten und der eingetragenen Kräfte nur Beanspruchungen zur Folge haben, welche von den verwendeten Materialien dauernd mit Sicherheit ertragen werden können. In den häufigsten Fällen erzeugen die eingetragenen, dauernd wirkenden Kräfte vorwiegend Druck in der Konstruktion; sie können jedoch infolge ihrer Exzentrizität auch Biegung und Schub verursachen ...".
Diese Definition ist sehr allgemein gehalten und macht keine Einschränkungen, wie wir sie heute in den Begriffen volle Vorspannung, beschränkte Vorspannung usw. haben. Viel mehr ist gesagt, daß man durch Eintragung eines „künstlich erzeugten und dauerhaft wirkenden Kraftsystems" gewisse Eigenschaften einer Konstruktion verbessern kann. Der letzte Satz in dieser Definition spricht von dauernd wirkenden Kräften. Diese Kräfte können Biegung und Schub verursachen, wenn sie exzentrisch angreifen. Schubkräfte können jedoch nur hervorgerufen werden, wenn die Kräfte eine veränderliche Exzentrizität zur Konstruktionsachse haben und dabei auch veränderliche Biegemomente hervorrufen. Somit wirken auch quer zur Achse innere Kräfte, welche man auch als Belastungen deuten kann.

Une révolution dans l'art de bâtir

LES CONSTRUCTIONS PRÉCONTRAINTES

Par M.E. Freyssinet

Ancien Ingénieur des Ponts et Chaussées

On est ainsi amené à la définition des constructions précontraintes dont je vais vous entretenir à présent.

J'entends par là, des constructions soumises avant application des charges qu'elles auront à supporter, ou, s'il s'agit de charges permanentes, en même temps que celles-ci, à un système d'efforts permanents, créé artificiellement, dans le but de déterminer des contraintes supplémentaires, de préférence de sens opposé à celles dues aux charges; et telles que les contraintes résultant de l'ensemble des forces appliquées — charges et efforts permanents ainsi créés — ne comportent aucun effort que le ou les matériaux utilisés ne puissent supporter indéfiniment en toute sécurité.

Dans la très grande majorité des cas, le système des efforts permanents à créer ne comprend que des compressions. Celles-ci peuvent toutefois par leur excentration, déterminer des flexions et des cisaillements.

Bild 2
Definition des Spannbetons nach FREYSSINET [2].

Die Wirkung der Vorspannung kann entweder als exzentrisch angreifende Kraft oder als zentrische Normalkraft und innere Belastung betrachtet werden.
FREYSSINET selber hat *Kräfte* sehr betont. So hatte er auf dem zweiten Kongreß der F.I.P. in 1955 in Amsterdam sehr klar darüber gesprochen und als Beispiel, das Andrehen der Bolzen, womit die Räder von Rennwagen an die Antriebachse verbunden werden,

erwähnt. „Jeder Monteur weiß, daß er diese Bolzen so stark wie möglich festdrehen muß" – Also „kräftige" Anschlüsse!

In statisch bestimmten Trägern kann man sehr gut auskommen, wenn man nur die Vorspann*kräfte* betrachtet, weil in jedem Querschnitt die Vorspannkraft allein auf den Querschnitt im Schwerpunkt der Spannglieder angreift, vorausgesetzt, daß die Stahlspannungen im Querschnitt in allen Spanngliedern gleich groß sind. Die Resultante der Druckkräfte im Betonquerschnitt, ausgeübt durch die Vorspannung allein, fällt zusammen mit der Resultante der Zugkraft in den Spanngliedern in diesem Querschnitt.

In einem statisch unbestimmten Tragwerk ist das Zusammenfallen der beiden Resultanten jedoch ausnahmsweise der Fall. Hier braucht man andere Betrachtungen und Berechnungsverfahren, um den Einfluß der Vorspannung allein zu bestimmen.

1957 hat MEHMEL in seinem Buch [8] den Lastfall Vorspannung ausführlich dargestellt und dabei den Zustand Vorspannung in Normalkraft, Endmoment und Umlenkkräfte aufgeteilt und die einzelnen Einflüsse getrennt betrachtet.

Anfangs der sechziger Jahre hatte T. Y. LIN die „Load balancing method" vorgeschlagen [3]. Die „load balancing" ist die Belastung, welche von den Umlenkkräften der Vorspannung, d. h. durch die Umlenkkräfte der Spannglieder, getragen wird. Der Anteil der Belastung, der nicht „balanced" ist, erzeugt Druck- und Zugspannungen in den Querschnitten, welche, zusammen mit den durch Vorspannen erzeugten Druckspannungen, entweder keine oder sehr kleine resultierenden Zugspannungen ergeben.

Der Autor hat 1963 [4], ebenso wie MEHMEL, vorgeschlagen, für die Berechnung von Spannungen aus Vorspannung in statisch unbestimmte Kontruktionen von einer Berechnungsmethode auszugehen, bei welcher die Wirkung der Vorspannung in drei Anteilen aufgeteilt wird.

In Bild 3 sind diese Anteile einfachshalber für einen Einfeldträger dargestellt.

Diese Anteile sind:

1. die in die Achse des Trägers in der Verankerungszone angreifende Normalkraft, welche nur eine zentrische Verkürzung der Konstruktion gibt – keine Krümmung –;
2. die Umlenkkräfte, erzeugt durch die Krümmung der Spannglieder;

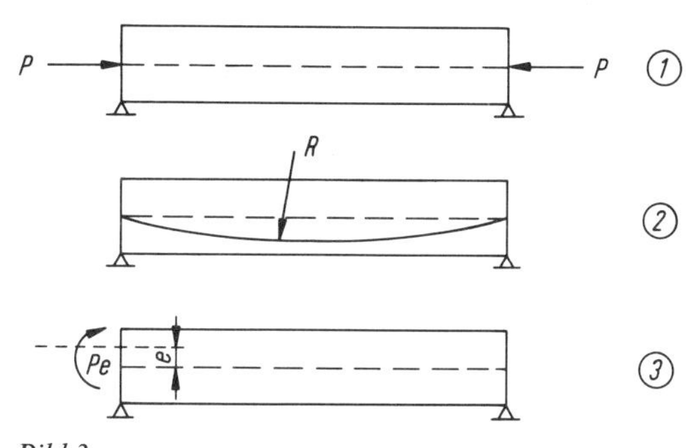

Bild 3
Aufteilung des Lastfalles Vorspannung.

3. das Biegemoment, das im Träger entsteht, wenn die resultierende Kraft in den Verankerungen exzentrisch auf den Querschnitt angreift.

Die Normalkraft (1) erzeugt Verkürzungen in Träger, welche z. B. durch Reibung in den Auflagern mehr oder weniger behindert werden, je nach Ausbildung dieser Auflager. Der von den Auflagern übernommene Anteil der Normalkraft wird weiter in das Bauwerk geführt. Diese anfänglichen Verkürzungen werden durch Schwinden und Kriechen des Betons noch verstärkt, wodurch der im Tragwerk verbleibende Anteil der Normalkraft (1) noch mehr reduziert wird. Temperaturdehnungen können diese Effekte vergrößern, jedoch auch kurzfristig abmindern.

Die Umlenkkräfte (2) und die damit zusammenhängenden Schubkräfte sind eigentlich eine künstlich erregte Belastung, welche normalerweise die Belastungen, welche auf die Konstruktion wirken, entgegengesetzt sind. Das bedeutet, daß die auf die Betonkonstruktion resultierende Belastung abgemindert wird, sogar eine teilweise negative resultierende Belastung erzeugt werden kann.

Diese über die Länge des Trägers vorhandene „Vorbelastung" durch Umlenkkräfte wird nur durch die Reibung der Spannglieder während des Vorspannens beeinflußt sowie durch die zeitabhängigen Einflüsse, welche eine Abminderung der Zugkraft in den Spanngliedern bewirken und mögliche Unterschiede durch Reibung teilweise ausgleichen.

3 Die Bedeutung der Vorbelastung für Stahlbetontragwerke

An einigen Beispielen wird nachfolgend gezeigt, was Vorbelasten für das Verhalten und die Bewehrung von Stahlbetonkonstruktionen bedeuten kann.

Vorbelasten wird dabei wie folgt definiert: „Vorbelasten einer Konstruktion bedeutet das Aufbringen einer Belastung, welche oft in entgegengesetzter Richtung wie das Eigengewicht oder der gleichzeitig oder später aufzubringenden Belastung wirkt.“
In verschiedenen Fällen wird die Vorbelastung gezielt aufgebracht und damit das Verhalten der Konstruktion unter Gebrauchsbelastung oder Belastung bei Auftreten von Kalamitäten in günstigem Sinne beeinflußt.

3.1 Kellerboden

In den Niederlanden befinden sich viele Kellerböden unterhalb des Grundwasserspiegels. In einem Polder wird die Höhe des Grundwasserspiegels reguliert und kann dabei schwanken zwischen zwei Niveaus: ein niedriges Niveau, das nie unterschritten wird, und ein hohes Niveau, das nur in Ausnahmefällen auftreten kann. Außerhalb der Deiche hängt insbesonders das Hochwasserniveau von natürlichen Einflüssen ab (Ausuferung). In diesen Fällen wird z. B. in Rotterdam vorgeschrieben, daß auf einem bestimmten Niveau Öffnungen in der Kellerwand vorgesehen werden müssen, damit durch Überbelastung bei zu hohen Wasserständen kein Aufbrechen des Kellerbodens möglich ist. Auch dann ist deshalb der Höchstwasserstand gewährleistet, weil jetzt der Keller mit Wasser gefüllt wird.
Wenn man den Kellerboden von einem auf Rammpfähle gegründeten Betonbauwerk betrachtet, sieht man, daß die folgenden extremen Lastfälle auftreten können (siehe Bild 4):

1. Hochwasser ($q_{w.max}$) und Eigengewicht (q_g) des Kellerbodens;

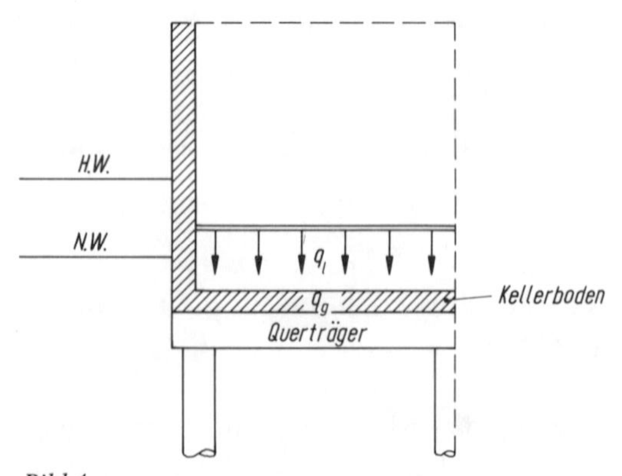

Bild 4
Lastfälle, welche auf einem Kellerboden auftreten können.

2. Niedrigwasser ($q_{w.min}$) und Vollbelastung (q_{max}).

In dem ersten Fall wird die Decke oft – abhängig vom Hochwasserstand – durch eine aufwärts wirkende resultierende Belastung ($q_g - q_{w.max}$) belastet. Im zweiten Fall wird die Decke belastet durch eine resultierende Belastung ($q_{max} - q_{w.min}$). Diese Belastung kann in Abhängigkeit des Niedrigwasserstandes entweder positiv oder negativ sein. Nebenbei wirkt in der Decke noch in zwei horizontalen Richtungen eine, in der Größe wechselnde Normalkraft; welcher Einfluß oft vernachlässigbar klein ist.
Die Bewehrung der Decke muß für die zwei erwähnten resultierenden Belastungen dimensioniert werden. Die Kontrolle der Bruchlast findet in der Regel für den Fall 2 statt: das bedeutet $q_u = \gamma \cdot q_{max} - q_{w.min}$ (γ = Sicherheitsbeiwert). Man erkennt daraus, daß in diesen Fällen, wo der Mindestwert der aufwärts wirkenden Wasserbelastung groß ist, die Bruchlast q_u wesentlich niedriger gehalten werden kann wie in dem Fall *ohne* Wasserbelastung, wenn man davon ausgeht, daß der Mindestwert nie unterschritten werden kann.
Das bedeutet, daß durch Vorbelasten, in diesem Fall durch natürliche Einflüsse, die Bewehrung eines Kellerbodens mit großer Nutzlast, wesentlich reduziert werden kann.

3.2 Sicherheitswand eines Flüssiggasbehälters

Ein interessantes, jedoch außergewöhnliches, Beispiel von Vorbelasten einer Behälterkonstruktion bietet die Anwendung der Erdbelastung [5]. Flüssiggas wird in Stahlbehältern von Ni-Stahl gelagert. Weil die Lagerung unter atmosphärischem Druck stattfindet, ist eine Lagerung der Flüssigkeit nur möglich, wenn sie tief gekühlt ist, z. B. bis −160 °C für Erdgas. Unter diesen Umständen hält man in vielen Fällen Sicherheitswände und sogar Sicherheitsbehälter aus Beton für notwendig, nicht nur, um das Lagergut zu schützen (gegen Explosion, Sabotage usw.), sondern auch die Umgebung (Sprödbruch des Stahles) [6].
Obwohl Spannbeton große Vorteile für die Betonkonstruktion der Sicherheitsbehälter bietet, hat man aus bestimmten Gründen Behälterwände aus Stahlbeton gewählt. Weil eine bestimmte Vorbelastung dieser Stahlbetonwände notwendig ist, hat man eine Bodenanschüttung angebracht (siehe Bild 5). Damit ist diese Sicherheitswand in Zusammenhang mit dieser Anschüttung in der Lage, auch große Horizontalkräfte (z. B. durch stoßartige Belastung) aufzunehmen. Ob-

Bild 5
Flüssiggasbehälter Abu-Dhabi.

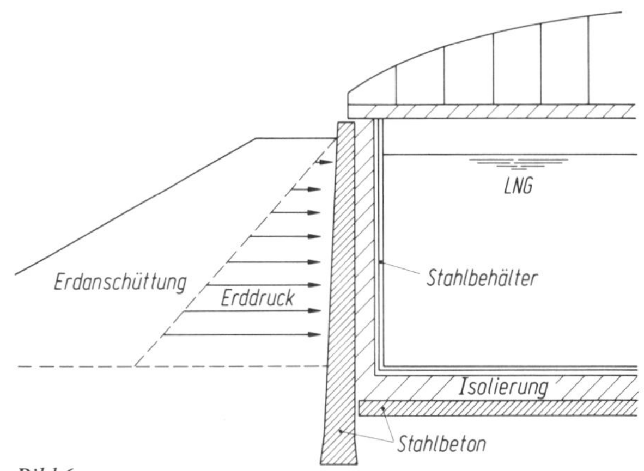

Bild 6
Stahlbetonsicherheitswand mit Erdanschüttung.

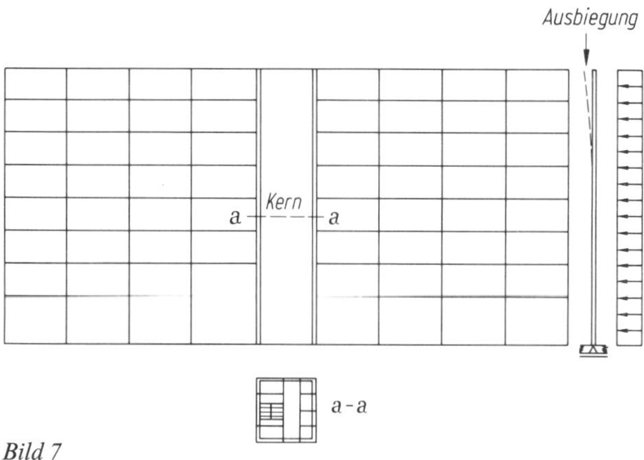

Bild 7
Rahmenwerk durch einen Kern ausgesteift.

wohl diese Anwendung der Vorbelastung durch Erddruck in den meisten Fällen nicht wirtschaftlich sein wird, zeigt sie doch, was man mit Vorbelasten erreichen kann.

Wird ein Mittelwert des Bodendruckes in Rechnung gestellt und sorgfältig gebaut, damit die Behälterwand wenig vom Kreiszylinder abweicht (sogenannte Unrundheit), so stehen die vertikalen Flächen der Stahlbetonwand unter Druckspannungen. Sie sind dann in der Lage, bestimmte Temperaturgradienten aufzunehmen, ohne unzulässige durchgehende Querrisse zu bilden (siehe Bild 6).

Ein anderes Beispiel von Vorbelastung durch Wasserdruck sind die großen Meeresbauten für die Gewinnung von Öl und Gas. In den Pylonen dieser Betonkonstruktionen werden durch Unterdruck diese Kreiszylinderpylonen von dem Wasserdruck vorgespannt. Auch dadurch verhalten sich die im Nordmeer installierte Bauten so außerordentlich gut [9].

3.3 Kern, zur Aussteifung eines mehrstöckigen Gebäudes

Gebäude werden oft als Rahmentragwerke ausgeführt. Die Aussteifung solcher Tragwerke wird oft mit einer Stabilitätskonstruktion (sogenannter Kern) ausgeführt, welche nicht nur die Windkräfte aufnimmt, sondern auch die aus der horizontalen Ausweichung hervorgerufene Horizontalkräfte erster und zweiter Ordnung (siehe Bild 7). Dabei wird jedoch diese Stabilitätskonstruktion stark auf Biegung beansprucht. Die vertikale Belastung des Kernes ist beschränkt auf sein Eigengewicht und auf einen geringen Anteil der Gebäudebelastung. Dadurch ist der untere Teil dieses Kernes im Zustand II. Durch Rißbildung in der Zugzone wird dabei die Biegesteifheit stark reduziert, die Horizontalverschiebungen vergrößert und die Horizontalkräfte zweiter Ordnung größer.

Wenn man das Bauwerk als Hängewerk ausbildet, wobei alle Belastungen des Gebäudes über ein Kopftragwerk in den Kern eingeleitet werden, dann wird diese Stabilitätskonstruktion auf zentrischen Druck vorbelastet. Abgesehen davon, daß die Horizontalkräfte zweiter Ordnung damit auch teilweise verschwinden und nur die Windbelastung aufgenommen werden muß, bleiben die maßgebenden Teile des Kernes im Zustand I. Das bedeutet, daß keine Rißbildung auftritt und die Biegesteifigkeit deshalb nicht dadurch reduziert wird, siehe Bild 8.

Ein sehr interessantes Beispiel ist das Kesseltraggerüst eines Kohlenkraftwerkes in Dürnrohr (Österreich) [7].

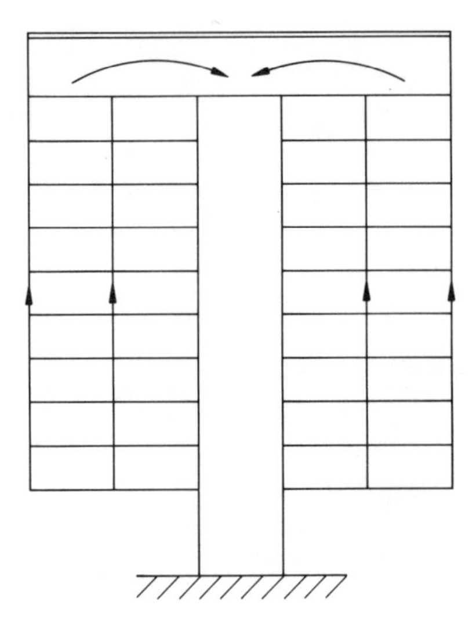

Bild 8
Hängewerk
für ein Bürogebäude.

In dem Tragwerk wird der Kessel mit einer Höhe von 90 m in vertikaler Richtung an ein Kopftragwerk aufgehängt. Das Kopftragwerk wird von vier Eckpylonen getragen. Weil das Kopftragwerk biegesteif mit den Pylonen verbunden ist, entsteht eine räumliche Rahmenkonstruktion, siehe Bild 9.

Mit dem Gewicht der angehängten Fassaden, dem Eigengewicht des Kopftragwerks und des Kessels wird auf einer Höhe von über 100 m eine Belastung von rund 2000 Tonnen in jeden Pylon eingeführt. Mit dem Eigengewicht der Pylonen und den Installationen von rund 2500 Tonnen ist die Vorbelastung in den maßgebenden Querschnitten 4500 Tonnen. Die von dieser Belastung hervorgerufene Druckspannung ist dann $\pm$ 4 N/mm^2 in der Unterseite der Pylonen und $\pm$ 2 N/mm^2 in der Oberseite der Pylonen. Damit wird das Stahlbetontragwerk im Zustand I bleiben, und es können außerdem Zugspannungen durch Temperaturgradienten aufgenommen werden.

Diese drei Beispiele zeigen, was Vorbelasten eines Stahlbetontragwerkes bedeuten kann, nämlich:

1. Reduktion des Bewehrungsgehaltes. Selbstverständlich ist eine bestimmte Mindestbewehrung notwendig. Dadurch ist das Ausmaß der Reduktion beschränkt.
2. Vergrößerung der Biegesteifigkeit entweder durch Beschränkung der maximal auftretenden Biegemomenten oder durch Einfluß einer Normalkraft. In den meisten Fällen bedeutet das auch Beschränkung der Rißbreiten.
3. Vergrößerung der Höhe der Druckzone durch den Einfluß einer Normalkraft, falls eine Stahlbetonwand durch eine, ungleichmäßig über den Querschnitt verteilte Temperatur beeinflußt wird. Wenn die Konstruktion dadurch auf Zwang beansprucht wird, d.h. sich nicht frei dehnen kann, dann entstehen Biegemomente. Vergrößerung der Normalkraft bedeutet die Vergrößerung der Druckzone, siehe Bild 10.
4. In den bisherigen Beispielen wurde noch nicht über zeitabhängige Einflüsse gesprochen, z.B. Zuwachs der Durchbiegung wegen Kriechen und Schwinden im Laufe der Zeit. Wenn die Vorbelastung so groß ist, daß in Zusammenspiel mit der Dauerbelastung auf die Konstruktion die resultierende Belastung nahezu Null ist, dann sind die Durchbiegungen einer Betondecke gering und dadurch auch der Zuwachs der Durchbiegung im Laufe der Zeit.

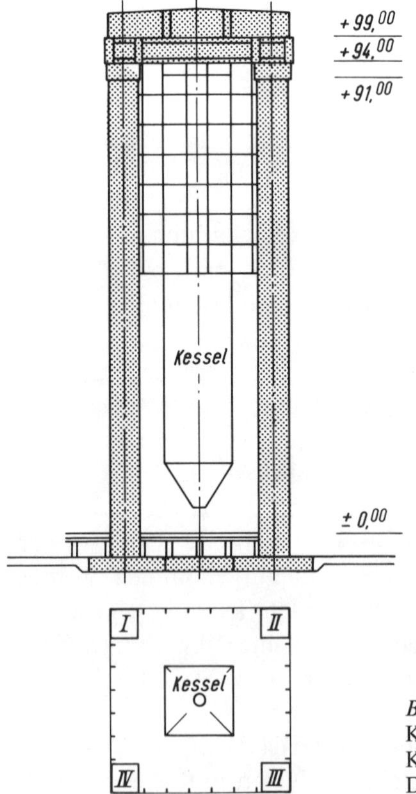

Bild 9
Kesseltraggerüst
Kohlenkraftwerk
Dürnrohr.

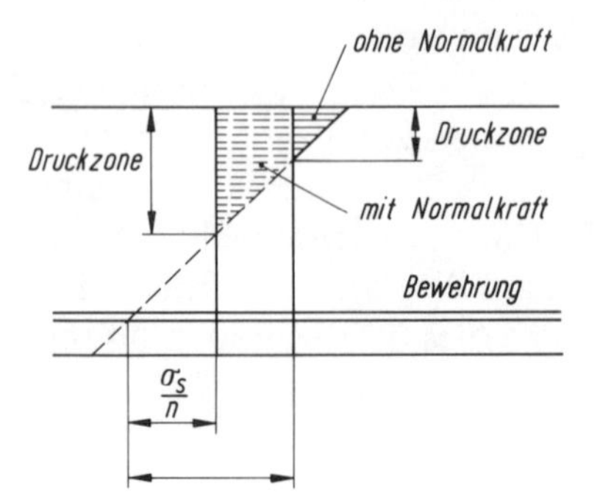

Bild 10
Einfluß der Normalkraft
auf die Höhe der
Druckzone.

4 Vorbelasten von Stahlbetonkonstruktionen mittels Spannverfahren

Aus den erwähnten Beispielen wird es deutlich, daß Vorbelasten von Stahlbetonkonstruktionen von wesentlicher Bedeutung für das Verhalten dieser Konstruktionen sein kann. Auch ist deutlich geworden, daß Vorbelasten sich in vielen Fällen praktisch und wirtschaftlich nicht durchführen läßt. Man braucht Wasserdruck, Erddruck, Kopftragwerke usw., um eine Vorbelastung zu ermöglichen.
Man kann deshalb auch die Vorspannung wie eine Vorbelastung betrachten, d.h. eine zentrische Druckkraft mit/ohne quer zur Konstruktionsachse wirkender Umlenkkräfte durch gekrümmte Spannglieder. Manchmal ist die Normalkraft von geringer Bedeutung, und es spielt die Belastung durch gekrümmte Spannglieder eine große Rolle. Manchmal auch ist das Umgekehrte der Fall. Betrachten wir wieder die unter 3 gegebenen Beispiele.

4.1 Stahlbetondecke oder Stahlbetonträger

In Bild 11 ist in einem Stahlbetonträger der Verlauf eines Spannkabels über mehrere Stützen angegeben. Wenn die Vorspannkraft in diesem Spannkabel die Größe P hat, dann ist der Krümmungsdruck in dem Feld P/R_f und über den Stützpunkten P/R_s. Die Gleichgewichtsbedingung erfordert:

$$P/R_f \cdot l_f + P/R_s \cdot l_s = 0\,.$$

Hierbei sind Reibungsverluste nicht in Betracht gezogen. Die maßgebenden Belastungen auf die Stahlbetonkonstruktion sind:

1. Anfangsvorspannung P_0/R_f und Eigengewicht (q_g);
2. Vorspannung im Gebrauchszustand (P_∞/R_f) und Vollbelastung ($q_{\max}$);
3. Lastfall $q_b = P_\infty/R_f$; in diesem Fall wirkt im Feld nur eine zentrische Normalkraft.

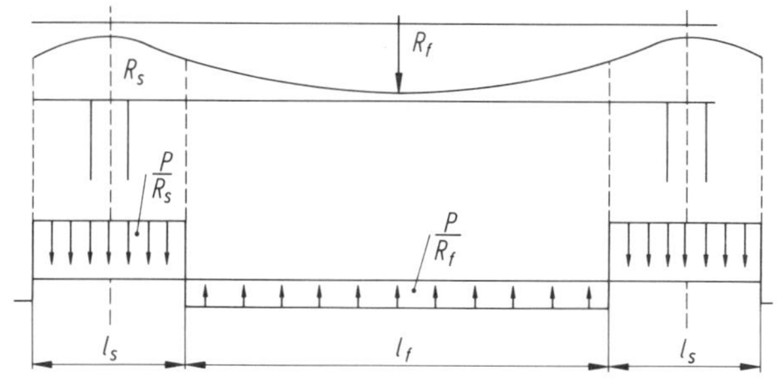

Bild 11
Verlauf von Spannkabel in einem Stahlbetonträger – oder Decke.

Die Belastung ($q_{\max} - q_b$) wird von der Stahlbetonbewehrung getragen. Dazu wird der Träger oder die Decke wie ein nomales Stahlbetontragwerk bewehrt. In vielen Fällen kann der Einfluß der Normalkraft bei der Bemessung der Bewehrung vernachlässigt werden, weil diese oft stark reduziert wird, da die Normalkraft oft von biegesteifen Säulen aufgenommen wird. Die Stahlbetondecke (oder Träger) ist dann, durch die Spannkabel, nahezu auf dieselbe Art vorbelastet wie der Kellerboden (Beispiel 3.1) durch Wasserdruck. Dabei ist Hochwasser zu vergleichen mit Anfangsvorspannung und Niedrigwasser mit Vorspannung im Gebrauchszustand. Wenn man Spannkabel ohne Verbund benützt, ist im Bruchzustand der Zuwachs der Vorspannung nahezu zu vernachlässigen. Bei großen Durchbiegungen nehmen im Bruchzustand die Umlenkkräfte P/R zu. Dieser Zuwachs kann in der Berechnung mit in Betracht gezogen werden, wobei $q_u = \gamma \cdot q_{\max} - r \cdot P_\infty/R_f (r > 1)$.
Die Stahlbetonkonstruktion muß so bewehrt werden, daß die Bruchlast q_u aufgenommen werden kann. Selbstverständlich muß eine Mindestbewehrung eingehalten werden, d.h. das Betontragwerk muß nach den Regeln von Stahlbeton bewehrt werden.
Wenn Spannkabel mit Verbund benutzt werden, kann oft der Vorspannstahl in der Feldmitte und über den Stützpunkt für die Berechnung des Bruchmoments mitberücksichtigt werden. Durch die Wirkung des Krümmungsdrucks der Spannkabel wird nicht nur die Durchbiegung der Decke oder des Trägers reduziert sowie auch der zeitabhängige Zuwachs der Durchbiegung. Vorbelasten von Stahlbetondecken kann deshalb von großer Bedeutung sein für das Verhalten der Konstruktion und auch für die Wirtschaftlichkeit.

4.2 Behälterwand

Es ist klar, daß man durch das Vorspannen einer Behälterwand die Zugspannungen, erzeugt vom Flüssigkeitsdruck, einfach aufnehmen kann. Wenn starke Temperaturgradienten über die Dicke der Wand auftreten, z.B. „cold spot“ Belastung einer Sicherheitswand, dann ist eine Bewehrung sehr wichtig, um die Rißbreiten zu beschränken und eine bestimmte Höhe der Druckzone zu gewährleisten (siehe Bild 10). Das bedeutet, daß man, im Grunde genommen, mittels Vorspannkabel oder einem Wickelverfahren den Stahlbetonbehälter künstlich vorbelastet. Es ergeben sich auch Beispiele, worin man Bodenplatten und Wände von Wasserbehälter mit rechteckigem Grundriß durch gewölbte Spannkabel vorbelastet, um die Wasserdichtigkeit (Rißbreiten, Höhe der Druckzone) zu gewährleisten.

4.3 Kern eines mehrstöckigen Gebäudes

Der aussteifende Kern des Gebäudes von Bild 8 kann auch mittels Vorspannkabel in jenen Zonen vorbelastet werden, welche durch Wind und Stabilitätskräfte in den Zustand II geraten. Man kann eine solche zentrische Normalkraft in den Kern einleiten, daß unter allen Umständen im Gebrauchszustand keine Rißbildungen mehr zu erwarten sind. Damit wird dann auch die Biegesteifigkeit vergrößert und die Effekte zweiter Ordnung wesentlich reduziert. Auch hier kann also Vorbelasten, mittels Vorspannverfahren, eine gute Konstruktion und eine wirtschaftliche Lösung ergeben, ohne daß man gezwungen ist, wesentliche Änderungen in der ursprünglich vorgeschlagenen Konstruktion durchführen zu müssen.
Diese Beispiele zeigen deutlich, daß Vorbelasten, mittels Vorspannverfahren, von Stahlbetonkonstruktionen zu wesentlichen Möglichkeiten führt. Man kann dabei folgende Tatsache feststellen: durch Vorspannung wird das Verhalten einer Konstruktion im Bereich des Gebrauchszustandes eindeutig verbessert, und zwar durch:

1. Einfache Bewehrung des Stahlbetontragwerkes;
2. Beschränkung von Rißbreiten, insbesondere auch in Fällen, wo Teile eines Stahlbetontragwerkes unter Temperatureinflüsse sich nicht unbehindert dehnen lassen;
3. Erhöhung der Biegesteifigkeit und Beschränkung der (zeitabhängigen) Durchbiegung.

Die Anwendung der Vorspannung in Stahlbetontragwerken ist nur möglich, wenn die folgenden Bedingungen erfüllt sind:

1. Das Stahlbetontragwerk muß dimensioniert werden nach den bekannten Regeln für Stahlbetonbau; z. B. in bezug auf Mindestbewehrung, Beschränkung der Rißbreiten, Durchbiegung, Betonüberdeckung usw.
2. Die Anwendung der Vorspanntechnik findet unter gleichen Bedingungen statt wie im Spannbetonbau üblich, nämlich:
   2.1. Spanngliedführung – möglichst einfache Führung der Spannglieder ist einzuhalten;
   2.2. Bewehrung der Verankerungszonen, Schutz gegen Korrosion der Verankerungen nach dem Auspressen der Hüllrohren;
   2.3. in Rechnung stellen von Verluste in der Vorspannkraft durch Reibung (während des Vorspannens der Spannglieder), durch zeitabhängige Einflüsse (Relaxation, Schwinden, Kriechen) und durch den Einfluß von Verkürzungen auf der Kraftverteilung in dem gesamten Tragwerk.
3. Die Dauerfestigkeit der Tragwerke muß gewährleistet sein.

In dieser Hinsicht ist die Betonüberdeckung der Spannglieder sehr wichtig, jedoch auch die Gewährleistung des Korrosionsschutzes in den Rissen der Zugzone.
Daher sind Einzellitzen in Kunststoffrohren, wie in Vorspannung ohne Verbund, sehr geeignet. Sie sind jedoch nur für kleine Vorspannkräfte anzuwenden.
Spannkabel in Hüllrohren sind, bei gut ausgeführtem Auspressen dieser Rohre nach dem Vorspannen, auch gut geschützt, wie dies Versuche an der TH Aachen [10] gezeigt haben. Die ausgepreßten Hüllrohre mit Spannstahl überbrücken die Risse in der Zugzone. Übrigens soll dieses Anwendungsgebiet in der Zukunft theoretisch und experimentell noch besser erfaßt werden. Man muß dabei der Praxis *einfache* Methoden anbieten, mit welchen man die Verbesserung der Eigenschaften eines Stahlbetonwerks durch Vorspannung bestimmen kann. Auf vielen Gebieten sind die Kriterien, d. h. Aussagen über Rißbreiten, Verformungen und Steifigkeiten, schon vorhanden. Diese stützen sich auf dasjenige, was auf dem Gebiet des Stahlbetonbaues schon erforscht worden ist.
Heute sind schon die Resultate verschiedener Versuchsreihen bekannt, bei welchen Stahlbetonträger mit verschiedenen hohen Vorspannungen geprüft wurden [11], [12].

5 Anwendungsbeispiele

5.1 Beschränken der Durchbiegung

Schon vor 30 Jahren ist in der Schweiz eine weitgespannte Stahlbeton-Massivplatte für ein Kaufhaus ausgeführt worden, bei welcher zusätzlich eine Anzahl parabolisch gekrümmte Spannkabel eingebaut wurden. Dadurch wurde verhindert, daß größere Durchbiegungen, die durch Schwinden und Kriechen des

Betons, auftreten konnten. Ähnliche Anwendungen liegen auch aus den Niederlanden vor. Sie beziehen sich auf Rahmentragwerke von Bürogebäuden. Auf dem Gebiet der Flachdecken hat das Vorspannen von Stahlbetondecken sich schon verschiedene Jahre in der Praxis bewehrt. Insbesonders erwähnenswert ist das Vorspannen mit Spanngliedern ohne Verbund in den Säulenstreifen der Flachdecken. Dadurch konnte die Durchbiegung dieser Decken beschränkt werden. Mit Ausnahme der Zonen der Stützpunkte kann man in diesen Stahlbetonflachdecken mit einer einfachen Mindestbewehrung auskommen.
In der Pyhrn-Autobahn in Österreich sind zur Zeit zwei Moorbrücken bei Trieben in Bau. Insgesamt 4,5 km Moor wird dabei überbrückt. Dazu sind Stahlbetonträger vorgefertigt mit einer Spannweite von 15 Meter. Nach der Montage dieser I-förmigen Träger wird eine Betondecke mit einer Breite von 12 m je Fahrrichtung an Ort und Stelle betoniert. Diese Betondecke ist mit 4 Stahlbetonträger je Fahrrichtung verbunden. Damit während dem Zusammenbau und dem Betonieren der Brückendecke keine zu große Durchbiegung auftreten wird, sind die Träger mittels parabolisch gekrümmte Spannglieder aufwärts vorbelastet.

5.2 Vermeiden von Dilatationsfugen

Die Nachklärbecken und Belüftungsbecken der Kläranlage Werdhölzli in Zürich werden vorgespannt, damit die Stahlbetonkonstruktionen ohne Dilatationsfugen gebaut werden konnte. Die Nachklärbecken mit 76 500 m^3 Inhalt stehen über einen benutzten Grundwasserstrom, und daher werden an die Dichtigkeit der Becken hohe Anforderungen gestellt. Gewählt wurde eine vorgespannte Stahlbetonkonstruktion ohne Dilatationsfugen, wobei je 6 Becken in einem Block zusammengefaßt sind.
Die zwei Nachklärbecken – die Blöcke haben die Abmessungen $145 \times 66 \times 6{,}4$ m^3.
Die zwölf Belüftungsbecken werden auch je 6 Becken in einem Block zusammengefaßt mit Abmessungen von $91 \times 80 \times 6{,}1$ m^3. Insgesamt werden hier mehr als 600 Tonnen Spannstahl verwendet.

5.3 Beschränken der Rißbildung und Rißbreiten

Ein interessantes Beispiel ist die Beschränkung der Rißbreiten in Tunnelbau.
Es handelt sich hier um einen Straßentunnel unter einem Fluß mit zwei Rohren je 10 m breit. Diese Tunnels werden abschnittsweise im Trockendock hergestellt und schwimmend transportiert. Dann werden die ca. 24 m breiten und ca. 100 m langen Tunnelelemente in richtiger Lage in eine ausgebaggerte Grube im Fluß abgesenkt. Diese Grube wird später wieder zugeschüttet.
Ein weiteres Beispiel ist der Benelux-Tunnel in der Nähe von Rotterdam.
Bei Höchstwasser kann eine Belastung von 24 m Wasser auf dem Tunneldach und von 32 m auf dem Tunnelboden wirken. Im Trockendock hat das Tunneldach nur das Eigengewicht zu tragen.
Jeder, der mit dieser Bauweise vertraut ist, weiß, daß das Dach und der Boden dieser Tunnelelemente sehr stark bewehrt sind, weil ihre Abmessungen beschränkt sind wegen dem Schwimmvermögen beim Transport. Das bedeutet auch, daß die Rißbreiten unter Vollbelastung oft über 0,4 mm betragen.
Durch parabolisch gekrümmte Spannglieder kann man diese Decke und Böden so vorbelasten, daß im Trockendock eine Belastung von rund 12 m Wasser aufwärts wirkt und bei Höchstwasser im Fluß eine Belastung von rund 12 m Wasser nach unten wirkt. Ähnliche Zahlen können für den Tunnelboden angegeben werden. Das bedeutet, daß die Belastung auf dem Tunneldach, einschließlich die Vorbelastung, nur noch die Hälfte beträgt und dabei entweder aufwärts (Trokkendock) oder nach unten (endgültige Lage) wirkt.
Das Tunneldach kann daher viel einfacher mit einer symmetrischen Bewehrung an Ober- und Unterseite der Platte bewehrt werden. Die Rißbreiten können dann im Gebrauchszustand unter 0,2 mm gehalten werden [13]. Es ist noch zu erwähnen, daß im Gebrauchszustand der Tunnel in Längsrichtung durch den Wasserdruck vorgespannt ist.

Die angeführten Beispiele machen es klar, daß durch die Vorspannung das Verhalten einer Stahlbetonkonstruktion im Bereich des Gebrauchszustandes eindeutig verbessert werden kann. Man kann daraus deutlich erkennen, daß die Anwendungsgebiete der Vorspanntechnik im Sinne von FREYSSINET noch nicht alle erfaßt sind und daß auf dem Gebiet des Stahlbetonbaues noch viele Möglichkeiten zur Anwendung der Vorspannung sich anbieten.

6 Zusammenfassung

In diesem Beitrag wird die Frage gestellt, ob die 1941 von FREYSSINET in der Zeitschrift „Travaux" gemachte Aussage, daß Vorspannen „eine Revolution für die Kunst des Bauens" bedeutete, richtig war. Diese Frage kann heute ohne weiteres bejaht werden. Man muß jedoch klar festhalten, was FREYSSINET mit dem Grundprinzip des Vorspannens sagen wollte. Dazu wurde die Definition von FREYSSINET genauer betrachtet, und es hat sich gezeigt, daß es sich nicht nur um das Einführen von künstlich erzeugten Belastungen handelt.

Anhand von drei Beispielen aus der Praxis wurde gezeigt, was Vorbelasten für eine Stahlbetonkonstruktion bedeutet. Daraus konnten interessante Folgerungen gezogen werden. Wenn man dann wieder die Aussage von FREYSSINET und seine Definition vom Vorspannen betrachtet, wird es klar, daß es sich lohnt, mittels Spannbetontechniken Stahlbetonkonstruktionen künstlich vorzubelasten. Das bedeutet, daß die von FREYSSINET angedeutete „Revolution" wieder einen neuen Impuls bekommen kann und dadurch neue Gebiete für die Vorspannung eröffnet werden. Zum Schluß wurde anhand einiger Beispiele gezeigt, welche neue Möglichkeiten sich anbieten und wie man dabei die Vorspanntechnik benutzen kann.

7 Literatur

[1] FREYSSINET, E.: Une révolution dans l'art de bâtir: les constructions précontrainte. In: Travaux (1941) Novembre.

[2] BIRKENMAIER, M.: Über einige Begriffe im Spannbeton. Institut für Bauwissenschaftliche Forschung. Zürich, Verlag Leeman.

[3] LIN, T. Y.: Load balancing method for design and analysis of prestressed concrete structures. In: Journal of the ACI (1963) January.

[4] BRUGGELING, A. S. G.: Theorie en praktijk van het voorgespannen beton. Amsterdam, Verlag „Het Koggeschip", 1963.

[5] CUPERUS, M. J.: Cryogenic storage facilities for LNG and NGL. Bucarest, World Petroleum Congress, 1979.

[6] BREUGEL, K. VAN: Minimizing of the vulnerability of storage systems for liquefied energy gases. Amsterdam, Second International Conference on Cryogenic Concrete, 1983.

[7] NUSSBAUMER, H.: Bautechnische Gesichtspunkte zum Entwurf und Bau von Stahlbetontraggerüsten für kohlegefeuerte Dampferzeuger. In: Elektrizitätswirtschaft (1983), S. 390–400.

[8] MEHMEL, A.: Vorgespannten Beton. 1957.

[9] FJELD, S. and ROLAND, B.: In-service experience with eleven offshore concrete structures. OTC 4358 – Offshore Technology Conference.

[10] THORMÄHLEN, U.: Zum Einfluß von Spanngliedern mit nachträglichem Verbund auf Rißbildung und Rißbreitenbeschränkung bei teilweiser vorgespannten Betonkonstruktionen. Dissertation Aachen, 1979.

[11] BACHMANN, H. und FLÜCKIGER, P.: 5-Jahres-Resultate der Langzeitversuche an teilweiser vorgespannten Leichtbetonbalken. Bericht des Instituts für Baustatik und Konstruktion der ETH-Zürich, Nr. 6504-7, 1976.

[12] BRUGGELING, A. S. G., BRUNEKREEF, S. H. und WALRAVEN, J. C.: Partially prestressed concrete. Theory and experiments. HERON 23 (1978) No. 1.

[13] BRUGGELING, A. S. G.: Partially prestressed concrete: a challenge for concrete designers. In: Proceedings of the International Symposium „Nonlinearity and continuity in prestressed concrete", University of Waterloo, Canada, 1983.

Geschweißte Betonstahlmatten

Franz Herkommer

1 Einleitung

Im Laufe von etwa 50 Jahren seit Einführung der geschweißten Betonstahlmatte als statische Bewehrung in Deutschland hat dieses rationelle Bewehrungselement in nahezu allen Bereichen des Stahlbetonbaus Eingang gefunden. Jährlich werden bei einem Anteil von rd. 40% am gesamten Bewehrungsmarkt 1,1–1,2 Millionen Tonnen Betonstahlmatten in flächenartigen und stabförmigen Bauteilen eingebaut.
Der Durchbruch im Markt erfolgte in den 60er und 70er Jahren und war der Erfolg eines marktgerechten Programmangebots und einer engen Zusammenarbeit mit Ingenieuren in Forschung, Baubehörden, Ingenieurbüros, Bauindustrie sowie mit dem Handel.
In den vergangenen Jahren wurde eine Reihe von theoretischen und praktischen Untersuchungen durchgeführt, deren Ergebnisse in DIN-Normen sowie allgemeinen Vorschriften, in der Fachliteratur und in Veröffentlichungen der Mattenhersteller ihren Niederschlag gefunden haben. Über einige Schwerpunkte der Forschungs- und Entwicklungstendenzen der geschweißten Betonstahlmatten in den letzten 2 Jahrzehnten soll im Folgenden aus der Sicht eines in der Mattenindustrie tätigen Bauingenieurs berichtet werden. Dabei muß verständlicherweise hier auf Einzelheiten verzichtet und auf die angegebene Fachliteratur verwiesen werden.

2 Verbund, Verankerung und Übergreifungsstöße

2.1 Verankerung und Verbund

Die Stäbe für geschweißte Betonstahlmatten wurden bis zum Jahre 1957 mit glatter und von 1957 bis 1968 mit profilierter Oberfläche, Betonstabstähle seit Mitte der 50er Jahre mit gerippter Oberfläche hergestellt.
Verankerungen an Mattenenden mußten entweder durch angeschweißte Querstäbe oder durch Endhaken vorgenommen werden. Eine Verankerung nur durch Haftverbund war nicht möglich, nachdem sich in Ausziehversuchen gezeigt hatte, daß bei solchen Verankerungen der Schlupf an den Stabenden zu groß wird. Für die Sicherstellung eines ausreichenden Verbundes zwischen Stahl und Beton mußten auch bei Matten aus profilierten Stäben noch eine ausreichende Anzahl scherfest verschweißter Querstäbe in den Matten angeordnet werden. Das Rißbild wurde durch die Lage der Querstäbe bestimmt. Die „bezogene Rippenfläche“ von profilierten Stäben entsprach nur etwa einem Drittel des Wertes von gerippten Stabstählen. Somit waren die Erkenntnisse aus [1] sowie die theoretischen und praktischen Untersuchungen zur Vorausberechnung der Rißabstände und Rißbreiten bzw. zur Beschränkung der Rißbreite unter Gebrauchslast für die Betonstahlmatte praktisch nicht anwendbar. Durch zunehmende Anwendung von Matten im Ingenieurbau – besonders Tiefbau – sowie Einsatz von Doppelstabmatten mit bis zu 12,0 mm Stabdurchmesser zeichnete sich die Notwendigkeit ab, die Staboberflächen viel stärker zu profilieren, de facto also die Stäbe zu rippen. Die BAUSTAHLGEWEBE GMBH entwickelte daraufhin mit dem KARI-Verfahren eine wirtschaftliche Methode der Herstellung von kaltgewalztem Betonrippenstahl und stellte im Jahre 1968 ihre gesamte Produktion auf Matten aus gerippten Stäben um. Diesen Schritt vollzogen in gleicher Weise die anderen Hersteller, so daß seit etwa 1970 im deutschen Baumarkt nur noch Matten aus gerippten Stäben Verwendung finden, ein Vorteil für Anwender und Hersteller.
Den Unterschied in den zulässigen Verbundspannungen bei glatten, profilierten, gerippten Stäben von Betonstahlmatten BSt 500 gemäß derzeitiger DIN 1045 zeigt beispielhaft Bild 1.
Die bezogene Rippenfläche für gerippte Stäbe von Betonstahlmatten wurde den Werten von gerippten Betonstabstählen angepaßt.

Oberflächengestaltung	zul. Verbundspannung MN/m²
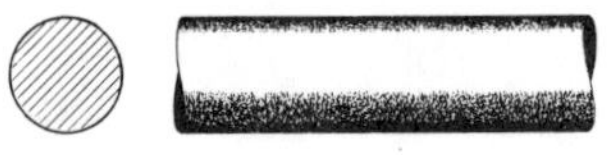 glatt	0,7
 profiliert	1,0
 gerippt	1,8

Bild 1
Oberflächengestaltung der Stäbe von Betonstahlmatten und zulässige Verbundspannungen (Beispiel: Grundwerte für Beton B 25, Verbundbereich I nach DIN 1045, Ausgabe Dezember 1978).

Haftverbund oder Haftverbund + Scherverbund durch Querstäbe?
Diese Frage stellte sich natürlich jetzt dem in Forschung und Anwendung tätigen Bauingenieur bei der Einführung von gerippten Stäben in Betonstahlmatten.
Aus Versuchen mit profilierten Stäben [2] war bekannt, daß die Scherfestigkeit von verankernden, einbetonierten Querstäben um mindestens 30% höher lag als im nicht einbetonierten Zustand.
Das Zusammenwirken von Schweißknoten und Verbund bei gerippten Stäben hat sich in Ausziehversuchen, über die in [3] berichtet wird, deutlich herausgestellt. Gegenüber dem reinen Haftverbund wird die Verankerung durch aufgeschweißte Querstäbe noch verbessert. Die Verankerungswirkung hängt wesentlich von dem Verhältnis $\varnothing_{quer}/Fe_L$ sowie bei mehreren Verankerungselementen auch von deren Abständen ab.

Die Erkenntnisse aus diesen Forschungen haben sich in DIN 1045, Dezember, 1978, Abschnitt 18 – Tabelle 20 – niedergeschlagen; dem Wunsch der Praxis nach kurzen Verankerungsmöglichkeiten konnte mit diesen Regeln voll entsprochen werden.

2.2 Übergreifungsstöße (Stöße von Tragstäben)

Mit gerippten Stäben in Betonstahlmatten ergaben sich für Übergreifungsstöße neue Konstruktionsmöglichkeiten, z. B. Einebenenstöße mit langen Überständen nach Bild 2. Auch war zu erwarten, daß bei Zweiebenenstößen weniger „wirksame" Querstäbe erforderlich werden.
Deshalb wurden am Institut für Baustoffkunde und Stahlbetonbau der Technischen Universität Braunschweig 4 Versuchsprogramme durchgeführt [4] und dabei viele Varianten mit unterschiedlichen Anzahlen von Querstäben im Stoßbereich überprüft.
Die Versuche ergaben, daß der Einfluß der Querstäbe im Stoß unbedeutend ist; bei Zweiebenenstößen wird jedoch zur Verhinderung des Abklappeffektes an jedem Mattenende im Stoßbereich mindestens ein „wirksamer" Querstab erforderlich. Dagegen ist der Einfluß des vorhandenen Stahlquerschnitts – also Stababstände und Stabdurchmesser – auf die erforderliche Übergreifungslänge erheblich.
Nach derzeitiger Regelung gemäß DIN 1045, Dezember 1978, kann die Übergreifungslänge von Betonstahlmatten im Einebenenstoß wie für Betonstabstahl ermittelt werden; für Zweiebenenstöße mußte aufgrund der Versuchsergebnisse [4] eine differenziertere Regelung erfolgen. Der Einebenenstoß wird im oberen Querschnittsbereich deutlich kürzer als der Zweiebenenstoß.

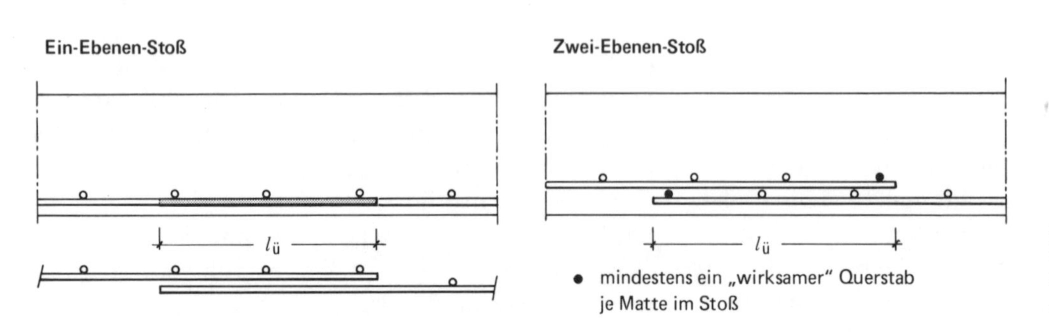

Bild 2
Übergreifungsstöße von Betonstahlmatten nach DIN 1045, Ausgabe Dezember 1978.

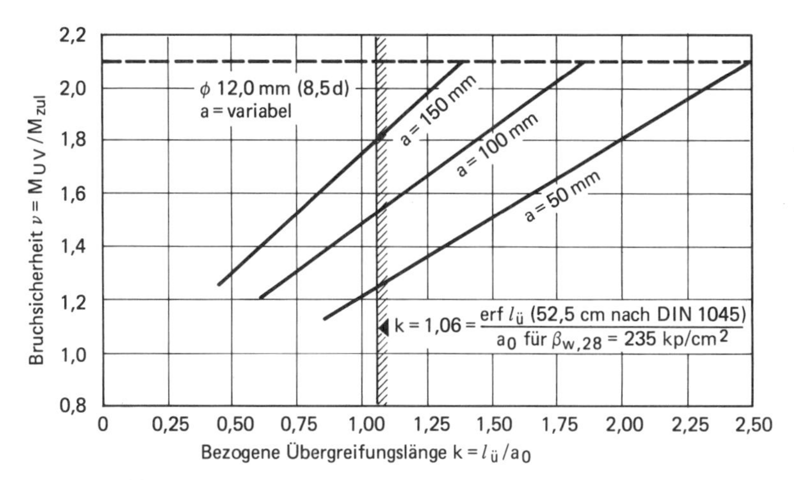

Bild 3a
Einfluß der Übergreifungslänge auf die Bruchsicherheit im Stoß in Abhängigkeit vom gestoßenen Stahlquerschnitt nach [4].

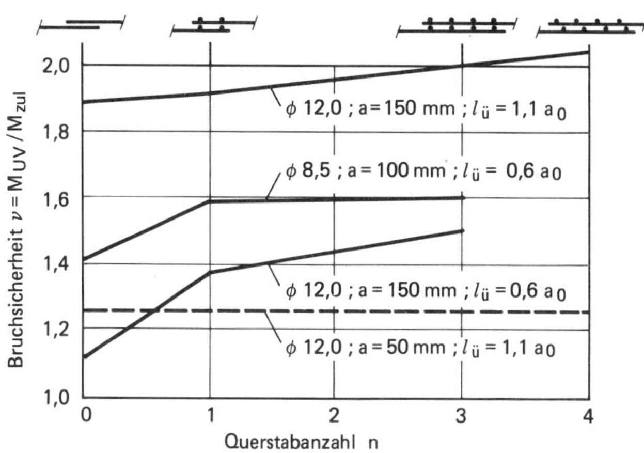

Bild 3b
Einfluß der Anzahl der sich abstützenden Querstäbe auf die Bruchsicherheit im Stoß nach [4].

3 Schubtragverhalten von mattenbewehrten flächenartigen und stabförmigen Bauteilen

3.1 Flächenartige Bauteile ohne/mit Schubbewehrung

Stahlbetonplatten ohne Schubbewehrung, ein hauptsächliches Anwendungsgebiet der Betonstahlmatten, konnten bis in die 60er Jahre ohne Beachtung einer speziellen Vorschrift hinsichtlich Begrenzung von Schubspannungen in Abhängigkeit von der Bewehrungsführung ausgeführt werden.

In höherbelasteten Decken des Industriebaus sowie im Tiefbau – speziell U-Bahn-Bau – wurde auch nach den damaligen Vorschriften oft die Schubspannungsgrenze für den Nachweis der Schubsicherung überschritten und dann eine volle Schubdeckung erforderlich, während dies für Platten im Wohnungs- und Verwaltungsbau wegen der relativ niedrigen Nutzlasten nicht notwendig wurde.

Um Stahl zu sparen, wurden Betonstahlmatten in Platten fast immer zweilagig „gestaffelt“ vorgesehen, in

– Zulagestaffelung oder
– verschränkter Staffelung.

Vorversuche der BAUSTAHLGEWEBE GMBH an Stahlbetonplatten mit durchgehender oder gestaffelter Bewehrung ließen die Notwendigkeit differenzierter Anwendungsregeln erkennen.

Im Hinblick auf die Neufassung Deutscher Stahlbetonbestimmungen und unter Auswertung der Ergebnisse der Stuttgarter Schubversuche – 1961 bis 1964 – war es angezeigt, die Schubtragfähigkeit von Platten ohne Schubbewehrung und mit abgestufter Biegezugbewehrung genauer zu untersuchen.

Über diese Versuche, die noch mit Matten aus profilierten Stäben – im Gegensatz dazu heute gerippte Stäbe – durchgeführt wurden, ist in [5] berichtet worden.

Es hat sich gezeigt, daß bei Platten ohne Schubbewehrung durch

- Vergrößerung der Verankerungslängen der abgestuften Bewehrung
- Begrenzung der Schubspannungen bei gestaffelter (abgestufter) Feldbewehrung
- Begrenzung der Schubspannungen bei durchgehender Feldbewehrung

ausreichende Schubbruchsicherheit erreicht wird. Auf dem Umweg über Schubspannungen wird somit von der Zugfestigkeit des Betons Gebrauch gemacht.

Die Staffelung der Stützbewehrung hatte keinen Einfluß auf diese „Schubtragwirkung“ und kann daher immer ausgeführt werden.

Im Zulassungsbescheid über „Geschweißte Betonstahlmatten BAUSTAHLGEWEBE“ vom 8. Dezember 1967 fanden diese Versuchsergebnisse erstmalig ihre offizielle Berücksichtigung [7]; in den danach folgenden Ausgaben der Stahlbetonnorm DIN 1045 – Januar 1972 und Dezember 1978 – wurden diese Regelungen in modifizierter Form beibehalten und damit dem planenden Bauingenieur sichere Konstruktionsregeln gegeben.

Heute kann die gestaffelte Bewehrung einlagig mit Feldsparmatten, bei denen verschiedene lange Stäbe in einer Mattenrichtung angeordnet sind, stahl- und arbeitszeitsparend ausgeführt werden [8], [25].

Bei niedrigen Bewehrungsquerschnitten – bis zu etwa 3 cm^2/m – hat die gestaffelte Bewehrung aus wirtschaftlichen Gründen an Bedeutung verloren.
Auch für Platten, in denen Schubbewehrung erforderlich wird, stehen geeignete Bewehrungselemente aus gebogenen Betonstahlmatten zur Verfügung [8].

3.2 Betonstahlmatten als Schubbewehrung in stabförmigen Bauteilen

Die technischen und wirtschaftlichen Vorteile in der Verwendung von räumlichen Bewehrungselementen aus Betonstahlmatten hat die Bauindustrie schnell erkannt. Eine Vielzahl von praxisgerechten Schneide- und Biegemaschinen – bis heute etwa 4000 Stück – finden in der Baupraxis Verwendung. Darüber hinaus verfügen die Mattenhersteller selbst noch über speziellere Verarbeitungsmöglichkeiten von Matten.
Der Schwerpunkt der Verwendung wie auch der noch zu klärenden Fragen für stabförmige Bauteile lag bei Stahlbetonbalken; die Anwendung von Betonstahlmatten in Stützen war demgegenüber relativ problemlos und wirtschaftlich weniger bedeutend.
Durch die umfangreichen Stuttgarter Schubversuche wurden neben der Klärung einer Vielzahl von Problemen der Schubdeckung und -bewehrung auch speziell den technischen und wirtschaftlichen Besonderheiten und Vorteilen der Betonstahlmatten Rechnung getragen. In [6] wurde auf die technischen und wirtschaftlichen Vorteile von gebogenen Betonstahlmatten hingewiesen.
Daraufhin wurden speziell „Geschweißte Bewehrungsmatten als Bügelbewehrung" untersucht und „Schubversuche an Plattenbalken und Verankerungsversuche" durchgeführt [9]. Das Ziel dieser Untersuchungen lag darin, die Festigkeitseigenschaften der Betonstahlmatten auch als Schubbewehrung in Balken voll ausnützen zu können und ggf. dafür entsprechende Regelungen für die Verwendung als Bügel festzulegen. Allein zur Untersuchung der Verankerung wurden dabei 21 Variationen geprüft [9]. Die Stäbe waren profiliert.
Als Ergebnisse dieser Untersuchungen wurde in [7] eine zulässige Stahlspannung für Bügel von 2400 kp/cm^2 genehmigt; dieser Wert war an besondere Verankerungen gebunden.
Am OTTO-GRAF-Institut der Universität Stuttgart wurden im Jahre 1974 weitere Untersuchungen an Stahlbetonbalken, die mit Bügelkörben aus gerippten Stäben bewehrt waren, vorgenommen [10] und zur Klärung der Frage „Mögliche Erhöhung der Bemessungsspannung für Bügel der Güte BSt 50/55 R auf $\sigma_e = 2860$ kp/cm^2" zusätzlich theoretische Untersuchungen angestellt.
Als Ergebnis wurde festgestellt, daß die Bemessungsspannung von $\sigma_e = 2860$ kp/cm^2 möglich ist. Gleichzeitig wurden dafür in Abhängigkeit von den Schubspannungen höchstzulässige – gegenüber bisher kleinere – Bügelabstände gefordert. Die entsprechenden Regeln wurden erstmalig in die „Ergänzenden Bestimmungen zu DIN 1045, Fassung April 1975" aufgenommen.
Weitere Untersuchungen [12] zur konstruktiven Gestaltung der Bügelbewehrung – z. B. offene Bügel im Bereich negativer Momente in Plattenbalken – führten auch zu neuen Bewehrungslösungen [13].
In Analogie zu den Regeln für Verankerung und Stöße

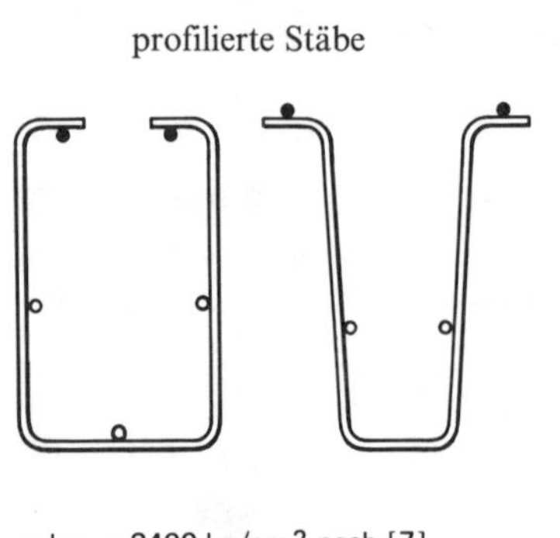

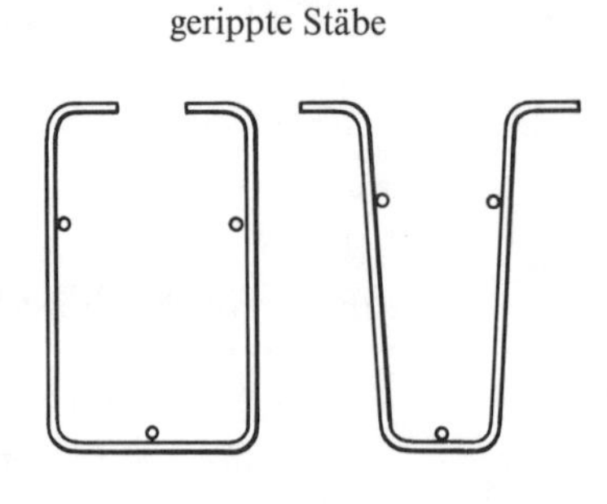

Bild 4
Bügelkörbe aus Betonstahlmatten (Beispiele): zulässige Stahlspannungen bei profilierten und gerippten Stäben.

von Betonstahlmatten in flächenartigen Bauteilen bringt auch hier in stabförmigen Bauteilen die Verwendung gerippter Stäbe die Möglichkeit, die Bügel mit oder ohne Querstäbe zu verankern.
Diese Freiheiten in der konstruktiven Gestaltung von Bewehrungselementen bedeuten in Verbindung mit der hohen zulässigen Bügelspannung erhebliche technische und wirtschaftliche Vorteile.

4 Verhalten unter „nicht vorwiegend ruhender Belastung“

4.1 Allgemeines

Die Anwendung von Betonstahlmatten unter „nicht vorwiegend ruhender Belastung“ war bis zum Jahre 1972 nicht zugelassen. Das Qualitätsmerkmal „Dauerschwingfestigkeit“ spielte erst eine wesentliche Rolle, als Matten im städtischen Tiefbau – z. B. U-Bahn-Bau – und im gewerblichen Bau eingesetzt werden sollten, deren Bauteile rollenden Verkehrsbelastungen aus Straßen- und Schienenfahrzeugen unterlagen.
Für stoßartige Belastungen, die niedrigen Lastspielzahlen unterliegen, wie z. B. im Schutzraumbau, sind entsprechend den dafür geltenden speziellen Richtlinien keine Beschränkungen auferlegt.
Beim Einsatz von Matten in selten befahrenen Decken – wie Hofkellerdecken, Einfahrten für Feuerwehr- oder Müllfahrzeuge etc. – oder in Tunnels, die neben Fahrzeuglasten anteilmäßig hohe ständige Lasten zu tragen hatten, stießen Mattenhersteller und Verwender von den Baubestimmungen her auf Schwierigkeiten, wenn es darum ging, arbeitszeitsparende Mattenbewehrungen zu verwenden.

4.2 Dauerschwingverhalten von Betonstahlmatten aus profilierten und gerippten Stäben

Ein erster, großer Schritt war getan, als die U-Bahn-Ämter großer Städte in ihren „U-Bahn-Richtlinien“ bei Verwendung von Matten unter nicht vorwiegend ruhender Belastung eine Erdüberdeckung von mindestens 1,50 m vorschrieben. Basis für diese Festlegung waren Ergebnisse von Dauerschwingversuchen, die mit Matten aus profilierten Stäben am OTTO-GRAF-Institut der Universität Stuttgart in den Jahren 1965/66 durchgeführt wurden. Damals wurden bei einer Grenzlastspielzahl 2×10^6 Schwingbreiten von 12 bis 14 kp/mm^2 ermittelt.
Versuche in den Jahren 1967/68 an der Technischen Universität München [14] haben gezeigt, daß bei verschweißten, gerippten Stäben gegenüber verschweißten, profilierten Stäben praktisch kein Unterschied in der Dauerschwingfestigkeit besteht; ausschlaggebend für die Dauerschwingfestigkeit ist also nicht die Rippenform, sondern die Schweißstelle.
Auch Untersuchungen an der TH Braunschweig [15] brachten das Ergebnis, daß eine Schwingbreite von 12 bis 16 kp/mm^2 bei einer Lastspielzahl von 2×10^6 erreicht wird.
Über weitere umfangreiche Dauerschwingversuche wurde im Hinblick auf die Stahlzusammensetzung und zusätzliche Bewertungen der Brüche nach metallographischen Untersuchungen in [16] berichtet.

4.3 „dyn-Matten“ in Bestimmungen

Wenn auch die Dauerschwingfestigkeit durch das Schweißen wesentlich niedriger wird als z. B. bei ungeschweißten Betonstählen, so war doch die Regelung einer „zulässigen Schwingbreite“ als ein wichtiges Ziel der Mattenindustrie zu sehen.
In der Stahlbetonnorm DIN 1045, Ausgabe Januar 1972 – Abschnitt 17 – wurde erstmalig für geschweißte Betonstahlmatten eine Anwendungsregel für „nicht vorwiegend ruhende Belastung“ aufgenommen. Die „zulässige Schwingbreite“ von $2\sigma_A = 800$ kp/cm^2 (80 MN/m^2) wurde auch in DIN 1045, Ausgabe Dezember 1978, übernommen.
Die Betonstahlnorm DIN 488 regelte erstmalig in der Fassung April 1972 die Höhe der nachzuweisenden „Versuchsschwingbreite“ von $2\sigma_A = 1200$ kp/cm^2 bei einer Lastspielzahl 2×10^6. Die Prüfbedingungen sind dabei so festgelegt worden, daß ein ungünstiger Merkmalswert – z. B. durch Prüfung am dünneren Querstab – ermittelt wird.
In dem Erfahrungsbericht [17] aus Überwachungsprüfungen wurde 3 Jahre nach Einführung dieser Normen bezüglich Dauerschwingfestigkeit von Betonstahlmatten festgestellt, daß die in DIN 488 im Jahre 1972 erstmalig festgelegten Werte zutreffend sind und eingehalten werden können.

4.4 Zeitfestigkeit, Betriebsfestigkeit

Bei vielen Anwendungsfällen, die gemäß DIN 1055 als „nicht vorwiegend ruhend“ belastet gelten, kann bei vorsichtiger Abschätzung eine überschaubare Last-

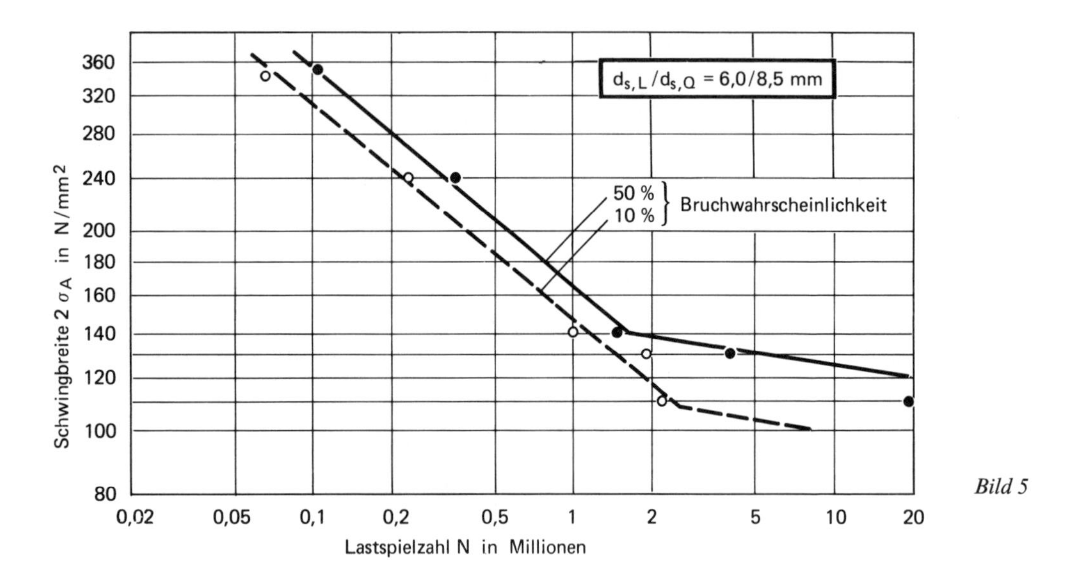

Bild 5

wechselzahl – besonders hinsichtlich der Höchstlast z. B. aus dem Schwerstfahrzeug nach DIN 1072 o. a. – innerhalb einer angenommenen Lebensdauer der Bauwerke ermittelt werden, die u. U. weit unter 2 Millionen liegt.

Über erste Versuche zur Beurteilung des Zeitfestigkeitsbereiches und mögliche Wege zur praktischen Nutzung wird in [19] berichtet.

In [20] wird am Beispiel der verschweißten Stabkombination 6,0/8,5 mm eine normierte Wöhlerlinie als Vorschlag für eine „Bemessungs-WÖHLER-Linie" für Betonstahlmatten aufgezeigt und über umfangreiche Untersuchungen im Zeitschwingfestigkeitsbereich berichtet.

Am gleichen Institut werden derzeit Forschungen über „Schadensakkumulationshypothesen beim Betriebsfestigkeitsnachweis für geschweißte Betonstahlmatten" durchgeführt.

Der Sinn dieser Untersuchungen liegt nicht nur darin, den Zeitfestigkeitsbereich über einen zweiten Wert zul. $2\sigma_A$ zu erschließen, sondern auch bei zukünftigen, neuen Belastungsvorschriften sichere Aussagen machen zu können.

In der z. Z. in Neubearbeitung befindlichen DIN 488 [18] wird daher neben der bisherigen Regel ein zusätzlicher Prüfhorizont $2\sigma_A = 200$ N/mm² bei 2×10^5 Lastwechsel aufgenommen, um zukünftig mit diesem Merkmal das Schwingverhalten im Zeitfestigkeitsbereich zu überwachen.

4.5 Betonstahlmatten für erhöhte dynamische Beanspruchung

In Bereichen ohne Schweißstellen weisen Betonstahlmatten dieselben dynamischen Eigenschaften auf wie ungeschweißte Betonstabstähle und dürfen im nichtgeschweißten Bereich gemäß Zulassungsbescheid [21] auch mit gleichen Schwingbreiten beansprucht werden.

$2\sigma_A = 180$ MN/m² im geraden und schwachgekrümmten Bereich

$2\sigma_A = 140$ MN/m² im Bereich von Abbiegungen und Bügeln.

Solche „Sonderdyn-Matten" müssen als Zeichnungsmatten speziell gefertigt werden.

5 Ausnutzung höherer Streckgrenzen

Betonstahlmatten werden im deutschen Baumarkt als BSt 500/550 RK (IVR) – zukünftig BSt 500 M (IV M) gemäß DIN 488 neu – mit Stabdurchmessern bis 12,0 mm und nur mit gerippten Stäben verwendet.

Mit gleicher Streckgrenze 500 N/mm², aber Stabdurchmesser bis 28 mm sind Betonstabstähle der Sorte IV zugelassen und werden zukünftig in der Qualität IVs genormt.

Für die Mattenindustrie stellte sich die Frage, ob für die von ihr verwendeten kleineren Stabdurchmesser höhere Streckgrenzen ausnutzbar sind.

Stähle höherer Festigkeit sind auf die Streckgrenzenlast bzw. Tragkraft bezogen in der Herstellung billiger

als weniger feste. Ein wirtschaftlicher Einsatz kann natürlich nur lohnend sein, wenn beim Verwender Ersparnisse an Material- und Lohnkosten aus einem kleineren Stahlbedarf entstehen.
Eine Steigerung der Stahlfestigkeiten und deren volle Ausnutzung warf hinsichtlich des Verbundwerkstoffes Stahlbeton eine Reihe von Fragen auf, die theoretisch und durch Versuche an verschiedenen Forschungsinstituten (TU München, TU Stuttgart, IBS München) sehr genau untersucht wurden. Frühzeitig wurde auch der Sachverständigenausschuß Betonstähle des Instituts für Bautechnik mit der Problematik befaßt.

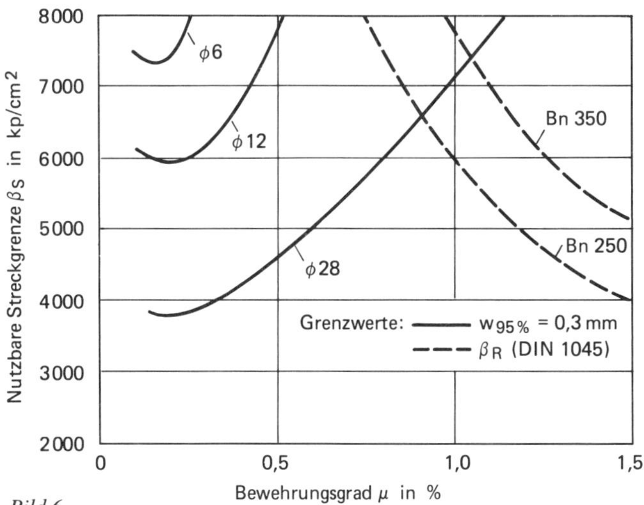

Bild 6
Einfluß von Bewehrungsgrad und Stabdurchmesser auf die nutzbare Streckgrenze nach [22].

Tabelle 1
Gegenüberstellung der Vorschriften über „Beschränkung der Durchbiegung unter Gebrauchslast" für Betonstahlmatten in normaler und höherwertiger Qualität

BSt 500/550 RK nach DIN 1045			BSt 630/700 RK nach [23]		
Nutzlast P	zul l_i/h bei Bauteilen ohne	mit Trennwänden	Nutzlast P kN/m² (kp/m²)	zul l_i/h bei Bauteilen ohne	mit Trennwänden
keine Regelung	35	$150/l_i$	≤ 2,75 (≤ 275)	35	$150/l_i$
			> 2,75 (> 275)	32	$135/l_i$
			≤ 5,00 (≤ 500)		
			> 5,00 (> 500)	30	$120/l_i$

Aus Veröffentlichungen und Untersuchungen ist erkennbar, daß die Ausnutzung einer höheren Streckgrenze nach den heutigen Maßstäben des normalen Stahlbetons nur mit dünnen Stabdurchmessern möglich ist, wie sie in Betonstahlmatten verwendet werden. Um die Gebrauchseigenschaften des Stahlbetons bei z. B. einer Streckgrenze der Mattenbewehrung von 630 N/mm² gegenüber 500 N/mm² nicht zu verschlechtern, wurden in [23] bezüglich der Beschränkung der Durchbiegung und Rißbreite unter Gebrauchslast gegenüber DIN 1045 – Dezember 1978 – einengende und differenziertere Regeln festgelegt.
Für die Bemessung auf Querkraft und Torsion wurden aufgrund von Versuchsergebnissen die zulässigen Schubspannungen in [23] gegenüber DIN 1045 etwas abgemindert.
Abschließend ist festzustellen, daß einerseits in einer Reihe von Anwendungsfällen eine höhere Streckgrenze wirtschaftlich vorteilhaft wäre, andererseits aber Anwendungsfälle bestehen, bei denen ein Kostenvorteil nicht gegeben ist.

6 Herstellung von Betonstahlmatten und ihre Verwendung im Baumarkt

6.1 Mattenarten, Mattenherstellung

Die etwa 1,1–1,2 Millionen Tonnen Betonstahlmatten der deutschen Hersteller werden zu etwa
- 75% als Lagermatten (standardisiert)
- 25% als Listen- und Zeichnungsmatten (bauteilbezogen variabel)

gefertigt.
Die Herstellung erfolgt auf Vielpunktschweißanlagen, auf denen mit 60–100 „Schweißtakten/Minute" bei jedem Takt bis zu 30 Stabkreuzungen im elektrischen Widerstandspunktschweißverfahren scherfest verbunden werden.
Der größte Rationalisierungseffekt in der Herstellung liegt bei diesen Schweißmaschinen. Demgegenüber hatten andere Fertigungsverfahren – wie automatisch arbeitende Knüpfmaschinen oder Maschinen, mit denen Kunststoffverbindungen an den Kreuzungsstellen hergestellt wurden – keine wirtschaftlichen Chancen.

6.2 Betonstahlmatten im Baumarkt

Betonstahlmatten finden vorzugsweise in „flächenartigen Bauteilen", zunehmend auch seit vielen Jahren in „stabförmigen Bauteilen" Verwendung.

Bild 7

In Tabelle 2 wird eine grobe Schätzung der Anteile von Betonstahlmatte und Betonstabstahl im deutschen Markt gegeben.

In flächenartigen Bauteilen findet die Matte ihre Grenze durch den nach oben festgelegten Querschnittsbereich (max. Stabdurchmesser 12 mm) sowie in der gegenüber ungeschweißten Betonstählen kleineren dynamischen Belastbarkeit.

In stabförmigen Bauteilen ist eine Zunahme der Mattenanteile in Form von Bügeln und Schubzulagen [24] denkbar. Eine Typisierung von Betonstahlmatten – lagerfähig – für stabförmige Bauteile setzt eine stärkere Vereinheitlichung voraus, die von den Querschnittsformen und der Schalung her erfolgen müßte. Versuche der Mattenindustrie brachten bisher nur Anfangserfolge.

Tabelle 2
Anteile (geschätzt) von Betonstahlmatten und Betonstabstahl im deutschen Baumarkt

Bauteil-arten	Betonstahl in Bauteilen %	Anteile	
		Beton-stabstahl %	Betonstahl-matten %
flächenartige Bauteile	50	15	35
stabförmige Bauteile	50	45	5
	100	60	40

6.3 Betonstahlmatten in flächenartigen Bauteilen

Über den Arbeitszeitbedarf für das Verlegen an der Baustelle sind in verschiedenen Veröffentlichungen Angaben gemacht, die hinsichtlich der absoluten Werte Stunden/Tonne in Abhängigkeit von Bewehrungsgewicht kg/m^2 unterschiedlich sind [25]. Einheitlich ergibt sich jedoch daraus ein Unterschied zwischen Matte und Stabstahl von 10 bis 18 Stunden/Tonne; einheitlich ist auch eine Relation der Verlegezeit von Matte : Stabstahl von etwa 1 : 2.

Wie aus [26] hervorgeht, werden für einlagige – und damit einfache – Bewehrungsanordnungen 1,5 bis 5 Stunden/Tonnen weniger Verlegezeit als für zweilagige

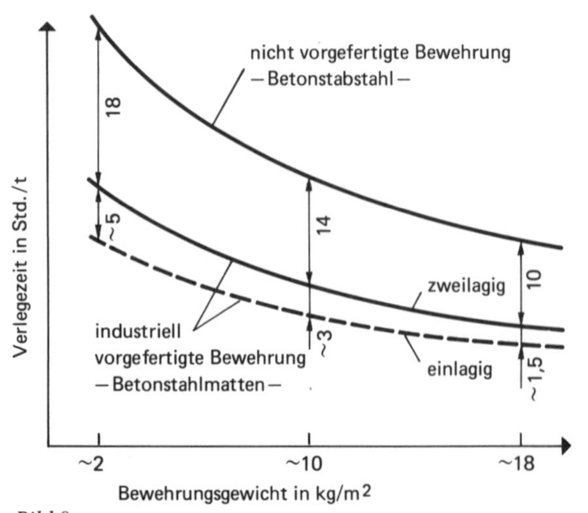

Bild 8
Zeitunterschied für das Verlegen von Betonstahlmatten und Betonstabstahl.

Bewehrungen benötigt. Mit Feldsparmatten kann dabei noch zusätzlich Stahl eingespart werden [25].
Für die Konstruktion der Mattenbewehrung stehen der Praxis verschiedene Möglichkeiten zur Verfügung [8]. Zwei grundsätzlich verschiedene Mattentypen stellen „Zweiachsmatten“ – tragende Stäbe in beiden Richtungen – und „Einachsmatten“ – tragende Stäbe nur in einer Richtung, in der anderen Richtung nur sogenannte Positionsdrähte – dar.
„Einachsmatten“ werden über Kreuz – also zweilagig – und ohne seitliche Überdeckung verlegt. Ein vorteilhaftes Anwendungsgebiet stellen Flachdecken dar [27].

6.4 Betonstahlmatten in stabförmigen Bauteilen

Der Arbeitszeitvergleich Matte: Stabstahl fällt in solchen Bauteilen mit etwa 1 : 2,5 noch günstiger aus als bei flächenartigen Bauteilen. Zeit- und Kostenvergleiche für das Schneiden von Betonstahlmatten und das Biegen zu Körben sind in [30] dargestellt.
Diese stets dreidimensional zu bewehrenden Bauteile – Balken- und Plattenbalken, Stützen u. a. – stellen hinsichtlich der Typisierung von Bewehrungselementen für eine Lagerhaltung eine wesentlich größere Problematik als flächenartige Bauteile dar.
Nach erfolgreicher Einführung der Betonstahlmatte hat die BAUSTAHLGEWEBE GMBH auch spezielle Untersuchungen mit zusammengesetzten, mattenartigen Bewehrungselementen durchführen lassen [28]. Das Ziel lag darin, dem Markt Elemente anzubieten, die für stabförmige Bauteile baukastenartig zusammengesetzt und ab Lager bezogen werden können.
Die Ergebnisse dieser sowie weiterer Untersuchungen [29] sind bei der Neufassung von DIN 1045, Abschnitt 18, berücksichtigt worden und boten verschiedene Möglichkeiten der Gestaltung der Schubbewehrung, z. B. Bügelkörbe und Schubzulagen sowie deren Verankerung, an.
Wenn die Fortschritte in der Rationalisierung der Bewehrungstechnik von stabförmigen Bauteilen auch für die Mattenhersteller nicht befriedigend sind, so kann doch festgestellt werden, daß jährlich beachtliche Mengen ihrer Produktion dort Verwendung finden.

6.5 Zeichnen von Betonstahlmatten in Verlegeplänen

Um auch die Bearbeitung von Matten-Bewehrungsplänen zu vereinfachen und edv-gesteuerten Zeichenanlagen zugänglicher zu machen, wurden in den letzten Jahren verschiedene Vorschläge in Veröffentlichungen [8], [31], [32] und in der Norm DIN 1356, Teil 10, gemacht. Diese Vorschläge stellen zunächst nur Anregungen für die Konstruktionsbüros dar; die Bedingungen der Praxis werden entscheiden, ob sie sich in breiter Form durchsetzen.

Bild 9

Bild 10

Bild 11

Bild 12

7 Schlußbemerkung

Mit den vorstehenden Ausführungen sollte ein Überblick über Schwerpunkte der Entwicklungsarbeiten in den vergangenen 15–20 Jahren für die Bewehrungstechnik mit geschweißten Betonstahlmatten gegeben werden.
Vieles, was heute global als Markt bezeichnet wird, ist auch im Falle der Betonstahlmatte und ihres großen Marktanteils in unserem Lande eine Folge von intensiver Forschung und danach Umsetzung in die Praxis über Baubestimmungen, Informationsmittel und Beratung. In 5 Jahrzehnten wurden in Deutschland rund 20 Millionen Tonnen dieser vorgefertigten Bewehrung in Stahlbetonbauteilen eingebaut.
Die Betonstahlmatte trug damit entscheidend zur Rationalisierung der Bewehrungstechnik bei.

8 Literatur

[1] Rehm, G.: Über die Grundlagen des Verbundes zwischen Stahl und Beton. (1961) Heft 138 des DAfSt.

[2] TH München, Bericht Nr. 7992, (1966): Scherversuche an einbetonierten und nicht einbetonierten geschweißten Baustahlmatten – nicht veröffentlicht.

[3] TH München, Bericht Nr. 6112 und 6198 (1969): Ausziehversuche zur Ermittlung des Zusammenwirkens zwischen Schweißknoten und Verbund bei Baustahlmatten aus Betonstahl KARI und Verankerungsversuche mit verschweißten KARI-Stäben. Veröffentlicht in „BAUSTAHLGEWEBE. Berichte aus Forschung und Technik“ Heft 3 und 4.

[4] Rehm, G., Tewes, R. und Eligehausen, R.: Übergreifungsstöße geschweißter Betonstahlmatten. Heft 291 des DAfSt und Beton- und Stahlbetonbau Heft 4/1976.

[5] Leonhardt, F. und Walther, R.: Beiträge zur Behandlung der Schubprobleme im Stahlbetonbau. Beton- und Stahlbetonbau 4/1964 und 5/1965.

[6] Leonhardt, F.: Die verminderte Schubdeckung bei Stahlbeton-Tragwerken. Der Bauingenieur 1/1965.

[7] Zulassungsbescheid „Geschweißte Betonstahlmatten BAUSTAHLGEWEBE“, Der Minister für Wohnungsbau und öffentliche Arbeiten des Landes Nordrhein-Westfalen 8. Dezember 1967.

[8] BAUSTAHLGEWEBE Konstruktionspraxis nach DIN 1045.

[9] Leonhardt, F. und Walther, R.: Geschweißte Bewehrungsmatten als Bügelbewehrung, Schubversuche an Plattenbalken und Verankerungsversuche. Die Bautechnik 10/1965.

[10] Untersuchungen „Zulässige Beanspruchung der Bügelbewehrung aus Baustahlgewebe in Balken.“ Rehm und Neubert, Bericht Universität Stuttgart, Oktober 1974.

[11] Gutachten Rehm, Universität Stuttgart, Oktober 1974.

[12] Rehm, G., Eligehausen, R. und Neubert, B.: Erläuterungen der Bewehrungsrichtlinien. (1979) DAfSt Heft 300.

[13] Die rationelle Bewehrung für Plattenbalken. (1982) BAUSTAHLGEWEBE GMBH.

[14] TH München, Bericht Nr. 310 (1968). Versuche an Stäben und geschweißten Matten aus Stäben mit neuartiger Profilierung. Veröffentlicht in „BAUSTAHLGEWEBE Berichte aus Forschung und Technik“ Heft 1.

[15] TH Braunschweig, Bericht Nr. 692620 (1969). Dauerschwingversuche im Zug-Schellbereich mit BAUSTAHLGEWEBE aus KARI-Stahl. Veröffentlicht in „BAUSTAHLGEWEBE Berichte aus Forschung und Technik“ Heft 6.

[16] Rehm, G. und Nürnberger, U.: Dauerschwingverhalten von widerstandspunktgeschweißten Baustahlmatten aus kaltgewalztem Betonrippenstahl. Schweißen + Schneiden, Bericht 1/1974.

[17] Rehm, G. und Russwurm, D.: Die Eigenschaften von geschweißten Betonstahlmatten. Betonwerk + Fertigteil-Technik 7/1975.

[18] DIN 488, Teil 1 (Entwurf Februar 1983, Gelbdruck).
[19] RUSSWURM, D. und REHM, G.: Dauerschwingfestigkeit (Betriebsfestigkeit) von Betonstahlmatten. Beton + Fertigteil-Technik 3/1979.
[20] MARTIN, H. und SCHIESSL, P.: Zeitschwingfestigkeit von geschweißten Betonstahlmatten. Betonwerk + Fertigteil-Technik 12/1981 und 1/1982.
[21] Geschweißte Betonstahlmatten BSt 500/550 RK für erhöhte dynamische Beanspruchung in Bereichen ohne Schweißstellen Zulassungsbescheid des Instituts für Bautechnik für die BAUSTAHLGEWEBE GMBH, 15.2. 1975ff.
[22] REHM, G.: Entwicklungstendenzen auf dem Gebiete der Beton- und Spannstähle. Vortrag Betontag 1975.
[23] Zulassungsbescheid „Geschweißte Betonstahlmatten BSt 630/700 RK (VR)“ des Instituts für Bautechnik.
[24] REHM, G., ELIGEHAUSEN, R. und NEUBERT, B.: Rationalisierung der Bewehrungstechnik, Vereinfachte Schubbewehrung in Balken. Betonwerk + Fertigteil-Technik 3/1978 und 4/1978.
[25] HERKOMMER, F.: Arbeitszeitrichtwerte mit geschweißten Betonstahlmatten und Stahlersparnis. Die Bautechnik 10/1977.
[26] Institut für Arbeits- und Baubetriebswissenschaft Dr. G. DRESSEL: Arbeitszeitrichtwerte für das Verlegen von Betonstahlmatten in Decken, Prüfbericht Nr. 116, Baugewerbe 6/1977.
[27] HÜTTEN, P. und HERKOMMER, F.: Bewehren von Flachdecken mit geschweißten Betonstahlmatten. Hoch- und Tiefbau 9/1981.
[28] TU Braunschweig: Bericht über Schubversuche an Stahlbetonbalken mit unterschiedlicher Ausbildung der Schubbewehrung (1973).
[29] REHM, G., ELIGEHAUSEN, R. und NEUBERT, B.: Rationalisierung der Bewehrungstechnik. Zwischenberichte über Balkenversuche. Universität Stuttgart Februar 1974 und Januar 1975.
[30] Institut für Arbeits- und Baubetriebswissenschaft Dr. G. DRESSEL: Zeit- und Kostenvergleiche für das Schneiden von Betonstahlmatten und das Biegen zu Körben. Überarbeitete Auflage 1975.
[31] Vereinfachte Darstellung von geschweißten Betonstahlmatten auf Verlegeplänen. Fachverband Betonstahlmatten e.V., Mitteilungen Heft 2.
[32] REHM, G., ELIGEHAUSEN, R. und MALLEÉ, R.: Rationalisierung der Zeichenarbeit im Stahlbetonbau. Betonwerk + Fertigteil-Technik 6/1975.

Rißfeld und Druckfeld in Stahlbetonscheiben

Herbert Kupfer
und Reinhard Mang

1 Stand der Forschung

Beim Bewehren von Flächentragwerken (Scheiben, Schalen, Platten) ist eine den gekrümmten Hauptzugspannungstrajektorien folgende Bewehrung, d. h. eine sogenannte Trajektorienbewehrung oft wegen des Arbeitsaufwandes nicht wirtschaftlich. Aus Gründen einer einfachen und rationellen Bewehrungsführung wird man vielmehr in der Regel orthogonale Bewehrungsnetze anordnen.

Die in diesem Fall erforderliche Bewehrung kann auf der Grundlage einer Fachwerkanalogie oder besser gesagt einer Druckfeldwirkung ermittelt werden:

Betrachtet man ein rechteckiges Scheibenelement mit den Seitenlängen $\Delta x = 1$ und $\Delta y = \tan\alpha$, dessen Schnittkanten parallel zu den orthogonalen Bewehrungsscharen verlaufen, so wirken an den Schnittkanten je Längeneinheit die äußeren Kräfte N_x, N_y und T (Bild 1).

Positive Kräfte N_x und N_y können unmittelbar der Bewehrung zugewiesen werden. Die Schubkraft T erzeugt – da der Beton keine Zugspannungen aufnehmen kann – neben Zugkräften im Stahl auch Betondruckspannungen, die unter dem Winkel α diagonal im Element gerichtet sind.

Ist die Richtung α dieser Druckdiagonalen bekannt, lassen sich die Zugkräfte in der Bewehrung ermitteln:

$$Z_x = N_x + |T| \cdot \cot\alpha \qquad (1\,a)$$

$$Z_y = N_y + |T| \cdot \tan\alpha \qquad (1\,b)$$

Obwohl es sich hierbei um ein baumechanisches Modell im Bruchzustand handelt, können die Zugkräfte Z_x und Z_y aus den Gebrauchsschnittgrößen N_x, N_y und T ermittelt werden; der erforderliche Stahlquerschnitt darf dann mit einer Stahlspannung von $\beta_s/1{,}75$ errechnet werden.

Wegen des Bruches des Betons ohne Vorankündigung ist die Druckfeldspannung im rechnerischen Bruchzustand, die der Druckfeldfestigkeit gegenüber zu stellen ist, jedoch unter 2,1facher Gebrauchslast zu ermitteln; es gilt somit

$$\sigma^{\mathrm{II}}_{2,u} = \sigma_{d,u} = \frac{2{,}1 \cdot \tau}{\sin\alpha \cdot \cos\alpha} \qquad (1\,c)$$

Die Neigung α des Druckfeldes wird in der Literatur unterschiedlich angenommen. So gehen LEITZ [1] und SCHOLZ [2] davon aus, daß das Druckfeld in Richtung der Winkelhalbierenden der Bewehrungsscharen verläuft; bei PETER [3] und EBNER [4] entspricht α der Richtung der Hauptdruckspannungen im ungerissenen Beton, bei KUYT [5] und THÜRLIMANN [6] ist der Winkel frei wählbar bzw. ergibt sich aus der eingelegten Bewehrung. Während LEITZ, SCHOLZ und KUYT auf die Rißrichtung nicht weiter eingehen oder sie der Druckfeldrichtung gleichsetzen, rechnen PETER und EBNER mit Schubspannungen im Riß infolge Rißverzahnung und Verdübelungswirkung der Bewehrung. BAUMANN [7] ermittelt den Winkel der Druckfeldrichtung nach dem Prinzip des Minimums der Formänderungsarbeit, wobei er ebenfalls Schubkräfte an den Rißufern berücksichtigt.

VECCHIO/COLLINS [8] errechnen den Winkel der Druckfeldneigung aus der – wie in [9], [10] gezeigt, nicht immer erfüllten – Bedingung, daß die sich unter Ein-

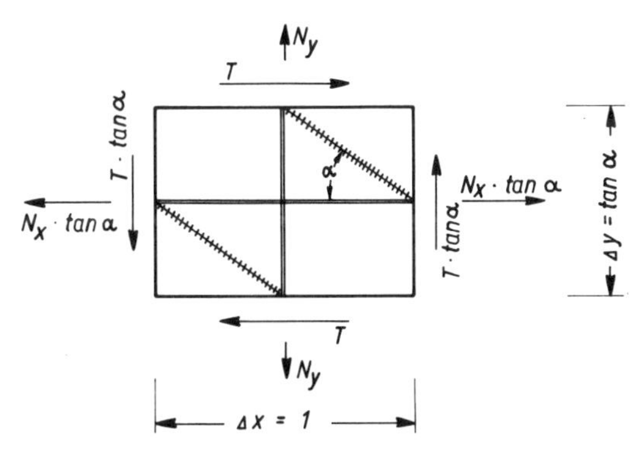

Bild 1
Element aus einer Stahlbetonscheibe mit orthogonaler Bewehrung.

schluß der verschmierten gegenseitigen Rißuferverschiebungen ergebenden Hauptdehnungsrichtungen mit den Hauptspannungsrichtungen übereinstimmen. Eine ihrer Versuchsauswertungen, wonach die Druckfeldfestigkeit mit zunehmendem Verhältnis von (positiver) Hauptdehnung zu Hauptstauchung sehr stark abnimmt, ist in der Fachwelt umstritten.
Obwohl die Richtung des Druckfeldes bei den unterschiedlichen Autoren variiert, haben sich diese Bemessungsverfahren in der Praxis doch bewährt. Die Erklärung ist in der Übertragung von Schubkräften in den Rissen des Betons zu finden, die erst in jüngster Zeit eingehend untersucht wurde [11], [12].
Im folgenden soll der Einfluß der Rißverzahnung für den Fall eines orthogonalen Bewehrungsnetzes dargestellt werden, wobei sich die Untersuchungen auf den Fall beschränken, daß in beiden Bewehrungsscharen Zugkräfte auftreten. Gemäß Gln. (1a) und (1b) ist dies immer dann der Fall, wenn

$$N_x + |T| \cdot \cot\alpha > 0 \quad \text{und} \qquad (1\,d)$$

$$N_y + |T| \cdot \tan\alpha > 0 \qquad (1\,e)$$

Im folgenden wird außerdem im Hinblick auf die Vertauschbarkeit von x und y unterstellt, daß

$$N_y > N_x \qquad (2)$$

Die Bedeutung der Rißverzahnung wurde in [9] am Beispiel des Schubtragverhaltens von Stahlbeton- und Spannbetonträgern aufgezeigt.

2 Bezeichnungen

Kräfte pro Längeneinheit:
N, T auf das untersuchte Scheibenelement einwirkende Normal- bzw. Schubkräfte
Z Zugkraft in der Bewehrung

Spannungen:
σ, τ Normal- und Schubspannungen
τ_{rd}, σ_r Rißverzahnungsspannungen parallel und senkrecht zum Riß

Dehnungen:
ε_s Stahldehnung

Verschiebungen:
v gegenseitige Verschiebung der Rißufer parallel zum Riß
w Rißbreite

Richtungen:
x, y Bewehrungsrichtung
φ Rißrichtung (gegenüber x-Achse)
α Druckfeldrichtung (gegenüber x-Achse)
$\delta = \alpha - \varphi$ Abweichung der Druckfeldrichtung von der Rißrichtung

Festigkeiten:
β_S Streckgrenze der Bewehrung
β_W Nennfestigkeit des Betons
β_d Druckfestigkeit des Betons im Druckfeld

Weitere Bezeichnungen werden im Text erläutert.

3 Umlagerung der Spannungen bei der Rißbildung

Beim Auftreten von Rissen im Beton, d. h. beim Übergang von Zustand I in Zustand II, kommt es zu einer Umlagerung der inneren Kräfte, wobei die inneren Zugkräfte im wesentlichen von der Bewehrung aufgenommen werden.
Bei einem schlanken Balken kann diese Umlagerung auch über die Balkenhöhe erfolgen. Daher ist es bekanntlich in diesen Fällen nicht notwendig, in der Schubzone von Balken mit vertikalen Bügeln eine horizontale Bewehrungsschar anzuordnen. Die horizontale Zugkraft der Schubzone wird vielmehr der Biegedruck- und Biegezugzone des Balkens zugewiesen (Versatzmaß).
Bei Scheiben, die aufgrund ihrer Form nicht mehr als Balken bemessen werden können, wird dagegen in der Regel davon ausgegangen, daß die Umlagerung der inneren Kräfte beim Übergang in Zustand II nur innerhalb eines Elementes bei Aufrechterhaltung der Elementkräfte erfolgt. Den folgenden Ableitungen liegt diese Annahme zugrunde.
Die Risse verlaufen im wesentlichen senkrecht zur Richtung der die Betonzugfestigkeit überschreitenden Hauptzugspannungen. Selbst wenn beide Hauptspannungen Zugspannungen sind, deren Richtungen von denen des Bewehrungsnetzes deutlich abweichen, entstehen keine rechtwinklig kreuzenden Risse, sondern nur ein einziges Rißfeld senkrecht zur größeren der beiden Hauptzugspannungen; sofort mit dem Entstehen dieses Rißfeldes muß sich zur Aufrechterhaltung des Gleichgewichtes ein Druckfeld ausbilden, das je nach der Rißverzahnungswirkung mehr oder weniger,

und zwar um den Winkel δ von der Rißrichtung abweicht. Maßgebend für die Rißrichtung ist jene Laststufe, unter der die Hauptzugspannung bzw. die größere Hauptzugspannung erstmals die Betonzugfestigkeit überschreitet. Dieser Laststufe wird der Index R zugeordnet, so daß

$$\tan 2\varphi = \frac{2\,T_R}{N_{yR} - N_{xR}} \tag{3}$$

wobei wie erläutert

$$\sigma_{1R} = \frac{N_{xR} + N_{yR}}{2} + \sqrt{\left(\frac{N_{xR} - N_{yR}}{2}\right)^2 + T_R^2} \approx \beta_{bZ} \tag{4}$$

Bei einem Lastpfad, bei dem das Verhältnis $N_x : N_y : T$ konstant bleibt, ist die Kenntnis der Kräfte N_{xR}, N_{yR} und T_R nicht nötig, da in diesem Fall das Verhältnis $T/(N_y - N_x)$ von der Laststufe unabhängig bleibt, so daß $\tan 2\varphi$ aus den Bemessungsschnittkräften N_x, N_y und T bestimmt werden kann. Lastpfade, die zu einer Rißfeldneigung $\varphi > 45^\circ$ führen ($N_{xR} > N_{yR}$), werden bei der weiteren Untersuchung ausgeschlossen.

Die Abweichung δ der Druckfeldrichtung α von der Rißrichtung φ bedingt, daß über die Risse hinweg Druck- und Schubkräfte übertragen werden können. Dies ist aufgrund der Verzahnung der Rißufer möglich.

Die parallel zum Riß übertragenen Schubspannungen τ_{rd} und die senkrecht zum Riß übertragenen Druckspannungen σ_r sind von der Rißbreite w und der Parallelverschiebung v der Rißufer abhängig. Von Bedeutung ist, daß die Übertragung der Druckfeldspannungen über den Riß hinweg nicht nur zu Schubspannungen τ_{rd} parallel zum Riß, sondern auch zu Druckspannungen σ_r senkrecht zum Riß führt.

Die Änderung des Spannungszustandes im Beton infolge des Auftretens von Rissen läßt sich mit Hilfe zweier Spannungskreise für den ungerissenen (rechten Kreis) und den gerissenen Beton (linker Kreis) veranschaulichen (Bild 2). Dabei ist zur besseren Verständlichkeit die Scheibendicke $d = 1$ gesetzt, so daß $\sigma_x = N_x$, $\sigma_y = N_y$ und $T = \tau_{xy} = \tau$.

Die Darstellung berücksichtigt folgende Bedingungen:

a) Beim Übergang in den Zustand II bleibt die Schubspannung τ bzw. die Schubkraft unverändert.

b) Nach der Rißbildung besteht ein einachsiger Druckspannungszustand (Druckfeld).

c) Die Richtung der Druckfeldspannung weicht um den Winkel δ von der Rißrichtung φ ab, wobei angenommen ist, daß sich letztere aus den Bemessungsschnittgrößen ergibt (proportionaler Lastpfad).

d) Die Zugkräfte Z_x und Z_y entsprechen den Gln. (1 a) und (1 b).

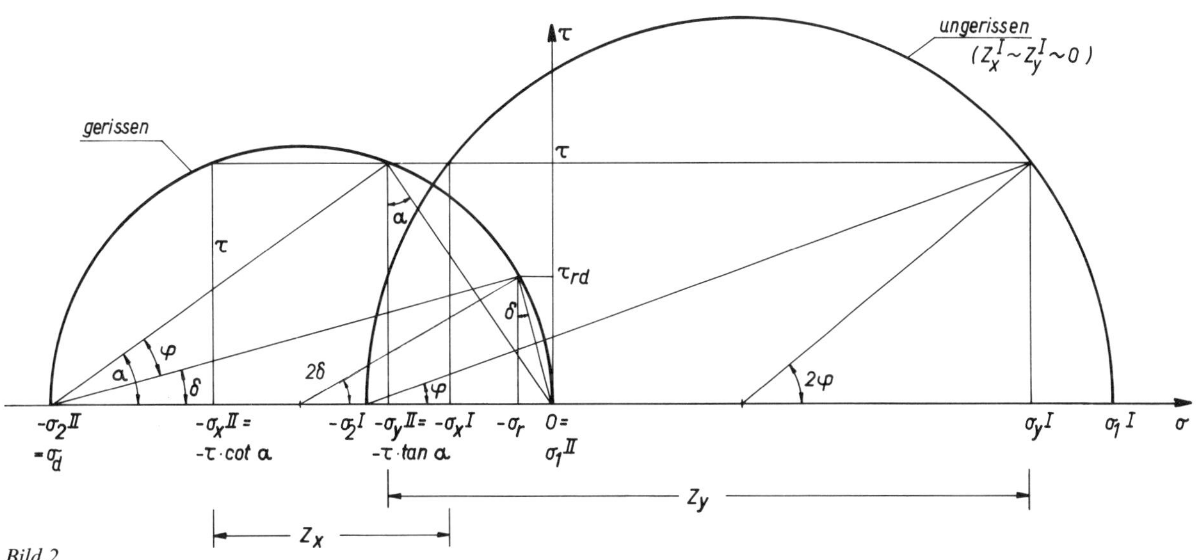

Bild 2
Darstellung der Betonspannungen vor der Rißbildung sowie der Betonspannungen und Stahlzugkräfte (Z_x, Z_y) nach der Rißbildung im MOHRschen Spannungskreisen.

4 Rißuferverschiebungen *v* und *w*

Die Größe der Rißverzahnungsspannungen τ_{rd} und σ_r ist von der Größe der parallelen und senkrechten Rißuferverschiebungen v und w abhängig. Einen entscheidenden Einfluß hat dabei das Verhältnis v/w von Parallelverschiebung zu Rißbreite.

Näherungsweise wird angenommen, daß in beiden Bewehrungsscharen die gleiche Dehnung ε_s auftritt, die Dehnungsverminderung zwischen den Rissen vernachlässigbar ist und die Betondehnungen gegenüber ε_s klein bleiben. Unter dieser Annahme ergibt sich eine eindeutige Abhängigkeit des Verhältnisses v/w von der Rißneigung φ (Bild 3):

$$\frac{v}{w} = \tan\psi = \tan(90 - 2\varphi) = \cot 2\varphi \qquad (5)$$

Diese Abhängigkeit ist in der unteren Hälfte des Bildes 4 dargestellt.

Eine genauere, die Verträglichkeit der Dehnungen des Druckfeldes unter Einschluß der verschmierten Rißuferverschiebungen mit den Dehnungen der Bewehrung berücksichtigende Untersuchung analog zu [9] ist vorgesehen. Wegen der geringen Empfindlichkeit der Rißverzahnungskräfte gegenüber Änderungen der Rißuferverschiebungen im interessierenden Bereich, dürfte jedoch das Ergebnis, nämlich die gesuchte Druckfeldrichtung α, nur wenig von dem hier gefundenen abweichen.

5 Rißverzahnungsspannungen

Die Rißverzahnung kann durch die Werte τ_{rd} und σ_r oder durch einen dieser beiden Werte und den Winkel δ zwischen Druckfeld- und Rißfeldrichtung beschrieben werden. Im folgenden wird die zweite Darstellung mit τ_{rd} und δ gewählt.

Eine Auswertung der Versuche von DASCHNER/KUPFER [11] gibt die in Bild 4 dargestellte Abhängigkeit des Winkels $\delta = \arctan(\sigma_r/\tau_{rd})$ vom Verhältnis v/w der gegenseitigen Verschiebungen für Rißbreiten von 0,05 bis 0,4 mm (entnommen aus [13]). Bei den Versuchen an Stahlbetonscheiben von VECCHIO/COLLINS [8] wurden die Rißbreiten im Bruchzustand in der gleichen Größenordnung gemessen.

Bild 5 zeigt die Abhängigkeit der Schubspannung τ_{rd} von v/w ebenfalls für konstante Rißbreiten. Die Schubspannung τ_{rd} ist dabei auf die Betonfestigkeit β_d des Druckfeldes bezogen. Sie kann bei geeigneter Beschränkung der Stahldehnungen bzw. Rißbreiten im Bruchzustand zu $\beta_d = 0,7 \cdot 0,8\,\beta_w = 0,56\,\beta_w$ angenommen werden, wobei der Faktor 0,7 den Gestalts- und Dauerlasteinfluß berücksichtigt, der Faktor 0,8 wurde in Versuchen [14] zur Bestimmung der einachsigen Druckfestigkeit in Rißrichtung (Druckfeldfestigkeit) ermittelt.

Diese Annahme trifft auch BAUMANN [7]. Sie wurde neuerdings durch Versuche von SCHLAICH/SCHÄFER [15] bestätigt.

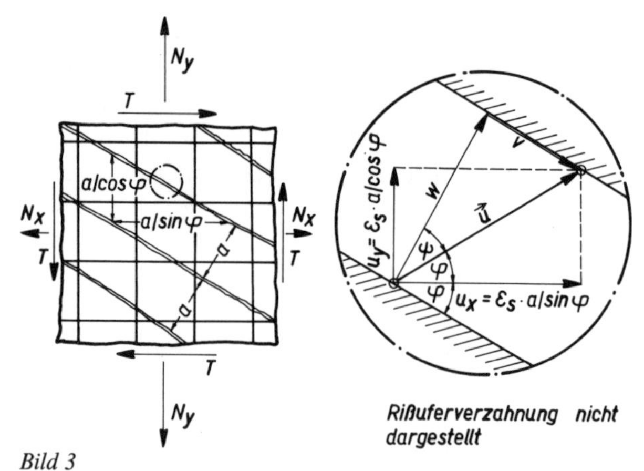

Bild 3
Ableitung der gegenseitigen parallelen (v) und senkrechten (w) Rißuferverschiebung in Abhängigkeit von der Rißrichtung.

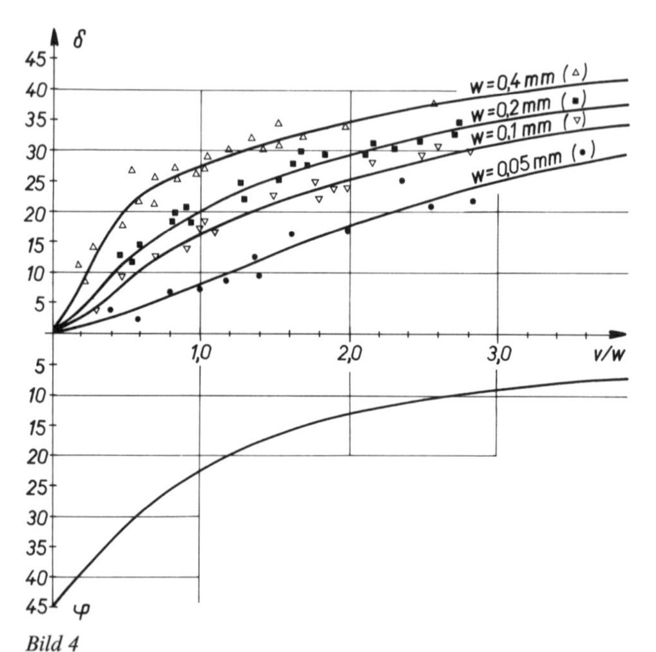

Bild 4
Winkel φ der Rißrichtung (untere Bildhälfte) und Winkel δ zwischen Rißrichtung und Druckfeldrichtung (obere Bildhälfte) in Abhängigkeit von v/w (parallele/senkrechte Rißuferverschiebung) mit der Rißbreite als Parameter.

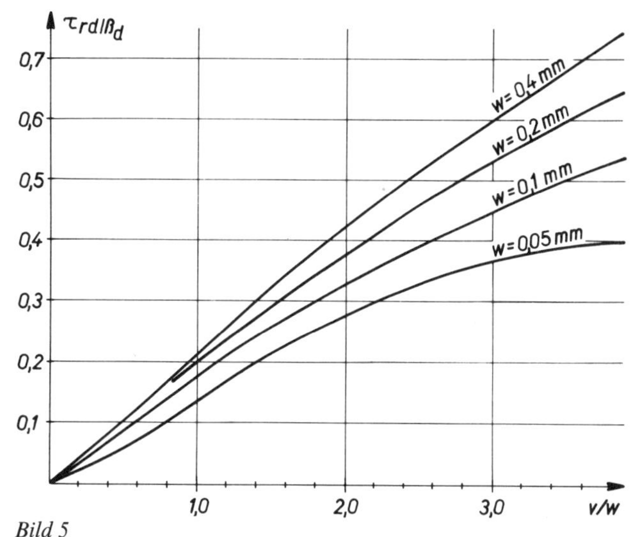

Bild 5
Bezogene Rißschubspannung τ_{rd}/β_d in Abhängigkeit von v/w mit der Rißbreite w als Parameter.

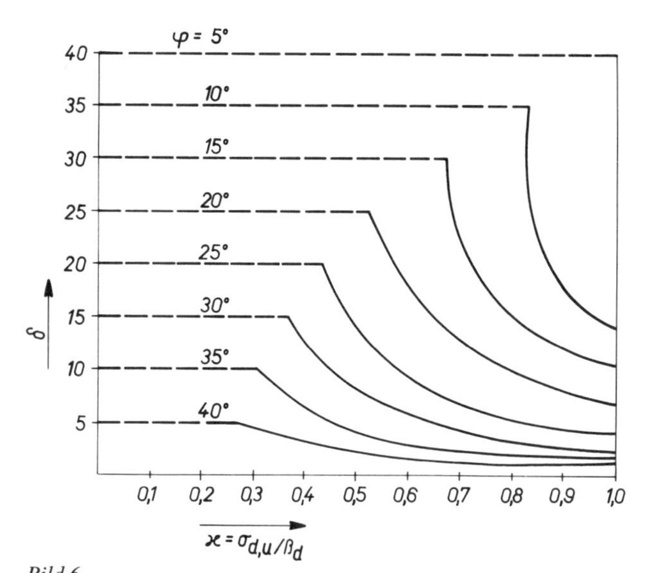

Bild 6
Möglicher Winkel δ zwischen Druckfeld und Rißfeld in Abhängigkeit von der bezogenen Hauptdruckspannung κ und der Rißfeldneigung φ.

Die Auswertung der Versuche erfolgte für einen Beton B25 mit 8 mm Größtkorndurchmesser. Für Betone mit größerem Größtkorndurchmesser ergeben sich größere Verzahnungskräfte, für Betone mit höherer Festigkeit geringfügig kleinere bezogene Rißschubspannungen. Die in Bild 5 angegebenen Rißschubspannungen sind Mittelwerte der Versuchsergebnisse.

6 Die mögliche Richtungsabweichung zwischen Druckfeld und Rißfeld

Die infolge der Rißverzahnung mögliche Abweichung des Druckfeldes vom Rißfeld wird im folgenden unter der Annahme ausgewertet, daß die experimentell festgestellten mittleren Rißschubspannungen nach Bild 5 im Hinblick auf den Dauerstandeinfluß und das Verhältnis der 5%-Fraktile zum Mittelwert mit dem Faktor 0,8 abzumindern sind. Dabei kann für den Dauerstandeinfluß wegen der langen Versuchsdauer ein relativ hoher Abminderungsfaktor von 0,95 angesetzt werden, so daß für die 5%-Fraktile ein Abminderungsfaktor von $0{,}8/0{,}95 = 0{,}84$ verbleibt. Die Rißschubspannungen nach Bild 5 beruhen auf relativ wenigen Versuchen. Daher war es nicht möglich, die 5%-Fraktilenwerte anzugeben. Es ist nicht auszuschließen, daß diese Fraktilenwerte – falls sich die Streuungen als groß erweisen sollten – noch niedriger liegen als hier berücksichtigt. Hierzu sind aber noch weitere Versuche nötig.

Bei der Auswertung der Bilder 4 und 5 wurden die Werte v/w, δ, w und τ_{rd}/β_d in Abhängigkeit von der Rißrichtung φ abgelesen und die zugehörigen bezogenen Druckfeldspannungen κ berechnet:

$$\kappa = \frac{\sigma_{d,u}}{\beta_d} = \frac{\tau_{rd}}{\beta_d} \cdot \frac{2}{\sin 2\varphi} \tag{6}$$

Bei der Auswertung ergeben sich zum Teil unrealistisch kleine Rißbreiten w. Bei einer Stahldehnung ε_s von 2‰ oder mehr (Bruchzustand) und einem Rißabstand von 15 cm oder mehr müßte w in der Größenordnung von 0,3 mm oder mehr liegen. Trotzdem erscheinen die δ-Kurven nach Bild 6 brauchbar, weil die Beziehung $v/w = \cot 2\varphi$ nur eine Mittelwertkurve für einen Streubereich darstellt, der sich bei $\varepsilon_{sx} \neq \varepsilon_{sy}$, d. h. bei unterschiedlichen Fließlaststufen der beiden Bewehrungsscharen einstellt und weil vor allem gemäß Bild 5 die Rißverzahnungsschubspannung τ_{rd} gegenüber einer Änderung der Rißbreite w nicht empfindlich ist.

7 Bemessungsdiagramm

Auf Bild 7 ist die Druckfeldrichtung α als Summe der Rißrichtung φ und der möglichen Winkelabweichung δ wiederum in Abhängigkeit von der Größe der Hauptdruckspannung dargestellt. Ferner sind die Kurven für die Größe der Schubbeanspruchung

$$\tau' = 2 \cdot \gamma \cdot \tau/\beta_d = 4{,}2\,\tau/\beta_d \tag{7}$$

angegeben.

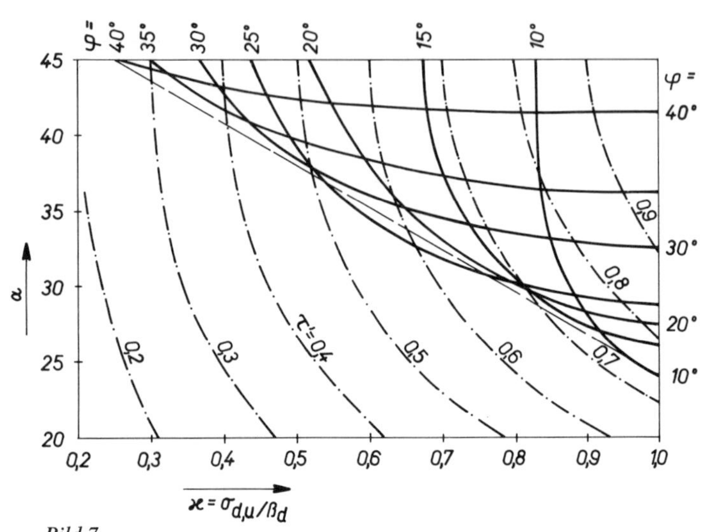

Bild 7
Bemessungsdiagramm für $\tau' = 4{,}2\,\tau/\beta_d$.

Dabei bedeutet:

τ die Schubspannung in Netzrichtung unter Gebrauchslast,

β_d die Druckfeldfestigkeit unter Dauerlast,

$\gamma = 2{,}1$ den Sicherheitsbeiwert für unangekündigten Bruch des Betons.

Das Diagramm ist wie folgt anzuwenden:

a) Aus der Lastkombination N_{xR}, N_{yR} und T_R wird die Rißrichtung φ nach Gl. (2) errechnet. (Bei Lastpfaden mit konstantem Verhältnis $N_x : N_y : T$ kann φ aus den Bemessungsschnittgrößen ermittelt werden.) Ebenso ist die bezogene Schubkraft τ' nach Gl. (7) bekannt. Aus dem Schnittpunkt der τ'- und φ-Linien läßt sich auf der Ordinate die Druckfeldrichtung α, auf der Abszisse die Größe der bezogenen Druckfeldspannung κ ablesen.

b) Die Rißrichtung φ ist nicht bekannt, weil der Lastpfad unsicher ist. In diesem Fall läßt sich ein auf der sicheren Seite liegendes Ergebnis erzielen, wenn der Winkel α aus der einhüllenden Geraden auf Bild 7 bestimmt wird.

Mit der so ermittelten Druckfeldrichtung läßt sich mit den Gln. (1a) und (1b) die erforderliche Bewehrung bestimmen.

Wenn, wie vorausgesetzt wurde, in beiden Bewehrungsscharen Zug auftritt, ergäbe sich nach den Gln. (1a) und (1b) der geringste Bewehrungsbedarf und nach Gl. (1c) auch die geringste Beanspruchung des Betons (kleinste Hauptdruckspannung) für eine Druckfeldrichtung von 45°. Wie aus der Darstellung ersichtlich, gestattet die Rißverzahnung aber nicht immer die Druckfeldrichtung mit 45° anzunehmen. Kleine Winkel α von etwa 25°, d.h. relativ flache Druckfelder, treten bei kleinen Rißneigungen φ und großen bezogenen Druckfeldspannungen κ auf, um die Rißverzahnungskräfte nicht zu groß werden zu lassen. Aber auch bei mittleren und niedrigen Druckfeldspannungen (κ-Werten) und Rißneigungen über 30° wird die Druckfeldneigung α deutlich kleiner als 45°, weil aufgrund kleiner Werte v/w nur kleine Rißverzahnungskräfte geweckt werden.

Abschließend sei nochmals darauf hingewiesen, daß einige günstige Wirkungen, wie Kräfteumlagerungen zwischen den Scheibenelementen und Verdübelungswirkungen der Bewehrung nicht erfaßt wurden. Außerdem ist nicht berücksichtigt, daß durch unterschiedliche Dehnungen der Bewehrungsscharen das Verhältnis v/w geändert werden kann, wodurch abweichende Rißverzahnungskräfte geweckt werden können.

8 Zusammenfassung

Es wurde gezeigt, daß in orthogonal bewehrten Stahlbetonscheiben, in denen die Bewehrung von den Hauptzugspannungen abweicht, das zur Aufnahme von Schubkräften erforderliche Druckfeld aufgrund der Verzahnung der Rißufer erheblich von der Rißfeldrichtung abweichen kann. Die Rißverzahnungskräfte wurden aus Versuchen [11] ermittelt.

Die Abweichung der Druckfeldrichtung von der Richtung der Risse ist von der Rißrichtung selbst und von der Höhe der Schubbeanspruchung abhängig.

Für den Fall, daß in beiden Bewehrungsscharen Zugkräfte auftreten, wurde ein Bemessungsdiagramm entwickelt, das erlaubt, bei Kenntnis der Rißrichtung die zum geringsten Bewehrungsbedarf führende Druckfeldneigung unter Beachtung der Rißverzahnung zu ermitteln. Ist die Rißrichtung wegen der Unsicherheit des Lastpfades nicht bekannt, kann aus dem Diagramm ein auf der sicheren Seite liegender Winkel der Druckfeldneigung entnommen werden.

Die Auswertung zeigt, daß die Druckfeldrichtung nur bei hoher Schubbeanspruchung wesentlich vom optimalen Winkel $\alpha = 45°$ abweicht. Dies erklärt, warum bekannte Bemessungsverfahren, denen eine von der Rißrichtung weitgehend unabhängige Druckfeldrichtung zugrunde liegt, in weiten Bereichen zu brauchbaren Ergebnissen führen.

9 Literatur

[1] LEITZ, H.: Über die Anwendung der Elastizitätstheorie auf kreuzweise bewehrten Beton. Beton und Eisen 25 (1926), S. 240–245.

[2] SCHOLZ, G.: Zur Frage der Netzbewehrung von Flächentragwerken. Beton- und Stahlbetonbau 53 (1958), S. 250–255.

[3] PETER, I.: Zur Bewehrung von Scheiben und Schalen für Hauptspannungen schiefwinklig zur Bewehrungsrichtung. Die Bautechnik 43 (1966), S. 149–154.

[4] EBNER, F.: Zur Bemessung von Stahlbetonplatten mit von der Richtung der Hauptzugspannung abweichenden Bewehrungsrichtung. Aus Theorie und Praxis des Stahlbetonbaus, Berlin, Wilh. Ernst & Sohn, 1969, S. 127–134.

[5] KUYT, B.: Zur Frage der Netzbewehrung von Flächentragwerken. Beton- und Stahlbetonbau (59) 1964, S. 158–163.

[6] THÜRLIMANN, B. et al.: Anwendung der Plastizitätstheorie auf Stahlbeton. Vorlesung zum Fortbildungskurs für Bauingenieure 1983. Institut für Baustatik und Konstruktion, ETH Zürich.

[7] BAUMANN, TH.: Tragwirkung orthogonaler Bewehrungsnetze beliebiger Richtung in Flächentragwerken aus Stahlbeton. Diss. TH München, 1971.

[8] VECCHIO, F. und COLLINS, M.P.: The response of reinforced concrete to in-plane shear and normal stresses. Publication 82-03 University of Toronto Dep. Civ. Eng. 1982.

[9] KUPFER, H., MANG, R. und KARAVESYROGLOU, M.: Bruchzustand der Schubzone von Stahlbeton- und Spannbetonträgern – Eine Analyse unter Berücksichtigung der Rißverzahnung. Bauingenieur 58 (1983), S. 143–149.

[10] KUPFER, H.: Zuschrift zu MEHLHORN, G. et al.: Schubbemessung für kombinierte Beanspruchung nach der Druckfeldtheorie von COLLINS/MITCHELL, Beton- und Stahlbetonbau 5, 1983. Beton- und Stahlbetonbau 79 (1984), S. 140.

[11] DASCHNER, F. und KUPFER, H.: Versuche zur Schubkraftübertragung in Rissen von Normal- und Leichtbeton. Bauingenieur 57 (1982) S. 51–55.

[12] WALRAVEN, I.C.: Experiments on shear transfer in cracks in concrete. Part II: Analysis of results. Report 5-79-10, Delft Universit of Technology.

[13] MANG, R.: Auswirkung der Verzahnung von Rissen in Beton auf das Tragverhalten der Schubzonen von Balken und Scheiben. In Vorbereitung.

[14] ROBINSON, J.R. und DEMORIEUX, J.M.: Essais de modéles d'âme de poutres en double Té. Annales de l'Institut Technique du Bâtiment et des Travaux Publics 354, Okt. 1977, S. 77–95.

[15] SCHLAICH, J. und SCHÄFER, K.: Druck-Querzugfestigkeit des Stahlbetons. Beton- und Stahlbetonbau (68) 1983, S. 73–78.

Bewehrung feuerwiderstandsfähiger Bauteile mit Stabbündeln

Karl Kordina
und Ulrich Quast

1 Einleitung

Bekanntlich ist ein Vorteil einer Bewehrungsanordnung in Form von Stabbündeln darin zu sehen, daß sich der Größtwert der Gesamtbewehrungsmenge auch bei Verwendung der größten zulässigen Stabdurchmesser noch weiter steigern läßt, wenn Stabbündel aus zwei oder drei Einzelstäben gebildet werden. DIN 1045, Beton und Stahlbeton – Bemessung und Ausführung, Ausgabe Dezember 78, enthält erstmalig Regelungen zur Anwendung von Stabbündelbewehrungen. Weitere Erläuterungen werden in [1] gegeben. Hiernach können die Bewehrungsrichtlinien für Einzelstäbe auf Stabbündel übertragen werden, wenn anstelle des Einzelstabdurchmessers d_s der Vergleichsdurchmesser d_{sV} verwendet wird, wobei sich der Vergleichsdurchmesser als Durchmesser eines dem Bündel flächengleichen Einzelstabes ergibt. Für ein Bündel aus n Einzelstäben gleichen Durchmessers d_s ergibt sich folglich $d_{sV} = d_s \cdot \sqrt{n}$.
Entsprechend dieser Überlegungen wurden Regelungen für Stabbündelbewehrungen auch in DIN 4102 Teil 4, Brandverhalten von Bauteilen, Ausgabe März 81, vorgenommen, obwohl zu diesem Zeitpunkt noch keine Erfahrungen über das Brandverhalten von Bauteilen mit Bewehrung aus Stabbündeln vorlagen. Dies war deshalb möglich, weil die Forschungsarbeiten im Sonderforschungsbereich 148, Brandverhalten von Baustoffen und Bauteilen, der TU Braunschweig so weit vorangeschritten waren, daß das Brandverhalten von Stahlbetonbauteilen mit Stabbündelbewehrung rechnerisch-theoretisch beurteilt werden konnte. Zur Bestätigung waren aber experimentelle Untersuchungen erforderlich, die inzwischen abgeschlossen sind [2]. Über die wichtigsten Zusammenhänge wird nachfolgend berichtet.

2 Bauteilerwärmung

Betonbauteile werden in Schadensfeuern im allgemeinen nur in ihren äußeren Randzonen so weit erwärmt, daß der Beton seine Festigkeit verliert und für die im Randbereich liegende Bewehrung die Gefahr besteht, unter Gebrauchslast in den Fließzustand zu geraten. Für Versuchszwecke wird die Einwirkung eines Schadensfeuers durch eine Beflammung nach einer Einheits-Temperaturzeitkurve (ETK) ersetzt, wobei nach 90 Minuten Brandraumtemperaturen von ca. 1000 °C vorliegen. Bild 1 zeigt den Verlauf der 500 °C-Isotherme nach 90 Minuten ETK-Einwirkung für einen 150 mm und einen 400 mm breiten Stahlbetonbalken. Bei üblicher Ausnutzung der schlaffen Bewehrung gerät diese – weitgehend unabhängig von ihrer Festigkeit – in den Fließzustand, wenn sie auf annähernd 500 °C erwärmt ist. Sofern die Zugfestigkeit der Bewehrung für die Tragfähigkeit eines biegebeanspruchten Bauteils maßgebend ist, wird hiermit der Versagenszustand erreicht. Aus diesem Grunde hat sich der Begriff der

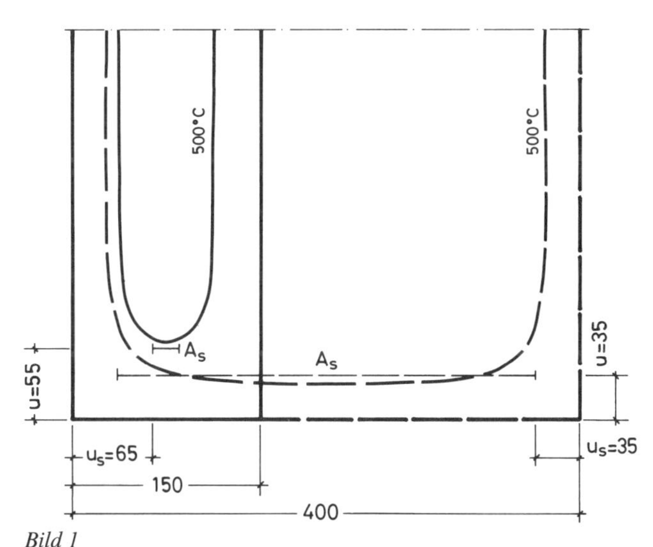

Bild 1
500 °C-Isothermen nach 90 Minuten ETK-Einwirkung in unterschiedlich breiten Balken.

Tafel 1
Bewehrungsrichtlinien aus DIN 4102 Teil 4, Ausgabe März 1981. (Mindestachsabstände sowie Mindeststabzahl der Zugbewehrung von 1- bis 4seitig beanspruchten, statisch bestimmt gelagerten Stahlbetonbalken[4]) aus Normalbeton)

Zeile	Konstruktionsmerkmale	Feuerwiderstandsklasse				
		F 30	F 60	F 90	F 120	F 180
1	Mindestachsabstände u[1]) und u_s[1]) sowie Mindeststabzahl n[2]) der Zugbewehrung unbekleideter, einlagig bewehrter Balken					
1.1	bei einer Balkenbreite b in mm von	80	≤120	≤150	≤200	≤240
1.1.1	u in mm	25	40	55[3])	65[3])	80[3])
1.1.2	u_s in mm	35	50	65	75	90
1.1.3	n	1	2	2	2	2
1.2	bei einer Balkenbreite b in mm von	120	160	200	240	300
1.2.1	u in mm	15	35	45	55[3])	70[3])
1.2.2	u_s in mm	25	45	55	65	80
1.2.3	n	2	2	3	3	3
1.3	bei einer Balkenbreite b in mm von	160	200	250	300	400
1.3.1	u in mm	12	30	40	50	65[3])
1.3.2	u_s in mm	22	40	50	60	75
1.3.3	n	2	3	4	4	4
1.4	bei einer Balkenbreite b in mm von	≥200	≥300	≥400	≥500	≥600
1.4.1	$u = u_s$ in mm	12	25	35	45	60[3])
1.4.2	n	3	4	5	5	5
2	Mindestachsabstände u, u_m und u_s sowie Mindeststabzahl n der Zugbewehrung bei unbekleideten, mehrlagig bewehrten Balken					
2.1	u_m nach Gleichung (3)	$u_m \geq u$ nach Zeile 1				
2.2	u und u_s	u und $u_s \geq u_{F30}$ nach Zeile 1 sowie u und $u_s \geq 0{,}5\,u$ nach Zeile 1				
2.3	Mindeststabzahl n	keine Anforderungen				
3	Mindestachsabstände u und u_s bzw. u_m von Balken mit Bekleidungen aus					
3.1	Putzen nach den Abschnitten 3.1.5.1 bis 3.1.5.5	u, u_m und u_s nach den Zeilen 1 und 2, Abminderungen nach Tabelle 2 sind möglich, u jedoch nicht kleiner als für F 30				
3.2	Unterdecken	u und $u_s \geq 12$ Konstruktion nach Abschnitt 6.5				

[1]) Zwischen den u- und u_s-Werten von Zeile 1 darf in Abhängigkeit von der Balkenbreite b geradlinig interpoliert werden.
[2]) Die geforderte Mindeststabzahl n darf unterschritten werden, wenn der seitliche Achsabstand u_s pro entfallendem Stab jeweils um 10 mm vergrößert wird; Stabbündel gelten in diesem Falle als ein Stab.
[3]) Bei einer Betondeckung $c > 40$ mm ist eine Schutzbewehrung nach Abschnitt 3.1.4 erforderlich.
[4]) Siehe Seite 43.

kritischen Temperatur crit T eingebürgert. Wenn Schnittgrößenumlagerungen nicht möglich sind, muß zur Gewährleistung ausreichender Feuerwiderstandsdauer sichergestellt sein, daß die Bewehrung nicht vor Ablauf dieser Zeit die kritische Erwärmung erfährt. Hierzu ist es erforderlich, daß nach Maßgabe des Verlaufs der 500 °C-Isotherme seitliche Abstände u_s und Abstände zur Unterseite u in bezug auf die Achse der Bewehrungsstäbe eingehalten werden. Diese Achsabstände sind für jede Feuerwiderstandszeit von der Balkenbreite abhängig, aber auch von der Anzahl der Bewehrungsstäbe in der untersten Lage. Die Erwärmung der Eckstäbe über 500 °C hinaus kann ohne Nachteil zugelassen werden, wenn dafür eine hinreichende Zahl innenliegender Stäbe noch nicht bis 500 °C erwärmt ist. Auf diese Weise kann sichergestellt werden, daß die erforderliche Zugkraft ausreichend lange von den unterschiedlich erwärmten Bewehrungsstäben aufgenommen wird. Die für $b = 150$ mm zu ersehenden Abweichungen dürfen nicht überwertet werden und wurden im Hinblick auf vereinfachte Festlegungen akzeptiert [3].

Entsprechende Regelungen enthält DIN 4102 Teil 4, aus der beispielhaft die Tabelle 5 hier als Tafel 1 wiedergegeben wird. Ausführliche Erläuterungen hierzu sind in [3] nachzulesen.

3 Stabbündelerwärmung

Die Durchwärmung eines Stahlbetonquerschnittes wird praktisch nicht von der Bewehrung beeinflußt, zumindest nicht solange praxisübliche Bewehrungen betrachtet werden. Wegen der guten Wärmeleitung von Stahl herrscht in Bewehrungsstäben näherungsweise ein einheitlicher Temperaturzustand. Entsprechend der Darstellung in Bild 2 ergibt sich dabei die Temperatur des Bewehrungsstabes so, wie sie im reinen Betonquerschnitt an der Stelle der Stabachse auftreten würde [4]. Diese Zusammenhänge sind sowohl experimentell als auch in rechnerisch-theoretischer Hinsicht weitgehend abgeklärt. Die Erwärmungsvorgänge können berechnet werden, wenn Wärmeleitzahl, Materialdichte, Betoneigenfeuchte und spezifische Wärmekapazität temperaturabhängig angesetzt werden. Rißbildungen im Beton und darin stattfindende Dampfströme haben Einfluß auf die Durchwärmung, wenn kleine Ausschnitte betrachtet werden. Die Zusammenhänge sind aber von zufälliger Natur und somit einfachen Regelungen nicht zugänglich. Die bisherigen Erfahrungen zeigen auch, daß diese Einflüsse zur Beurteilung der Bauteilsicherheit bei Brandeinwirkung nicht von entscheidender Bedeutung sind.

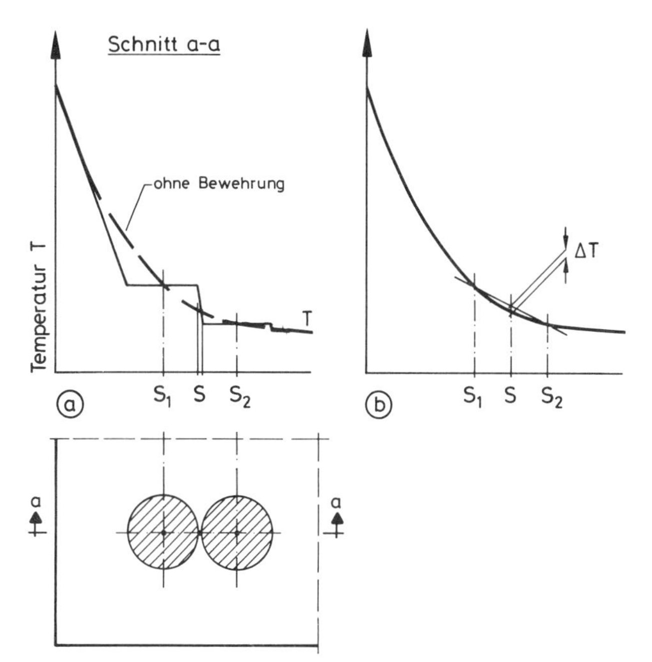

Bild 2
ⓐ Prinzipielle Veränderung der Temperaturverläufe durch ein Stabbündel und
ⓑ Abweichung ΔT des Mittelwertes der Temperaturen in den Einzelschwerpunkten S_1 und S_2 von der Temperatur im Gesamtschwerpunkt S.

Für das Erwärmungsverhalten von Stabbündeln können neben den erwähnten, im allgemeinen nicht berücksichtigten Rißbildungen noch weitere Einflüsse hinzukommen, dadurch, daß der Beton in den Zwikkeln zwischen den Stäben der Stabbündel Abweichungen von der Grundstruktur dieses Baustoffs zeigt; außerdem ist eine Rißbildung infolge behinderter Temperaturdehnung der Stabbündel vorstellbar. Aus diesem Grunde wurde es für unerläßlich gehalten, die Erwärmung von einbetonierten Stabbündeln experimentell zu untersuchen.

[4]) Die Tabellenwerte gelten auch für Spannbetonbalken; die Mindestachsabstände u, u_m und u_s sind jedoch entsprechend den Angaben von Tabelle 1 um die folgenden Δu-Werte zu erhöhen:
Bei vergüteten Drähten mit crit $T = 450$ °C um $\Delta u = 5$ mm,
bei kaltgezogenen Drähten und Litzen mit crit $T = 350$ °C um $\Delta u = 15$ mm.
Sofern die Mindestachsabstände u oder u_s mit 12 mm angegeben sind, dürfen die Δu-Werte um 2 mm verringert werden.

Die erwarteten Temperaturdifferenzen in den einzelnen Stäben eines Bündels wurden im wesentlichen bestätigt. Je nach Abstand von der Bauteiloberfläche und dem Stabdurchmesser wurden unterschiedlich große Temperaturdifferenzen zwischen den einzelnen Stäben gemessen. Diese nahmen, wie erwartet, mit der Zeit zu. Bei den einzelnen Versuchen wurden durchaus unterschiedliche Verhaltensweisen beobachtet. In Einzelfällen nahmen die Temperaturdifferenzen über den gesamten Versuchszeitraum zu, so wie es auch eine idealisierende Modellrechnung zeigt. In Einzelfällen wurde die größte Differenz bereits in der ersten Hälfte des Versuchszeitraumes erreicht und blieb danach konstant oder verringerte sich auch gelegentlich. Diese Beobachtungen können nur so erklärt werden, daß die Erwärmung eines Stabbündels von Zufälligkeiten beeinflußt sein kann, über die keine eindeutige Vorhersage möglich ist. Die größte Temperaturdifferenz wurde in einem Stabbündel aus drei Stäben mit d_s = 12 mm nach 40 Minuten Brandeinwirkungszeit zu 100 K beobachtet.

Für die Übertragung der brandschutztechnischen Richtlinien für Einzelstäbe auf Stabbündel ist zunächst nicht die Temperaturdifferenz zwischen den einzelnen Stäben eines Stabbündels maßgebend. Entscheidend ist vielmehr die mittlere Temperatur der Stäbe eines Stabbündels. Wie in Bild 2 dargestellt ist, ergibt sich aber in Stabbündeln eine um ΔT größere mittlere Temperatur als sie im – unbewehrten – Beton im Schwerpunkt des Bündels auftreten würde. Die Abweichung ΔT wird um so größer, je näher das Stabbündel an der Bauteiloberfläche liegt und je größer die Durchmesser der einzelnen Stäbe sind. Um also beurteilen zu können, ob die Festlegungen der Achsabstände für Einzelstäbe auch auf den Vergleichsstab eines Stabbündels angewendet werden können, ist es erforderlich, die auftretenden Abweichungen ΔT zu kennen. Diese Werte wurden rechnerisch ermittelt. Bild 3 zeigt die Ergebnisse für eine bestimmte Balkenbreite und die zugehörigen Achsabstände, die zu einem freiaufliegenden Balken der Feuerwiderstandsklasse F 90 nach Tafel 1 gehören. Die entsprechenden Zusammenhänge wurden nur für Stabdurchmesser aufgestellt, die Vergleichsdurchmesser der Stabbündel von höchstens d_{sV} = 36 mm ergeben, so daß keine zusätzliche Hautbewehrung nach DIN 1045, Abschnitt 18.11.3 erforderlich ist.

Die zusätzlich erforderliche Hautbewehrung bei größerem Vergleichsdurchmesser beeinträchtigt mit Sicherheit die Wirtschaftlichkeit der Stabbündelbewehrung im üblichen Hochbau, weswegen hier nur Stabbündel betrachtet werden, die keine zusätzliche Hautbewehrung erfordern. In dem 200 mm breiten Balken können drei Stäbe d_s = 22 mm, aber nur zwei Stabbündel aus je drei Stäben Durchmesser d_s = 18 mm untergebracht werden. Die Anordnung von drei Stabbündeln aus je drei Einzelstäben würde nur so geringe Stabdurchmesser zulassen, daß kein größerer Gesamtquerschnitt untergebracht werden kann als bei Verwendung von Einzelstäben.

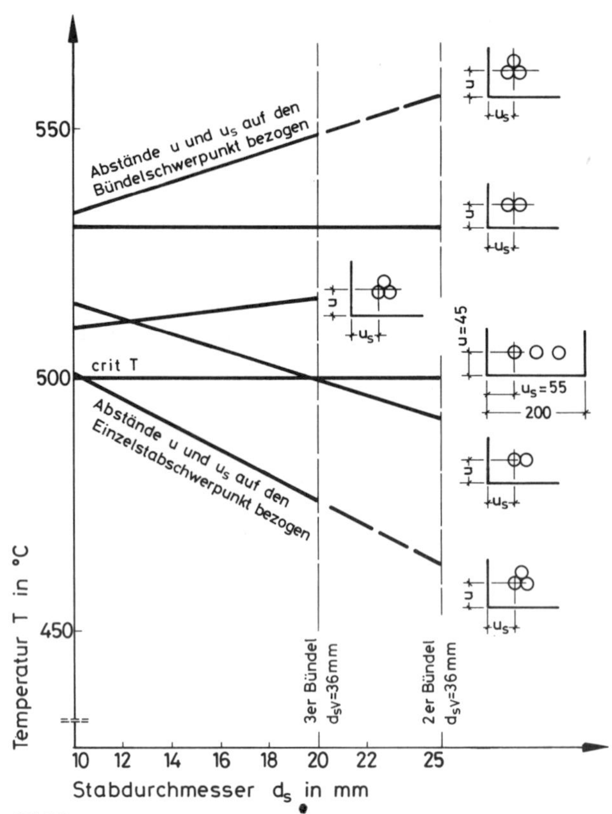

Bild 3
Mittelwerte der Temperaturen nach 90 Minuten ETK-Einwirkung bei Einzelstab- und bei Stabbündelbewehrung.

Aus diesem Grunde interessiert die Frage, welche mittleren Temperaturen in Stabbündeln auftreten, wenn im Querschnitt nur in den Ecken je ein Stabbündel angeordnet wird. Die Ergebnisse sind in Bild 3 dargestellt: Werden die gleichen Achsabstände wie sie für Einzelstäbe gelten auf den Schwerpunkt der Stabbündel angewendet, dann ergeben sich bedeutende Überschreitungen der kritischen Temperatur crit T. Werden hingegen die Achsabstände jeweils auf den der Ecke am nächsten liegenden Einzelstab des Stabbündels bezogen, ergeben sich überwiegend günstigere Verhält-

nisse, als dies für den Vergleichsfall mit Einzelstäben der Fall ist.
Soweit Stabbündel aus drei Einzelstäben verwendet werden, ergibt sich noch eine weitere Möglichkeit, die ebenfalls in Bild 3 dargestellt ist. Hierbei wird der seitliche Achsabstand u_s auf den maßgebenden Einzelstab bezogen, der Achsabstand u zur Unterseite aber auf den Schwerpunkt des Stabbündels. In diesem Fall ergibt sich zwar ebenfalls eine Überschreitung der kritischen Temperatur; jedoch bleibt die Überschreitung so klein, daß sie in Anbetracht der angesprochenen Zufälligkeiten, die bei der Erwärmung von Stabbündeln eintreten können, nicht überbewertet werden sollte.

4 Regeln für Achsabstände

Die in Bild 3 dargestellten und im vorangegangenen Abschnitt besprochenen Zusammenhänge waren in [2] zum Anlaß genommen worden, eine Änderung der diesbezüglichen Regeln in DIN 4102 Teil 4 vorzuschlagen. Die in DIN 4102 Teil 4 derzeit enthaltene Regelung ist auf den gemäß DIN 1045, Abschn. 18.11, definierten Vergleichsdurchmesser für das Stabbündel abgestellt. Bei Behandlung des Bündels als Einzelstab ergeben sich zunächst Überschreitungen der kritischen Temperatur soweit die erforderliche Bewehrung nach DIN 4102 Teil 4, Tabelle 5 aus nur zwei oder einem Einzelstab besteht. Soweit drei oder mehr Einzelstäbe erforderlich werden, läßt sich durch Anordnung gleichvieler Stabbündel kein Vorteil erzielen, weil der für die Unterbringung der Stabbündel erforderliche Platz bei gleicher Balkenbreite nicht zur Verfügung steht. Deshalb wurde in [2] der Vorschlag gemacht, die erforderliche Mindestzahl von Stabbündeln gegenüber der nach DIN 4102 erforderlichen Zahl von Einzelstäben grundsätzlich um eins zu verringern und die Achsabstände nicht auf den Schwerpunkt des Stabbündels, sondern auf den Schwerpunkt des maßgebenden Einzelstabes im Stabbündel zu beziehen. Die in [2] durchgeführten rechnerisch-theoretischen und experimentellen Untersuchungen ergaben für diesen Vorschlag ausreichende Sicherheiten.
Die geforderte Mindestzahl an Bewehrungsstäben darf nach DIN 4102 ebenfalls um eins vermindert werden, wenn der seitliche Achsabstand u_s um 10 mm vergrößert wird. Soweit Stabbündel aus je drei Stäben $d_s =$ 20 mm betrachtet werden, entspricht dieses Maß von 10 mm etwa dem Abstand des Schwerpunktes des Eckstabes im Stabbündel vom Schwerpunkt des Bündels. Beide Regelungen laufen also auf das gleiche hinaus, lediglich der untere Achsabstand u unterscheidet sich in beiden Vorgehensweisen. Aus Bild 3 ist jedoch zu ersehen, daß keine nennenswerte Überschreitung der kritischen Temperatur eintritt, wenn der untere Achsabstand u auf den Schwerpunkt des Stabbündels bezogen wird und der seitliche Achsabstand u_s auf den maßgebenden Einzelstab bzw. bei vergrößertem seitlichem Achsabstand auf den Bündelschwerpunkt.
Neben den Achsabständen nach DIN 4102 sind auch die Überdeckungsmaße der Längsbewehrungen bzw. der in Balken immer vorhandenen Bügel nach DIN 1045 zu beachten. In der Regel ist für die Bewehrungsanordnung immer nur ein Kriterium maßgebend. Bei Anwendung von Stabbündeln ist häufig der nach DIN 4102 geforderte Achsabstand u zum unteren Rand aufgrund anderer Festlegungen in DIN 1045 überschritten. Auch dies spricht dafür, die aus Bild 3 zu ersehende geringfügige Überschreitung der kritischen Temperatur bei Bezug des Achsabstandes u auf den Bündelschwerpunkt nicht zu hoch zu bewerten, weil häufig nicht in jedem Einzelfall zu berücksichtigende günstige Einflüsse vorliegen können. Es spricht deshalb nichts dagegen, es bei der augenblicklichen Regelung in DIN 4102 zu belassen und auf die in [2] vorgeschlagene Änderung vorläufig zu verzichten, wenn sich ergibt, daß sich auch ohne Änderung die erwünschten Vorteile durch die Anwendung von Stabbündeln erzielen lassen.
Bekanntlich lassen sich mehrparametrige Zusammenhänge bezüglich optimaler Zustände nur schwer veranschaulichen. Dies trifft auch für den hier zu besprechenden Zusammenhang zu. Deshalb wird die Auswirkung der möglichen Regeln für die Achsabstände nur für die wichtigsten Anwendungen untersucht. Als wichtig werden in diesem Zusammenhang die Vergrößerung der Gesamtbewehrung durch Anwendung von Stabbündeln angesehen, wobei der größte Vorteil bei Anwendung von Bündeln aus drei Einzelstäben erwartet wird.

5 Vergrößerung der Gesamtbewehrung

Die erforderliche Mindestbreite von Balken ergibt sich für Einzelstabbewehrungen unter Beachtung der erforderlichen Stababstände a_{sb} nach DIN 1045 und der Achsabstände u_s nach DIN 4102 sowie der seitlichen Betondeckung c_{sb} der Längsbewehrung nach DIN

1045 entsprechend den Gleichungen (1) und (2) in Bild 4. Die verwendeten Bezeichnungen sind in Bild 4 erläutert. Mit Gl. (3.1) in Bild 5 ergibt sich die erforderliche Mindestbreite bei Beachtung der seitlichen Achsabstände nach derzeitiger DIN 4102 und nach Gl. (3.2) nach dem Änderungsvorschlag in [2] zu DIN 4102. Nach Gl. (4) ergibt sich die Mindestbreite bei Beachtung der Betondeckung der Längsbewehrung nach DIN 1045. Die Bezeichnungen bei Verwendung von Stabbündeln sind in Bild 5 erläutert.

Mit den Gl. (1) bis (4) wurden die größtmöglichen Bewehrungsquerschnitte für die in Tafel 1 zu den einzelnen Feuerwiderstandsklassen aufgeführten Balkenbreiten bei Verwendung von Einzelstäben und Stabbündeln ermittelt. Die Ergebnisse sind in Tafel 2 zusammengestellt. Zur besseren Übersichtlichkeit wurde jeweils nur eine der beiden Zeilen für Einzelstäbe bzw. Stabbündel ausgefüllt und der jeweils nicht maßgebende Wert weniger auffällig dargestellt.

Für die Feuerwiderstandsklassen F 30 und F 60 wurden durchweg die Festlegungen der Mindestüberdekkung nach DIN 1045 maßgebend und nicht die Regelungen der Achsabstände nach DIN 4102. Ferner zeigte sich, daß sich nicht in allen Fällen die unterzubringende Bewehrungsmenge bei Verwendung von Stabbündeln steigern läßt. Dies liegt daran, daß zum einen nur ausgewählte Balkenbreiten behandelt sind, und zum anderen auch daran, daß die Wahl einheitlicher Stabdurchmesser gelegentlich allzu sprunghafte Veränderungen bewirkt.

Für die Feuerwiderstandsklassen F 120 und F 180 sind durchweg die Achsabstände nach DIN 4102 für die erreichbare Größtbewehrung maßgebend. Diese beiden Feuerwiderstandsklassen spielen nur für Gebäude besonderer Art und Nutzung eine Rolle, im wesentlichen dabei auch nur die Feuerwiderstandsklasse F 120 in Zusammenhang mit den Regelungen für Hochhäuser. Bei der Feuerwiderstandsklasse F 90 sind bei den behandelten Balkenbreiten wechselweise sowohl die Regelung nach DIN 1045 als auch die nach DIN 4102 maßgebend. In fast allen untersuchten Fällen lassen sich Steigerungen der Größtbewehrung durch Stabbündel erreichen. Häufig ist dabei die Anzahl der Stabbündel kleiner als die der Einzelstäbe. Nur in einzelnen Fällen ergibt sich die größte Bewehrungsmenge nicht bei Verwendung von Stabdurchmessern $d_s = 20$ mm. Soweit kleinere Durchmesser als $d_s = 20$ mm für Stabbündel verwendet werden, ergeben sich mit der Regelung nach DIN 4102 entsprechend Gl. (3.1) geringfü-

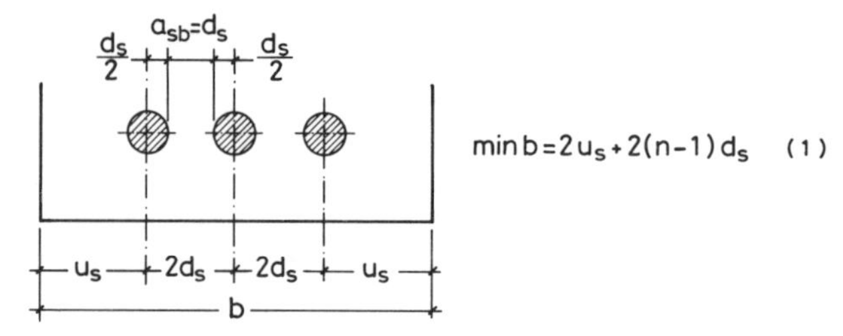

Mindestbalkenbreite min b für Einzelstabbewehrung und Oberflächenachsabstände nach DIN 4102, Teil 4

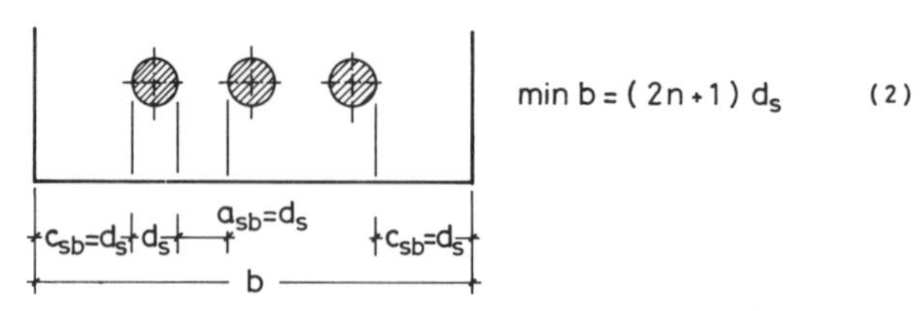

Mindestbalkenbreite min b für Einzelstabbewehrung und Betondeckung der Längsbewehrung nach DIN 1045

Bild 4

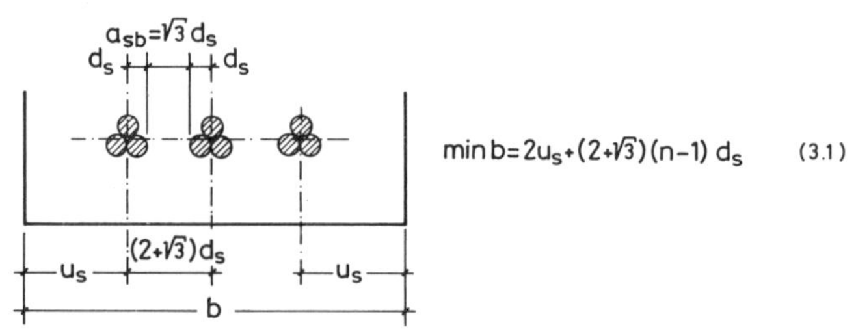

Mindestbalkenbreite min b für Stabbündelbewehrung und Oberflächenachsabstände nach DIN 4102

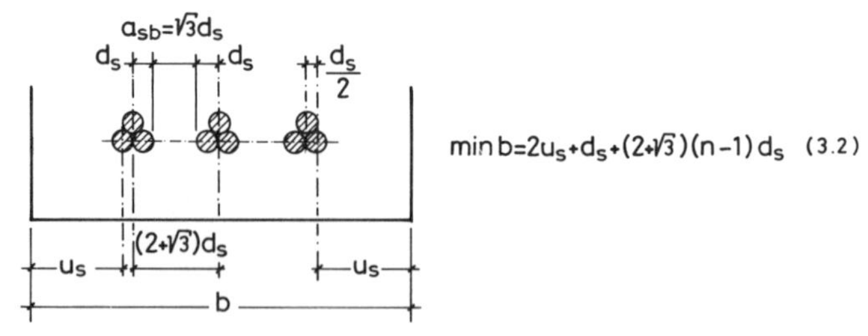

Mindestbalkenbreite min b für Stabbündelbewehrung und Oberflächenachsabstände nach Änderungsvorschlag zu DIN 4102

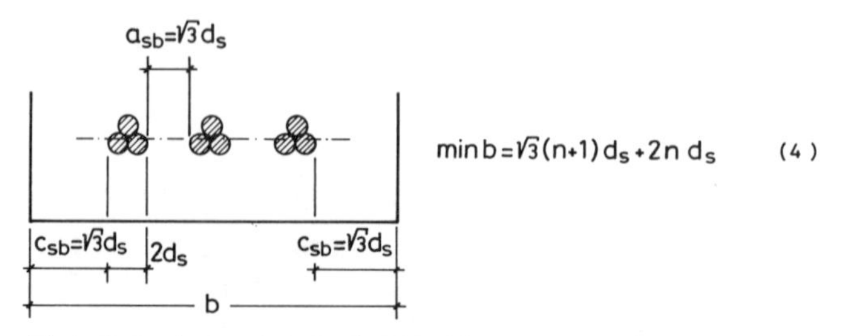

Mindestbalkenbreite min b für Stabbündelbewehrung und Betondeckung der Längsbewehrung nach DIN 1045

Bild 5

Tafel 2
Maximale Bewehrung mit zugehöriger Querschnittsfläche A_s in cm^2 für Einzelstab- und Stabbündelbewehrung. Bei Eintragung in der jeweils oberen Zeile ist DIN 4102 maßgebend, bei Eintragung in der jeweils unteren Zeile DIN 1045

	Feuerwiderstandsklasse				
	F 30	F 60	F 90	F 120	F 180
b in mm	80	120	150	200	240
Einzel- stäbe	 1⌀25 4,91	 *1⌀28 6,16*	 *1⌀28 6,16*	2⌀25 9,82 	2⌀28 12,3
Stab- bündel	 *1 · 3⌀14 4,62*	 1 · 3⌀20 9,42	 1 · 3⌀20 9,42	 *1 · 3⌀20 9,42*	*2 · 3⌀16 12,1*
b in mm	120	160	200	240	300
Einzel- stäbe	 *2⌀22 7,60*	 2⌀28 12,3	*2⌀28 12,3* 	*3⌀25 14,7* 	*3⌀28 18,5*
Stab- bündel	 1 · 3⌀20 9,42	 *2 · 3⌀16 12,1*	2 · 3⌀18 15,3 	2 · 3⌀20 18,9 	3 · 3⌀18 22,9
b in mm	160	200	250	300	400
Einzel- stäbe	 2⌀28 12,3	 *3⌀28 18,5*	 4⌀25 19,6	*4⌀28 24,6* 	*5⌀28 30,8*
Stab- bündel	 *2 · 3⌀16 12,1*	 2 · 3⌀20 18,9	 *2 · 3⌀20 18,9*	3 · 3⌀20 28,3 	4 · 3⌀20 37,7
b in mm	200	300	400	500	600
Einzel- stäbe	 *3⌀28 18,5*	 *4⌀28 24,6*	 *6⌀28 37,0*	*8⌀28 49,3* 	*9⌀28 55,4*
Stab- bündel	 2 · 3⌀20 18,9	 4 · 3⌀18 30,5	 5 · 3⌀18 38,2	 6 · 3⌀20 56,6	7 · 3⌀20 66,0

gig größere Bewehrungen, als dies aufgrund des Änderungsvorschlages in [2] entsprechend Gl. (3.2) der Fall wäre. Bei Verwendung von Stabbündeln aus Einzelstäben mit Durchmesser $d_s = 20$ mm ergeben sich keine Unterschiede zwischen Gl. (3.1) und Gl. (3.2). Die in [2] vorgeschlagene Änderung ist deshalb entbehrlich, weil sich bei vollständigerer Untersuchung bezüglich der optimalen Verhältnisse stets eine Vergrößerung des seitlichen Achsabstandes u_s nach DIN 4102 als erforderlich herausstellt und sich nur durch Verringerung der Anzahl der Stabbündel eine Vergrößerung der Gesamtbewehrungsfläche gegenüber der Anwendung von Einzelstabbewehrungen ergibt. Ohne Vergrößerung der Bewehrungsfläche erscheint aber die Anwendung von Stabbündeln nicht wirtschaftlicher als die Anwendung von Einzelstäben, weil erfahrungsgemäß die Kosten der Verlegearbeit durch die zu verarbeitenden Stab-Stückzahlen beeinflußt werden, auch wenn nach wie vor die Vorkalkulation und Abrechnung der Verlegearbeiten über Tonnenpreise üblich ist. Dies führt später häufig zu den hinlänglich bekannten Reibereien, weil der Unternehmer meint, aufgrund des abgegebenen Preises nur Bewehrungen geringer Stückzahl auskömmlich verarbeiten zu können, weswegen er sich gegen sinnvolle Bewehrungskonzepte wendet, die der entwerfende Ingenieur nach sachlichen Gesichtspunkten anders festlegt.

6 Zusammenfassung

Entsprechend der Vorgehensweise in DIN 1045 lassen sich auch die Festlegungen für Einzelstäbe in DIN 4102 Teil 4 auf Stabbündel anwenden, wenn anstelle des Einzelstabdurchmessers d_s der Vergleichsdurchmesser d_{sV} eingesetzt wird. Rechnerisch-theoretische und experimentelle Untersuchungen haben zwar ergeben, daß bei solcher Regelung Überschreitungen der kritischen Temperaturen in Stabbündeln nicht ausgeschlossen sind. Geht man jedoch davon aus, daß Stabbündel immer nur dann angewendet werden, wenn die erforderliche Bewehrungsmenge nicht durch Einzelstäbe untergebracht werden kann, so ergibt sich, daß – denkbare – unsichere Anordnungen nicht ausgeführt werden. Die in DIN 4102 Teil 4, Ausgabe März 81, allein auf rechnerisch-theoretischen Untersuchungen abgestützten Untersuchungen konnten nachträglich durch experimentelle Untersuchungen und ergänzende Betrachtungen über die Wirtschaftlichkeit von Bewehrungen aus Stabbündeln abgesichert werden.

7 Literatur

[1] REHM, G., ELIGEHAUSEN, R. und NEUBERT, B.: Erläuterungen der Bewehrungsrichtlinien, Schriftenreihe Deutscher Ausschuß für Stahlbeton, Heft 300, 1979.

[2] KORDINA, K., WIESE, J. und KLINGSCH, W.: Brandverhalten von Stahlbetonbauteilen mit Stabbündelbewehrung, Forschungsbericht des Instituts für Baustoffe, Massivbau und Brandschutz der TU Braunschweig, März 1983.

[3] KORDINA, K. und MEYER-OTTENS, C.: Beton-Brandschutz-Handbuch, Beton-Verlag, Düsseldorf, 1981.

[4] EHM, H.: Rechnerische Ermittlung der Erwärmungsvorgänge von brandbeanspruchten, balkenartigen Stahlbetonbauteilen, Schriftenreihe des Deutschen Ausschusses für Stahlbeton, Heft 230, Teil 2, 1975.

Zur Berechnung schiefer Plattenbrücken

Christian Menn

1 Grundsätzliches

Die Berechnung von Stahlbetonplatten kann mit zwei grundsätzlich verschiedenen Methoden durchgeführt werden.

1. Statische Methode: Ermittlung der Schnittkräfte und anschließend Bemessung der maßgebenden Plattenpunkte.
2. Kinematische Methode: Ermittlung der Tragsicherheit aufgrund der maßgebenden Bruchlinienkonfiguration.

Die statische Methode ergibt bei jeder Schnittkraftverteilung, die einem Gleichgewichtszustand entspricht, immer einen unteren Grenzwert für die Tragsicherheit. Die kinematische Methode hat den schwerwiegenden Nachteil, daß sie grundsätzlich einen oberen Grenzwert der Tragsicherheit ergibt, weil die maßgebende Bruchlinienkonfiguration nicht eindeutig ermittelt werden kann. Diese Methode ist deshalb zu vermeiden oder höchstens für Kontrollzwecke zu verwenden. Sie trägt auch in keiner Weise den heute zur Verfügung stehenden Möglichkeiten der Computerstatik Rechnung.

Bei der Ermittlung der Schnittkräfte spielt die Wahl des statischen Modells eine entscheidende Rolle. Je nach Größe und Bedeutung der zur berechnenden Platte werden einfache oder komplexe Modelle und Berechnungsverfahren verwendet. Die Verträglichkeitsbedingungen, denen bestimmte Annahmen über die Querschnittssteifigkeit zugrunde liegen, werden mehr oder weniger stark verletzt; der Gleichgewichtszustand ist aber immer erfüllt.

- Die Plattentheorie ergibt eine Schnittkraftverteilung, die die Verträglichkeit in allen Plattenpunkten erfüllt; Grundlage ist ein linear-elastisches Materialverhalten sowie eine dem homogenen Querschnitt entsprechende Steifigkeit.
- Bei Trägerrostmodellen wird die Platte durch einen Trägerrost nachgebildet. Die Schnittkräfte werden nach der Stabstatik ermittelt, die Verträglichkeit ist nur beschränkt und in einzelnen Punkten erfüllt.
- Beim Lastaufteilungsverfahren, das sich besonders für Rechteckplatten eignet, wird die Last in gewissen Plattenbereichen jeweils einer bestimmten Tragrichtung zugewiesen. Verträglichkeitsbedingungen werden nur generell berücksichtigt.
- Momentenverteil-Verfahren eignen sich für die Berechnung von Pilz- und Flachdecken. Die Momente werden zunächst global ermittelt und dann unterschiedlich auf bestimmte Plattenbereiche (Gurt- und Feldstreifen) verteilt.
- Das Momentenumlagerungs-Verfahren, das im folgenden näher erläutert wird, besteht im Prinzip darin, daß der Momentenverteilung gemäß Plattentheorie „lastfreie Spannungszustände“ überlagert werden. Das Ziel dieser Spannungsüberlagerung beziehungsweise Momentenumlagerung besteht darin, daß konzentrierte Spannungsspitzen abgebaut und die Tragfähigkeit der konstruktiv vernünftig angeordneten Bewehrung voll ausgenützt werden kann. Das Verfahren entspricht im wesentlichen der bekannten Momentenumlagerung bei statisch unbestimmten Stabtragwerken und nützt die Möglichkeiten der modernen Computerstatik.

Bei einfachen, statisch unbestimmten Stabtragwerken setzt die Momentenumlagerung (Stütze – Feld) im wesentlichen erst bei der Ausbildung eines plastischen Gelenkes ein. Steifigkeitsänderungen infolge Rißbildung haben normalerweise einen kleinen Einfluß auf die Momentenumlagerung.

Bei verformungsgekoppelten Systemen (Trägerroste, Platten) wirkt sich dagegen eine Steifigkeitsänderung direkt und sehr stark auf die Momentenumlagerung aus; d. h. die Momentenumlagerung erfolgt nicht wie üblich im plastischen, sondern bereits im elastischen Beanspruchungsbereich. Deshalb können auch bei Platten ohne Bedenken Tragmodelle und Verfahren gewählt werden, die eine von

der Elastizitätstheorie stark abweichende Schnittkraftverteilung ergeben, und andererseits führt die Elastizitätstheorie je nach Bewehrungsanordnung nicht unbedingt zu einer realistischen Lösung.
Am rationellsten und zweckmäßigsten ist deshalb das Momentenumlagerungs-Verfahren; d.h. die Kombination der elastischen Lösung mit lastfreien Spannungszuständen.

2 Lastfreie Spannungszustände

Lastfreie Spannungszustände sind Lösungen der homogenen Gleichgewichtsdifferentialgleichung.

Balken: $\frac{d^4w}{dx^4} = 0$

Platten: $\frac{\partial^4 w}{\partial x^4} + 2\frac{\partial^4 w}{\partial x^2 \partial y^2} + \frac{\partial^4 w}{\partial y^4} = 0$

Die Anzahl der unabhängigen lastfreien Spannungszustände entspricht dem Grad der statischen Unbestimmtheit des Systems. Da man Platten als unendlichfach statisch unbestimmtes Trägerrostsystem auffassen kann, existieren bei Platten unendlich viele unabhängige lastfreie Spannungszustände.
Lastfreie Spannungszustände können auf verschiedene Arten ermittelt werden; z.B.

- durch Auflagerverschiebungen
- durch Belastung von zwei verschiedenartigen Tragsystemen einerseits mit $+q$ und andererseits mit $-q$.

Die Zwängungsmomente infolge Vorspannung entsprechen immer einem lastfreien Spannungszustand.
Bei Platten lassen sich Zwängungsmomente (m_{Zp}) infolge Vorspannung sehr einfach folgendermaßen ermitteln:
Aus

$$m_p = m_{op} + m_{Zp}$$

folgt

$$m_{Zp} = m_p - m_{op}$$

m_p sind die Momente infolge Vorspannung, die bei Platten aus der „Plattenbelastung" durch Umlenk- und Ankerkräfte ermittelt werden.
m_{op} sind die „statisch bestimmten Momente":

$$m_{op} = -p \cdot e$$

(e = Exzentrizität des Spanngliedes)

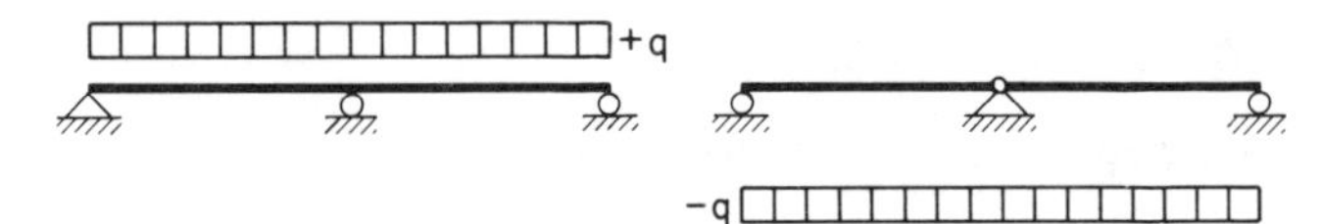

Bild 1
Belastungen $+q$, $-q$ an zwei verschiedenen Tragsystemen angreifend

Dieser lastfreie Spannungszustand kann nun im Prinzip in beliebiger Größe der elastischen Lösung aus Lasteinwirkung so überlagert werden, daß die Bewehrung optimal ausgenutzt ist.

3 Beispiel

Durchlaufende Platte mit Stützstreifenvorspannung
Belastung: $q = 20$ kN/m²
Vorspannung: $p = 5000$ kN/m²
Momente gemäß Plattentheorie mit Finite-Elemente Programm gerechnet.

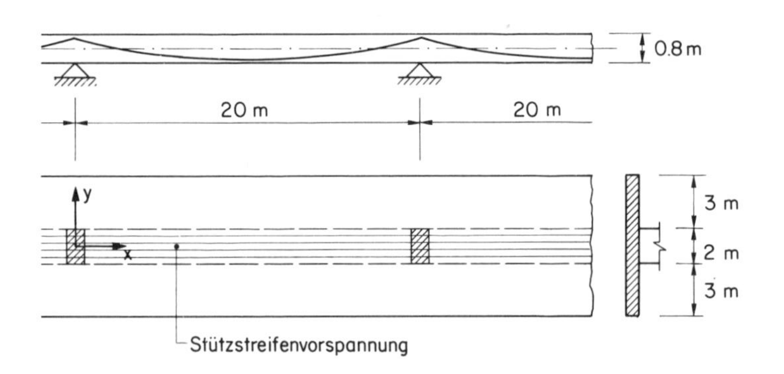

Bild 2
Längsschnitt und Grundriß

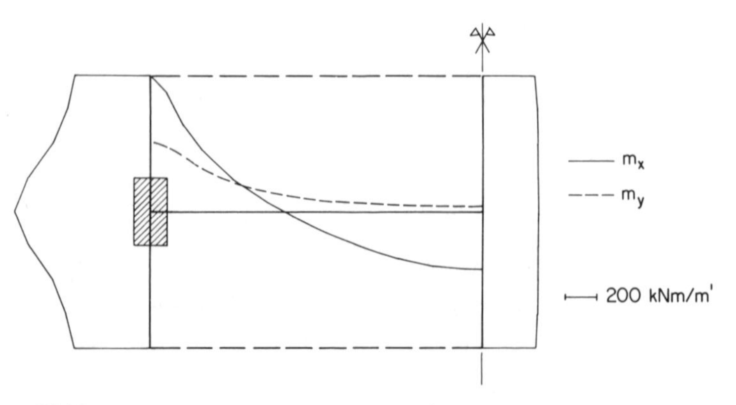

Bild 3
m_q: Momente infolge gleichmäßig verteilter Belastung q

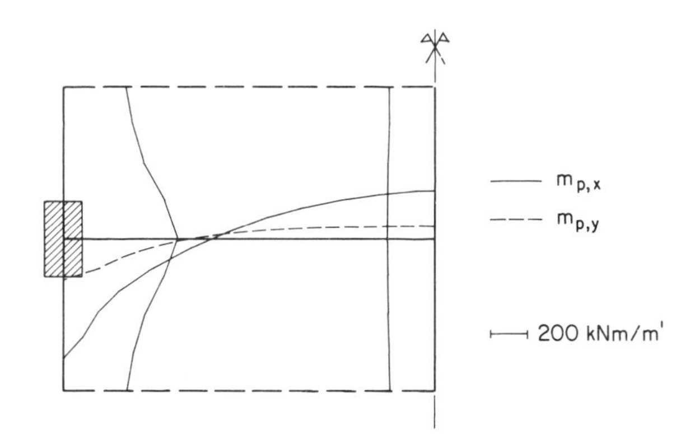

Bild 4
m_p: Momente infolge Vorspannung

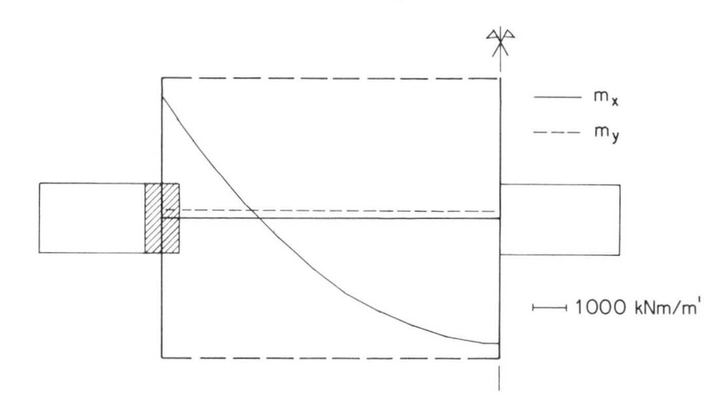

Bild 7
m^*: Momentenverteilung für Tragsicherheitsnachweis:
$m^* = 1{,}8 \cdot m_q + 1{,}8 \cdot 1{,}3 \cdot m_{Zp}$, Faktor $1{,}8 \cdot 1{,}3$ frei gewählt

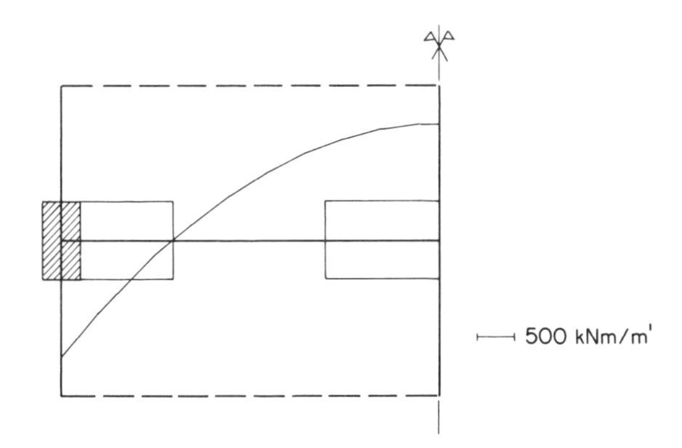

Bild 5
$m_{op} = -p \cdot e$

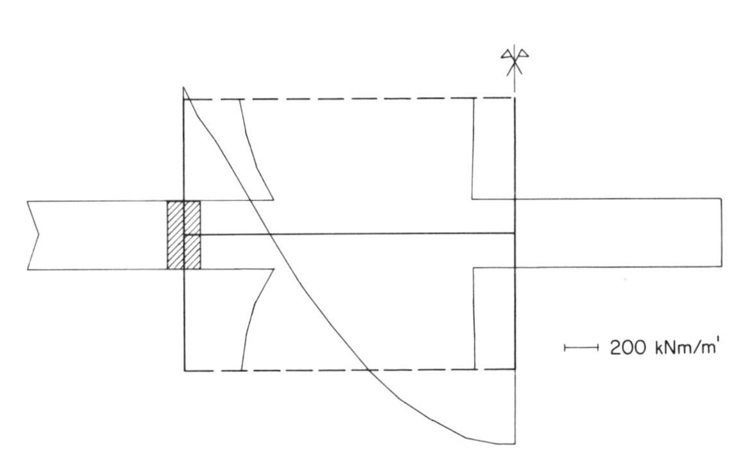

Bild 6
m_{Zp}: Lastfreier Spannungszustand, in der Größe frei wählbar

4 Bemessung von Platten bei beliebiger Bewehrungsanordnung [1], [2], [3]

Bei schiefen Platten weicht die Bewehrungsrichtung sehr oft beträchtlich von der Richtung der Hauptmomente ab, und überdies muß die Bewehrung oft in drei Richtungen angeordnet werden. Eine zweckmäßige Berechnung läßt sich in diesen Fällen nur aufgrund der Plastizitätstheorie durchführen. Im Folgenden wird das Verfahren kurz aufgezeigt.
Grundsätzlich lassen sich aus den mechanischen Bewehrungsgehalten ω_i (Bewehrungsrichtung i) über ω_x und ω_y und ω_{xy} äquivalente orthogonale Bewehrungsgehalte ω_1 und ω_2 in den Hauptrichtungen bestimmen. Mit

$$\omega_x = \sum_i \omega_i \cdot \cos^2\gamma_i$$

$$\omega_y = \sum_i \omega_i \cdot \sin^2\gamma_i$$

$$\omega_{xy} = \sum_i \omega_i \cdot \sin\gamma_i \cdot \cos\gamma_i$$

ergeben sich die äquivalenten Bewehrungsgehalte in den Hauptrichtungen ξ und η:

$$\omega_1 = \omega_\xi = \omega_x \cdot \cos^2\varphi_1 + \omega_y \cdot \sin^2\varphi_1 + \omega_{xy} \cdot \sin 2 \cdot \varphi_1$$

$$\omega_2 = \omega_\eta = \omega_x \cdot \sin^2\varphi_1 + \omega_y \cdot \cos^2\varphi_1 - \omega_{xy} \cdot \sin 2 \cdot \varphi_1$$

wobei

$$\operatorname{tg} 2 \cdot \varphi_1 = \frac{2 \cdot \omega_{xy}}{\omega_x - \omega_y}$$

Dieser „Ersatzbewehrung“ entsprechen die in der üblichen Weise zu bestimmenden plastischen Momente $m_{u\xi}$ und $m_{u\eta}$

$$m_{u\xi} = d^2 \cdot f_{cu} \cdot \omega_\xi \cdot \left(1 - \frac{\omega_\xi}{2}\right)$$

$$m_{u\eta} = d^2 \cdot f_{cu} \cdot \omega_\eta \cdot \left(1 - \frac{\omega_\eta}{2}\right)$$

Die auf die Richtung der Ersatzbewehrung transformierten Bemessungsmomente ($m^* = \gamma \cdot m; \gamma =$ globaler Sicherheitskoeffizient)

$$m^*_\xi = m^*_x \cdot \cos^2\varphi_1 + m^*_y \cdot \sin^2\varphi_1 + m^*_{xy} \cdot \sin 2 \cdot \varphi_1$$

$$m^*_\eta = m^*_x \cdot \sin^2\varphi_1 + m^*_y \cdot \cos^2\varphi_1 - m^*_{xy} \cdot \sin 2 \cdot \varphi_1$$

$$m^*_{\xi\eta} = \frac{1}{2} \cdot (m^*_y - m^*_x) \cdot \sin 2 \cdot \varphi_1 + m^*_{xy} \cdot \cos 2 \cdot \varphi_1$$

können nun in die linearisierten Fließbedingungen eingesetzt werden.

Für positive Momente:

$$(m_{u\xi} - m^*_\xi) - |m^*_{\xi\eta}| \geq 0$$
$$(m_{u\eta} - m^*_\eta) - |m^*_{\xi\eta}| \geq 0$$

und für negative Momente:

$$(m'_{u\xi'} + m^*_{\xi'}) - |m^*_{\xi'\eta'}| \geq 0$$
$$(m'_{u\eta'} + m^*_{\eta'}) - |m^*_{\eta'\eta'}| \geq 0$$

Im allgemeinen sind die Richtungen der Ersatzbewehrung für die positiven und die negativen Bruch- und Bemessungsmomente verschieden, das heißt

$$\varphi_1 \neq \varphi'_1, \quad \xi \neq \xi', \quad \eta \neq \eta'$$

5 Beispiel

Bemessung einer schiefen Plattenbrücke

– System, Abmessungen und Belastung
  Die Berechnung erfolgt der Einfachheit halber nur für einen Lastfall: $g = 22{,}5\ \text{kN/m}^2$, $q = 4{,}5\ \text{kN/m}^2$, $Q = 200\ \text{kN}$.

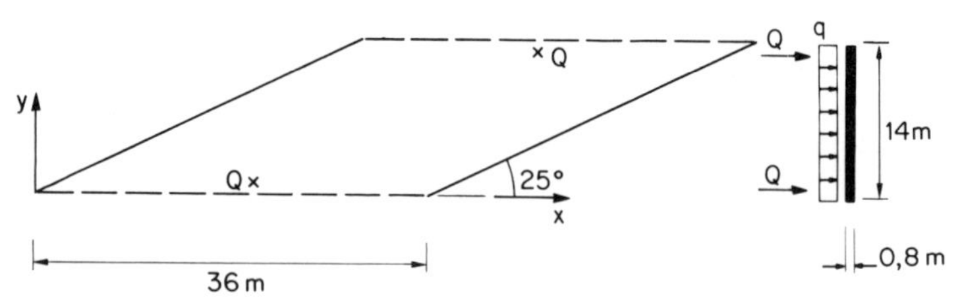

Bild 8
System, Abmessungen und Belastung

– Material

Beton:	f_{cu}	= 24 N/mm²
Stahl:	f_{sy}	= 460 N/mm²
Spannstahl:	f_{py}	= 1520 N/mm²

– Anordnung der Vorspannung
  Kabelfeld I:
  $l = 19{,}0$ m $f = 0{,}3$ m, f = Pfeilhöhe
  $b = 9{,}5$ m $p = 5480$ kN/m′ $a_p = 4696$ mm²/m′
  Kabelfeld II:
  $l = 36{,}0$ m $f = 0{,}3$ m,
  $b = 2 \cdot 3{,}0$ m $p = 6120$ kN/m′ $a_p = 5232$ mm²/m′

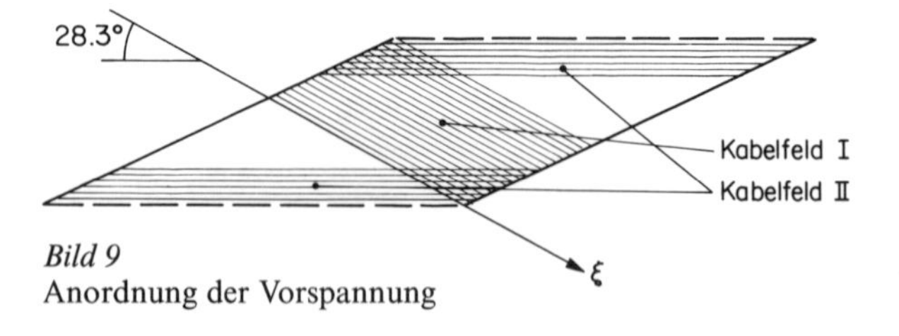

Bild 9
Anordnung der Vorspannung

– Momente infolge ständiger Last und Verkehrslast

Tabelle 1
Momente infolge ständiger Last und Verkehrslast

Pkt	m_x	m_y	m_{xy}
1	−1428	−87	−600
2	55	544	−855
3	585	122	−378
4	781	83	−424

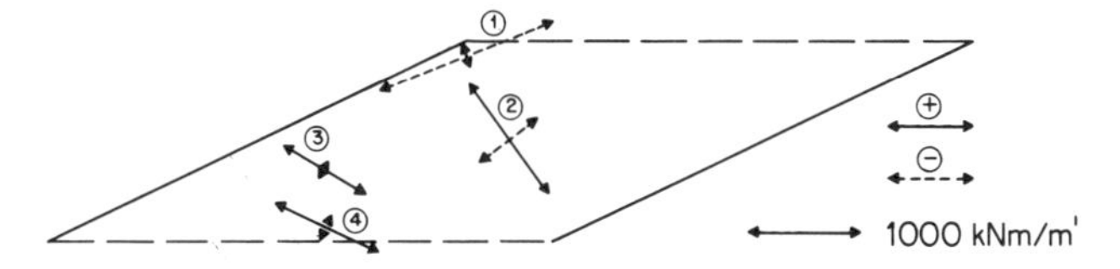

Bild 10
Hauptmomente infolge ständiger Last und Verkehrslast

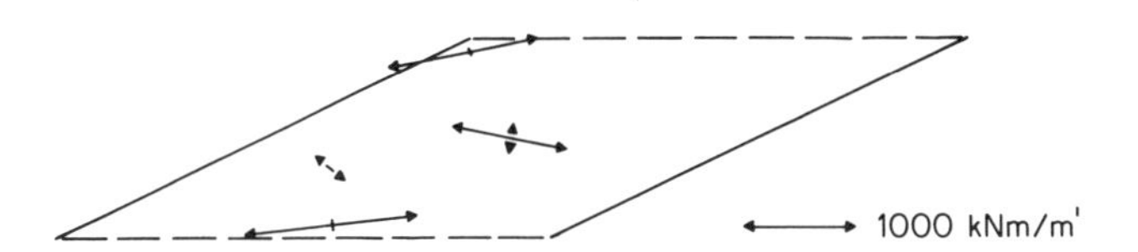

Bild 12
Hauptmomente infolge Zwängung aus Vorspannung

- Momente infolge Vorspannung
  Berechnet aus den Umlenk- und Ankerkräften; die Kabel sind bei den Auflagern in der Schwerachse verankert.

Tabelle 2
Momente infolge Vorspannung

Pkt	$m_{p,x}$	$m_{p,y}$	$m_{p,xy}$
1	715	−54	380
2	−288	−567	475
3	−162	−82	142
4	−288	−38	180

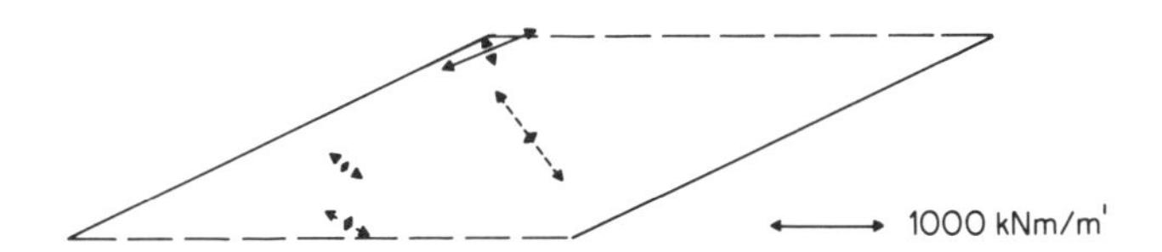

Bild 11
Hauptmomente infolge Vorspannung

- Zwängungsmomente infolge Vorspannung
  $m_{Zp} = m_p - m_{op}$, wobei $m_{op} = -p \cdot e$

Tabelle 3
Zwängungsmomente

Pkt	$m_{Zp,x}$	$m_{Zp,y}$	$m_{Zp,xy}$
1	1368	23	236
2	986	−198	−211
3	−162	−82	142
4	1548	−38	180

- Bemessungsmomente $m^* = 1{,}8\,(m_{g+q} + m_{Zp})$
  Die Größe der Vorspannung wurde so angesetzt, daß die Zwängungsmomente mit dem gleichen Lastfaktor wie die Momente aus der Belastung berücksichtigt werden können, um einen günstigen Spannungszustand zu erhalten.

Tabelle 4
Bemessungsmomente

Pkt	m^*_x	m^*_y	m^*_{xy}
1	−108	−115	−655
2	1874	623	−1919
3	761	72	−425
4	4192	81	−439

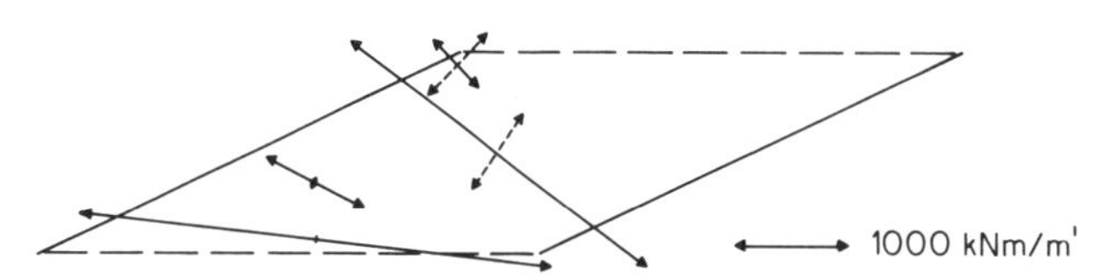

Bild 13
Bemessungsmomente in den Hauptrichtungen

- Bewehrungsanordnung
  $\varrho_{min} = 0{,}15\%$

Tabelle 5
Oben liegende Bewehrung

Pkt	a'_x	a'_y
1	∅ 18, $t = 10$	∅ 18, $t = 10$
2	∅ 12, $t = 10$	∅ 26, $t = 12{,}5$
3	∅ 12, $t = 10$	∅ 12, $t = 10$
4	∅ 12, $t = 10$	∅ 12, $t = 10$

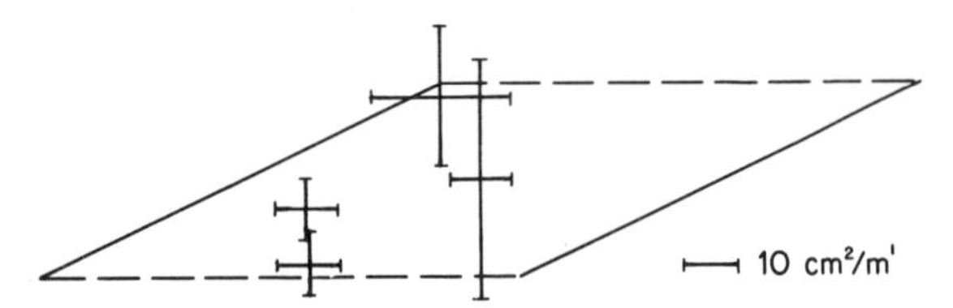

Bild 14
Oben liegende Bewehrung

Tabelle 6
Unten liegende Bewehrung

Pkt	a_x	a_y
1	∅ 12, t = 10	∅ 12, t = 10
2	∅ 12, t = 10	∅ 12, t = 10
3	∅ 22, t = 10	∅ 14, t = 10
4	∅ 12, t = 10	∅ 14, t = 10

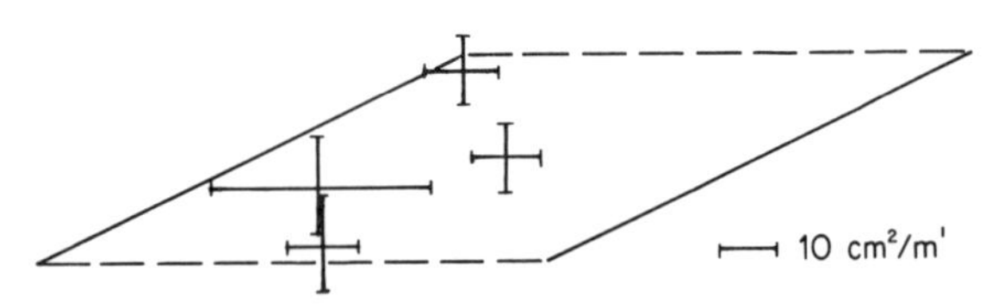

Bild 15
Unten liegende Bewehrung

– Bruchmomente

Tabelle 7
Bruchmomente entsprechend der „Ersatzbewehrung"

Pkt	φ	$m_{u\xi}$	$m_{u\eta}$	φ'	$m'_{u\xi}$	$m'_{u\eta}$
1	−13,3°	2332	535	0°	849	849
2	−28,3°	4121	348	0°	386	1391
3	0°	1246	521	0°	386	386
4	0°	4633	521	0°	386	386

– Nachweise

Tabelle 8
Auf die Richtung der Ersatzbewehrung transformierte Bemessungsmomente

Pkt	m^*_ξ	m^*_η	$m^*_{\xi\eta}$	$m^*_{\xi'}$	$m^*_{\eta'}$	$m^*_{\xi'\eta'}$
1	185	−408	−584	−108	−115	−655
2	3195	−698	−534	1874	623	−1919
3	761	72	−425	761	72	−425
4	4192	81	−439	4192	81	−439

Die Fließbedingungen

$$(m_{u\xi} - m^*_\xi) - |m^*_{\xi\eta}| \geq 0$$
$$(m_{u\eta} - m^*_\eta) - |m^*_{\xi\eta}| \geq 0$$
$$(m'_{u\xi'} + m^*_{\xi'}) - |m^*_{\xi'\eta'}| \geq 0$$
$$(m'_{u\eta'} + m^*_{\eta'}) - |m^*_{\xi'\eta'}| \geq 0$$

sind für alle Punkte erfüllt.

6 Zusammenfassung

Bei Platten und insbesondere bei vorgespannten Platten haben konstruktive und ausführungstechnische Gesichtspunkte einen wesentlichen Einfluß auf die Bewehrungsanordnung. Das vorgestellte Verfahren zeigt wie die Schnittkräfte eines günstigen Gleichgewichtszustandes in einer Platte unter Berücksichtigung der Bewehrungsanordnung ermittelt werden können und wie die Bemessung durchzuführen ist.

7 Literatur

[1] Menn, C.: „Brückenbau II", Institut für Baustatik und Konstruktion ETH Zürich, Autographie zur Vorlesung Brückenbau AK, Zürich, 1981.
[2] Marti, P.: „Plastische Berechnungsmethoden", Institut für Baustatik und Konstruktion ETH Zürich, Autographie zur Vorlesung Baustatik III, Zürich, 1978.
[3] Wolfensberger, R.: „Traglast und optimale Berechnung von Platten", Technische Forschungs- und Beratungsstelle der Schweiz. Zementindustrie, Wildegg, 1964.

Untersuchungen über das Verhalten von rechteckigen und kranzförmigen Konsolen insbesondere auch unter exzentrischer Belastung

Heinrich Paschen
und Hermann Malonn

1 Problemstellung

Im Fertigteilbau insbesondere im Skelettbau kommt es vor, daß die Träger (Unterzüge) im Bauzustand, aber auch nach Fertigstellung infolge einseitiger Belastung durch das Deckengewicht bzw. durch Verkehrslast Torsion erhalten. Infolgedessen geben sie an ihren Auflagern, also an die Stützenkonsolen exzentrisch wirkende Auflagerdrücke ab (s. Bild 1). Exzentrisch angreifende Auflagerkräfte erzeugen in den Konsolen Torsion. Die Frage ist, ob hierdurch im kurzen Kragarm dieselben Beanspruchungen hervorgerufen werden wie in einem stabförmigen Bauteil und ob daher die übliche Torsionsbewehrung auch hier optimal zur Aufnahme der inneren Kräfte geeignet ist. In der Literatur wurde bislang auf diese Frage noch nicht eingegangen, und in der Praxis ist die mögliche Exzentrizität oft genug ganz unbeachtet geblieben. Um so wichtiger war es zu prüfen, welche Einbußen an Tragfähigkeit bei Konsolen durch exzentrische Belastung hervorgerufen

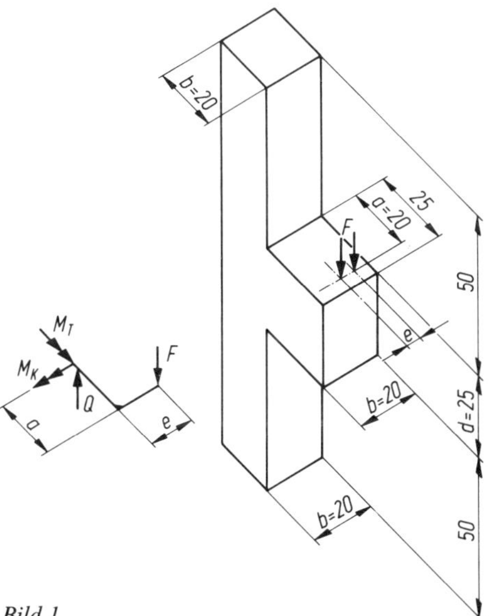

Bild 1
Versuchskörperabmessungen. Zentrische bzw. exzentrische Belastung

werden und wie ihnen durch Bewehrung zweckmäßig begegnet werden kann.
Zur Klärung dieser Frage sind vom Institut für Baukonstruktion und Vorfertigung der Technischen Universität Braunschweig Versuche durchgeführt worden, die vom niedersächsischen Kultusministerium und von der Deutschen Forschungsgemeinschaft finanziert worden sind. Die Versuchsdurchführung erfolgte am Institut für Baustoffe, Massivbau und Brandschutz der Technischen Universität Braunschweig unter aktiver Mitarbeit der Herren Klinkert und Ege.
Im Rahmen dieser Versuche sollte auch das Verhalten von „Kranzkonsolen“, insbesondere auch unter exzentrischer Belastung untersucht werden. Bei Innenstützen sog. „gerichteter“ Skelettsysteme werden Konsolen in der Regel paarweise und einander gegenüberliegend angebracht, weil Unterzüge hier nur in einer Gebäude-Achs-Richtung verlegt werden. Es kommen jedoch auch Konstruktionen zum Einsatz, bei welchen die Anordnung von Unterzügen in beiden Achsrichtungen vorgesehen ist bzw. möglich sein soll. In solchen Fällen werden auch „Kranzkonsolen“ verwendet, Konsolen also, die mit konstanter Auskragung die Stütze kranzartig, also auch im Eckbereich umgeben. Derartige Konsolen werden vielfach z. B. durch hutförmige Unterzüge in den Eckbereichen belastet, bzw. es erfolgt aus den vorgenannten Gründen eine exzentrische Einleitung der Unterzugs-Auflagerkräfte (s. Bild 2). Auch hier stellt sich die Frage nach der wirksamsten Bewehrungsform. In das Versuchsprogramm wurden deshalb auch einige Kranzkonsolen aufgenommen.

2 Vorhandene Untersuchungen

Neben älteren und ausländischen Arbeiten sind die Arbeiten von Franz, Mehmel-Becker und neuerdings wieder von Franz grundlegend. Beachtenswert

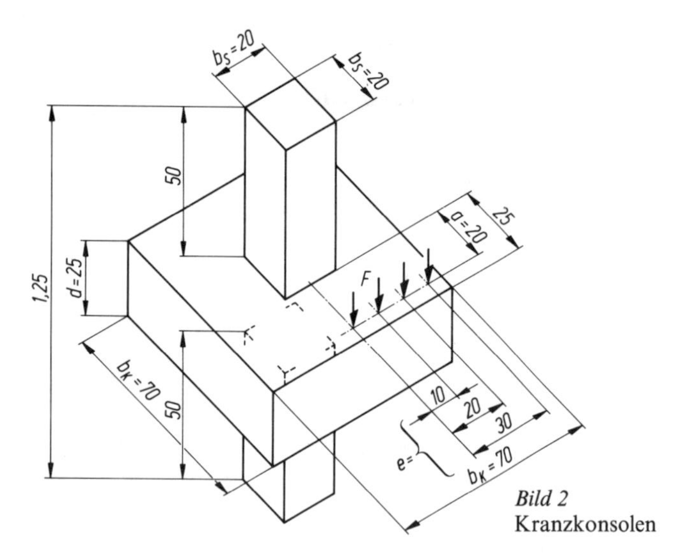

Bild 2
Kranzkonsolen

ist auch eine vergleichende Studie von Steinle. In seiner neuesten Arbeit empfiehlt Franz wiederum eine ausschließlich horizontal liegende Bewehrung im Gegensatz zu Mehmel, der eine Kombination von Horizontal- und Schärgankerbewehrung als optimal ansah. FRANZ fordert außerdem eine Begrenzung des Bewehrungsverhältnisses $\mu = As/b \cdot h$.

Bemerkenswerterweise sind alle bisherigen Versuche an symmetrischen Versuchskörpern (also an Doppelkonsolen) unter symmetrischer Last durchgeführt worden. Bei einseitigen Konsolen ist jedoch der Kräftefluß Konsole–Stütze anders.

STEINLE hat in seiner Untersuchung insbesondere auf die Divergenz der verschiedenen Ansätze und auf die Unstetigkeit an der Übergangsstelle zur Biegebemessung ($a/h > 1$) hingewiesen (s. Bild 3). Alle Veröffentlichungen stimmen darin überein, daß bei der Bemessung nicht nur der Stahlquerschnitt bestimmt, sondern auch nachgewiesen werden muß, daß die Betonabmessungen ausreichen.

Keine der vorhandenen Veröffentlichungen befaßt sich mit der exzentrisch belasteten bzw. mit der Kranzkonsole.

3 Versuchsparameter

Als Versuchsparameter, deren Einfluß auf das Konsoltragverhalten durch geeignete Variation im Rahmen des Forschungsvorhabens untersucht werden sollte, wurden vorgesehen:

- die Konsolform: Einzel- oder Kranzkonsole,
- die Bewehrungsform und -menge,
- die Exzentrizität der Last in Querrichtung.

Konstant sollten dagegen gehalten werden:

- die Konsolabmessungen,
- der Lasthebel *a*,
- die Beton- und Stahlgüte.

Die Abmessungen der beiden Konsoltypen gehen aus Bild 1 und 2 hervor. Als Betongüte wurde B 25, hier mit einer 28-Tage-Festigkeit von 25 N/mm² als Serienfestigkeit angestrebt, um eine unmittelbare Beziehung zwischen tatsächlich vorhandener Würfelfestigkeit und den Versuchsergebnissen zu erhalten. Als Bewehrungsstahl wurde einheitlich Betonstahl BSt 420/500 eingesetzt.

Bei der Bewehrungsform sollte es in erster Linie um Klärung der Frage gehen, ob beim kurzen Kragarm eine Torsionsbewehrung, bestehend aus Längsstäben und Bügeln gegenüber exzentrischem Lastangriff wirksam ist, oder ob eine schräg angeordnete Hauptbewehrung (obere Horizontalbewehrung, Schrägbügel) die Last eher zu zentrieren vermag. Die schlechten Ergebnisse der ersten Versuche gaben Veranlassung, die von MEHMEL-BECKER und FRANZ vorgeschlagenen

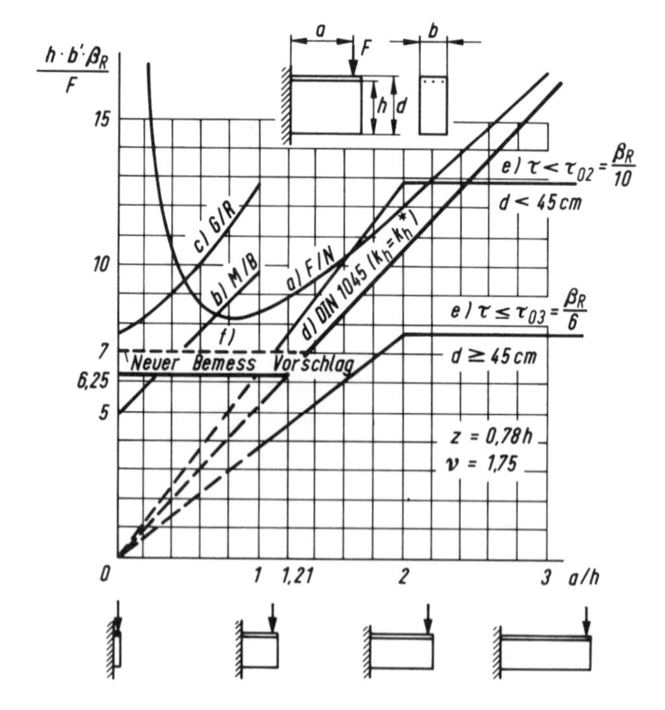

Bild 3
Vergleich verschiedener Bemessungsvorschläge nach [8] von:
a) FRANZ/NIEDENHOFF [5]
b) MEHMEL/BECKER [6]
c) GRASSER/RUHNAU [10]
d) DIN 1045 k_h-Verfahren
e) DIN 1045 Schubspannungsnachweis
f) STEINLE [8]

Bewehrungsformen nochmals hinsichtlich ihrer Wirksamkeit vergleichend zu testen. Das geschah insbesondere mit Hilfe von Dehnungsmessungen an der Bewehrung, mit welchen bei vergleichbaren Verhältnissen sehr aufschlußreiche Ergebnisse erhalten wurden. Sie lassen unmittelbar erkennen, welche Bewehrungsstäbe sich bei welchem Belastungsniveau stärker und welche sich weniger an der Lastabtragung beteiligen, und liefern daher Hinweise darauf, wie optimal bewehrt werden kann. Ein neuerliches Eingehen auf die Frage, ob ergänzende Schrägbewehrung zu empfehlen ist, erschien auch deshalb wünschenswert, weil eine Reihe von Untersuchungen in den letzten Jahren verdeutlicht hat, um wieviel wirkungsvoller bei einspringenden Ekken den Zugtrajektorien folgende Bewehrungen sind als solche, die tangential zu den Winkelschenkeln verlaufen [11], [12], [13] u. a. Im Laufe der Versuche tauchten noch weitere Fragestellungen auf. So störte die frühzeitige, meist schon unter Gebrauchslast einsetzende Rißbildung in der einspringenden Ecke und die überwiegend vertikal, d. h. stützenparallel verlaufenden Risse, die sich bei höheren Laststufen bis tief in den Stützenquerschnitt hinein entwickelten (s. Bild 4). Es wurde vermutet, daß die Nachgiebigkeit der Zugbewehrungsverankerung an der Umlenkstelle der wie üblich in der Stütze nach unten abgebogenen horizontalen Zuggurtbewehrung dafür mit ursächlich ist. Deshalb wurde ein Teil dieser Bewehrung bei einigen Versuchen in Form von mit Überdeckung geschlossenen Bügeln ausgeführt. Außerdem wurden zusätzliche, auf den Tragbewehrungsquerschnitt nicht angerechnete, auf Konsolhöhe verteilte Horizontalbügel getestet.

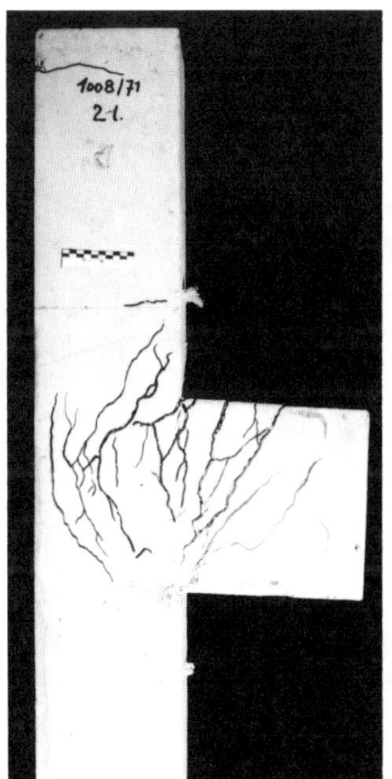

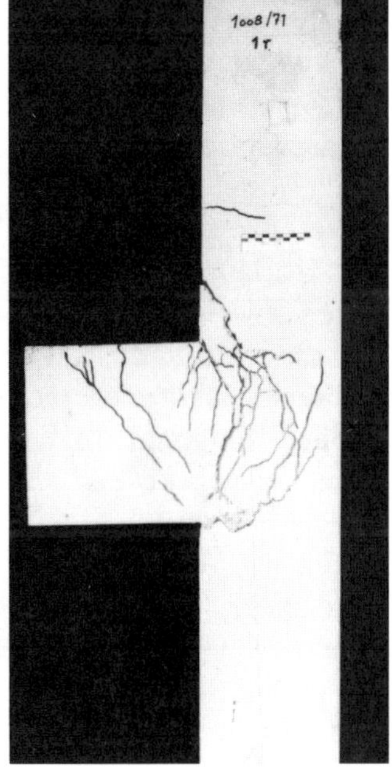

Bild 4

Bezüglich der zweckmäßigsten Bewehrung von Kranzkonsolen insbesondere zur Aufnahme exzentrischer Lasten lagen noch keine Untersuchungen vor. Hier sollten daher wenigstens zwei verschiedene Bewehrungsanordnungen getestet werden. Bild 7 bis 10 verdeutlichen im Schema die untersuchten Bewehrungsvarianten.

Eine systematische Untersuchung des Einflusses der Bewehrungsmenge $\left(\mu = \frac{A_s}{b \cdot h}\right)$ war im Rahmen dieser Arbeit nicht möglich. Dennoch schien eine Variation der Bewehrungsmenge zumindest in dem von Franz [9] gesteckten Rahmen sinnvoll. Höhere Bewehrungsprozentsätze mit zu untersuchen, wäre zwar höchst wünschenswert zur Klärung u. a. der Frage, ob die Bruch-Vorankündigung verloren geht, so daß ab einem bestimmten μ mit höherer Sicherheit gerechnet werden müßte, war aber im Rahmen der zur Verfügung stehenden Mittel bei diesem Forschungsvorhaben nicht möglich. Die Größe praktisch auftretender Exzentrizitäten kann bei Einzelkonsolen im Gegensatz zu Kranzkonsolen nur 20% bis 30% der Stützenbreite erreichen [1]. Für die Versuche mit Einzelkonsolen wurden daher die Exzentrizitäten $e/b = 0{,}15$, $0{,}25$ und $e/b = 0{,}3$ vorgesehen. Bei Kranzkonsolen wurden dagegen Exzentrizitäten von $e/b = 0{,}5$, $1{,}0$ und $1{,}5$ getestet.

4 Versuchseinrichtung

Das Schema der Versuchseinrichtung ist aus Bild 5 ersichtlich. Der Stützenabschnitt wird am oberen und unteren Ende in horizontaler Richtung gehalten. Die Einleitung einer zusätzlichen Normalkraft in den Stützenkopf war nicht vorgesehen. Die Konsolbelastung selbst wurde über eine jeweils aufgeklebte Stahlplatte, mitteils einer Halbwelle zentriert am vorderen Konsolrand eingeleitet. Der äußere Lasthebel a wurde extrem groß gewählt, indem die Lasteinleitung im vorderen Drittel der Konsole, beginnend am vorderen Rand erfolgte, was bei praktischer Anwendung vermieden werden muß.

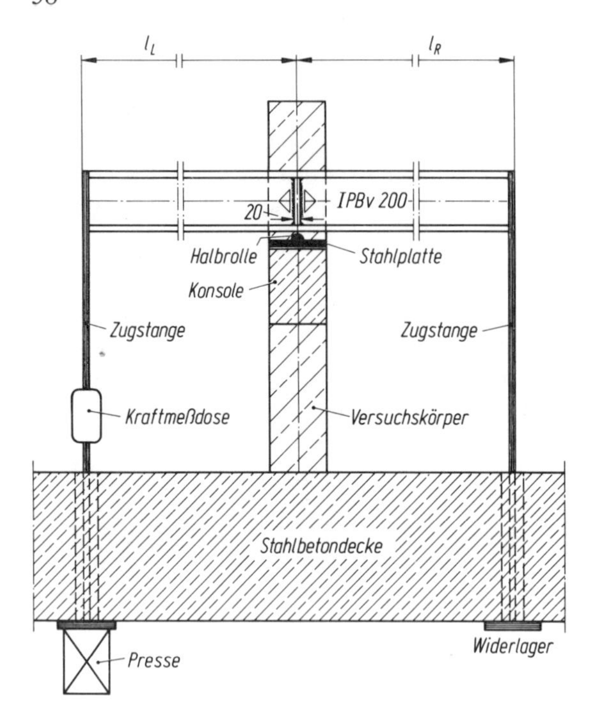

Bild 5
Belastungsschema der Konsole bei $e = 0$

5 Die Versuchskörper

Die Stützenabschnitte wurden liegend und so betoniert, daß die Konsolen oben in der Schalung lagen. Damit ergaben sich in der Konsole symmetrische Verhältnisse, der Konsolbeton als solcher war der qualitativ weniger gute, die Bewehrung ist in ihren wesentlichen Bestandteilen allerdings dem Verbundbereich I zuzurechnen. Die Versuchskörper wurden wie hierbei üblich gelagert, die Versuchsdurchführung erfolgte im allgemeinen nach 28 Tagen. Die Betonrezeptur aller Versuchskörper war einheitlich. Von jeder Betoncharge wurden mindestens 6 Probewürfel entnommen und deren 28-Tage-Festigkeit sowie deren Festigkeit am Tage der Versuchsdurchführung bestimmt. Ferner wurde von jeder Stahllieferung die σ-ε-Linie der für die Hauptbewehrung verwendeten Durchmesser ermittelt. Ausbildung und Bewehrung der Versuchskörper im einzelnen sind für die Einzelkonsolen aus Bild 6 bzw. 13 und 14 ersichtlich. Zur Ausbildung der Versuchskörper ist folgendes zu sagen:

Am oberen Konsolrand angeordnete Bewehrung wird im Folgenden vereinfachend als „Horizontalbewehrung“, Schräganker, die mit 60° oder 45° Neigung in die Konsole eingreifen, werden als „Schrägbügel“ oder „Schrägbewehrung“ bezeichnet.

Die Horizontalbewehrung kann aus durch Abbiegen in die Stützen in diesen verankerten Schlaufen oder aus Bügeln bestehen, die hinter der Tragbewehrung der Stütze geschlossen sind. Die Schlaufen können gekreuzt (Versuch 8, 9, 10, 11) bzw. diagonal angeordnet sein (Versuch 20). Dasselbe trifft für die Schrägbügel zu. Zusätzlich sind Horizontal- und Vertikalbügel in unterschiedlichen Kombinationen vorhanden.

Die Versuchskörper 1 und 4 erhielten als Horizontalbewehrung Bügel, deren Biegeradien entsprechend der bei Versuchsdurchführung noch gültigen Fassung der DIN 1045 von 1972 ausgebildet waren. Obwohl der Bewehrungsanteil der Horizontalbewehrung nur 0,8% beträgt, ergibt sich dann die statische Höhe mit $h/d = \frac{18{,}5}{25} = 0{,}74$ sehr ungünstig. Die Versuchskörper erhielten außerdem auch eine dichte, konstruktive Horizontal- und Vertikalverbügelung vornehmlich, um mit Hilfe von Dehnungsmessungen deren Wirksamkeit erfassen zu können.

Der Versuchskörper 2 erhielt im Gegensatz dazu horizontale Bügelschlaufen mit Biegeradien, wie sie damals nur für Haken zulässig waren, jetzt allerdings auch für Schlaufen zulässig sind. Damit ergab sich die statische Höhe zu

$$h/d = \frac{21{,}3}{25} = 0{,}852\,.$$

Die Tragebewehrung und sonstige Bewehrung stimmten mit Versuchskörper 1 überein.

Die Bewehrung der weiteren Versuche mit Einfachkonsolen ist aus Bild 6 ersichtlich.

Für die Versuche mit Kranzkonsolen ist die Bewehrungsausbildung und -anordnung in den Bildern 7 bis 10 dargestellt.

Die Tragbewehrung dieser Versuche I, II, III, IV wurde von Horizontal- und Schrägbügeln gebildet. Letztere waren jeweils von den unteren Ecken der Kranzkonsole diagonal in die Stütze geführt worden.

Bei Versuchskörper I und III waren auch die Horizontalbügel diagonal im Konsolkranz angeordnet. Die beim Versuchskörper I vorhandene orthogonale, die Stützenbewehrung tangierende vertikale Verbügelung entfiel bei Versuchskörper III, da sie sich als wenig wirksam erwies, sie wurde dort jedoch durch zwei sich kreuzende Horizontalbügel ersetzt.

Bei Versuchskörper II und IV waren die Horizontalbügel parallel zu den Seiten der Kranzkonsole angeordnet. Der Horizontalbewehrungsquerschnitt war bei Versuch IV gegenüber Versuch II reduziert auf 62%,

1	2	3	4	5	6	7	8	9	10	11	12	13	14	15	16	17	18	19	20	21	22
Versuch		Zugbewehrung										stat. Höhe	Bügelbewehrung								Bewehrungsgehalt
		horizontale Schlaufen					schräge Schlaufen						Horizontalbügel				Vertikalbügel				
		Form	lagig	Ø	Anzahl	A	Form	lagig	Ø	Anzahl	A		Form	Ø	Anzahl	A	Form	Ø	Anzahl	A	μ
Nr.	Anlage	A_0		mm	Stck.	mm²	A_s		mm	Stck.	mm²	mm		mm	Stck.	mm²		mm	Stck.	mm²	%
1	4		4	8	4	402						185		6 8	2 1	113 100		6 8	4 1	226 100	1,08
2	5		2	8	4	402						213		6 8	2 1	113 100		6 8	4 1	226 100	0,94
3	6		4	8	4	402		3	6	3	170	185		6 8	2 1	113 100		6 8	4 1	226 100	1,55
4	4		4	8	4	402						185		6 8	2 1	113 100		6 8	4 1	226 100	1,08
5	7		2	8	4	402		3	6	3	170	213		6 8	1 2	56 201		6 8	3 2	170 201	1,34
6	8		2	8	3	302		3	8	3	302	218		6 8	1 2	56 201		6 8	3 2	170 201	1,39
7	7		2	8	4	402		3	6	3	170	213		6 8	1 2	56 201		8	5	503	1,34
8	9		4	8 8	2 2	201 201		2	8	2	201	185		8	3	302		8	5	503	1,63
9	10		4	8 8	2 2	201 201		2	8	2	201	185		–	–			8	1	100	1,63
10	11		4	8 8	2 2	201 201		2	10	2	314	185		8	3	302		8	5	503	1,94
11	12		4	8 10	2 2	201 314		2	8	2	201	185		8	3	302		8	5	503	1,94
12	13		2	8 8	2 2	201 201		2	10	2	314	213		8	2	201		8	3	302	1,68
13 +17	13		2	8 8	2 2	201 201		2	10	2	314	213		8	2	201		8	3	302	1,68
14 +16	14		2	8 8	2 2	201 201		2	10	2	314	213		8	2	201		8	3	302	1,68
15	15		4	10	4	628						185		6 8	2 1	113 100		6	5	283	1,70
18 +19	16		2	8 8	2 2	201 201		1	12	1	226	213		8	3	302		6	3	170	1,48
20	17		3 2	10 8	3 2	471 201		1	10	1	157	213		8	2	201		8	3	302	1,94

Bild 6
Bewehrungsanordnung und -querschnitte der geprüften Versuchskörper

da sich die innen liegende Bewehrung bei Versuch II als nicht ausgenutzt erwies. Außerdem war bei den Versuchen I und III eine obere horizontale Ringbewehrung sowie bei allen Versuchen – vornehmlich als Montagebewehrung – eine vertikale Steckbügelbewehrung eingebaut.

Die Bemessung der Versuchskörper mit Einfachkonsole erfolgte nach den Formeln aus [5] und [6]:
Die danach rechnerisch zu erwartenden Bruchlasten betrugen bei den Einfachkonsolen bei zentrischer Last nach [5] aufgrund der eingelegten Horizontalbewehrung:

$$F_{uR} = A_s \cdot \frac{0{,}85 \cdot h}{a} \cdot \beta_s \quad (a \text{ vgl. Bild 1})$$

aus der Betonfestigkeit ergibt sich nach [5]:

$$F_{uR} = \frac{b \cdot h^2 \cdot \beta_w}{2{,}97 \cdot a} \cdot \frac{1}{\sqrt{1 + 0{,}61 \frac{h^2}{a^2}}}$$

Nach [6] betrugen die rechnerischen Bruchlasten aus der eingelegten Horizontalbewehrung A_{sh}:

$$F_{uR} = A_{sh} \cdot \frac{h}{0{,}9 \cdot a} \cdot \beta_s$$

aus der vorhandenen Schrägbewehrung A_{ss}:

$$F_{uR} = A_{ss} \cdot \frac{h}{0{,}4 \cdot a} \cdot \beta_s$$

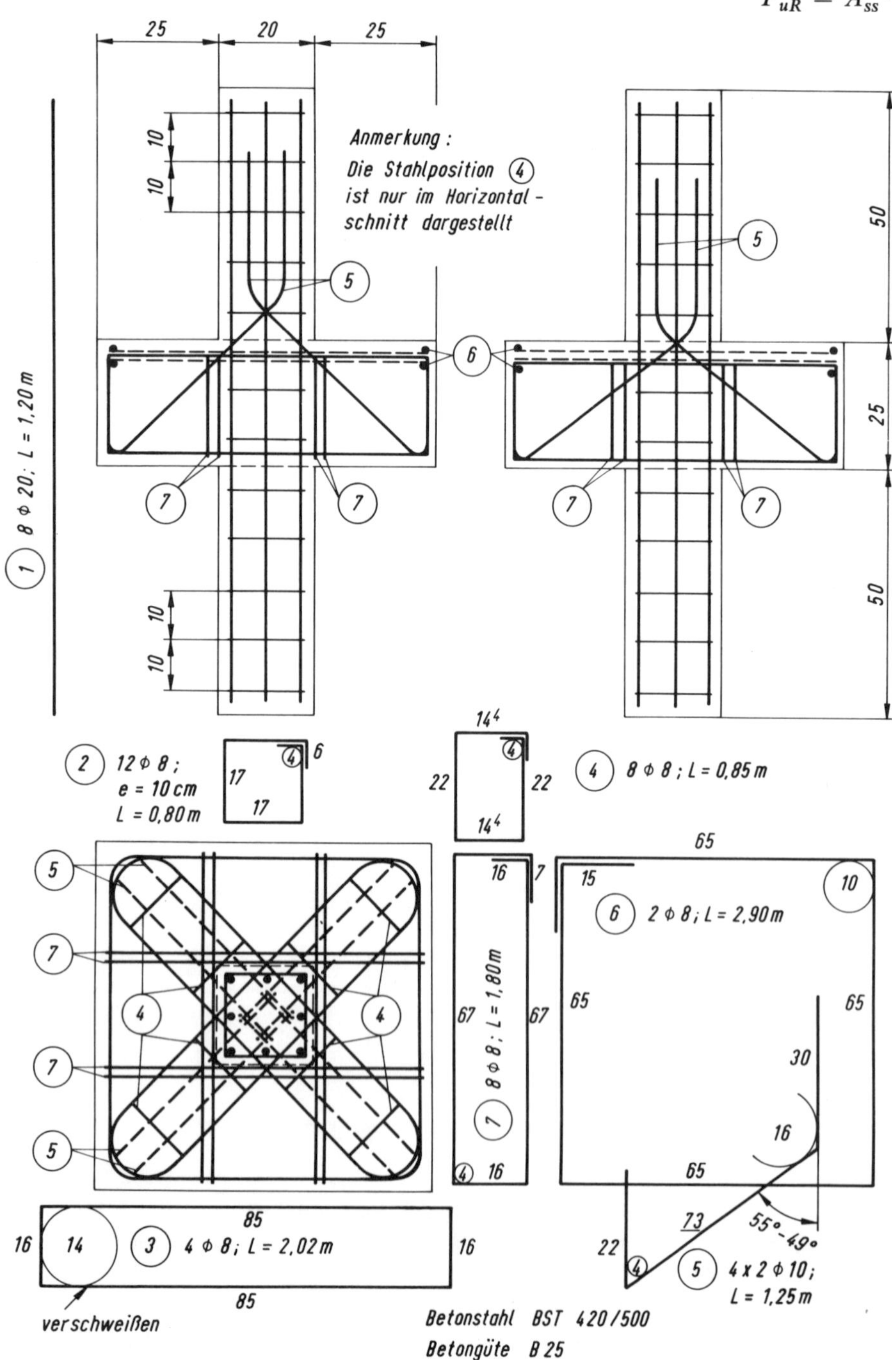

Bild 7
Versuchskörperbewehrung Versuch I

und aus der Betonfestigkeit:

$$F_{uR} = \frac{b \cdot h \cdot \beta_w}{2{,}8} \cdot \frac{1}{1 + \frac{a}{h}}$$

Bei der Bemessung der Kranzkonsolen wurde darauf verzichtet, die vom Beton aufnehmbare Konsollast zu bestimmen, weil sich hier ein größerer mitwirkender Bereich des Betons an der Lastabtragung beteiligen kann. Da die Versuchseinrichtung nur eine Belastung in begrenzter Größe zuließ, wurde die Bewehrung so gewählt, daß mit Hilfe der maximal erreichbaren Pressenkraft auch das Versagen des Versuchskörpers herbeizuführen war.

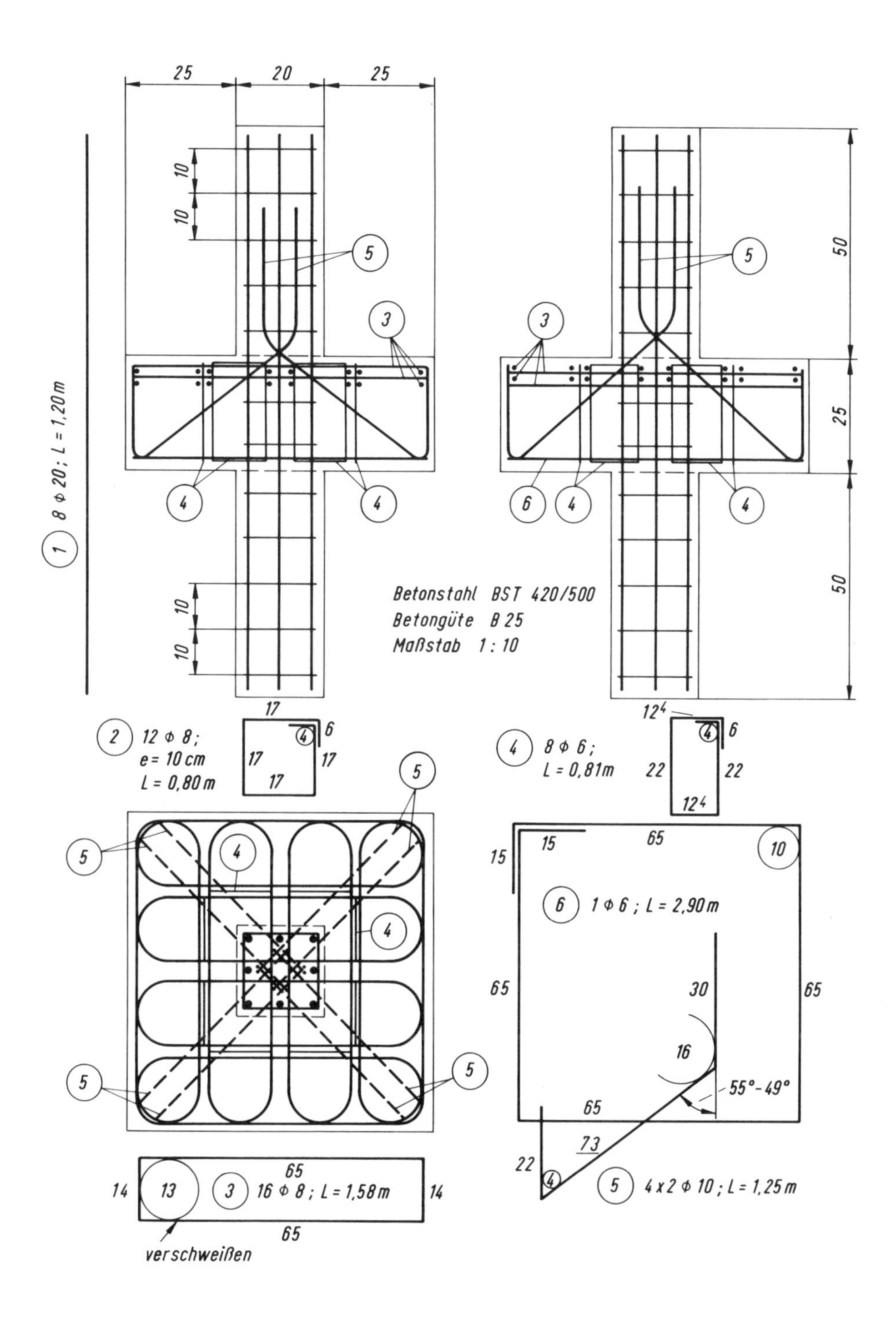

Bild 8
Versuchskörperbewehrung Versuch II

6 Meßstellenanordnung und Messungen

Es wurden sowohl Dehnungsmessungen an der Bewehrung als auch Stauchungsmessungen am Beton durchgeführt. Die Dehnungsmessungen erfolgten mit paarweise angebrachten Dehnmeßstreifen (DMS), die auf die Bewehrungsstäbe jeweils seitlich aufgeklebt wurden (auf Bild 12 und 13 angedeutet). Zuvor wurden die Rippen abgeschliffen, um glatte Klebeflächen zu erhalten. Die Querschnittsminderung wurde bei der Auswertung berücksichtigt.

Mit Hilfe von Stauchungsmessungen in der Betondruckzone wurde versucht, deren Verlauf zu erfassen. Diese Messungen wurden mit Meßbrücken von der

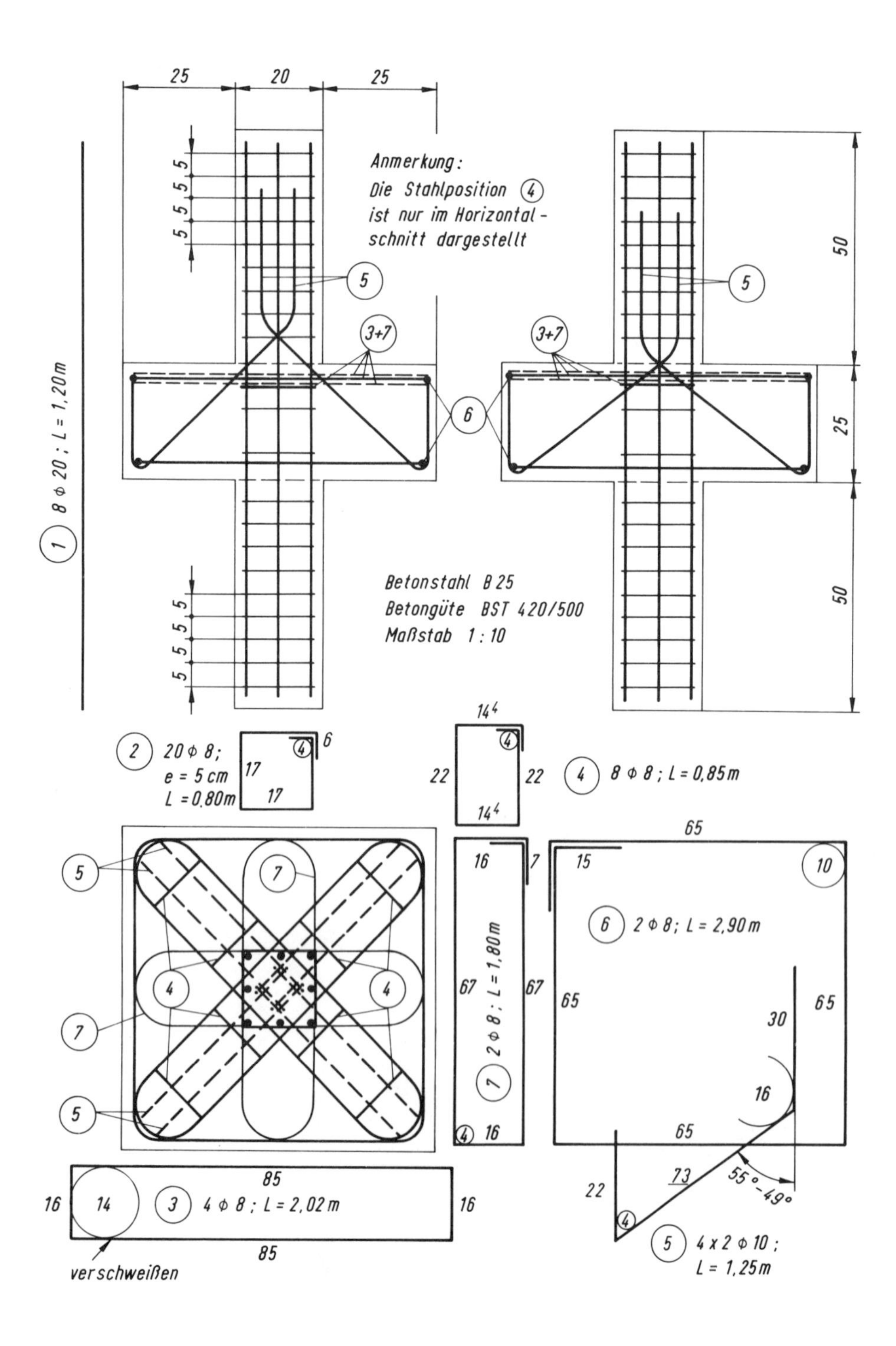

Bild 9
Bewehrung einer Kranzkonsole – Versuchskörper III

Basislänge 60 mm bzw. 100 mm vorgenommen (s. Bild 11 und 12). Außerdem wurde mit Hilfe mehrerer 1/100 Uhren versucht, das Verformungsverhalten, insbesondere die Verdrillung der Konsolen zu erfassen.
An den Kranzkonsolen wurden ebenfalls Stahldehnungen, Betonstauchungen und Verformungen gemessen.

7 Belastung

Alle Versuche, ausgenommen die Versuche 9, 16 und 17 wurden mit zentrischer Belastung bzw. mit einer konstanten Exzentrizität gefahren. Die Exzentrizität e/b betrug:

$e/b = 0$	bei Versuch	1, 2, 3 und 12
$= 0{,}25$	bei Versuch	4
$= 0{,}3$	bei Versuch	5, 6, 7, 8, 10, 11, 13, 14, 15, 18, 19, 20

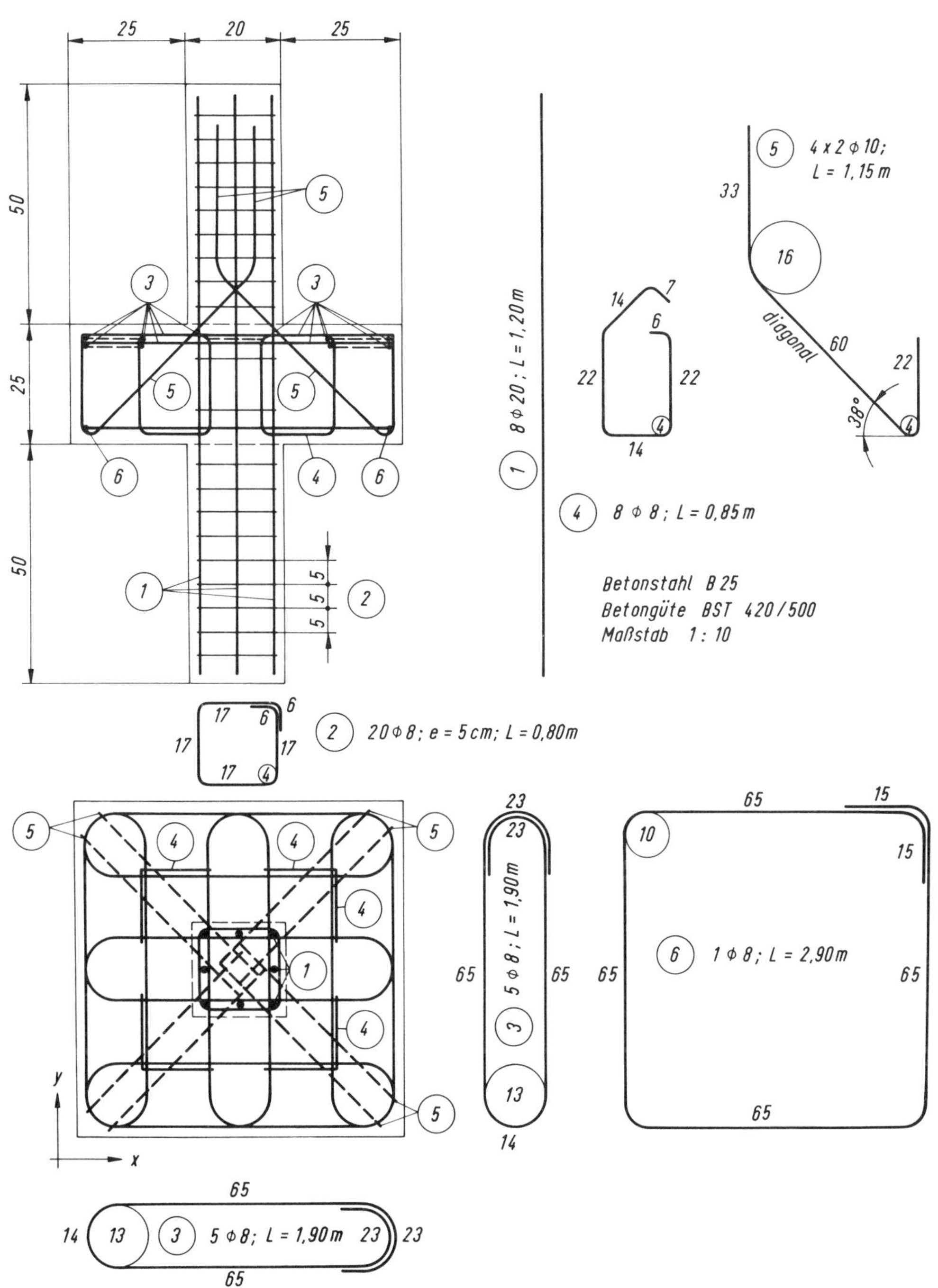

Bild 10
Bewehrung einer Kranzkonsole – Versuchskörper IV

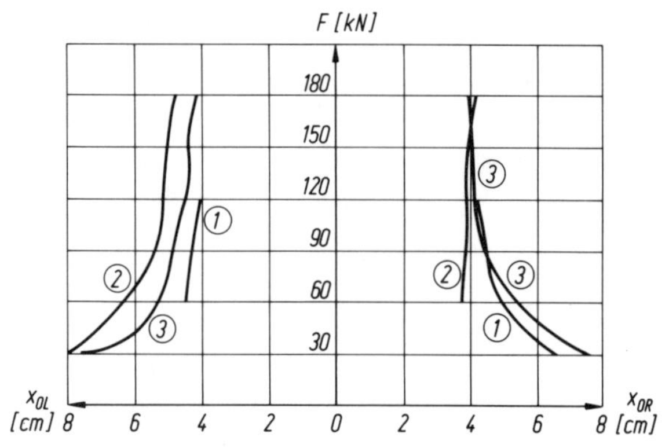

Bild 11
Nullinienänderungen in der Winkelhalbierenden der unteren einspringenden Ecke der zentrisch belasteten Versuchskörper bei Versuchskörper 1, 2, 3

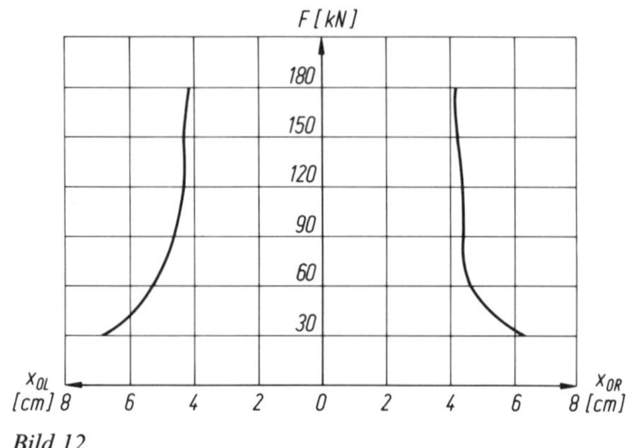

Bild 12
Nullinienänderungen in der Winkelhalbierenden der unteren einspringenden Ecke der zentrisch belasteten Versuchskörper als Mittelwerte aus Versuchskörper 1, 2, 3, 9.0, 12, 16.0, 17.0

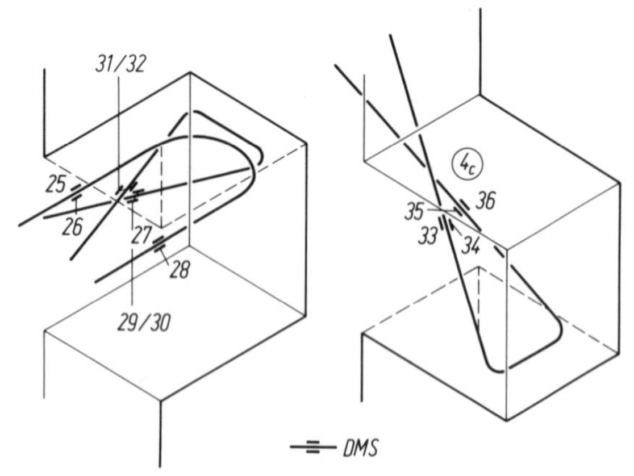

Bild 13
Bewehrungsanordnung der Tragbewehrung mit Meßstellen bei Versuch 10

Bei den Versuchen 9, 16 und 17 wurde nacheinander $e/b = 0$, 0,15 und 0,3 gefahren. Bei Versuch 9 wurde die Last bei $e/b = 0$ bis 120 kN und bei $e/b = 0{,}15$ bis 110 kN gesteigert. Bei $e/b = 0{,}3$ wurde der Versuch bis zum Bruch gefahren. Die Belastungsgrenze bei Versuch 16 und 17 wurde für $e/b = 0$ und $e/b = 0{,}15$ als diejenige Last festgelegt, bei welcher an der höchstbelasteten Stelle eine Stahldehnung von ca. 1,8‰ gemessen wurde.

Bei den Kranzkonsolen wurden die Versuchskörper I und II nacheinander mit der Exzentrizitäten $e = 0$, $e = 10$ cm, $e = 20$ cm und $e = 30$ cm, die Versuchskörper III und IV nur jeweils mit $e = 30$ cm belastet. Der Lasthebel bis zum Konsolanschnitt betrug nach wie vor 20 cm (s. Bild 2). Mit der größten Exzentrizität wurde die Konsole bis zum Versagen belastet, während bei den kleineren Exzentrizitäten ($e = 0$, 10, 20 cm) die Konsole auch nur so weit belastet wurde, bis am stärksten beanspruchten Bewehrungsstab eine Dehnung von ca. 1,8‰ auftrat.

Die Versuche wurden abgebrochen, sobald bei laufenden Meßuhren bzw. Meßwerten keine Laststeigerung mehr möglich war.

8 Ergebnisse der Versuche mit Einfachkonsolen

8.1 Für zentrische Belastung hat sich folgendes ergeben:

Wie sich schon bei einer Reihe von Untersuchungen der letzten Jahre an einspringenden Ecken herausgestellt hat, ist eine Bewehrung senkrecht zur Winkelhalbierenden besonders wirksam. Deshalb sollte auch hier eine Schrägbewehrung vorhanden sein, die mehr oder weniger senkrecht zu den Zugtrajektorien verläuft und einen Teil der Last nach oben in die Stütze verhängt (Mehmelscher Vorschlag).

Besteht die Horizontal- bzw. Schrägbewegung aus mehreren Lagen, so nahmen die Stahldehnungen von Lage zu Lage stark ab, wie das bei den geringen Querschnittsabmessungen und der dadurch bedingten schnellen Abnahme des Nullinienabstands von Lage zu Lage zu erwarten war. Insbesondere bei kleinen Abmessungen sind mehrere Bewehrungslagen daher uneffektiv, die unteren Lagen lassen sich nicht ausnutzen, und von daher ist der Franzsche Vorschlag einer Begrenzung von μ sicher mitbegründet. Nach der ergänzten Fassung der DIN 1045 vom Dez. 1978 Tab. 18 sind allerdings für Schlaufen $\varnothing < 20$ mm Biegedurchmesser von $4d_s$ zugelassen. Damit lassen sich immer mehrere Schlaufen in eine Lage bringen.

Die Versuche haben ergeben, daß:
frühzeitig, d. h. schon unter Gebrauchslast Risse in der einspringenden Ecke auftreten und daß die Risse z. T. dem Verlauf der Abbiegungen der Horizontalschlaufen folgen.

Betrachtet man auf Bild 4 noch einmal das typische Rißbild, so fällt die starke, überwiegend lotrecht orientierte Rißbildung in Stütze und Konsolanschnitt auf. Eine zusätzliche Horizontalverbügelung ist daher sinnvoll. Die Messungen haben auch gezeigt, daß solche Bügel bei wachsender Last zunehmende Dehnung erfahren. Sie bringen zwar für die Bruchlast wenig, beeinflussen aber das Rißverhalten günstig.

Im Gegensatz dazu erfahren Vertikalbügel kaum eine Beanspruchung. Sie können daher zum Tragverhalten wenig beitragen, sind eigentlich nur für die Montage nützlich und können daher rein konstruktiv aus dünneren Stäben (∅ 6) hergestellt werden. Zu empfehlen ist (im Grundriß gesehen) die folgende Anordnung:

äußerste Lage: Vertikalbügel
nächste Lage: Horizontalbügel (in der gleichen Lage wie Stützenbügel)
dann Stützenlängsstäbe.

8.2 Ergebnisse unter exzentrischer Belastung

Bei exzentrischer Belastung sind die Dehnungen der Tragbewehrung auf der belasteten Seite erwartungsgemäß höher und am höchsten in dem am weitesten außen am belasteten Rand liegenden Stab. Lagen vier Stäbe in einer Lage, so war in signifikanter Weise der auf der lastabgewandten Seite innen liegende Stab stets geringer beansprucht als der dort außen liegende Stab. Bei möglicher exzentrischer Last sollte daher die Tragbewehrung – auf beiden Seiten – so weit wie möglich außen liegen, da innen liegende Stäbe geringere Wirksamkeit besitzen. Auch hier erwies sich eine Schrägankerbewehrung als besonders wirksam.

Um die exzentrisch angreifende Last möglichst in das Zentrum des gefährdeten Querschnitts am Stützenanschnitt abzuleiten, wurden auch gekreuzte Bewehrungen verwendet s. Bild 13. Bei einigen Versuchen ergab die Messung der Betonstauchungen über der unteren einspringenden Ecke günstigere, d.h. gleichmäßiger über den Querschnitt verteilte Stauchungen, während bei üblicher Bewehrungsanordnung auf der belasteten Seite sehr viel schneller ansteigende Stauchungen als auf der unbelasteten registriert werden konnten. Insbesondere dann, wenn die Bewehrung insgesamt im Grundriß diagonal zur belasteten Ecke hinweisend angeordnet werden kann (s. Bild 14) tritt eine Entlastung des Betons auf der belasteten Seite ein. Bei gekreuzter Bewehrungsanordnung sind natürlich die zur belasteten Ecke hinweisenden Stäbe sehr hoch, die diagonal dazu verlaufenden Stäbe nur gering beansprucht. Praktisch läßt sich diese Bewehrung unter exzentrischer Last also nur zur Hälfte voll ausnutzen, während ihre Wirksamkeit unter zentrischer Last der der üblichen Bewehrungsanordnung entspricht.

Diagonal zu einer Ecke hinweisende Tragbewehrung kann folglich nur dann empfohlen werden, wenn die Last ausschließlich exzentrisch über dieser Ecke angreift. Gekreuzte Bewehrung bringt also keine nennenswerten Vorteile.

Was die Bügel anbelangt, so ist wohl besonders bemerkenswert, daß ein Versuchskörper mit zusätzlicher Schrägankerbewehrung aber ohne Vertikal- und ohne Horizontalbügel dieselbe Traglast erreichte wie im Mittel normal – also mit Bügeln – bewehrte Versuchskörper, wohingegen zwei Versuchskörper mit kräftiger Torsionsbewehrung sogar ungünstigere Resultate erbrachten. Die Messungen ergaben auch durchweg nur geringe Stahldehnungen in der Torsionsbewehrung, solange die Fließgrenze in der Tragbewehrung noch nicht überschritten war.

Es kann mithin auch bei exzentrischem Lastangriff die Empfehlung beibehalten werden, konstruktiv zusätzliche – geschlossene – Horizontalbügel und leichte Vertikalbügel vorzusehen. Eine klassische Torsionsbemessung und die Anordnung einer Torsionsbewehrung ist überflüssig, weil wirkungslos.

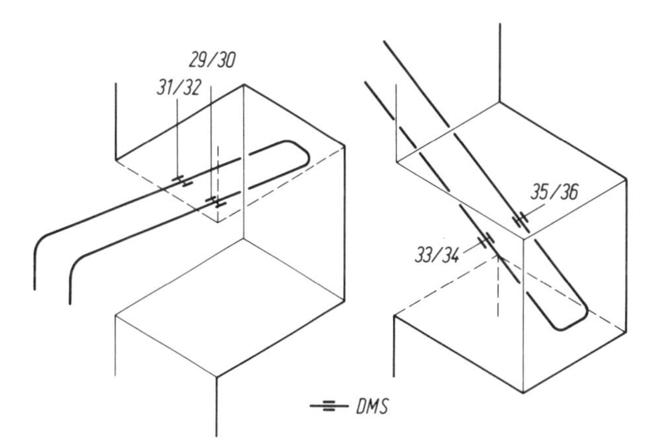

Bild 14
Bewehrungsanordnung der Tragbewehrung mit Meßstellen bei Versuch 20

8.3 Bruchlasten bei zentrischer und exzentrischer Belastung

Der Bruch wird eingeleitet durch das Fließen der Tragbewehrung, das seinerseits zur Einschnürung der Betondruckzone und dann zum Betonversagen führt. Bei exzentrischer Last setzt die Rißbildung auf der belasteten Seite früher ein, entwickelt sich ausgeprägter als auf der unbelasteten Seite und führt schließlich auch früher zum Versagen. Die exzentrisch belastete Konsole besitzt also unter sonst gleichen Verhältnissen eine geringere Tragfähigkeit als die zentrisch belastete. Die erreichbare Traglast hängt von der Exzentrizität e/b ab, s. Bild 15. Man kann mithin wie für zentrische Belastung bemessen – zumal sich an der Bewehrungsanordnung im Normalfall nichts ändert – und dann die zulässige Last in Abhängigkeit von e/b abmindern bzw. vorh F vergrößern und damit bemessen.
Bei den Versuchen sind die Lastplatten jeweils bündig mit dem Konsolrand, also absichtlich extrem weit nach außen gelegt worden. Die Folge waren vorzeitige Abplatzungen bei mehreren Versuchen. Das bestätigt, daß die Lasteinleitung genügend weit vom Rande weg erfolgen sollte.

8.4 Betonstauchungen bzw. -dehnungen

Die Stauchungsmessungen senkrecht zur Winkelhalbierenden an der unteren, einspringenden Ecke ermöglichten die Bestimmung der Nullinienlage auf der Winkelhalbierenden in Abhängigkeit von der Last. In Bild 11 oben ist deren Entwicklung für die Versuche 1 bis 3 für beide Seiten der Konsolen aufgetragen. Die Übereinstimmung zwischen links und rechts ist für Lasten ≥ 90 kN bei Versuch 1 und 3 ganz gut, bei Versuch 2 weniger gut. Signifikante Unterschiede bestehen hier nicht, der Nullinienabstand x_0 tendiert für höhere Belastung gegen 4 cm. Nimmt man Versuch 12 und die zentrischen Belastungszustände bei Versuch 9, 16 und 17 hinzu, so ergibt sich bei Ausmittlung der jeweils links und rechts gemessenen Werte das Bild 12. Man erkennt, daß die Nullinie bei geringer Last 6–8 cm Abstand von der einspringenden Ecke hat und daß dieser Abstand für höhere Lasten ziemlich einheitlich gegen 4 cm konvergiert.
Bezüglich des Stauchungsverlaufs in der Druckzone ist der Vergleich zwischen den Versuchen 1, 2 und 3 aufschlußreich. Bei gleicher Laststufe sind die Stauchungen bei Versuch 2 erheblich geringer als bei Versuch 1, was auf die günstigere Lage der Bewehrung mit größerem, innerem Hebelarm zurückzuführen ist. Sehr viel günstiger noch ist das Bild bei Versuch 3. Hier wurde bei 150 kN nur 3‰ Randstauchung erreicht, während bei Versuch 2 bereits 5,5‰ gemessen wurden. Die Zusatzbewehrung in Form der Schrägbügel hat sich also vorteilhaft ausgewirkt. Ebenso konnte man unter exzentrischer Last auf der Lastseite bei Versuch 6 günstigere Stauchungen vermerken als bei Versuch 5. Die Bewehrung war hier gleich groß, bei Versuch 6 der Schrägbewehrungsanteil jedoch höher. Versuch 7 mit geringerem Schrägbewehrungsanteil zeigt wieder höhere Betonstauchungen. Auch der Stauchungsverlauf bei den Versuchen 10, 13, 14 ist günstiger als bei Versuch 11. Das spricht für einen angemessenen Anteil von Schrägbewehrung. Versuch 14 mit gekreuzter Bewehrung zeigt etwas günstigere Resultate als Versuch 13 (Parallelbewehrung). Dasselbe war beim Vergleich von Versuch 8 mit Versuch 7 schon festgestellt worden. Das schien für eine gekreuzte Bewehrung zu sprechen, hat sich aber bei den Versuchen 16 bis 19 nicht bestätigt.

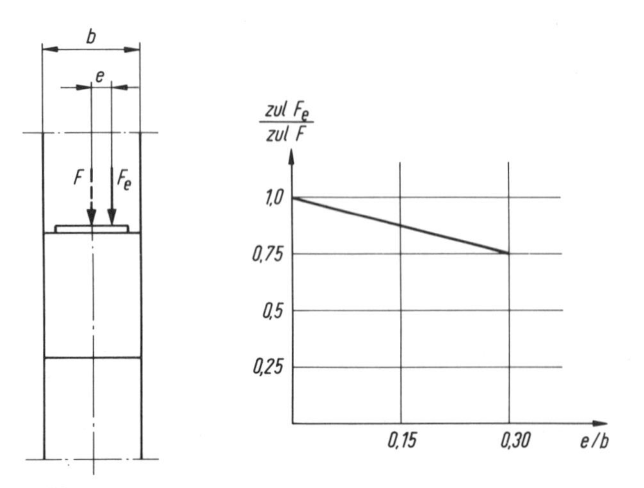

Bild 15
Zusammenhang zwischen zulässiger exzentrischer Last F_e und zulässiger zentrischer Last F bei gleichen Konsolabmessungen und gleicher Bewehrung

9 Ergebnisse der Versuche mit Kranzkonsolen

Die Versuche mit Kranzkonsolen wurden mit zwei verschiedenen Bewehrungsanordnungen durchgeführt. Bild 7 zeigt die Bewehrungsform 1. Sie ist gekennzeichnet durch verbügelte Diagonalschlaufen Pos. 3 und diagonal angeordnete Schräganker Pos. 5. Neben einem Kranzbügel Pos. 6 sind außerdem stehende Parallelbügel Pos. 7 vorhanden.

Die Bewehrungsform 2 (Bild 8) dagegen zeigt Parallelschlaufen Pos. 3 in Verbindung mit Vertikalbügeln Pos. 4, aber auch die Schräganker Pos. 5 und den Kranzbügel Pos. 6.
Da sich bei Form 1 die stehenden Parallelbügel als wenig wirksam erwiesen, wurde diese Bewehrungsform durch Wegfall dieser Bügel modifiziert. Statt dessen wurden mittig Parallelschlaufen angeordnet, s. Bild 9. Bei Form 2 wurde prinzipiell nichts geändert, nur wurden statt je $2 \cdot 4$ Parallelschlaufen Pos. 3 nur $2 \cdot 2 + 1$ Schlaufen eingebaut, dafür aber wurde der Kranzbügel Pos. 6 verstärkt, s. Bild 10.
Die Dehnungsmessungen an der Bewehrung ergaben folgendes:

Bewehrungsform 1

Bei zentrischer Last wiesen der Kranzbügel, dicht gefolgt von den Schrägankern die größten Dehnungen auf. Die Diagonalschlaufen waren geringer, die stehenden Parallelbügel am geringsten beansprucht. Bei großer Exzentrizität dagegen ($e = 30$ cm) wurden die Diagonalschlaufen zur belasteten Ecke am höchsten beansprucht, dicht gefolgt von den Schrägankern und dem Kranzbügel. Die Querbügel erwiesen sich somit insgesamt als am wenigsten wirksam und wurden deshalb bei der modifizierten Bewehrungsform 1 weggelassen.

Bewehrungsform 2

Bei zentrischer Belastung waren die Parallelschlaufen Pos. 3 in der oberen Lage alle hoch beansprucht, am höchsten die außen liegenden Schenkel einschließlich des Kranzbügels Pos. 6. Die Schräganker waren dagegen weniger beansprucht. Bei gleichem Bewehrungsgewicht erwies sich die Form 2 unter zentrischer Last als etwas günstiger als die Form 1.
Auf exzentrische Belastung reagiert die Form 2 dagegen ungünstiger. Für $e = 30$ cm wiesen nur noch die außen unter der Last liegenden Parallelschlaufen hohe Beanspruchung auf (für $F = 135$ kN $\varepsilon = 4{,}62‰$), während bereits der nächste Schlaufenschenkel nur noch geringe ($\varepsilon = 1{,}13‰$) Dehnung zeigte. Der Kranzbügel erwies sich auch hier wieder als besonders wirksam und natürlich auch die zur belasteten Ecke gerichteten Schräganker. Bei der modifizierten Form 2 wurde daher die Anzahl der Parallelschlaufen bei gleichzeitiger Konzentration an den Konsolrändern reduziert.
Die Vertikalbügel Pos. 4 waren nur gering beansprucht.
Bei stark exzentrischer Last waren die Betonstauchungen an der belasteten Ecke bei Bewehrungsform 2 deutlich größer als bei Form 1.
Zusammenfassend kann gesagt werden:
Diagonalbewehrung (Form 1) erweist sich bei stark exzentrischer Belastung (Eckbelastung) als besonders günstig. Um auch bei zentrischer Belastung möglichst effektiv zu sein, sollte sie durch mittig angeordnete Parallelschlaufen ergänzt werden (modifizierte Form 1).
Orthogonale Bewehrung (Form 2) eignet sich bei zentrischer und gering exzentrischer ($e/b \leqq 0{,}5$) Belastung. Bei wachsender Exzentrizität beteiligen sich die innen liegenden Parallelschlaufen kaum mehr an der Lastaufnahme. Deshalb ist in solchen Fällen die modifizierte Bewehrungsform 2 zu empfehlen, bei welcher die Zahl der mittig liegenden Parallelschlaufen zugunsten der am Rande liegenden vermindert wird. Die innen liegenden Schlaufen sollten für starke Exzentrizität ebenso wie die im unbelasteten Bereich liegenden Schlaufen gar nicht in Ansatz gebracht werden.

10 Vorschläge für die Konsolbemessung

10.1 Einfachkonsolen

Die den folgenden Empfehlungen zugrunde liegenden theoretischen Betrachtungen und Vergleiche zwischen ihren Resultaten und den Versuchsergebnissen können hier aus Platzgründen nicht wiedergegeben werden. Sie können aber in einer demnächst erscheinenden Veröffentlichung in der Schriftenreihe des Deutschen Ausschuß für Stahlbeton nachgelesen werden.
Die Untersuchungen haben ergeben, daß für $a/h > 0{,}4$ (a vgl. Bild 1) der von STEINLE [8] vorgeschlagene Ansatz in der Form

$$\frac{b \cdot h \cdot \beta_R}{\operatorname{vorh} F} \geqq 7 \tag{1}$$

immer zu ausreichenden Betonabmessungen führt. Ein Nachweis der Schub- bzw. Scherspannung ist dann nicht mehr erforderlich.
Die Ermittlung der Bewehrungsquerschnitte ist mit Hilfe einer einfachen Kräftezerlegung möglich (s. Bild 16). Es gilt:

$$F_{\text{vorh}} = (A_h \cdot \tan\gamma_D + A_s \cdot \sin\gamma_s) \cdot \frac{\beta_s}{\gamma} \tag{2}$$

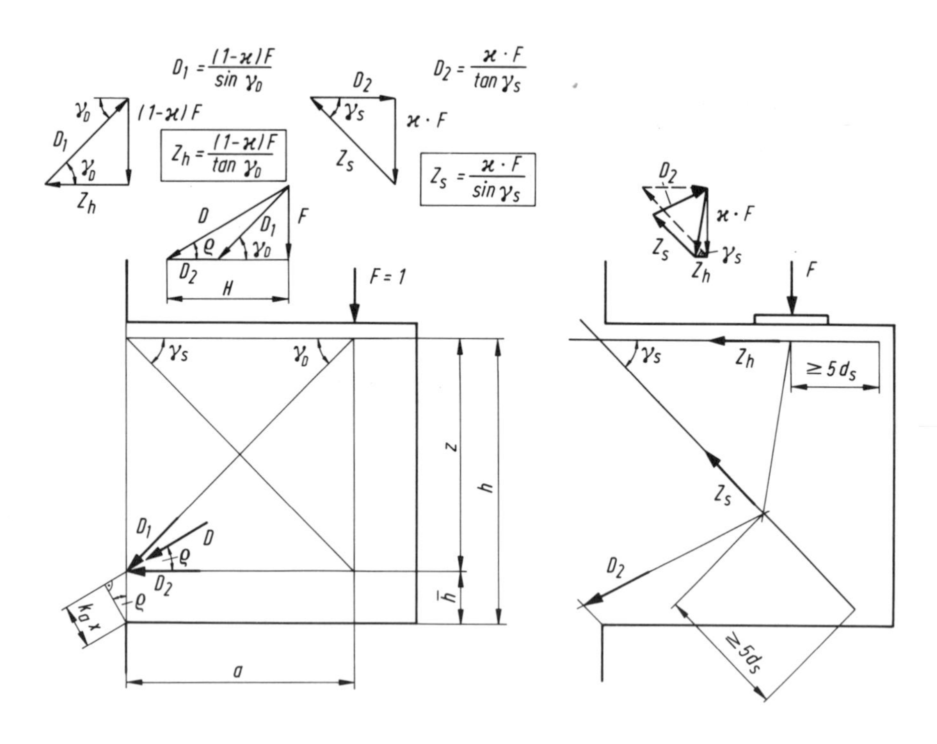

Bild 16 a (links)
Konsole mit Horizontal- und Schrägbewehrung

Bild 16 b (rechts)
Komponenten von F bei Berücksichtigung der Verankerungslänge $5\,d_s$ der Schrägbügelschlaufe

bzw.

$$A_h = \frac{\text{vorh}\,F \cdot (1 - \varkappa)}{\beta_{s/\gamma} \cdot \tan\gamma_D}, \qquad (3)$$

$$A_s = \frac{\text{vorh}\,F \cdot \varkappa}{\beta_{s/\gamma} \cdot \sin\gamma_s}$$

Darin ist:

γ_D, γ_s = Winkel gemäß Bild 16
A_h = Querschnitt der Horizontalbewehrung
A_s = Querschnitt der Schrägbewehrung
$\varkappa \cdot F$ = Anteil der Last der durch Schrägbewehrung aufgenommen wird
$\frac{\beta_s}{\gamma}$ = zulässige Stahlspannung hierbei sollte gesetzt werden für
$\gamma = 1{,}75$ für $\bar{\mu} \leqq 2\%$
$\gamma = 2{,}10$ für $\bar{\mu} > 2\%$

$$\mu = \mu_h + \mu_s \cdot \cos\gamma_s \qquad (4)$$

Der Winkel γ_D kann genau genug mit Hilfe der Annahme $z = 0{,}87 \cdot h$ gefunden werden (vgl. Bild 16). Eine immer wieder auftauchende Frage ist die nach den erforderlichen Verankerungslängen der Konsoltragbewehrung. Die eigenen aber ebenso auch die Versuchserfahrungen von ROSTÁSY-STEINLE [17] deuten darauf hin, daß infolge der herrschenden Querdrücke Verankerungslängen von $5\,d_s$ (d_s = Stahldurchmesser) in der Konsole ausreichend sind.

10.2 Kranzkonsolen

Eine ähnliche Betrachtung wie bei den Einfachkonsolen ist bei Kranzkonsolen kaum möglich, da die Einleitung der Konsol-Druckkraft in die Stütze bei zentrischer Last nicht nur über die lastzugewandte Leibung, sondern auch über die flankierenden Leibungen erfolgt, wie das Rißbild erkennen läßt. Bei exzentrischer Belastung ist es noch schwieriger, den bei der Druckübertragung mitwirkenden Bereich zu erfassen. Andererseits zeigen die Rißbilder der Kranzkonsolen gerade bei Eckbelastung deutliche Symptome eines Durchstanz-Versagens (weshalb die Schrägbewehrung sehr zweckmäßig und natürlich auch als Schubbewehrung anrechenbar ist).

Es wird deshalb für sinnvoll gehalten, ausreichende Betonabmessungen durch einen Durchstanz-Nachweis entsprechend DIN 1045 Abschnitt 22.5 sicherzustellen. Die sich daraus ergebenden Abmessungen stimmen mit denen der Versuchskörper gut überein, sind daher durch die Versuchsergebnisse gedeckt. Bei einem solchen Nachweis muß der mitwirkende Bereich des Rundschnitt-Umfangs allerdings sinnvoll festgelegt werden. Bei einseitiger Belastung einer im Grundriß quadratischen Stütze schwankt der mitwirkende Teil des Umfangs nach DIN 1045 Abschnitt 22.5.1.1 und entsprechend auch hier zwischen $0{,}6\,u_0$ und $0{,}3\,u_0$

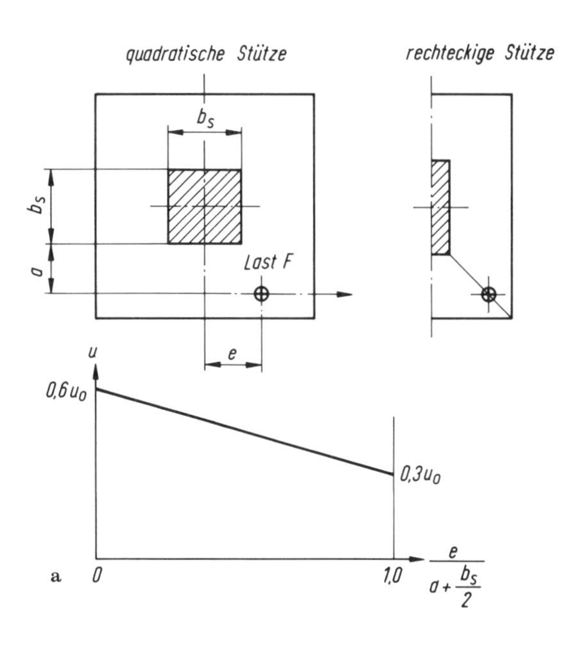

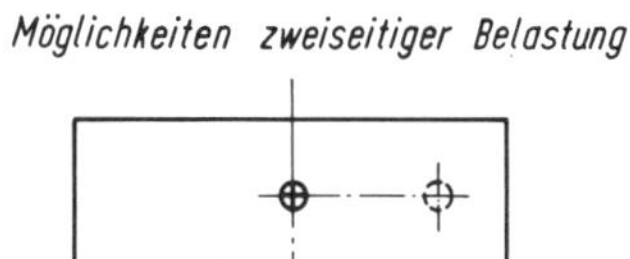

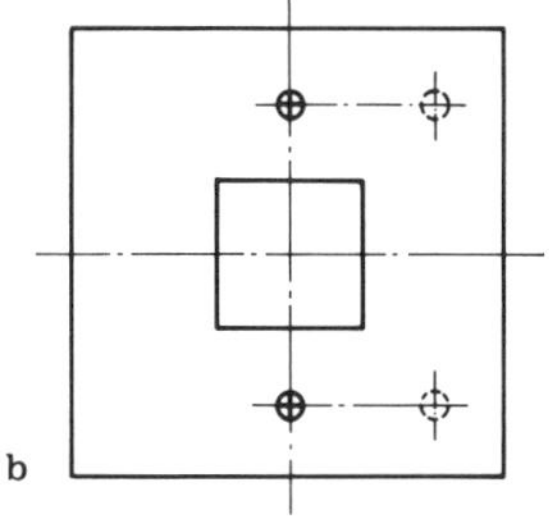

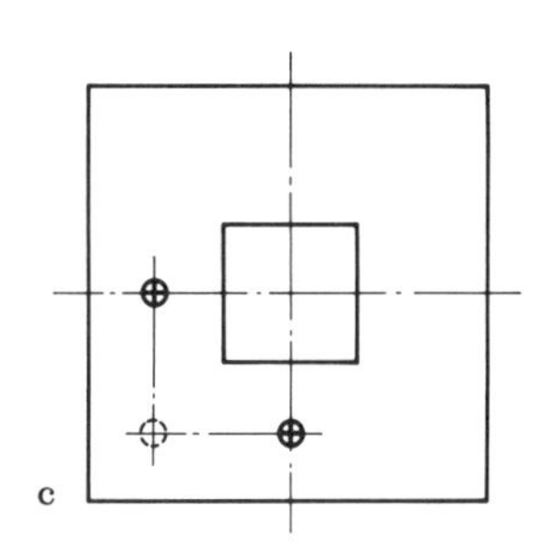

Bild 17a
Bestimmung von u (in Anlehnung an DIN 1045 Abs. 22.5.1.) bei quadratischer Stütze und einseitig exzentrischer Last in Abhängigkeit von $\frac{e}{a+\frac{b_s}{2}}$

Anmerkung:
Bei rechteckiger Stütze entspricht dem Wert $\frac{e}{a+\frac{b_s}{2}} = 1$ die Stellung der Last über den Eckdiagonalen

Bild 17b
Einachsig symmetrische Belastung
$u_0 \geq u \geq 0{,}6\,u_0$

Bild 17c
Unsymmetrische Belastung
$0{,}7\,u_0 \geq u \geq 0{,}3\,u_0$

(Randstütze und Eckstütze), wobei sich die Interpolationsgerade in Bild 17a in Abhängigkeit von

$$\frac{e}{a+\frac{b_s}{2}}$$

ergibt.
u_0 entspricht DIN 1045 Abschnitt 22.5.1.
Bei vierseitiger, annähernd gleichmäßiger Belastung ist $u = u_0$.
Bei zweiseitiger, symmetrischer Belastung entsprechend Bild 17b kann auch $u = u_0$ gesetzt werden. Rükken die Lasten nach außen, so liegt ein mit einseitig zentrischem Lastangriff vergleichbarer Fall vor, so daß $u = 0{,}6\ u_0$ sinnvoll ist.
Greifen die Lasten über Eck entsprechend Bild 17c an, so sind die Verhältnisse bei über den Querschnittsachsen stehenden Lasten günstiger als bei einseitig zentrischem Angriff, so daß u zu $0{,}7\ u_0$ angenommen werden kann. Nähern sich die Lasten beide einer Ecke, so liegt schließlich der Fall

$$\frac{e}{a+\frac{b_s}{2}} = 1$$

vor, der $u = 0{,}3\ u_0$ erfordert. Zur Interpolation kann sinngemäß und bei entsprechend angepaßten Koordinaten das Diagramm in Bild 17a verwendet werden. Die Bemessung der Stahleinlagen erfolgt zweckmäßig durch einfache Kräftezerlegung.

Für orthogonale Bewehrungen gilt gem. Bild 18a bei quadratischer Stütze:

$$\xi = \arctan \frac{e}{\frac{b_s}{2}+a}$$

$$Z_x = Z_h \cos\xi, \quad Z_y = Z_h \sin\xi \qquad (5)$$

$$r = \left(\frac{b_s}{2}+a\right)\cdot\frac{1}{\cos\xi}$$

Nimmt man an, daß der Punkt A (vgl. Bild 18b) im Grundriß auf einem Kreisbogen um die Stützenachse mit dem Radius

$$\frac{b_s}{2} - f$$

und im Abstand c oberhalb der Konsolunterkante liegt – was dem Umstand Rechnung trägt, daß A bei Lastangriff im Eckbereich am Stützenanschnitt höher liegt – und setzt man für f und c auf der sicheren Seite liegende Werte, wie sie sich bei 2,5‰ Randstauchung für die Einfachkonsole ergeben:

	a/h	f/h
	0,4	0,19
$c = 0{,}13\ h$	0,6	0,15
	0,8	0,12
	1,0	0,10

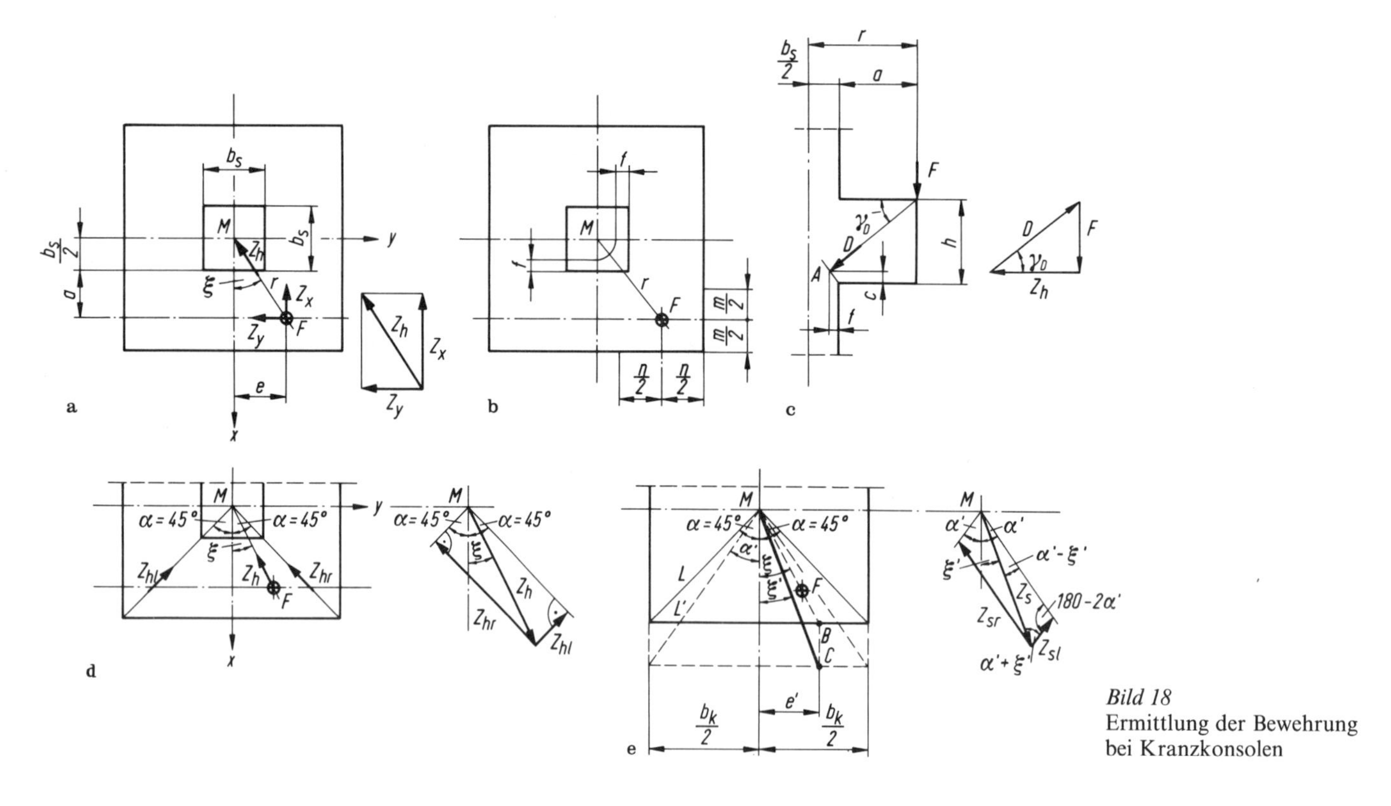

Bild 18
Ermittlung der Bewehrung bei Kranzkonsolen

So folgt aus Bild 18c:

$$Z_h = F \cdot \text{ctan}\,\gamma_D = F \frac{r - \frac{b_s}{2} + f}{h - c}$$

$$Z_h = F \cdot \frac{\left(\frac{b_s}{2} + a\right)\frac{1}{\cos\xi} - \frac{b_s}{2} + f}{0{,}87\,h} \tag{6}$$

Die sich aus Z_h bzw. Z_x und Z_y ergebende Bewehrung sollte im Bereich m und n (vgl. Bild 18b) angeordnet werden. Wenn die Laststellung sich nicht verändern kann, genügt im übrigen Konsolbereich eine konstruktive Bewehrung.
Bei diagonal zu den Konsolecken gerichteten Bewehrungen gilt entsprechend Bild 18d:

$$Z_{hr} = Z_h \sin(45 + \xi)$$
$$Z_{hl} = Z_h \sin(45 - \xi) \tag{7}$$

Bei diagonal zu den unteren Ecken gerichteten Schrägbewehrungen gilt vorstehende Kräftezerlegung im Prinzip auch, jedoch treten an die Stelle von α und ξ die Winkel α', ξ' und der Zusammenhang zwischen der nunmehr schrägen Zugkraft Z_s und F ergibt sich gemäß Bild 16a zu

$$Z_s = \frac{F}{\sin\gamma_s}$$

Dabei ist zu beachten, daß γ_s' hier der Neigungswinkel der gedachten, in die Komponenten Z_{sr} und Z_{sl} zu zerlegenden Schrägbewehrung in der zugehörigen Vertikalebene MB ist. Sind die Schrägbewehrungen unter γ_s zur Horizontalen geneigt, so ergibt sich für γ_s':

$$\tan\gamma_s' = \tan\gamma \cdot \sqrt{2} \cdot \cos\xi$$

für α' gilt:

$$\alpha' = \arcsin \frac{b_k}{2 \cdot l'}$$

wobei

$$l' = \sqrt{l^2 + d^2}$$

ist und für ξ' gilt:

$$\xi' = \arcsin \frac{e'}{MC} \quad (MC = \text{Strecke } M-C)$$

wobei

$$MC = \sqrt{MB^2 + d^2} \quad (MB = \text{Strecke } M-B)$$

ist.

Mit $\alpha' \neq 45°$ wird das in Bild 18 d dargestellte Krafteck zum schiefwinkligen Kräfteparallelogramm, und es ergibt sich daher (Bild 18 e):

$$Z_{sr} = Z_s \frac{\sin(\alpha' + \xi')}{\sin(180 - 2\alpha')}$$
$$Z_{sl} = Z_s \frac{\sin(\alpha' - \xi')}{\sin(180 - 2\alpha')} \tag{8}$$

11 Literatur

[1] PASCHEN, H. und MALONN, H.: Bemessung eines Randträgers im Falle exzentrischer Belastung durch Nebenträger. Betonwerk + Fertigteiltechnik 12/77.

[2] MÖRSCH, E.: Der Eisenbetonbau, 5. Aufl., Band I. 2. Stuttgart, K. Wittwer.

[3] RAUSCH, E.: Drillung, Schub und Scheren beim Stahlbetonbalken. VDI-Verlag Düsseldorf 1953.

[4] NIEDENHOFF, H.: Untersuchungen über das Tragverhalten von Konsolen und kurzen Kragarmen. Dissertation TH Karlsruhe 1961.

[5] FRANZ, G. und NIEDENHOFF, H.: Die Bewehrung von Konsolen und gedrungenen Balken. Beton- und Stahlbetonbau 5/1963.

[6] MEHMEL, A. und BECKER, G.: Zur Schubbemessung des kurzen Kragarms. Bauingenieur 6/1965.

[7] MEHMEL, A. und FREITAG, W.: Tragfähigkeitsversuche an Stahlbetonkonsolen. Bauingenieur 10/1967.

[8] STEINLE, A.: Zur Frage der Mindestabmessungen von Konsolen. Beton- und Stahlbetonbau 6/1975.

[9] FRANZ, G.: Stützenkonsolen. Beton- und Stahlbetonbau 6/1975.

[10] GRASSER, E. und RUHNAU, J.: Unveröffentlichte Untersuchungen über die „Begrenzung der Hauptdruckspannungen bei der Bemessung von Konsolen und kurzen Balken“.TU München.

[11] KORDINA, K.: Noch nicht veröffentlichter Institutsbericht. TU Braunschweig.

[12] REHM, G.: Erläuterungen zu den Stahlbetonbestimmungen. DAfStb., Heft 300 (1978).

[13] LEONHARDT, F.: Vorlesungen über Massivbau. Bd III. Springer-Verlag Berlin, 1974, „Rahmenecken“.

[14] GRASSER, E.: Die Bemessung der Stahlbetonbauteile. Beton-Kalender 1974, Band I, Berlin, W. Ernst & Sohn 1974.

[15] ROBINSON, J. R.: L'armature des Consoles Courtes. Aus Theorie und Praxis des Stahlbetonbaues. Festschrift zum 65. Geburtstag von Prof. FRANZ, S. 67. Berlin, W. Ernst & Sohn 1969.

[16] PASCHEN, H.: Das Bauen mit Beton-Stahlbeton und Spannbetonfertigbauteilen. Beton-Kalender 1982, Teil II, Berlin, W. Ernst & Sohn.

[17] STEINLE, A. und ROSTÁSY, F. S.: Zum Tragverhalten ausgeklinkter Trägerenden. Betonwerk + Fertigteiltechnik 6/1975.

[18] MALONN, H. und PASCHEN, H.: Untersuchungen über das Verhalten von rechteckigen und kranzförmigen Konsolen insbesondere auch unter exzentrischer Belastung. Unveröffentlichter Forschungsbericht 1982 DFG.

Bruchbedingung für zweiachsig beanspruchtes Mauerwerk

Bruno Thürlimann
und Hans R. Ganz

1 Einleitung

Zur Berechnung der Tragfähigkeit einer Konstruktion ist die Kenntnis der Bruch- respektive Fließ-Bedingung des verwendeten Baustoffes eine wesentliche Grundlage. Darunter versteht man das Versagen oder Fließen des Materials unter einem beliebigen Spannungszustand. Zeigt das Material eine genügende Duktilität, so kann die Traglast relativ einfach mit den Grenzwert-Sätzen der Plastizitätstheorie bestimmt werden.

Die Festkörperphysik kann heute das physikalische Verhalten von vielen Stoffen aus ihrer Atom-, Molekular- und Kristallstruktur grundlegend erklären. Es ist ihr bis heute jedoch noch nicht gelungen, die Festigkeit von Werkstoffen und im besonderen von Baustoffen quantitativ zu bestimmen. Dieser Umstand ist keineswegs erstaunlich, da die Baustoffe im allgemeinen äußerst komplex aufgebaute Systeme sind. So besteht Stahl aus einem Haufwerk von Kristallen mit einer Unzahl von Einschlüssen und Dislokationen. Die Kristallgrenzen und die Kontakte zwischen den Kristallen verursachen weitere Schwierigkeiten. Alle diese Unregelmäßigkeiten verunmöglichen bis heute eine quantitative theoretische Analyse. Beton ist ein noch viel komplizierter aufgebauter Baustoff aus mineralischen Zuschlagstoffen, Zementmatrix und Wasser. Für den Verbundbaustoff Stahlbeton verursacht das Zusammenwirken von Stahl und Beton weitere Schwierigkeiten.

Backsteine bestehen aus gebranntem Ton. Dieses Material hat selber einen sehr homogenen und isotropen Aufbau. Hingegen führt die geometrisch gerichtete Struktur des Steines bereits zu einem anisotropen Verhalten. Weiter beeinflussen das Fugenmaterial, die Ausbildung der Lager- und Stoßfugen und die Art des Verbandes die Festigkeit ganz entscheidend.

Aus dieser Sachlage ergibt sich, daß die Herleitung einer Bruch- oder Fließbedingung von gebräuchlichen Baustoffen theoretisch auf der Ebene der Kontinuumsmechanik erfolgen sollte. Neben mechanisch zwingenden Relationen können die übrigen Annahmen intuitiv getroffen oder aus experimentellen Beobachtungen begründet werden. Dadurch ist der ingenieurmäßigen Behandlung des Problems der nötige Spielraum gegeben. Theoretisches Wissen, Erfahrung, Beobachtung, angemessene Genauigkeit, Übersichtlichkeit, einfache Anwendung usw. sind alles Bestandteile, welche darin einfließen und schlußendlich zu einer adäquaten Lösung führen sollen.

Vor einigen Jahren wurde eine Fließbedingung für armierte Stahlbetonscheiben hergeleitet und veröffentlicht [1]. In dieser Arbeit wird mit einem analogen Vorgehen eine allgemeine Bruchbedingung für Mauerwerksscheiben aufgestellt. Die Bestimmung wird an einem Beispiel illustriert und auch quantitativ festgelegt.

2 Grundsätzliche Überlegungen zur Bruchbedingung von Mauerwerk

Mauerwerk ist aus Steinen und aus Fugenmörtel aufgebaut. Eine Bruchbedingung für Mauerwerk sollte deshalb Versagen in diesen beiden Komponenten berücksichtigen.

Die Steine für tragendes Mauerwerk sind normalerweise so gelocht, daß die Nettoquerschnittsfläche in einem Schnitt parallel zu den Lagerfugen bedeutend größer ist als in einem Schnitt senkrecht zu den Lagerfugen. Der Steinquerschnitt läßt sich generell in einen zweiachsig beanspruchten Teil (z. B. Längsscheiben) und in einen einachsig beanspruchten Teil (z. B. Querstege) aufteilen. Die zweiachsig beanspruchten Elemente ergeben einen isotropen Verlauf der Steinfestigkeit. Durch die Überlagerung mit einachsig beanspruchten Elementen wird die Festigkeit in der Richtung der Elemente erhöht, und es resultiert eine anisotrope Steinfestigkeit. Durch Teilvermörtelung der Fugen, im besonderen der Stoßfuge (Bild 6), wird die Aufteilung in ein- und zweiachsig beanspruchte Elemente weiter beeinflußt. Ist die Aufteilung des Steinquerschnittes und die Festigkeit des Steinmaterials bekannt, kann die Bruchbedingung des Steines aufgestellt werden. Es gelingt jedoch nur selten, diese beiden Steinanteile auch theoretisch sauber auseinander zu halten. Die effektive Verteilung kann deshalb nur anhand von Versuchen festgestellt werden.

In den Mörtelfugen wird nur Schub- und Zugversagen untersucht. Die Druckfestigkeit des Mörtels wird wegen des dreiachsigen Druckspannungszustandes nicht maßgebend. Werden für Mauerwerk im Läuferverband Scherbrüche durch Stoßfugen und Steine ausgeschlossen, so wird der Schubwiderstand der Stoßfugen unter zweiachsiger Druckbeanspruchung nie maßgebend.

Die Mörtelfugen haben auch einen starken Einfluß auf die Druckfestigkeit des Mauerwerks. Anstelle der Steindruckfestigkeit ist nur die z.T. bedeutend geringere Mauerwerksdruckfestigkeit erreichbar. Infolge ungleicher Querdehnung von Stein und Mörtel entstehen Querzugspannungen, welche zu dieser Abminderung führen. Da sich dieses Zusammenwirken kaum analytisch erfassen läßt, wird im Sinne einer Vereinfachung die Bruchbedingung des Steines auf die experimentell zu bestimmenden Mauerwerksdruckfestigkeiten abgemindert. Die Form der Bruchbedingung wird beibehalten.

3 Bruchbedingung von Mauerwerk

3.1 Annahmen

Es werden folgende Annahmen getroffen:

1. Für alle Materialien wird starr-idealplastisches Verhalten angenommen.
2. Die Zugfestigkeit des Mauerwerks (Steine und Fugen) wird vernachlässigt.
3. Die Bruchbedingungen der Mauerwerkskomponenten folgen den Beziehungen im Bild 1.
4. Die Bewehrung wird nur einachsig in Stabrichtung belastet. Ihre Wirkung auf Druck wird vernachlässigt.

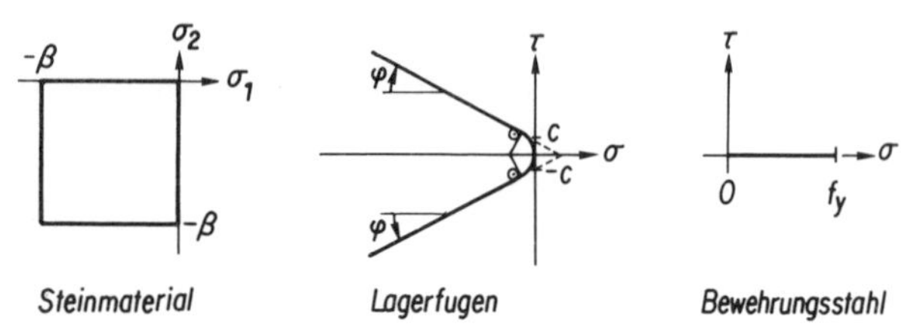

Bild 1
Bruchbedingungen der Mauerwerkskomponenten

3.2 Bruchbedingung bei Versagen der Steine

Im Bild 2 ist ein Element einer Mauerwerksscheibe mit den daran angreifenden Schnittkräften n_x, n_y und n_{xy} dargestellt. Werden Stoß- und Lagerfugen voll vermörtelt, so kann die Querschnittsfläche des gelochten Steines in den zweiachsig beanspruchten Teil mit der Fläche A_{xy} (Längsscheiben) und in den einachsig beanspruchten Teil mit der Fläche A_x (Querstege) unterteilt werden.

Im gewählten Koordinatensystem steht die x-Achse senkrecht zur Lagerfuge. Unter den getroffenen Annahmen lautet die Bruchbedingung für die isotropen, zweiachsig beanspruchten Querschnittsteile (vgl. Bild 1) in Hauptspannungen:

$$0 \geq n_{1,2} = \frac{1}{2}(n_x + n_y) \pm \sqrt{\frac{1}{4}(n_x - n_y)^2 + n_{xy}^2} \geq -\beta \cdot A_{xy} \tag{1 a}$$

oder gleichwertig

$$\begin{aligned} n_{xy}^2 - n_x \cdot n_y &\leq 0 \\ n_{xy}^2 - (n_x + \beta \cdot A_{xy}) \cdot (n_y + \beta \cdot A_{xy}) &\leq 0 \end{aligned} \tag{1 b}$$

Für die isotropen, einachsig beanspruchten Querschnittsteile ergibt sich

$$\begin{aligned} 0 \geq n_x &\geq -\beta \cdot A_x \\ n_y = n_{xy} &\equiv 0 \end{aligned} \tag{2}$$

Den Widerstand des zusammengesetzten Steinquerschnittes erhält man als Linearkombination der

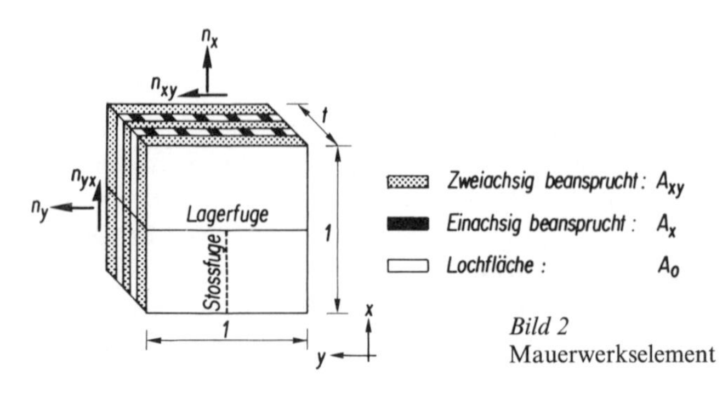

Bild 2
Mauerwerkselement

Bruchbedingungen der Elemente. Es ergeben sich drei Bereiche:

$$\begin{aligned} &n_{xy}^2 - n_x \cdot n_y \leq 0 \\ &n_{xy}^2 + n_y \cdot (n_y + \beta \cdot A_{xy}) \leq 0 \\ &n_{xy}^2 - (n_x + \beta \cdot (A_{xy} + A_x)) \\ &\qquad \cdot (n_y + \beta \cdot A_{xy}) \leq 0 \end{aligned} \tag{3}$$

Werden die Beziehungen (3) durch die Bruttoquerschnittsfläche A des Mauerwerks dividiert, so ergibt sich die Bruchbedingung bei Versagen der Steine in Bruttospannungen.

$$\begin{aligned} &\tau_{xy}^2 - \sigma_x \cdot \sigma_y \leq 0 \\ &\tau_{xy}^2 + \sigma_y \cdot \left(\sigma_y + \beta \cdot \frac{A_{xy}}{A}\right) \leq 0 \\ &\tau_{xy}^2 - \left(\sigma_x + \beta \cdot \left(\frac{A_{xy}}{A} + \frac{A_x}{A}\right)\right) \\ &\qquad \cdot \left(\sigma_y + \beta \cdot \frac{A_{xy}}{A}\right) \leq 0 \end{aligned} \tag{4}$$

Darin bedeuten

$$\begin{aligned} &\beta \cdot \left(\frac{A_{xy}}{A} + \frac{A_x}{A}\right) = f_{mx} \\ &\beta \cdot \left(\frac{A_{xy}}{A}\right) = f_{my} \end{aligned} \tag{5}$$

die einachsigen Druckfestigkeiten des Mauerwerks in x- bzw. y-Richtung bezogen auf die Bruttoquerschnittsfläche.

3.3 Bruchbedingung bei Gleiten der Lagerfugen

Unter den getroffenen Annahmen (vgl. Bild 1) lauten die Beziehungen für Fugenversagen im gewählten Koordinatensystem

$$\begin{aligned} &\tau_{xy}^2 - (c - \sigma_x \cdot \tan\varphi)^2 \leq 0 \\ &\tau_{xy}^2 + \sigma_x \cdot \left(\sigma_x + 2 \cdot c \cdot \tan\left(\frac{\pi}{4} + \frac{\varphi}{2}\right)\right) \leq 0 \end{aligned} \tag{6}$$

Die Spanngen sind wieder auf die Bruttoquerschnittsfläche bezogen. Die erste Bedingung entspricht der Geraden im Bild 1, die zweite gilt für den Kreis, der die erwähnte Gerade sowie die τ_{xy}-Achse berührt.

3.4 Bruchbedingung für unbewehrtes Mauerwerk

Für den Widerstand von Mauerwerk gelten die Beziehungen (4) und (6) gleichzeitig. Damit ergibt sich folgende Bruchbedingung für Mauerwerk:

$$\begin{aligned} &(1){:}\quad \tau_{xy}^2 - \sigma_x \cdot \sigma_y \leq 0 \\ &(2){:}\quad \tau_{xy}^2 - (\sigma_x + f_{mx}) \cdot (\sigma_y + f_{my}) \leq 0 \\ &(3){:}\quad \tau_{xy}^2 + \sigma_y \cdot (\sigma_y + f_{my}) \leq 0 \\ &(4\text{a}){:}\quad \tau_{xy}^2 - (c - \sigma_x \cdot \tan\varphi)^2 \leq 0 \\ &(4\text{b}){:}\quad \tau_{xy}^2 + \sigma_x \cdot \left(\sigma_x + 2 \cdot c \cdot \tan\left(\frac{\pi}{4} + \frac{\varphi}{2}\right)\right) \leq 0 \end{aligned} \tag{7}$$

Im dreidimensionalen Spannungsraum wird die Bruchbedingung von Mauerwerk durch zwei elliptische Kegel (1 und 2), zwei Kreiszylinder (3 und 4b) und eine Ebene (4a) dargestellt. Sie besitzt vier unabhängige Materialparameter. Bild 3 zeigt die Bruchbedingung in Höhenliniendarstellung. Die fünf Regimes der Bruchbedingung haben folgende Bedeutung:

(1): Zugversagen des Mauerwerks
(2): Druckfestigkeit des gesamten Querschnittes ausgenützt
(3): Druckfestigkeit der zweiachsig beanspruchten Querschnittsteile ausgenützt
(4a): Gleiten der Lagerfugen
(4b): Zugversagen (Trennbruch) in der Lagerfuge

Der Verlauf der Hauptspannungen als Funktion der Fugenneigung kann gewonnen werden, indem die Spannungstransformation

$$\begin{aligned} &\sigma_x = \sigma_2 \cdot \left(\cos^2\vartheta + \frac{\sigma_1}{\sigma_2} \cdot \sin^2\vartheta\right) \\ &\sigma_y = \sigma_2 \cdot \left(\sin^2\vartheta + \frac{\sigma_1}{\sigma_2} \cdot \cos^2\vartheta\right) \\ &\tau_{xy} = \sigma_2 \cdot \left(\frac{\sigma_1}{\sigma_2} - 1\right) \cdot \sin\vartheta \cdot \cos\vartheta \end{aligned} \tag{8}$$

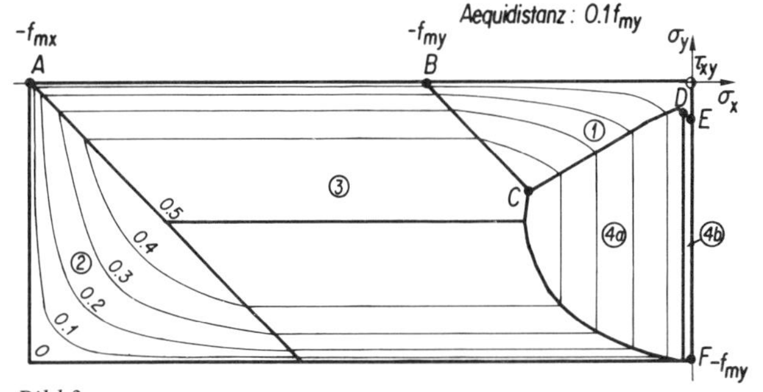

Bild 3
Bruchbedingung für unbewehrtes Mauerwerk

in die Bruchbedingungen (7) eingesetzt wird. Für drei Verhältnisse σ_1/σ_2 ist der Verlauf der kleineren Hauptspannung im Bild 4 dargestellt. Im speziellen ergibt sich der Verlauf der einachsigen Druckfestigkeit zu

$$\begin{aligned}
&\text{A:} && \vartheta = 0°: \\
& && \sigma_2 = -f_{mx} \\
&\text{BC, F:} && 0 < \vartheta \leq \vartheta_{Gr},\ \vartheta = \frac{\pi}{2}: \\
& && \sigma_2 = -f_{my} \\
&\text{CD:} && \vartheta_{Gr} \leq \vartheta \leq \frac{\pi}{4} + \frac{\varphi}{2}: \\
& && \sigma_2 = -\frac{c}{\sin\vartheta \cdot \cos\vartheta - \cos^2\vartheta \cdot \tan\varphi} \\
&\text{DE:} && \frac{\pi}{4} + \frac{\varphi}{2} \leq \vartheta < \frac{\pi}{2}: \\
& && \sigma_2 = -2 \cdot c \cdot \tan\left(\frac{\pi}{4} + \frac{\varphi}{2}\right)
\end{aligned} \tag{9}$$

Die Spannungspunkte, die einachsigen Druckversuchen entsprechen (A bis F), liegen im Bild 3 auf Kanten der Bruchbedingung.

Vier Merkmale der Bruchbedingung sollen besonders hervorgehoben werden:

- Unter einachsigen Druckspannungszuständen kann die kleinere Druckfestigkeit f_{my} – mit Ausnahme von $\vartheta = 0°$ – nicht überschritten werden. Ohne Berücksichtigung von Gleiten der Lagerfugen ergibt sich ein quasi isotropes Verhalten.
- Für Winkel $\vartheta > \vartheta_{Gr}$ fällt die einachsige Druckfestigkeit infolge Gleiten der Lagerfugen stark ab und bleibt bis $\vartheta = \frac{\pi}{2}$ auf dem Minimum.

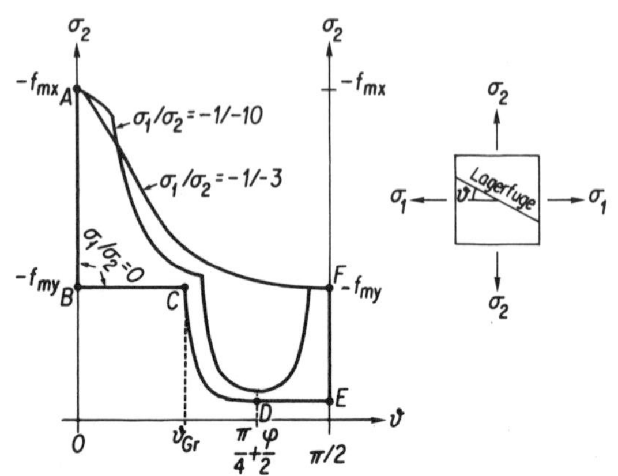

Bild 4
Druckfestigkeit in Funktion der Lagerfugenneigung

- Bereits unter geringer, zweiachsiger Druckbeanspruchung ergibt sich ein starker Festigkeitsanstieg.
- Für große Verhältnisse σ_1/σ_2 verschwindet die Bruchart „Gleiten der Lagerfugen" vollständig.

Es lassen sich zahlreiche Parameter-Kombinationen und dementsprechend unterschiedliche Bruchbedingungen denken. Hier sei nur auf den Spezialfall einer Bruchbedingung für Mauerwerk mit Vollsteinen (isotroper Stein) und voll vermörtelten Fugen hingewiesen. Bei gleichen Druckfestigkeiten $f_{mx} = f_{my}$ verschwindet der Zylinder (3) und die Bruchbedingung besteht einzig aus zwei Kegeln (1 und 2) und den Fugenbedingungen (4a, b).

Eine Bruchbedingung für Mauerwerk mit Zugfestigkeit kann ähnlich hergeleitet werden wie hier dargestellt. Für Mauerwerk mit Zugfestigkeit fällt die einachsige Druckfestigkeit kontinuierlich von der großen (f_{mx}) auf die kleine Druckfestigkeit (f_{my}) ab und steigt für Winkel $\vartheta > \frac{\pi}{4} + \frac{\varphi}{2}$ wieder an. Ihr Verlauf gleicht der im Bild 4 eingezeichneten Kurve $\frac{\sigma_1}{\sigma_2} = \frac{-1}{-10}$.

Das Steinformat erscheint in der Bruchbedingung ohne Zugfestigkeit nicht explizit. Es ist aber über die experimentell zu bestimmenden Festigkeitsparameter in der Bruchbedingung enthalten. Wird eine Zugfestigkeit angesetzt, so erscheint das Steinformat explizit in den Gleichungen.

Da keine Versuche vorhanden sind, wurde die Form der Bruchbedingung des Steinmaterials (Bild 1) möglichst einfach gewählt. Ein anderer Verlauf der zweiachsigen Druckfestigkeit hat in der dargestellten Bruchbedingung nur Einfluß auf das Regime (2).

3.5 Bruchbedingung für bewehrtes Mauerwerk

Im Mauerwerksbau werden die Bewehrungsstäbe in die Lagerfugen und/oder in die Steinlochung verlegt. Deshalb ergibt sich immer eine orthogonale Bewehrungsanordnung in x- und y-Richtung. Die am Scheibenelement angreifenden Kräfte (Bild 2) können in den Anteil des Mauerwerks und in den Anteil der Bewehrung aufgespalten werden:

$$\begin{aligned}
n_x &= \sigma_x + z_x \\
n_y &= \sigma_y + z_y \\
n_{xy} &= \tau_{xy} + z_{xy}
\end{aligned} \tag{10}$$

Da die Bewehrungsrichtungen senkrecht zueinander stehen, gilt

$$z_{xy} \equiv 0\,.$$

Allen Linearkombinationen von Spannungen $\{\sigma_x, \sigma_y, \tau_{xy}\}$ und $\{z_x, z_y\}$, die die Bruchbedingungen von Mauerwerk und Bewehrungsstahl nicht verletzen, entsprechen Bereiche der Bruchbedingung von bewehrtem Mauerwerk. Werden für den Stahl die mechanischen Bewehrungsgehalte

$$\omega_x = \mu_x \cdot \frac{f_y}{f_{my}} \quad \text{und} \quad \omega_y = \mu_y \cdot \frac{f_y}{f_{my}}$$

eingeführt, so ergeben sich folgende Ungleichungen als Bruchbedingung für bewehrtes Mauerwerk:

$$\begin{aligned}
&(1): && n_{xy}^2 - (\omega_x \cdot f_{my} - n_x) \cdot (\omega_y \cdot f_{my} - n_y) \le 0 \\
&(2): && n_{xy}^2 - (n_x + f_{mx}) \cdot (n_y + f_{my}) \le 0 \\
&(3\,a): && n_{xy}^2 + \left(n_y + f_{my} \cdot \left(\frac{1}{2} - \omega_y\right)\right)^2 - \left(\frac{1}{2} \cdot f_{my}\right)^2 \le 0 \\
&(3\,b): && n_{xy}^2 + \left(n_y + \frac{1}{2} \cdot f_{my}\right)^2 - \left(\frac{1}{2} \cdot f_{my}\right)^2 \le 0 \qquad (11) \\
&(4\,a): && n_{xy}^2 - (c - (n_x - \omega_x \cdot f_{my}) \cdot \tan\varphi)^2 \le 0 \\
&(4\,b): && n_{xy}^2 + (n_x - \omega_x \cdot f_{my}) \cdot \Bigg((n_x - \omega_x \cdot f_{my}) \\
& && \quad + 2 \cdot c \cdot \tan\left(\frac{\pi}{4} + \frac{\varphi}{2}\right)\Bigg) \le 0 \\
&(5): && n_{xy}^2 - \left(\frac{1}{2} \cdot f_{my}\right)^2 \le 0 \\
&(6): && n_{xy}^2 + \left(n_x + f_{mx} - \frac{1}{2} \cdot f_{my}\right)^2 \\
& && \quad - \left(\frac{1}{2} \cdot f_{my}\right)^2 \le 0
\end{aligned}$$

Durch Einlegen einer Lagerfugenbewehrung allein entstehen neu die Bruchregimes (5) und (6). Die anderen Regimes sind lediglich verschoben worden. Auf den einzelnen Regimes der Bruchbedingung von bewehrtem Mauerwerk (11) herrschen die gleichen Spannungszustände $\{\sigma_x, \sigma_y, \tau_{xy}\}$ wie auf den entsprechenden Regimes des unbewehrten Materials.
Bild 5 zeigt die Bruchbedingung des in der Anwendung häufigen Falles, bei dem die Bewehrung nur in die Lagerfugen verlegt wird.
Die obigen Ableitungen setzen starren Verbund zwischen Bewehrungsstahl und Mauerwerk voraus. In den Lagerfugen scheint dies näherungsweise gewährleistet zu sein. Dagegen sind die Verhältnisse bei der in die Steinlochung verlegten Bewehrung unsicher, und ihre Wirkung sollte deshalb vorsichtig angesetzt werden.
Eine detaillierte Behandlung und Diskussion der Bruchbedingung von Mauerwerk ist in Vorbereitung [2].

4 Experimentelle Untersuchungen

Zur Kontrolle der theoretisch entwickelten Bruchbedingung in einzelnen Punkten und zur Festlegung der Materialparameter wurden zwölf Versuche an Mauerwerkskörpern durchgeführt [3].
Die quadratischen Versuchskörper waren 1200 mm lang und 150 mm dick. Es wurden ein gelochter Modulbackstein normaler Qualität mit einer Druckfestigkeit von $f_s = 31$ N/mm² (Lochfläche 46%) und ein Zementmörtel mit einer Druckfestigkeit von $f_M = 28$ N/mm² verwendet. Die Lagerfugenbewehrung erreichte eine Fließspannung von $f_y = 550$ N/mm².
Die Lagerfugen waren voll vermörtelt. In den Stoßfugen wurde der Mörtel, wie im Bild 6 dargestellt, nur in die Steinnut eingebracht.
Die Mauerwerkskörper wurden mit den Hauptspannungen in unterschiedlichen Richtungen zu den Fugen

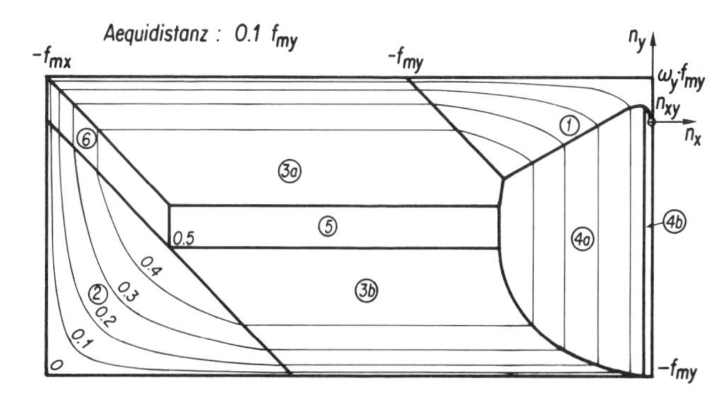

Bild 5
Bruchbedingung für Mauerwerk mit Lagerfugenbewehrung

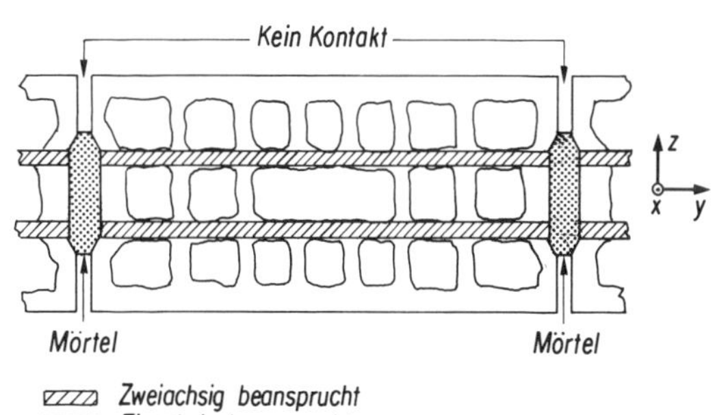

Bild 6
Lochbild des Steines und Stoßfugenausbildung

Tabelle 1
Bruchspannungen

Versuch	σ_1/σ_2	ϑ [Grad]	Experimentelle Bruchspannungen [N/mm²] σ_x	σ_y	τ_{xy}	Theoretische Bruchspannungen [N/mm²] σ_x	σ_y	τ_{xy}	Bruch-regime	Theorie / Versuch
K 3	0	0,0	−7,63	0,00	0,00	−7,60	0,00	0,00	2	1,00
K 4	0	90,0	0,00	−1,83	0,00	0,00	−2,70	0,00	3	1,48
K 6	0	45,0	−0,32	−0,32	0,32	−0,32	−0,32	0,32	4a	1,00
K 7	0	22,5	−2,25	−0,39	0,93	−2,30	−0,40	0,95	3	1,02
K 8	0	67,5	−0,04	−0,22	0,09	−0,04	−0,22	0,09	4a	1,00
K 9*)	−0,097	22,5	−2,37	−0,18	1,09	−2,51	−0,19	1,16	3a	1,06
K 10	0,328	0,0	−6,44	−2,11	0,00	−7,60	−2,49	0,00	2	1,18
K 11	0,306	22,5	−4,49	−2,04	1,23	−4,37	−1,98	1,19	3	0,97
K 12	0,304	45,0	−2,03	−2,03	1,08	−2,10	−2,10	1,12	3	1,03

*) Mit Lagerfugenbewehrung $\omega_y = 0{,}17$, Spannungen $\{n_x, n_y, n_{xy}\}$

belastet. Es wurden Zug-Druck-Versuche sowie einachsige und zweiachsige Druckversuche durchgeführt. Die Hauptspannung σ_2 war immer negativ. In einen Versuchskörper wurde vorgefertigte Lagerfugenbewehrung (2 ∅ 4 mm pro Lagerfuge) eingelegt. Die Belastung wurde proportional bis zum Bruch der Prüfkörper gesteigert. Im Bild 7 ist die Prüfeinrichtung für zweiachsige Druckbelastung schematisch dargestellt. Druckkräfte wurden über einzelne Stahl-Neopren-Stücke eingeleitet, um die Querverformungen der Wände nicht zu behindern.

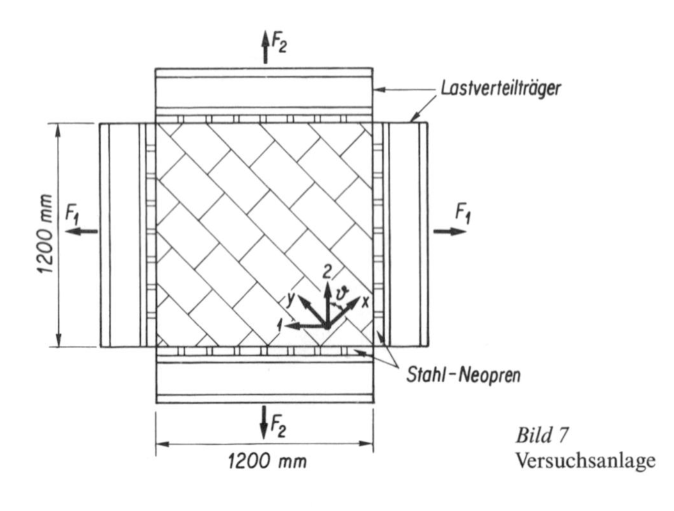

Bild 7
Versuchsanlage

In der Tabelle 1 sind die Resultate der ein- und zweiachsigen Druckversuche sowie die des bewehrten Versuchskörpers zusammengestellt. Die theoretischen Bruchlasten wurden mit den Beziehungen (7) und (11) berechnet. Als Materialparameter wurden folgende Werte verwendet:

$$f_{mx} = 7{,}6\,\text{N/mm}^2 \qquad c = 0{,}06\,\text{N/mm}^2$$
$$f_{my} = 2{,}7\,\text{N/mm}^2 \qquad \tan\varphi = 0{,}81$$

Die Übereinstimmung zwischen experimentellen und theoretischen Werten darf als zufriedenstellend bezeichnet werden. Anpassungen der Bruchbedingung sind einzig im Bereich des einachsigen Druck-Versuchs K 4 vorzunehmen, wo die spezielle Stoßfugenausbildung zu einem frühzeitigen Versagen führt.

5 Literatur

[1] MARTI, P., und THÜRLIMANN, B.: „Fließbedingung für Stahlbeton mit Berücksichtigung der Betonzugfestigkeit", Beton- und Stahlbetonbau 72 (1977), Heft 1.

[2] GANZ, H. R.: „Mauerwerksscheiben unter Normalkraft, Querkraft und Biegemoment", Institut für Baustatik und Konstruktion, ETH Zürich, Bericht in Vorbereitung.

[3] GANZ, H. R., und THÜRLIMANN, B.: „Versuche über die Festigkeit von zweiachsig beanspruchtem Mauerwerk", Institut für Baustatik und Konstruktion, ETH Zürich, Versuchsbericht Nr. 7502-3, Februar 1982, Birkhäuser Verlag Basel und Stuttgart.

Kunstharzgebundene Glasfaserstäbe – eine korrosionsbeständige Alternative zum Spannstahl

Martin Weiser
und Lothar Preis

1 Einführung

Die Entwicklung des Spannbetons stellte der Bauindustrie die Aufgabe, hochfeste Spannstähle sicher und dauerhaft gegen Korrosion zu schützen. Neben diesem auch heute noch aktuellen Anliegen fanden jedoch einzelne Forscher auf der Suche nach Alternativen zum Spannstahl schon sehr frühzeitig, daß bei Verbundstäben aus E-Glas-Fasern und einem geeigneten UP-Harz die bei Werkstoffen so seltene Kombination von hoher Zugfestigkeit und guter Korrosionsbeständigkeit erreicht werden kann.
In den 70er Jahren hat dann REHM in vier aufeinanderfolgenden systematischen Forschungsarbeiten [1], [2], [3], [4] die grundsätzliche Eignung kunstharzgebundener Glasfaserstäbe als Bewehrung im Betonbau nachgewiesen. Kurz darauf gelang der Bayer AG die Entwicklung eines im Labormaßstab kontinuierlich arbeitenden Herstellungsverfahrens für hochfeste Rundstäbe aus kunstharzgebundenen Glasfasern.
Auf dieser Basis aufbauend gründeten die Firmen Strabag Bau-AG als Anwender und Bayer AG als Hersteller eine Arbeitsgemeinschaft zur marktreifen Entwicklung der Glasfaservorspannung. Seit 1978 betreiben sie gemeinsam ein vom Bundesminister für Forschung und Technologie (BMFT) zu 50% gefördertes langfristiges Forschungsvorhaben, in dem die Eignung der von Bayer unter dem Markennamen Polystal® hergestellten Glasfaserstäbe für eine Verwendung als hochfeste Zugglieder im Bauwesen demonstriert werden soll. Als mögliche Anwendungsbereiche werden dabei alle Arten und Grade der Vorspannung, mit und ohne Verbund, sowie temporäre und permanente Erd- und Felsanker untersucht.
Das Programm dieses Vorhabens der beiden Partner, an dem auch REHM und andere namhafte Vertreter der Wissenschaft mitarbeiten [6], ist in [5] ausführlich beschrieben. Es umfaßt als Teilaufgaben die Ermittlung der Werkstoffkennwerte, ihre Auswirkung auf Bemessung und Sicherheit, die Herstellung des Stabmaterials, die Krafteinleitung im Verankerungsbereich sowie die praxisnahe Erprobung der Zugglieder in Bauteilversuchen und Probebauwerken.
Eine 1983 vom BMFT bewilligte Aufstockung des Vorhabens brachte als neue Teilaufgaben die Sicherstellung der Medienbeständigkeit sowie die Herstellung und Verankerung von Rechteckstäben, während bis dahin nur Rundstäbe untersucht worden waren.

2 Werkstoffkennwerte und ihre Auswirkung auf Bemessung und Sicherheit

Im Auftrag der Arbeitsgemeinschaft ermittelt das Institut für Werkstoffe im Bauwesen (REHM) die Kennwerte der Polystal-Stäbe. KÖNIG bearbeitet die Teilaufgabe „Bemessung und Sicherheit".
Durch den hohen Anteil der Glasfasern (ca. 68 Vol.%) und ihre streng unidirektionale Führung erreicht Polystal mit 1600 N/mm² die Längszugfestigkeit hochfester Spannstähle (Bild 1). Die Zeitstandfestigkeit beträgt ca. 70% dieses Wertes, die Querdruckfestigkeit jedoch nur ca. 10% (anisotropes Material).
Die niedrige Dichte von Polystal (2,1 g/cm³) erleichtert Transport und Einbau der Spannglieder. Einen Vergleich mit den mechanischen Eigenschaften anderer hochzugfester Stoffe gibt Tabelle 1.
Die Schubfestigkeit des Matrixharzes beträgt bei Raumtemperatur (RT) 32 N/mm² und kann durch Überlagerung von Querdruck wesentlich gesteigert werden. Bei 70 °C verliert sie gegenüber RT etwa 20% ihres Wertes (Bild 2). Die sehr gute Hitzebeständigkeit der tragenden Glasfasern zeigt Bild 3.
Infolge seines niedrigen E-Moduls (50 000 N/mm²) hat Polystal gegenüber vergleichbaren Spannstählen den 4fachen Spannweg und erleidet somit nur ein Viertel der entsprechenden Spannkraftverluste aus Kriechen und Schwinden des Betons. Wegen dieser großen Ver-

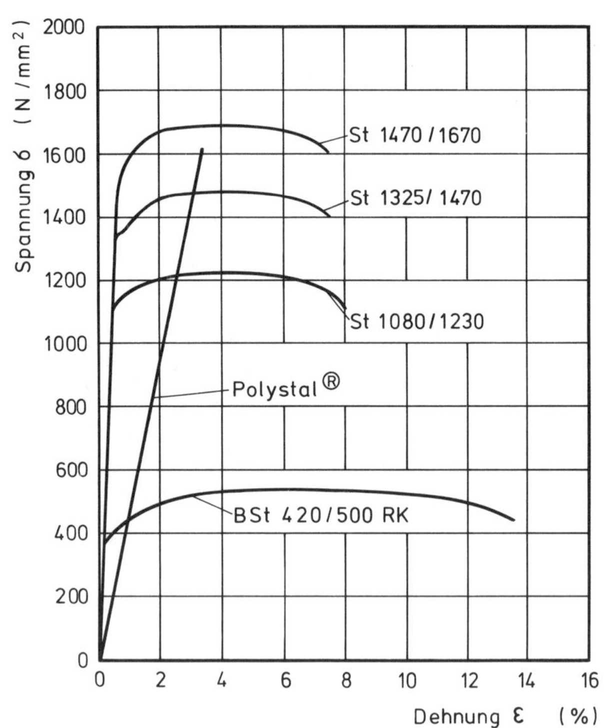

Bild 1
Spannungs-Dehnungs-Diagramm von Polystal®-Stäben im Vergleich zu verschiedenen Spannstählen und Betonstahl BSt 420/500

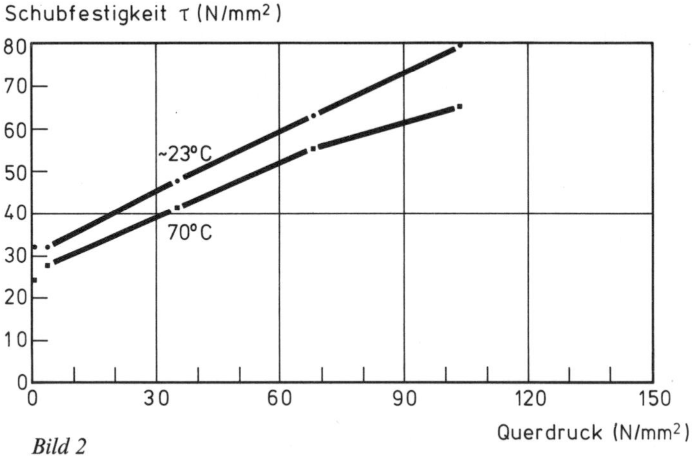

Bild 2
Schubfestigkeit des Matrixharzes in Abhängigkeit vom Querdruck

formungen eignet sich Polystal jedoch nur in Ausnahmefällen als schlaffe Bewehrung von Beton.
Durch das Fehlen einer Streckgrenze erfolgt bei Polystal der Bruch schlagartig mit dem charakteristischen Bruchbild eines Faserpinsels. Dennoch liegt hier kein Bruch ohne Vorankündigung vor, der einen höheren Sicherheitsbeiwert erfordern würde: anstelle der fehlenden plastischen Verformung wirkt nämlich bei Polystal die große elastische Verformung jenseits des Gebrauchszustandes als erwünschte Warnung vor dem Eintreten eines Bruches.

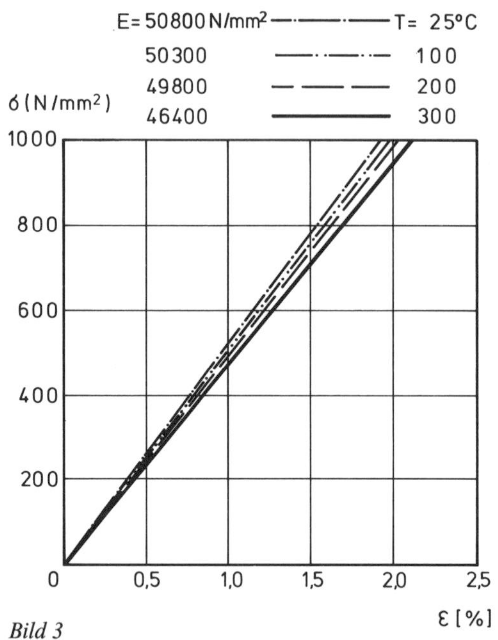

Bild 3
Spannungs-Dehnungs-Linien von Polystal® bei 25, 100, 200 und 300 °C

Tabelle 1
Vergleich der charakteristischen Eigenschaften von Polystal® mit anderen Werkstoffen

	Betonstahl BSt 420/500	Spannstahl St 1470/1670	Polystal® (68% Glasfasern)	Kohlenstoff-fasern
Zugfestigkeit (N/mm²)	> 500	> 1670	1600	2800
Streckgrenze (N/mm²)	> 420	> 1470	–	–
Bruchdehnung (%)	10	6	3,2	0,7
Elastizitätsmodul (N/mm²)	$21 \cdot 10^4$	$21 \cdot 10^4$	$5 \cdot 10^4$	$40 \cdot 10^4$
Reißlänge (km)	6,4	21,5	76	160
Dichte (g/cm³)	7,85	7,85	2,1	1,75

3 Herstellung der Polystal-Stäbe

3.1 Rundstäbe

Zu Beginn des FE-Vorhabens gelang es Bayer, sein schon vorher im Labormaßstab entwickeltes kontinuierliches Herstellungsverfahren zu einer funktionsechten Simulation eines industriellen Verfahrens für die wirtschaftliche Herstellung hochwertiger Rundstäbe auszubauen.
Auf der von Bayer neu entwickelten Fertigungsanlage erfolgt die Produktion von Rundstäben mit Durchmessern von 1 bis 25 mm. Die Ausgangsstoffe sind Stränge („Rovings") aus je ca. 2000 nur wenige µm starken Glasfasern sowie flüssiges UP-Harz. Ein Stab von 7,5 mm Durchmesser besteht z. B. aus etwa 60 000 Einzelfasern. Diese Glasfaserstränge werden im Tauchbad mit Harz getränkt, in einer Düse sowie durch die nachfolgende Umwicklung mit Spinnfasern zum Rundstab geformt und unter Wärmezufuhr gehärtet. Der endlos gefertigte Stab wird in Ringen von 1,50 m Durchmeser gelagert und befördert.
Der Querschnitt eines Polystal-Stabes enthält 68% Glasfasern und 32% Harz. Während das Glas dem Elastizitätsmodul-Verhältnis entsprechend praktisch die gesamten Zugkräfte aufnimmt, hält das Harz die Fasern zusammen und schützt sie gegen mechanische und chemische Schädigungen. An den Stabenden wird die Schubfestigkeit des Harzes beansprucht, um die dort an der Staboberfläche angreifenden Kräfte bis in die innersten Fasern des Querschnittes einzuleiten.

3.2 Rechteckstäbe

Aufbauend auf den Erfahrungen mit der Herstellung von Rundstäben werden z. Z. (im Rahmen der Aufstockung des Vorhabens) Fertigungsverfahren für Rechteckstäbe entwickelt: zunächst diskontinuierlich arbeitend, dann im Labormaßstab kontinuierlich als Vorstufe für die spätere Simulation eines industriellen Herstellungsverfahrens.
Polystal-Stäbe mit Rechteckquerschnitt versprechen eine wesentlich bessere Aufnahme der Querpressungen, die an den Umlenkstellen im Hüllrohr und insbesondere in den Klemmverankerungen auftreten (vgl. Abschnitt 5), sowie eine optimale Ausfüllung der Hüllrohrquerschnitte. Wenn die bei dieser Entwicklungsstufe des Verfahrens erzielbare Herstellungsgenauigkeit der Flächen und Kanten des Querschnittes für eine optimale Krafteinleitung noch nicht ausreichen sollte, werden die Ankerenden der Stäbe in einem zweiten Arbeitsgang durch Vergießen oder Verpressen vollends exakt geformt.
Da sich die Entwicklung dieses kontinuierlichen Herstellungsverfahrens auf den Labormaßstab und die hierbei erreichbare Genauigkeit beschränkt, bleibt der Kostenaufwand in vertretbaren Grenzen, ohne daß dadurch die Effektivität der Untersuchungen wesentlich verringert würde.

4 Sicherung der Medienbeständigkeit

Das für Faserverbundstoffe üblicherweise verwendete E-Glas besitzt, wie die UP-Harze, eine sehr gute Beständigkeit gegenüber den meisten Medien. Gegen die Alkalität von jungem Beton oder Zementmörtel bedürfen die Glasfasern jedoch eines besonderen Schutzes. Er kann im einfachsten Falle durch das Matrixharz des Stabes erfolgen, das den Glasfaserstrang durchdringt und ummantelt. Als weitere Schutzmöglichkeiten kommen eine zusätzliche Ummantelung des Stabes oder die Verwendung entsprechender Verpreßmörtel in Frage. Da die zweifelsfreie Sicherstellung der Medienbeständigkeit nur durch sehr zeitaufwendige Versuchsreihen erfolgen kann, werden in unserem Vorhaben die drei Möglichkeiten zeitlich parallel untersucht.

4.1 Medienschutz durch Matrixharz

Dieses Matrixharz muß, neben seiner hohen Schubfestigkeit zur Einleitung der Verankerungskräfte, eine so große Dehnfähigkeit besitzen, daß es trotz seiner starken Zugbeanspruchung – aus Schwinden nach dem Herstellprozeß und aus Vorspannung im Bauwerk – keine bis auf den Glaskern durchgehenden Risse bekommt, die seine Dichtigkeit gefährden würden.

4.2 Medienschutz durch Ummantelung

Eine zusätzliche Ummantelung muß medienbeständig und dehnfähig sein und zur Staboberfläche eine gute Haftung besitzen. Sie läßt sich durch eine sofortige Aufbringung im „on line-Verfahren" verbessern. Anzustreben sind „ankerfeste" Mäntel, die im Ankerbereich nicht abgeschält werden müssen.

Bild 4
Auf dem Versuchsstand wird ein Polystal-Spannglied zum Verpressen mit Kunststoff-Mörtel vorbereitet

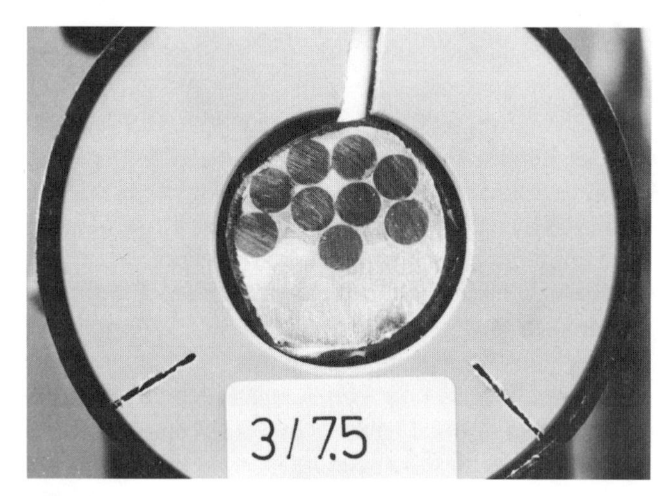

Bild 5
Querschnitt durch ein mit Kunststoff-Mörtel verpreßtes Polystal-Spannglied

4.3 Medienschutz durch Verpreßmörtel

Als Verpreßmörtel kommen besondere hydraulische Mörtel oder Mörtel auf Kunststoffbasis in Frage. Die von Bayer entwickelten Mörtel genügen, wie die Versuche zeigten, allen an Verpreßmörtel zu stellenden Anforderungen bezüglich Festigkeit und Fließfähigkeit (Bilder 4 und 5).

4.4 Tests zur Prüfung der Medienbeständigkeit

Im Standardversuch (Bild 6) wird ein Polystal-Stab auf 50% seiner Bruchlast vorgespannt und mit seinen Ankern gegen ein äußeres Stahlrohr als Widerlager festgelegt. Dann füllt man Zementmörtel als aggressives Medium in das innenliegende PE-Hüllrohr. Nach einer Einlagerungszeit zwischen 18 und 52 Wochen wird im Kurzzeitversuch die Restzugfestigkeit des Polystal-Stabes ermittelt. Liegt sie unter der Vergleichsfestigkeit von nicht in Zementmörtel gelagerten Stäben, so muß auf einen Alkalischaden geschlossen werden. Während der zeitaufwendige Standardversuch den endgültigen sicheren Nachweis der Medienbeständigkeit bringt, erlaubt der Schnelltest (Bild 7) eine kurzfristige Vorauswahl unter verschiedenen Materialien. Anstelle des Zementmörtels dient hier eine wässerige Zementlösung mit pH = 13,0 als aggressives Medium. Sie wird mit einem Thermostat auf 50 bis 60 °C erhitzt und zwischen 4 und 14 Tage lang fortlaufend umgewälzt. Dabei wirkt die erhöhte Temperatur als Zeitraffer. Einen eventuellen Medienschaden erkennt man wiederum am Abfall der Restzugfestigkeit.

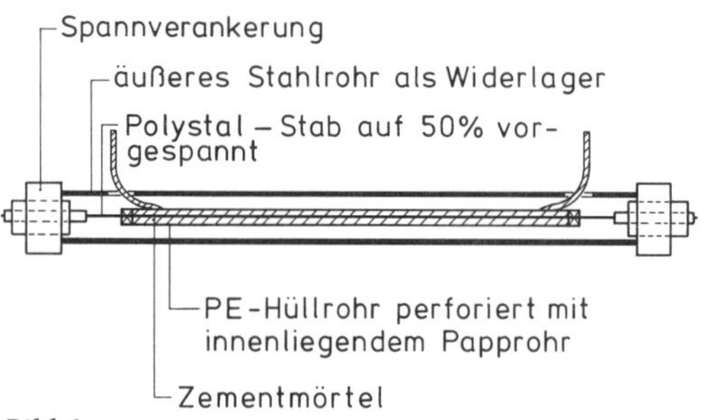

Bild 6
Standardversuch zur Medienbeständigkeit

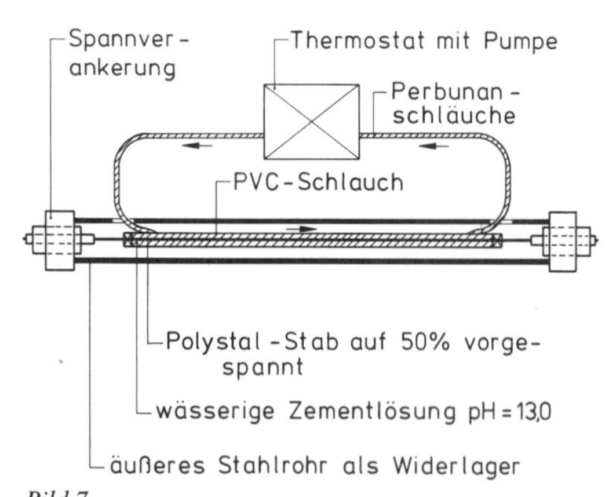

Bild 7
Schnelltest zur Medienbeständigkeit

Die Herstellung der wässerigen Zementlösung erfolgt nach Vorschlag des Instituts für Werkstoffe im Bauwesen Stuttgart durch 7 Tage dauerndes Umwälzen von Wasser durch gebrochenen Zementmörtel (Bild 8). Durch dieses Verfahren sollen alle im Zementmörtel enthaltenen, das Glas angreifenden Stoffe gelöst wer-

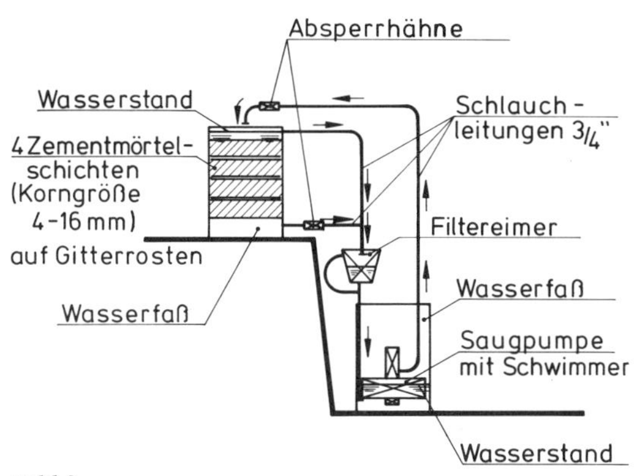

Bild 8
Herstellung der wässerigen Zementlösung für Medienschnelltests (pH-Wert = 13)

den, um später im Schnelltest zur Wirkung zu kommen.

5 Krafteinleitung

Die Verankerung von kunstharzgebundenen Glasfaserstäben ist gekennzeichnet durch

- das Fehlen der Möglichkeit einer Kaltverformung (aufgerollte Gewinde, aufgestauchte Köpfchen, harte Keile mit „scharfem Einbiß"),
- die Empfindlichkeit des anisotropen Werkstoffes gegen (insbesondere ungleichmäßigen) Querdruck,
- die relativ niedrige interlaminare Schubfestigkeit des Matrixharzes, die eine vergleichsweise große Ankerlänge erfordert.

Schon 1978 kam REHM (siehe [4], S. 49 und 54) zu der Feststellung, daß Vergußverankerungen durchaus voll wirksam sein können, und daß sich Klemmverankerungen bereits als geeignet erwiesen haben.

In unserem Forschungsvorhaben wird die Entwicklung werkstoffgerechter Verankerungen vom Partner Strabag in Zusammenarbeit mit REHM (Klemmanker) und ROSTÁSY (Vergußanker) betrieben. Eine ausführliche Beschreibung dieser Teilaufgabe wurde in [6] gegeben. Sie wird ergänzt durch die nachfolgenden Ausführungen über Konzeption und derzeitigen Stand dieser Ankerentwicklung.

Die Zwischenergebnisse dieser Entwicklungsarbeiten bestätigten immer deutlicher die Vorteile einer zweigleisigen Konzeption, die eine technisch und wirtschaftlich bessere Anpassung an die werkstoffspezifischen Gegebenheiten von Polystal ermöglicht.

Während die ursprüngliche Planung aus Kostengründen eingleisig ausgelegt und auf die preisgünstigeren und technisch einfacher herstellbaren Rundstäbe beschränkt war, erfordert die neue Konzeption die zusätzliche Entwicklung rechteckiger Stäbe und kann daher erst im Rahmen der 1983 bewilligten Aufstokkung des Vorhabens ihrer Realisierung zugeführt werden.

Nach dieser Konzeption werden „kurze und leichte" Spannglieder mit Längen bis ca. 15 m und Gebrauchslasten bis 650 kN (z. B. als Erdanker oder als Quervorspannung einer Brücke) aus Rundstäben, runden Hüllrohren und Vergußankern gefertigt, während „lange und schwere" Spannglieder mit Gebrauchslasten über 1000 kN (z. B. für die Längsvorspannung) aus Rechteckstäben, rechteckigen Hüllrohren und Keilklemmplattenankern hergestellt werden.

Die erstgenannte Kombination genügt den Anforderungen des kleineren Spanngliedtyps und bildet für ihn mit Abstand die wirtschaftlichste Lösung. Ihre Übertragung auf schwere Spannglieder wäre aber sowohl beim Anker als auch bei Stab und Hüllrohr technisch und wirtschaftlich zumindest unbefriedigend.

Hier erlaubt jedoch der für leichte Spannglieder zu teure Keilklemmplattenanker ohne entsprechenden Mehraufwand eine im praktischen Anwendungsbereich fast unbegrenzte Steigerung der Gebrauchslast und ist daher für schwere Spannglieder die wirtschaftlichste Lösung.

Auch der Rechteckstab bringt, in Kombination mit Rechteckhüllrohr und Klemmplattenanker, bei schweren Spanngliedern überzeugende technische und wirtschaftliche Vorteile. Er erlaubt eine kompaktere Ausbildung von Anker- und Spanngliedquerschnitt und erträgt die großen Querkräfte, die an den Umlenkstellen der viellagigen schweren Spannglieder und zwischen den Klemmplatten der Anker auftreten.

Bei der Entwicklung der Vergußanker wurde als erster Teilschritt ein Prototyp mit 260 kN Gebrauchslast in Kurzzeit-, Zeitstand- und Zugschwellversuchen optimiert. Nach den dabei gewonnenen Erkenntnissen konnte sodann ein 680-kN-Anker entworfen werden, der bereits gute Kurzzeitwerte gebracht hat.

Klemmplattenanker mit Rundstäben, die in halbkreisförmige Rillen der Stahlplatten eingelegt waren, brachten zwar durchaus befriedigende Ergebnisse, aber auch den deutlichen Hinweis auf technische und wirtschaftliche Verbesserungsmöglichkeiten bei Verwendung rechteckiger Stäbe.

Als Ersatz für die noch fehlenden Rechteck-Vollstäbe wurden daher zunächst 2 bis 4 Rundstäbe mit ihren

Verankerungsenden in stählerne Schalformen gelegt und durch Verguß mit Kunstharzmörtel auf eine rechteckige Form gebracht. In Kurzzeit- und Zeitstandversuchen können so schon vorab die gute Eignung des Rechteckquerschnittes für einen Klemmanker bewiesen und der zulässige Bereich der Querpressungen abgeschätzt werden. Der demnächst zur Verfügung stehende Rechteck-Vollstab läßt wegen seiner Homogenität eher noch bessere Ergebnisse erwarten. Mit ihm wird dann die endgültige Optimierung des Keilklemmplattenankers in Kurzzeit-, Zeitstand- und Zugschwellversuchen durchgeführt werden. An der Materialprüfungsanstalt (MPA) in Stuttgart wurden inzwischen die Reibungswerte ermittelt, die für die Dimensionierung der die Querpressung erzeugenden Keile benötigt werden.

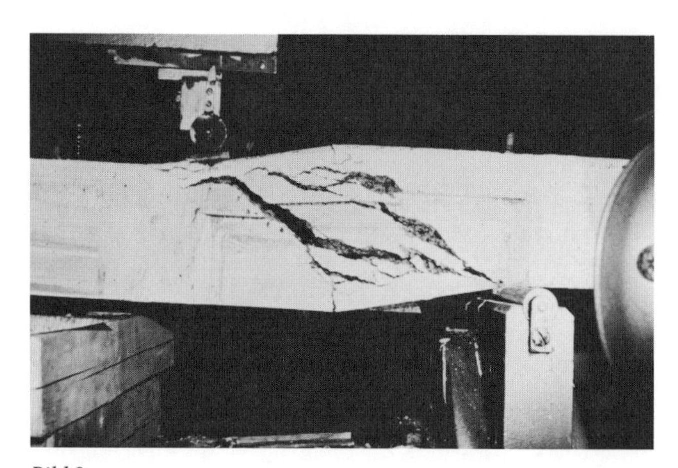

Bild 9
Schubbruchversuch an einem mit Polystal-Stäben vorgespannten Betonbalken

6 Praxisnahe Erprobung

6.1 Bauteilversuche

Die aus Detailversuchen an kleinen Prüfkörpern sowie aus Berechnungen gewonnenen Erkenntnisse über das Zusammenwirken von Vorspannung, schlaffer Bewehrung und Beton im Gebrauchs- und Bruchzustand müssen in sog. Bauteilversuchen an größeren Betonbalken oder -platten auf ihre Übereinstimmung mit der Wirklichkeit überprüft werden. Dabei unterscheidet man zwischen Bruch- und Brandversuchen.
Bei den Bruchversuchen interessieren insbesondere die Vorgänge im Verankerungsbereich, das Schubverhalten, die Bruchsicherheit bei Biegung, das Verhalten der Polystal-Stäbe im Bereich von Rissen (Reibung, Querpressung) sowie der Einfluß von Vorspanngrad und „nicht ruhender" Belastung. Die bei KORDINA geplanten Brandversuche sollen Aufschluß geben über Feuerwiderstandsdauer und Versagensart.
Die Bruchversuche werden von KÖNIG konzipiert und bei REHM durchgeführt. Zur ersten Absicherung seines neuen Bemessungskonzeptes hat KÖNIG bereits bei Strabag Biege- und Schubbruchversuche an kleineren vorgespannten Betonbalken durchführen lassen, die seine rechnerischen Ansätze voll bestätigt haben (Bild 9).

6.2 Probebauwerke

Forschungsvorhaben auf dem Bausektor enthalten, als unentbehrliches Bindeglied zwischen Entwicklung und

Bild 10
Antennenmast-Abspannung mit Polystal-Stäben. Fußpunkt einer Abspannseil-Gruppe mit Verankerungen und spiralförmigen Schwingungsdämpfern.

Anwendung ihrer neuen Technologie, stets auch die Durchführung von einem oder mehreren Probebauwerken. Diese werden – mit „bauaufsichtlicher Zustimmung im Einzelfall" – von unabhängigen Bauherren zur normalen Nutzung errichtet, ermöglichen jedoch darüber hinaus den Einbau und die laufende Überwachung der neu entwickelten Elemente unter den echten Bedingungen der praktischen Anwendung.
Bereits in den Jahren 1980 und 1981 wurden die beiden ersten Probebauwerke dieses Vorhabens errichtet: eine mit den elektromagnetisch neutralen Trag- und Abspannseilen aus Polystal ausgestattete Antennen-Anlage in Niedersachsen (Bild 10), und eine kleine, mit Polystal-Gliedern vorgespannte Straßenbrücke in Düsseldorf (Bild 11). Diese beiden Baumaßnahmen wurden in [7] ausführlich beschrieben.

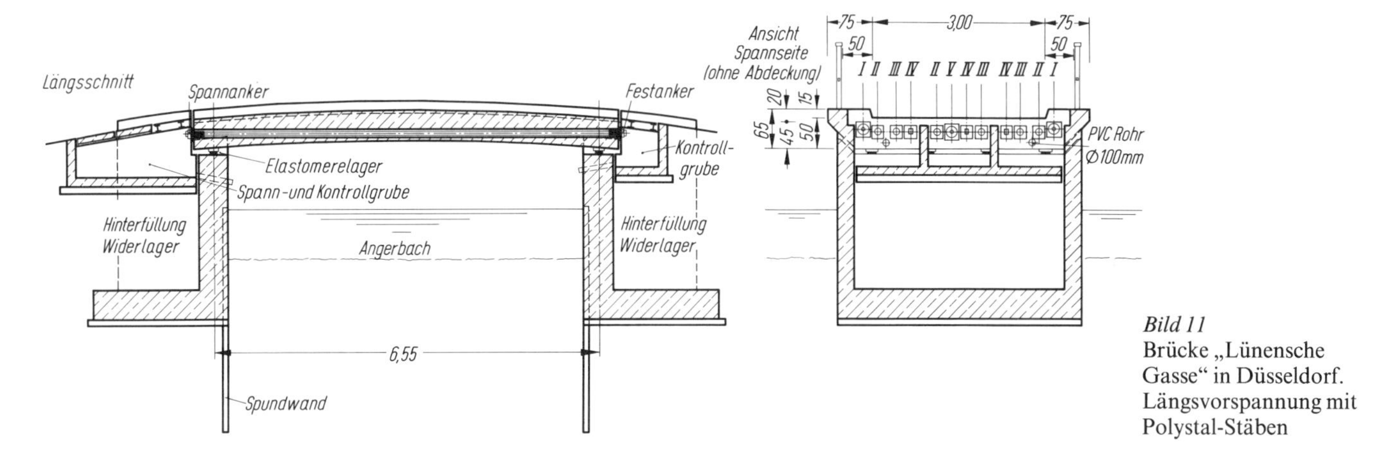

Bild 11
Brücke „Lünensche Gasse“ in Düsseldorf. Längsvorspannung mit Polystal-Stäben

Als drittes Probebauwerk wird eine 1986 zu erstellende Straßenbrücke geplant. Sie soll durch Polystal-Spannglieder mit Verguß- und Keilklemmplattenankern eine beschränkte Vorspannung erhalten. Zum Zwecke der Erprobung sind hier neben den Spanngliedern mit nachträglichem Verbund auch solche ohne Verbund vorgesehen.

Die Auswertung der an diesem Probebauwerk zu sammelnden Erfahrungen über den wirtschaftlichen Aufwand bei Herstellung und Einbau der Bewehrungselemente sowie die nur hier zu gewinnende baupraktische Bestätigung unserer Erkenntnisse über ihr Langzeitverhalten bilden sodann den letzten Schritt der im Rahmen dieses FE-Vorhabens geplanten Entwicklung der Polystal-Spannglieder bis zur Demonstrationsreife.

7 Literatur

[1] REHM, G., und FRANKE, L.: Kunstharzgebundene Glasfaserstäbe als Bewehrung im Betonbau. Forschungsbericht im Auftrage des Bundesministeriums für Städtebau und Wohnungswesen vom 30.6.1972.

[2] REHM, G., und FRANKE, L.: Kunstharzgebundene Glasfaserstäbe als Bewehrung im Betonbau, Stufe II. Forschungsbericht im Auftrage des Bundesministers für Raumordnung, Bauwesen und Städtebau vom 26.5.1975.

[3] REHM, G., und FRANKE, L.: Kunstharzgebundene Glasfaserstäbe als Bewehrung im Betonbau, Stufe III. Forschungsbericht im Auftrage des Bundesministers für Raumordnung, Bauwesen und Städtebau vom 7.11.1977.

[4] REHM, G., FRANKE, L., und PATZAK, M.: Untersuchungen zur Frage der Krafteinleitung in kunstharzgebundene Glasfaserstäbe. Forschungsbericht im Auftrage des Innenministers des Landes Nordrhein-Westfalen vom 8.5.1978.

[5] WEISER, M., und PREIS, L.: Glasfaser-Kunststoff-Elemente als Bewehrung im kommunalen Tiefbau. Tiefbau, Ingenieurbau, Sraßenbau 1980. H. 1, S. 14–18.

[6] WEISER, M., und PREIS, L.: Einsatz kunstharzgebundener Glasfaserstäbe als Bewehrung im Bauwesen. Bauwirtschaft 1982, H. 43, S. 1615–1621.

[7] WEISER, M.: Erste mit Glasfaser-Spanngliedern vorgespannte Betonbrücke. Beton- und Stahlbetonbau 1983, H. 2, S. 36–40.

[8] REHM, G., und FRANKE, L.: Persönliche Mitteilung an die Verfasser.

[9] REHM, G.: Glasfaser-Harz-Verbundstäbe als Bewehrung. Vortrag auf dem Betontag 1973, Deutscher Betonverein e.V.

[10] REHM, G.: GFK-Stäbe als Bewehrung. Betonwerk + Fertigteil-Technik 1973, H. 9.

[11] REHM, G., und FRANKE, L.: Kunstharzgebundene Glasfaserstäbe als Bewehrung im Betonbau. Die Bautechnik 1974, H. 4.

[12] REHM, G., und FRANKE, L.: Verhalten von kunstharzgebundenen Glasfaserstäben bei unterschiedlichen Beanspruchungszuständen. Die Bautechnik 1977, H. 4.

[13] REHM, G., und FRANKE, L.: Kunstharzgebundene Glasfaserstäbe als Bewehrung im Betonbau. Reihe „Deutscher Ausschuß für Stahlbeton“, Heft 304, Verlag Wilhelm Ernst & Sohn, Berlin 1979.

[14] REHM, G., und FRANKE, L.: Entwicklungstendenzen. VDI-Berichte Nr. 384 (1980), S. 133–140.

Werkstoff-Forschung, Dauerhaltbarkeit

Karbonatisierung und Dauerhaftigkeit von Beton

Hubert K. Hilsdorf, Jörg Kropp und Martin Günter

1 Einführung

Bei der Diskussion der Dauerhaftigkeit von Stahlbetonbauwerken wurde in der Vergangenheit die Karbonatisierung der Betonrandzonen fast ausschließlich unter dem Aspekt der Neutralisierung des Betons und der damit verbundenen Korrosionsgefahr für die äußeren Bewehrungslagen betrachtet.
Experimentelle Untersuchungen zeigten aber auch, daß die durch die Karbonatisierung des Zementsteines hervorgerufenen Veränderungen sich auf die Dauerhaftigkeit des Betons selbst auswirken können, indem die Betonrandzone verdichtet und damit das Eindringen von betonaggressiven und auch stahlaggressiven Medien behindert wird. Unter der Voraussetzung einer ausreichenden Betondeckung kann die Karbonatisierung somit auch einen Beitrag zum Korrosionsschutz der Bewehrung leisten.
Im folgenden sollen die Auswirkungen der Karbonatisierung auf jene Parameter des Betons diskutiert werden, die die Dauerhaftigkeit einer Stahlbetonkonstruktion maßgeblich beeinflussen. Abschließend wird über den Widerstand von karbonatisiertem Beton gegenüber chemischen und physikalischen Beanspruchungen berichtet. Dabei werden sowohl eigene als auch von anderen Forschern veröffentlichte Ergebnisse herangezogen.

2 Die Veränderung der Zementsteinstruktur durch Karbonatisierung

Das in der natürlichen Atmosphäre enthaltene Kohlendioxid dringt auf dem Weg der Diffusion in das Porensystem der Zementsteinmatrix ein und reagiert dort mit den alkalischen Hydratationsprodukten des Zementes. Nach STEINOUR [1] sind mit Ausnahme des Gipses alle kalziumhaltigen Hydratationsprodukte sowie die in geringen Mengen vorliegenden Hydroxide von Natrium und Kalium an der CO_2-Aufnahme beteiligt. Die Umlagerung der Hydratationsprodukte in Karbonate wird begleitet von einer Freisetzung chemisch gebundenen Wassers, die je nach zersetztem Hydratationsprodukt in unterschiedlich starkem Umfang abläuft. Im Falle des Portlandites $Ca(OH)_2$ steht der Aufnahme von 1 Mol CO_2 die Abgabe von 1 Mol H_2O gegenüber. Bei der Karbonatisierung von Zementstein aus Portlandzement wurde als Durchschnittswert jedoch eine Wasserabgabe von nur 0,4–0,6 Mol H_2O je aufgenommenem Mol CO_2 beobachtet, so daß die CO_2-Aufnahme die Wasserabgabe überwiegt [2].
Die Neubildung von Phasen führt neben der Veränderung der chemischen Zusammensetzung zu einem höheren Raumfüllungsgrad des Zementsteins. Der als stabile Endform von Kalziumkarbonat gebildete Kalzit nimmt rechnerisch gegenüber dem Ausgangsprodukt $Ca(OH)_2$ ein um 11% größeres Volumen ein [3]. Insbesondere jene Zementsteinphasen, die bei der Karbonatisierung kein chemisch gebundenes Wasser freisetzen, tragen zu einer weiteren Vergrößerung des Feststoffvolumens bei. In der Literatur wird über eine Reduktion der Gesamtporosität des Zementsteines infolge Karbonatisierung um ca. 20% berichtet [4], [5]. Die umfangreichen, an Zementen mit unterschiedlichem Hüttensandgehalt durchgeführten Untersuchungen von SMOLCZYK und ROMBERG [5] zeigen, daß sich diese Reduktion der Gesamtporosität weitgehend unabhängig vom Hüttensandgehalt des Zementes einstellt. Die Verringerung der Porosität ist jedoch um so deutlicher ausgeprägt, je geringer der Wasserzementwert der Proben.
Untersuchungen der Porenstruktur mit Hilfe der Quecksilberdruckporosimetrie zeigten, daß alle in der Zementsteinmatrix vorliegenden Porenräume, die durch diese Meßmethode erfaßt werden können, von der Karbonatisierung betroffen werden. An Mörtelproben, die nach einer einjährigen Vorlagerung unter Wasser karbonatisieren konnten, wurde gezeigt, daß

der Hüttensandgehalt des Zementes die als Folge einer Karbonatisierung eintretenden Strukturveränderungen beeinflußt [5].

Als Beispiel ist in Bild 1 die Porengrößenverteilung von Mörtelproben mit einem ω-Wert von 0,5 bzw. 0,7 aus einem Zement mit 75 M.% Hüttensand vor und nach der Karbonatisierung dargestellt [5]. Die Porosität ist dabei auf das Volumen des in der Probe enthaltenen Zementsteins umgerechnet. Aus Bild 1 folgt, daß über den ganzen Meßbereich des Porenradius $7{,}5 < r < 100$ nm die Karbonatisierung zu einer Erhöhung der Porosität führt. Die beobachtete Reduzierung der Gesamtporosität durch die Karbonatisierung muß daher auf eine Reduzierung der Gelporosität zurückzuführen sein.

Weitere Versuchsergebnisse, über die in [5] berichtet wird, zeigen, daß mit abnehmendem Hüttensandgehalt die Karbonatisierung nicht mehr zu einer Zunahme, sondern zu einer Abnahme der Porosität auch im Bereich $7{,}5 < r < 100$ nm führt. Auch der von PIHLAJAVAARA [6] beobachtete Abfall der spezifischen Oberfläche von karbonatisiertem Zementstein deutet an, daß die hochdisperse Gelstruktur im Zementstein durch die Karbonatisierung verändert wird.

In eigenen Versuchsreihen wurde mit unterschiedlichen Analysenmethoden die Zementsteinstruktur von karbonatisierten und nicht karbonatisierten Proben charakterisiert [7]. Die Zementsteinproben wurden aus einem Portlandzement PZ 45F und den Wasserzementwerten von $\omega = 0{,}5$ und $\omega = 0{,}65$ hergestellt. Um einen hohen Hydratationsgrad zu erzielen, wurden alle Proben zunächst bis zum Alter von 28 Tagen unter Wasser vorgelagert, ehe mit einer beschleunigten Karbonatisierung in einer Atmosphäre mit 2 Vol.% CO_2 bei 50% r. LF begonnen wurde. Parallelproben wurden in CO_2-freier Luft einer gleichwertigen Vorbehandlung unterzogen.

Die mit Hilfe der Wassersättigung bestimmte Gesamtporosität der Proben zeigte, daß durch die Karbonatisierung neuer Feststoff gebildet wird, der die Gesamtporosität des Zementsteins deutlich reduziert. Unter Berücksichtigung der veränderten Reindichte des karbonatisierten Zementsteines wurde eine Erhöhung des Feststoffvolumens um ca. 25% bei einem Karbonatisierungsgrad von $k \simeq 0{,}8$ berechnet. Entsprechend der Ausgangsporosität des Zementsteines wird dadurch die Gesamtporosität reduziert.

Messungen mit Hilfe der Quecksilberdruckporosimetrie zeigten, daß die neu gebildeten Karbonate bevorzugt in den Kapillarporen der Zementsteinmatrix abgelagert werden und damit zu einer Reduktion dieser Porenräume führen. Bild 2 zeigt die Verteilung der Porenräume über einen Meßbereich von 2,2 bis 1000 nm. Daraus folgt, daß die Verringerung der Poren mit Radien zwischen 10 und 100 nm am stärksten ausgeprägt ist.

Mit Hilfe der Stickstoffadsorption und der Röntgenkleinwinkelstreuung konnte nachgewiesen werden, daß auch die Mikrostruktur des Zementsteines durch die Karbonatisierung weitreichenden Veränderungen unterliegt.

Die aus der Stickstoffadsorption nach BET ermittelte spezifische Oberfläche des Zementsteines sinkt durch die Karbonatisierung auf ca. 20% des Ausgangswertes ab und zeigt damit die weitgehende Auflösung der

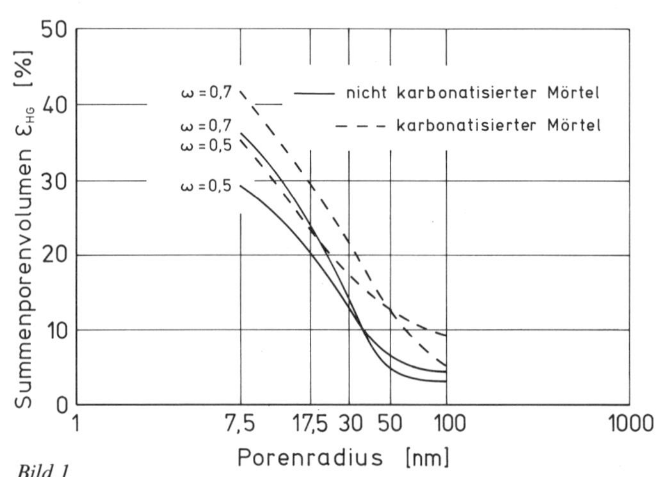

Bild 1
Porengrößenverteilung in der Zementsteinmatrix von Mörtel aus Hochofenzement mit 75 M% Hüttensand vor und nach der Karbonatisierung

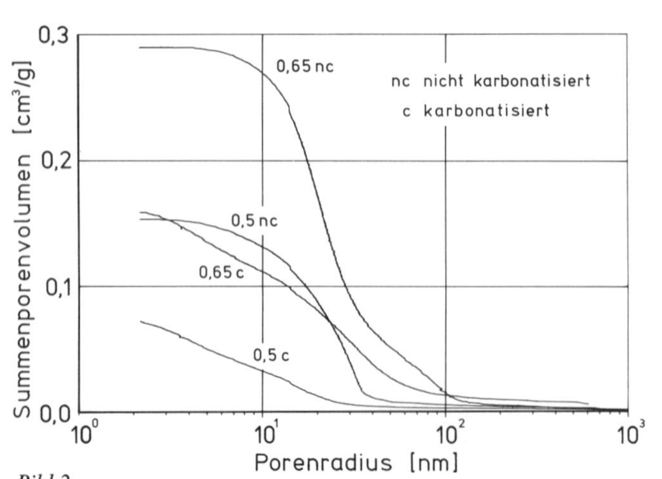

Bild 2
Porengrößenverteilung in Zementstein aus Portlandzement PZ 45 F vor und nach der Karbonatisierung

hochdispersen Gelstruktur an. Aus der mit Hilfe der BET-Methode errechneten Verteilung der Poren bis zu Radien von $r \sim 10$ nm, die in Bild 3 dargestellt ist, wird deutlich, daß die feinsten, im Zementstein vorliegenden Porenräume durch die Karbonatisierung aufgelöst werden.

Mit der Röntgenkleinwinkelstreuung (SAXS) kann im Gegensatz zur Stickstoffadsorption die spezifische Oberfläche des Zementsteines außer im trockenen Zustand auch im wassergesättigten Zustand ermittelt werden. Eine Schädigung der Zementsteinstruktur durch Probenpräparation wird damit vermieden. In Tabelle 1 sind zusammen mit der Gesamtporosität ε_0 und der spezifischen Oberfläche nach BET S_{N_2} die mit Hilfe der Röntgenkleinwinkelstreuung bestimmten spezifischen Oberflächen S_{SAXS} für trockenen und für wassergesättigten Zementstein angegeben. Die Meßergebnisse zeigen eine deutliche Abhängigkeit der spezifischen Oberfläche vom Wassergehalt der Probe. Der starke Anstieg der spezifischen Oberfläche der nicht karbonatisierten Proben bei Wassersättigung wird auf eine Aufweitung der Gelstruktur und Bildung neuer Grenzflächen durch eine Separation der Gelpartikel bei Wassereinlagerung zurückgeführt. Der Einfluß des Feuchtegehalts auf die spezifische Oberfläche ist bei den karbonatisierten Proben wesentlich geringer. Dies deutet darauf hin, daß die hochdisperse Gelstruktur des Zementsteins durch die Karbonatisierung teilweise aufgelöst wurde und daß die neu gebildeten kristallinen Karbonate zu einer hygrisch stabileren Struktur führen.

3 Der Wassertransport durch die Zementsteinmatrix

In der Literatur liegen nur wenige Aussagen über die Auswirkungen der Karbonatisierung auf den Transport von Wasser durch die Zementsteinmatrix vor. Gestützt auf die Verringerung der Porosität wird vielfach eine Behinderung des Stofftransportes durch die Zementsteinmatrix postuliert bzw. eine Verlangsamung des Feuchtetransportes aus der Beobachtung des Austrocknungsverlaufes von Probekörpern abgeleitet [8], [9]. In Laborversuchen an Zementmörtelproben $\omega = 0{,}45$–$0{,}6$, die in 100%iger CO_2-Atmosphäre karbonatisiert wurden, fanden Houst et al. [10] eine Erhöhung des Diffusionskoeffizienten für Wasserdampf als Folge der Karbonatisierung um bis zu 150%.

An Probekörpern, die aus den Randbereichen alter Bauwerksbetone entnommen wurden, konnte sowohl eine Erhöhung als auch eine Reduktion des Feuchtetransportes ermittelt werden [11], während VOLKWEIN [12] aus der deutlich höheren Kapillaraktivität des karbonatisierten Randzonenbereiches gegenüber dem nicht karbonatisierten Kernbeton auf eine Erhöhung der Durchlässigkeit durch die Karbonatisierung schloß. Bei der Betrachtung von Bauwerksbetonen ist jedoch zu berücksichtigen, daß die Randzonen von Betonbauteilen aufgrund der ungünstigeren Hydratationsbedingungen auch ohne Karbonatisierung eine höhere Durchlässigkeit aufweisen als die Kernbereiche [13].

In Laborversuchen wurde der Feuchtetransport durch die Zementsteinmatrix von Mörtelproben untersucht, indem die Diffusionskoeffizienten für Wasserdampf sowie die Permeabilität der Mörtel bei Druckwasser-

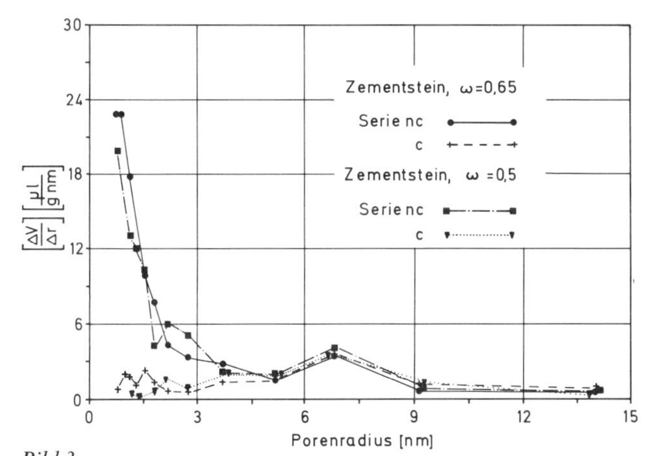

Bild 3
Porengrößenverteilung in Zementstein aus Portlandzement, ermittelt durch N_2-Adsorption

Tabelle 1
Porosität und spezifische Oberfläche von Zementstein aus Portlandzement

Zementsteinserie ω	Lagerung	Gesamtporosität ε_0 [%]	Spezifische Oberfläche [m²/g] BET	SAXS trocken	SAXS naß
0,5	nicht karbon.	41,9	52,6	160	450
0,5	karbonatisiert	28,1	10,2	90	160
0,65	nicht karbon.	50,2	69,5	125	510
0,65	karbonatisiert	39,1	17,7	95	200

beaufschlagung bestimmt wurden [7]. Die Mörtel wurden hergestellt mit einem PZ 45F und Quarzsand 0–2 mm. Das Mischungsverhältnis betrug 1 : 3 MT für den Wasserzementwert $\omega = 0{,}5$ bzw. 1 : 3,55 MT für den Wasserzementwert von $\omega = 0{,}65$. Die Vorlagerung der zylindrischen Mörtelproben $\varnothing 80 \times 150$ mm entsprach den in Abschnitt 2 dargestellten Lagerungsbedingungen der Zementsteinproben. Zur Bestimmung der Transportkoeffizienten wurden jeweils Scheiben von den karbonatisierten bzw. nicht karbonatisierten Stirnseiten der Zylinder abgetrennt und für die Untersuchungen herangezogen.

Die Diffusionskoeffizienten für Wasserdampf wurden in Anlehnung an DIN 52615 in 6 enggestaffelten Feuchteintervallen im Bereich von 31 bis ca. 98% rel. Luftfeuchte bestimmt. In den Diagrammen des Bildes 4 sind die gemessenen mittleren Diffusionskoeffizienten der einzelnen Feuchteintervalle als Balken dargestellt. Für Probenserien mit einem Wasserzementwert von $\omega = 0{,}5$ zeigten die Versuche, daß die Feuchtediffusion über den gesamten geprüften Feuchtebereich durch die Karbonatisierung deutlich behindert wird. An Probenserien mit dem höheren Wasserzementwert von $\omega = 0{,}65$ konnte eine Transportbehinderung bei karbonatisierten Proben nur bei mittleren und niedrigen relativen Feuchten beobachtet werden, während im Bereich hoher relativer Feuchte ein verstärkter Wassertransport auftrat. Diese Diffusionsversuche zeigten, daß die Porenraumverengung durch Karbonatisierung den Wassertransport in der Dampfphase behindert. Dieser Effekt wurde bei beiden Wasserzementwerten beobachtet. Bei hohem Feuchtegehalt der Proben gewinnt der Transport von flüssigem Wasser in den Kapillaren zunehmend an Bedeutung. Über eine Reduzierung der Porendurchmesser durch Karbonatablagerungen an den Porenwandungen wurde bei den Proben mit dem höheren ω-Wert dieser Transportmechanismus wegen der Verengung der Porendurchmesser verstärkt. Aufgrund der geringeren und diskontinuierlichen Kapillarporosität tritt bei Proben mit einem Wasserzementwert von $\omega = 0{,}5$ dieser Transportmechanismus wesentlich weniger auf.

Mit Hilfe der in die Diagramme von Bild 4 eingetragenen stetigen Funktionen wurde der experimentell bestimmte Trocknungsverlauf von Mörtelzylindern mit und ohne karbonatisierter Randschicht rechnerisch nachvollzogen. Sowohl die Trocknungsversuche wie

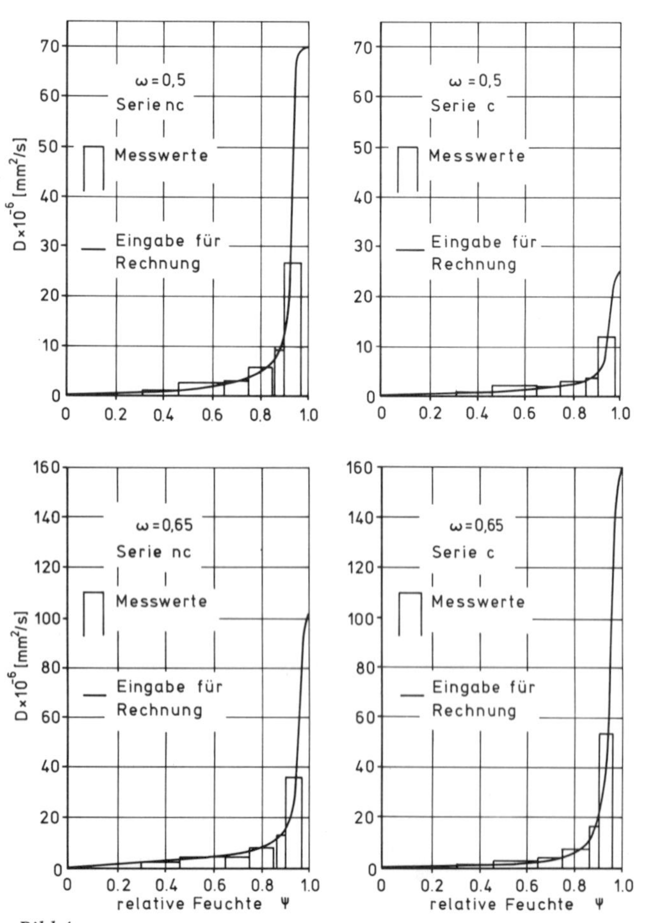

Bild 4
Diffusionskoeffizienten für den Feuchtetransport durch nicht karbonatisierte und karbonatisierte Zementmörtel

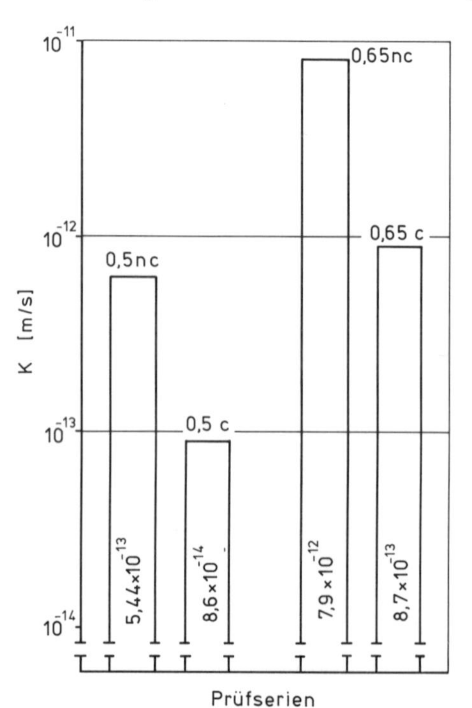

Bild 5
Permeabilitätskoeffizienten für Zementmörtel

auch die Berechnungen zeigten, daß eine karbonatisierte Randschicht den Trocknungsverlauf verzögert. Für Proben mit einem Wasserzementwert von $\omega =$ 0,65 tritt diese Verzögerung jedoch erst für Trocknungszeiten $t > 28$ Tage auf.
Auch die im Permeationsversuch vorliegende gesättigte Porenströmung wird durch die Karbonatisierung der Zementsteinmatrix der Mörtel reduziert. In Bild 5 sind die erzielten Ergebnisse grafisch dargestellt. Der Vergleich der einzelnen Prüfserien zeigt, daß die Strukturverdichtung der Karbonatisierung in ihrer Auswirkung auf die Permeabilität einer Reduktion des Wasserzementwertes entspricht. Der Permeabilitätskoeffizient der nicht karbonatisierten Mörtel mit einem Wasserzementwert von $\omega = 0{,}65$ wurde durch die Karbonatisierung auf das Niveau der Mörtel mit einem Wasserzementwert von $\omega = 0{,}5$ abgesenkt.

4 Die Beeinflussung der mechanischen Eigenschaften zementgebundener Werkstoffe durch die Karbonatisierung

In zahlreichen, in der Literatur veröffentlichten Untersuchungen wurde i. A. als positives Resultat der Karbonatisierung eine Steigerung der Druckfestigkeit von Zementstein- oder Zementmörtelproben auf Werte bis zu 210% der Ausgangsfestigkeit ermittelt. In systematischen Versuchsreihen über das Verhalten von Mörteln aus verschiedenen Zementen zeigte MANNS [2] jedoch, daß ein ausgeprägter Einfluß der Zementart auf die erzielbare Festigkeitssteigerung durch die Karbonatisierung existiert. Während bei Proben aus Zementen mit hohem Klinkergehalt eine Karbonatisierung zu bedeutenden Festigkeitssteigerungen führte, wurde an einem Sulfathüttenzement mit einem Klinkergehalt von 3 M% ein Festigkeitsabfall beobachtet. Auch SMOLCZYK und ROMBERG untersuchten den Einfluß des Hüttensandgehaltes und der Karbonatisierung auf die Druckfestigkeit von Mörtelproben [5]. In Bild 6 sind die von MANNS [2] und die von SMOLCZYK erzielten Ergebnisse zusammengefaßt. Dabei wurde die nach der Karbonatisierung beobachtete Druckfestigkeit β_{Dc} auf die Druckfestigkeit nicht karbonatisierter Proben β_{Dnc} bezogen und in Abhängigkeit vom Hüttensandgehalt der verwendeten Zemente aufgetragen. Als Bezugsgröße β_{Dnc} diente bei den Versuchsergebnissen von MANNS die 28-Tage-Festigkeit der nicht karbonatisierten Mörtel, während die Ergebnisse von

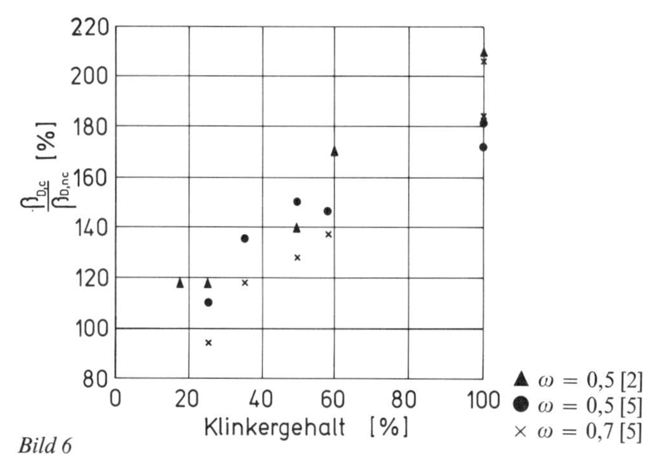

Bild 6
Relative Druckfestigkeit karbonatisierter Mörtel aus Portland- und Hochofenzement in Abhängigkeit vom Klinkergehalt des Zementes nach Ergebnissen von MANNS und SCHATZ [2] und SMOLCZYK und ROMBERG [5]

SMOLCZYK auf die Festigkeit von 1 Jahr unter Wasser gelagerten Proben bezogen wurden.
Auch für die Biegezugfestigkeit wurde eine der Druckfestigkeitsentwicklung vergleichbare Abhängigkeit gefunden. Die Festigkeitssteigerungen konnten jedoch nur an vollständig karbonatisierten Proben beobachtet werden, während Proben, die nur in den Randbereichen karbonatisiert waren, Festigkeitsminderungen aufwiesen [2]. Dieses Verhalten ist auf Schwindspannungen in den Randzonen zurückzuführen, die durch die Karbonatisierung der Randzonen verstärkt werden [8].
An den von uns untersuchten und in Abschnitt 3 beschriebenen Mörteln aus Portlandzement wurde auch die Oberflächenhärte von karbonatisierten und nicht karbonatisierten Proben qualitativ beurteilt. Dazu wurde an den plangeschliffenen Stirnflächen der Mörtelzylinder eine Härteprüfung nach BRINELL in Anlehnung an DIN 50351 durchgeführt und der Durchmesser des Kugeleindruckes bestimmt. Bei einem Kugeldurchmesser von 2,5 mm wurde der Durchmesser des Kugeleindruckes von 1,3 mm an nicht karbonatisiertem Mörtel, $\omega = 0{,}5$, durch die Karbonatisierung auf 1,0 mm reduziert. Für den höheren Wasserzementwert, $\omega = 0{,}65$, stellte sich eine Verringerung von 1,4 mm auf 1,0 mm ein.
Aus der Reduktion der Durchmesser des Kugeleindruckes von ca. 25% ist die Erhöhung der Oberflächenhärte der Mörtel als Folge der Karbonatisierung abzuleiten. Obwohl experimentell nicht überprüft, ist damit eine Erhöhung des Abriebwiderstandes durch die Karbonatisierung zu erwarten.

5 Die Dauerhaftigkeit der karbonatisierten Randzone

Die in der Zementsteinmatrix durch die Karbonatisierung vollzogenen Umwandlungen bewirken neben einer Reduzierung des charakteristischen Porenraumes auch die Überführung von chemisch reaktiven Hydratationsprodukten in stabilere Karbonate. Die Einwirkung von aggressiven Substanzen auf einen karbonatisierten Beton oder Mörtel kann somit in zweifacher Weise beeinflußt werden:

- Die Reduzierung des Porenraumes wirkt sich verlangsamend auf Transportvorgänge aus, so daß das Eindringen von aggressiven Stoffen behindert wird.
- Die in die Matrix eingedrungenen Stoffe können nicht mit den Karbonatisierungsprodukten reagieren.

Der Einfluß der Karbonatisierung auf die Dauerhaftigkeit zementgebundener Werkstoffe wurde in der Vergangenheit anhand des Eindringens von Chloriden in die Zementsteinmatrix sowie des Angriffes verschiedener Sulfate untersucht. Während sich die Aggressivität der Chloride vornehmlich auf die Korrosion der Bewehrung im Stahlbetonbau bezieht, führt die Reaktion von Sulfaten mit Kalziumhydroxid oder Kalziumaluminatphasen zu einer Zerstörung der Zementsteinmatrix selbst durch die Bildung expansiver Phasen. Sowohl bei der Einlagerung vorkarbonatisierter Proben in Meerwasser [14] als auch bei der Einwirkung einer aufstehenden Salzlösung [15] wurde beobachtet, daß die über die Eindringtiefe ermittelte Cl^--Konzentration in den karbonatisierten Probenbereichen stets deutlich geringer war als bei nicht karbonatisierter Zementsteinmatrix. Durch die Reduktion des Porenraumes wurde das Eindringen der stahlaggressiven Ionen behindert. Bei bereits chloridverseuchtem Beton verringert aber die Karbonatisierung der Hydratationsprodukte die Bindekapazität einzelner Hydratphasen für Chloride, so daß der Anteil des wasserlöslichen Chlorides am Gesamtchloridgehalt zunimmt [12].

In umfangreichen Untersuchungen wurde nachgewiesen, daß karbonatisierte Zementstein- und Zementmörtelproben einen erhöhten Widerstand gegen angreifende Sulfatlösungen aufweisen [16]. Das günstigere Verhalten der karbonatisierten Proben wird zurückgeführt auf die Umwandlung der reaktiven Zementsteinphasen in chemisch stabilere Karbonate. Da durch die vollständige Karbonatisierung die im Zementstein enthaltenen Kalziumaluminathydrate zersetzt wurden, ist bei Sulfateinwirkung die Bildung von Ettringit weitgehend ausgeschlossen.

An Proben mit karbonatisierter Randschicht, die in Sulfatlösungen gelagert waren, wurde Sulfattreiben erst mit einer erheblichen Zeitverzögerung gegenüber den nicht karbonatisierten Proben beobachtet. Diese Feststellungen deuten auf eine Transportbehinderung der angreifenden Stoffe durch die karbonatisierte Randschicht hin [16].

In eigenen experimentellen Untersuchungen wurde der Einfluß der Karbonatisierung auf den Widerstand von Beton gegen Frost-Tausalzbeanspruchung untersucht [17]. Zur Herstellung der Probewürfel $100 \times 100 \times 100$ mm^3 wurden zwei verschiedene Zemente, ein Portlandzement PZ 35F und ein Hochofenzement HOZ 35L NWHS mit einem Hüttensandgehalt von ca. 75 M% herangezogen. Der Zuschlag bestand aus Rheinkies entsprechend der Sieblinie A/B 16 bei einem Mischungsverhältnis von 1 : 5, 26 : 0,5 MT. Durch Zugabe eines Luftporenbildners wurde ein Luftgehalt von 4,5% im Frischbeton erzeugt.

Nach dem Ausschalen wurden die Betonwürfel zunächst einer 6- bzw. 48tägigen Wasserlagerung unterzogen und anschließend in Luft bei 65% relativer Feuchte und unterschiedlichen CO_2-Gehalten gelagert:

Serie 1: Lagerung in CO_2-freier Luft
Serie 2: Lagerung in normaler Laborluft (ca. 0,03 Vol.% CO_2)
Serie 3: Lagerung in Luft mit 2 Vol.% CO_2

Im Anschluß an diese Vorlagerung wurden die Probekörper entsprechend dem vom DBV vorgeschlagenen „Verfahren zur Prüfung des Frost- und Tausalzwiderstandes von Beton für Brückenkappen und ähnliche Bauteile“ [18] beansprucht.

Bild 7a zeigt die Ergebnisse der Frost-Tausalzprüfung von Proben, die nach einer 6tägigen Wasserlagerung 49 Tage an Luft gelagert wurden. In Bild 7a ist die nach 70 Frost-Tauwechseln in gesättigter NaCl-Lösung beobachtete Abwitterung in Abhängigkeit vom CO_2-Gehalt der Luft während der vorangegangenen Trocknung aufgetragen. Bild 7b zeigt die Abwitterung von Proben nach 35 Frost-Tauwechseln, die vorher 48 Tage in Wasser und anschließend 42 Tage an Luft gelagert waren. In Bild 7 sind auch die beobachteten Karbonatisierungstiefen angegeben. Bei den Proben mit karbonatisierten Randzonen war in keinem Fall die Abwitterung bis in den nicht karbonatisierten Bereich vorgedrungen.

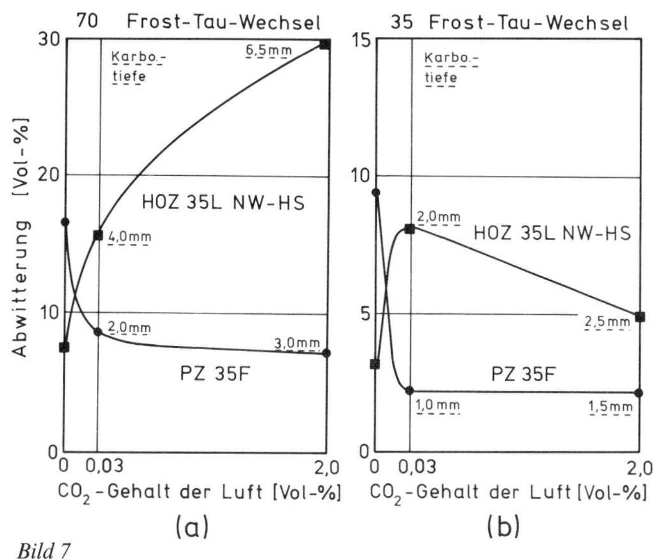

Bild 7
Abwitterung von Betonen aus Portlandzement und Hochofenzement durch Frost-Tausalzbeanspruchung, in Abhängigkeit vom CO_2-Gehalt der Luft während der Vorlagerung

Die erzielten Versuchsergebnisse zeigen, daß sowohl nach einer 6tägigen als auch nach einer 48tägigen Wasserlagerung der in CO_2-freier Luft gelagerte Beton aus Hochofenzement eine geringere Abwitterung aufweist als die entsprechenden Betone aus Portlandzement. Durch die Karbonatisierung der Randzonen, die während der Lagerung in CO_2-haltiger Luft einsetzen konnte, wurde die Abwitterung der Portlandzementbetone jedoch reduziert, während für Betone aus Hochofenzement eine Zunahme der Abwitterung beobachtet wurde. Diese Beeinflussung des Frost-Tausalzwiderstandes von Hochofenzementbetonen zeigte sich besonders deutlich für die Probenserien, die einer nur 6tägigen Wasserlagerung unterzogen wurden.
Der günstige Einfluß der Karbonatisierung auf den Frost-Tausalzwiderstand von Portlandzementbetonen kann durch die aufgetretene Strukturverdichtung der Zementsteinmatrix erklärt werden. Neben einer Erhöhung der Festigkeit führt die Verringerung des verfügbaren Porenraumes zu einer Eindringbehinderung der Salzlösung in die Randzone sowie zu einer Reduktion der Menge an gefrierbarem Wasser im Porensystem.
Diese positiven Einflüsse der Karbonatisierung sind offensichtlich bei Hochofenzementen mit hohem Hüttensandgehalt nicht gegeben. Die in Abschnitt 2 dargestellte Erhöhung der Kapillarporosität hüttensandreicher Mörtel durch Karbonatisierung erleichtert das Eindringen von Tausalzen und bewirkt eine Erhöhung der Menge an gefrierbarem Wasser im Vergleich zu nicht karbonatisierten Proben. Der experimentelle Nachweis dieser Hypothese steht allerdings noch aus. Von besonderer baupraktischer Relevanz ist die Beobachtung, daß eine lange Nachbehandlung bzw. ein hoher Hydratationsgrad die negativen Auswirkungen einer Karbonatisierung von Betonen aus hüttensandreichem Hochofenzement deutlich reduziert. Ein hoher Hydratationsgrad kann auch bei Wechsellagerung feucht/trocken oder bei Lagerung in Luft mit einer rel. Feuchte $> 80\%$ erreicht werden [17].

6 Zusammenfassung

6.1 Eine Karbonatisierung von Beton führt zu einer Reduktion des Kapillarporenvolumens und der inneren Oberfläche von Zementstein, die um so deutlicher ist, je geringer der ω-Wert und je geringer der Hüttensandgehalt des verwendeten Zementes. Bei hochgeschlackten HOZ-Zementen führt die Karbonatisierung zu einer Erhöhung der Kapillarporosität.

6.2 Als Folge dieser Strukturveränderungen kann die Karbonatisierung den Diffusionswiderstand von Beton aus Portlandzement erhöhen und seine Permeabilität gegen Wasser reduzieren.

6.3 Entsprechend bewirkt die Karbonatisierung eine Erhöhung der Druckfestigkeit und der Oberflächenhärte von Portlandzementbetonen.

6.4 Eine Karbonatisierung kann den Widerstand von Beton gegen chemische Angriffe erhöhen, und zwar durch Reduktion der Eindringgeschwindigkeit aggressiver Substanzen und durch Umwandlung der Hydratationsprodukte in chemisch stabilere Verbindungen. Eine Karbonatisierung reduziert jedoch die Bindekapazität der Hydratationsprodukte für Chloride.

6.5 Karbonatisierte Betone aus Portlandzement weisen einen höheren Frost-Tausalzwiderstand auf als nicht karbonatisierte Betone. Bei Proben aus hüttensandreichem Hochofenzement verringert die Karbonatisierung den Frost-Tausalzwiderstand. Auch diese Einflüsse können auf die beobachteten Strukturveränderungen als Folge der Karbonatisierung zurückgeführt werden.

7 Literatur

[1] Steinour, H. H.: Some effects of carbon dioxide on mortars and concrete. Journal of the ACI, 2, 1959.

[2] Manns, W., Schatz, O.: Über die Beeinflussung der Festigkeit von Zementmörteln durch die Karbonatisierung. Betonsteinzeitung, Heft 4, 1967.

[3] Smolczyk, H. G.: Physical and chemical phenomena of carbonation. Rilem International Symposium on Carbonation of Concrete, Fulmer Grange, 1976.

[4] Fernandez-Pena, O.: Action du CO_2 sur pates pures et sur mortier (1 : 2) élaborés avec differents ciments espagnols. Rilem International Symposium on Carbonation of Concrete, Fulmer Grange, 1976.

[5] Smolczyk, H. G., Romberg, H.: Der Einfluß der Nachbehandlung und der Lagerung auf die Nacherhärtung und Porenverteilung von Beton. Tonindustrie-Zeitung, 100, Heft 10, 11, 1976.

[6] Pihlajavaara, S. E.: Some results of the effect of carbonation on the porosity and pore size distribution of cement paste. Materiaux et Constructions, Nr. 6, 1968.

[7] Kropp, J.: Karbonatisierung und Transportvorgänge in Zementstein. Dissertation, Universität Karlsruhe, 1983.

[8] Verbeck, G. J.: Carbonation of hydrated portland cement. ASTM STP, Nr. 205, 1958.

[9] Hilsdorf, H. K.: Austrocknung und Schwinden von Beton. Stahlbetonbau – Berichte aus Forschung und Praxis – Festschrift Rüsch, Verlag Wilhelm Ernst und Sohn, 1969.

[10] Houst, Y. F., Roelfstra, P. E., Wittmann, F. H.: Ein Modell zur Vorhersage der Nutzungsdauer von Betonkonstruktionen. Berichtsband Internationales Kolloquium Werkstoffwissenschaften und Bausanierung, 6.–9. Sept. 1983, Esslingen.

[11] Meyer, A., Wierig, H. J., Husman, K.: Karbonatisierung von Schwerbeton. DAfStb, Heft 182, 1967.

[12] Volkwein, A., Springenschmid, R.: Corrosion of reinforcement in concrete bridges at different age due to carbonation and chloride penetration. Proc. Second International Conference on the Durability of Building Materials and Components, 14.–16. 9. 1983, Maryland.

[13] Pfeifer, D. W.: Steel corrosion damage and useful evaluation tests. Concrete Construction, Feb. 1981.

[14] Regourd, M.: Carbonatation accelerée et resistance des ciments aux eaux agressives. Rilem International Symposium on Carbonation of Concrete, Fulmer Grange, 1976.

[15] Rauen, A.: Chlorideindringtiefen in karbonatisiertem Beton. Bericht Nr. 45/17 zum FuE Vorhaben „Spätschäden an Spannbetonbauteilen“, BMFT, FKZ: Bau 7006.

[16] Arber, M. G., Vivian, H. E.: Carbonation as a means of inhibiting sulphate attack in mortar and concrete. Australien Journal of Applied Science, Vol. 12, No. 3.

[17] Günter, M., Hilsdorf, H. K.: Einfluß der Nachbehandlung auf die Widerstandsfähigkeit von Betonoberflächen. Schlußbericht zum Forschungsauftrag DBV-Nr. 88, 1983, Institut für Massivbau und Baustofftechnologie, Abt. Baustofftechnologie, Universität Karlsruhe (TH).

[18] Verfahren zur Prüfung des Frost- und Tausalzwiderstandes von Beton für Brückenkappen und ähnliche Bauteile. Betonwerk und Fertigteiltechnik, Heft 1, 1976.

Auffinden, Beurteilung und Folgen geschädigter Bewehrung bei zu erhaltenden Bauwerken

Dieter Jungwirth

1 Einführung

Staatsverschuldung, Mittelknappheit, Ausbau-Sättigung lassen Neubautrends zurückgehen. Die Stimmen nach Bauwerks- bzw. Substanzerhaltung werden lauter; es wird gar zur Mode, sich dieser Aufgaben anzunehmen. Mit etwa 15% Anteil am Bruttosozialprodukt, d. h. mit 250 Mrd. DM/Jahr, hat das Baugewerbe einschließlich Nebenbetrieben in den letzten 20 Jahren ca. 5 Bill. DM Bausubstanz geschaffen, die es zu erhalten gilt. Betonkosmetik und Versiegelung wird vorgenommen; Abplatzungen und Risse des Betons werden behoben.

Wesentliches zur Standsicherheit trägt jedoch die Bewehrung bei. Obgleich der Gesamtkostenanteil der Spannbewehrung und der schlaffen Bewehrung an der Gesamtbausubstanz nur ~9% (Tabelle 1) beträgt, steht und fällt das Langzeitverhalten mit der Zuverlässigkeit der Bewehrung. Die mit der Bewehrung verbundenen Aufgaben der Bauwerkserhaltung sind im Schrifttum untervertreten. Wenn vom Betonstahl, Verbundproblemen, Korrosionsverhalten u.a. die Rede ist, fehlt der Name Rehm nicht. Es ist daher naheliegend, in dieser Festschrift all die Erkenntnisse im Bereich der Bauwerkserhaltung zusammenzutragen, die mit Bewehrung in Verbindung stehen.

Bevor jedoch vom Erkennen, Beurteilen und Beheben von Bewehrungsschäden die Rede ist, bedarf es ein paar genereller Anmerkungen zur Schadenssituation [1]:

- Der Bauboom nach dem Zweiten Weltkrieg, Zeitdruck und Vergabe nach Billigstangeboten ließen die Ausführungsqualität stellenweise zu kurz kommen. Aber auch Fehler im Entwurfsbüro, fehlende Kenntnisse im Tragverhalten (z. B. Koppelfuge), technologisch nicht abschließend geklärte Fragen (z. B. Einpreßtechnik) und mangelhafte Bauwerksüberwachung bzw. unzulässige Änderung der Einwirkungen während der Nutzung sind zu verzeichnen, so daß alle am Bau Beteiligten sich angesprochen fühlen müssen.
- Offenen Erfahrungsaustausch zu betreiben, um – wegen der Schläue des Materials – verzögert in Erscheinung tretende Mängel baldmöglichst zu erkennen und zu beheben, ist der derzeit sicher richtige Weg.

Tabelle 1
Aufteilung der Rohbaukosten in %

	Allg. Hochbau	Großbrücken	U-Bahn	Kraftwerke	Mittel
Entwurf	5	2	2	2	3
Technische Bearbeitung	3	4	3	2	3
Lohn	43	35	36	42	39
Material ohne Bewehrung	24	29	28	27	27
Bewehrung	7	14	9	7	9
Geräte	8	6	12	10	9
Gemeinkosten	10	10	10	10	10
	100	100	100	100	100

- Ganz ohne Fehler wird es auch in Zukunft nicht gehen. Fehler zeigen die Grenzen des Machbaren. Ein Gesamtkostenminimum muß angegangen werden, das sich grob aus Herstellkosten, Risiko (möglichst abgedeckt durch Versicherungen) und Unterhaltungskosten ergibt.
- Grobe Schäden sind sehr selten. Kleinere bis mittlere Mängel treten häufiger auf. Sie gilt es rechtzeitig zu erkennen, und zu beheben, um langfristig daraus entstehende Schäden zu vermeiden. Am häufigsten auftretende Mängel sind ungenügende Betondeckung und nicht sachgemäß verpreßte Spannglieder.

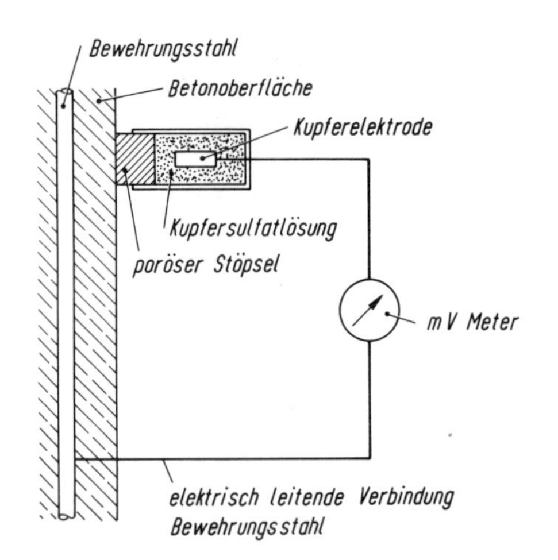

Bild 1
Prinzip der Potentialmessung am Stahlbeton [3]

2 Auffinden von Stahlschädigungen [2], [3]

2.1 Äußerlich wahrnehmbare Schäden

Betonabplatzungen und Risse, verursacht durch Volumenvergrößerung der Korrosionsprodukte, zeigen Stellen, an denen z. B. wegen mangelnder Betondeckung oder karbonatisierter poröser Zonen der Stahl rostet. Abplatzungen können aber auch durch Frostschäden in nichtverpreßten Spannkanälen entstehen und künden so Mängel an.

2.2 Auffinden der Bewehrung mit verschiedenen Hilfen, Beurteilung

Mittels Bewehrungssuchgeräten, die nach dem Prinzip der Änderung eines elektromagnetischen Wechselfeldes arbeiten, wird der Stahl gefunden. Es lassen sich Überdeckung und Durchmesser des Stahles feststellen. Die Reichweite geht nur bis 100 mm (Tiefensonden bis 150 mm). Bei mehrlagiger Bewehrung und engem Abstand treten schon vorher Meßprobleme auf. Aber auch über Durchstrahlung (siehe Abschn. 2.3.3) und thermographische Untersuchung [4] (Tiefenwirkung bis 100 mm) läßt sich Bewehrung orten. Schließlich kann ggf. über Einmessen (Planhilfe) ebenfalls Bewehrung gefunden werden.
Nun kann im nächsten Schritt, durch Freistemmen, Freilegen und Anbohren der Zustand der Bewehrung beurteilt werden. Im Zuge dieser Tätigkeit ist das Umfeld auf Auftretenswahrscheinlichkeit von Korrosion zu überprüfen. Gemeint ist die Beurteilung der die Korrosion (Voraussetzung: Elektrolyt, Sauerstoff, Potential) verursachenden probabilistischen Faktoren wie Porigkeit, Chloride u. ä., Feuchtigkeit, Karbonatisierung (mittels Phenolphthaleinlösung feststellbar). Zerstörungsfrei kann auch über Potentialmessungen (Bild 1) auf Stahlkorrosion geschlossen werden. Hier ist differenziertes Wissen erforderlich, das nur wenige Fachgruppen besitzen [5], [6].
Es ist stets fallweise zu prüfen, welches Verfahren am geeignetsten ist. Häufig müssen sich mehrere Verfahren ergänzen. Stets ist jedoch bei solchen Untersuchungen nach der Methode der kleinen Schritte vorzugehen. Wird man – von Stichproben ausgehend – fündig, wird das Untersuchungsnetz verdichtet. Schließlich soll nicht noch mehr Schaden angerichtet werden durch groß angesetzte Erhebungsbohrungen.

2.3 Mangelhaft verpreßte Spannkanäle

2.3.1 Allgemeines

Unausgereifte Technologien, ungeeignete Zemente und nicht sorgfältige Ausführung führten zu unterschiedlich gut verpreßten Spannkanälen. Um diese Fehlstellen zu erkennen, gibt es verschiedene Methoden, die, sich gegenseitig ergänzend, auch in Kombination eingesetzt werden können. Auch hier gilt das Vorgehen nach der Methode der kleinen Schritte.

2.3.2 Ultraschallprüfung [7], [8]

Mit diesem Verfahren können Spannstahlbrüche aufgefunden, ihre Lage exakt gemessen und außerdem qualitativ der Injektionszustand über den Energieverlust des Ultraschallimpulses festgestellt werden. Es gibt dafür zwei Meßmethoden, das Reflexionsverfahren und das Durchschallungsverfahren (Bild 2). Die optimale Prüffrequenz liegt bei 2 MHz. Für die Praxis eignet sich das Reflexionsverfahren besser, da die Messung nur von einem Spannstahlende ausgeht.

Die Grenzen der Anwendung sind:
nur bei glattem Stahl anwendbar
maximale Stablänge bei geraden Stäben < 20 m
bei stark gekrümmten Stäben < 8 m

Der kleinste, feststellbare Verpreßfehler, der jedoch nicht lokalisiert werden kann, beträgt 0,5 bis 1 m (Bild 3), ggf. sind Eichmessungen an Versuchsstäben durchzuführen.

Die Anwendung der Ultraschallprüfung erfordert hohe Vertrautheit mit dem Verfahren. Fehler in Höhe der Meßgenauigkeit sind sonst nicht selten. Allein die Schallübertragung vom Sender auf die Stirnseite des Stahles oder die Beschaffenheit des reflektierenden Endes können beachtliche Streuungen verursachen.

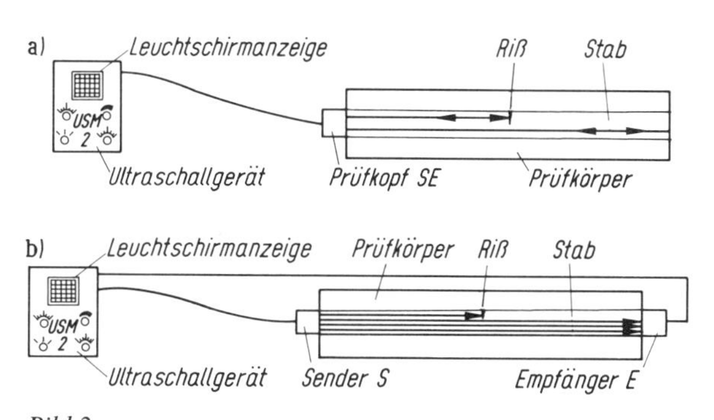

Bild 2
Ultraschallverfahren. a) Reflexionsverfahren, b) Durchschallverfahren

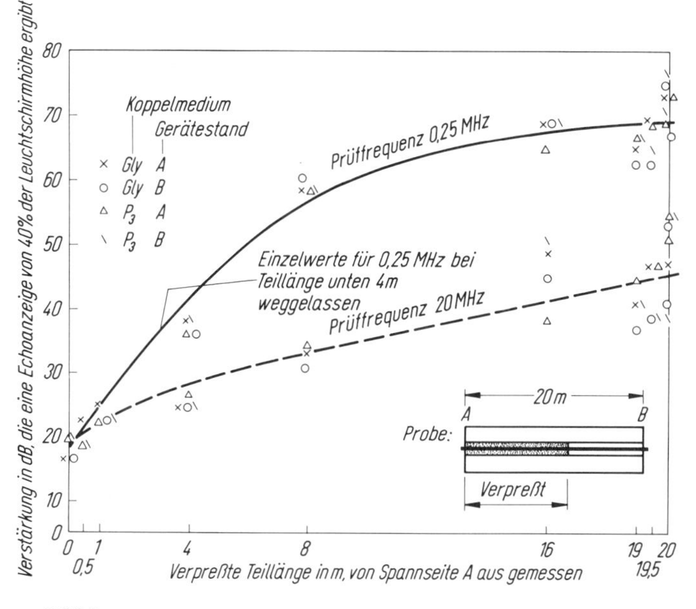

Bild 3
Ergebnis der Ultraschallprüfung

2.3.3 Durchstrahlungsprüfung (Gammagraphie)

Zur Feststellung von Verpreßmängeln und Spannstahlbrüchen kann das Bauwerk auch mit Gammastrahlen durchstrahlt werden. Die Verpreßmängel zeichnen sich durch unterschiedliche Helligkeitsgrade auf dem Film ab (Bild 4).

Filmqualität und Art der Strahlungsquelle beeinflussen die Belichtungszeit und die Bildqualität, d. h. das Auflösungsvermögen [8], [9]. Neben eigenen Untersuchungen und Baustellenauswertungen liegen seit kurzem auch Ergebnisse vom Betriebsforschungsinstitut Düsseldorf vor, welches die Möglichkeiten einer Optimierung dieser Prüftechnik untersuchte.

Die Leistungsgrenze bei einer Belichtungszeit von 60 Min. liegt zur Zeit für die (Bild 5)

Strahlungsquelle Ir 192 bei 30 cm Betondicke
und mit Strahlung Co 90 bei 60 cm Betondicke

Nachteilig sind außerdem die hohe Strahlungsgefahr (die Baustelle muß geräumt werden) und die rel. hohen Kosten bei kleinem Bildausschnitt. In einer Ebene liegende Spannglieder erschweren die Deutung.

Neuerdings wird mit beschleunigter Strahlung gearbeitet, was eine beachtliche Erweiterung des bisherigen Anwendungsbereiches erwarten läßt.

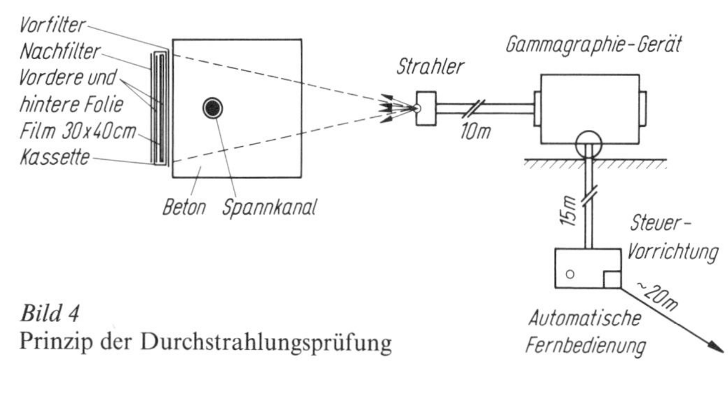

Bild 4
Prinzip der Durchstrahlungsprüfung

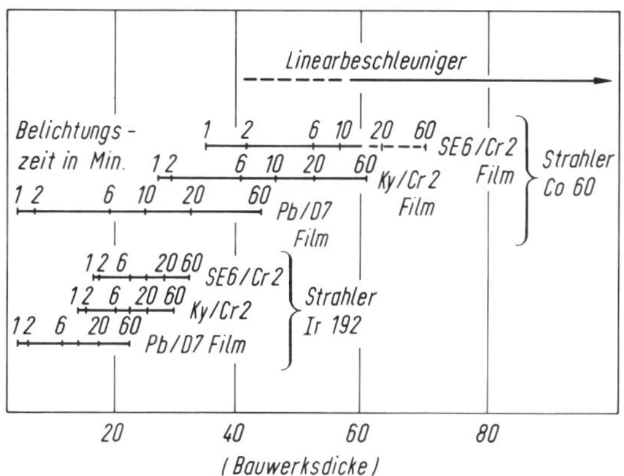

Bild 5
Leistungsfähigkeit der Durchstrahlungsprüfung

2.3.4 Anbohren des Spannkanals

Sprechen die örtlichen Verhältnisse gegen die Anwendung des Ultraschall- bzw. Durchstrahlungsverfahrens oder soll der Prüfaufwand gering gehalten werden, so kann schonend angebohrt werden, möglichst an Schwachstellen des Systems z. B. Hochpunkte, Koppelstellen usw. Die Bohrung dient gleichzeitig einer ggf. erforderlichen Sanierung, die dem Öffnen zur Vermeidung weiterer Korrosion rasch folgen sollte.
Für diese Untersuchungsvariante wurden geeignete Geräte (langsame Bohrgeschwindigkeit, Spezialbohrkopf, geringe Schlagkraft und Absaugen des Bohrstaubes waren die Kriterien) und günstigste, d. h. spannstahlschonende Bohrtechniken entwickelt. Weitere Verbesserungen, wie z. B. Abschaltautomaten, werden angeboten bzw. verfolgt (Bild 6).
Zur Erleichterung der Begutachtung wurde das aus der Medizin bekannte Endoskop (Bild 7) auf seine Brauchbarkeit geprüft. Mit diesem Gerät ist es einwandfrei möglich, auch in größeren Tiefen exakte Beurteilungen abzugeben. Außerdem kann der Befund ggf. fotografisch festgehalten werden. Protokollierung ist für die später vorzunehmende Beurteilung wichtig.

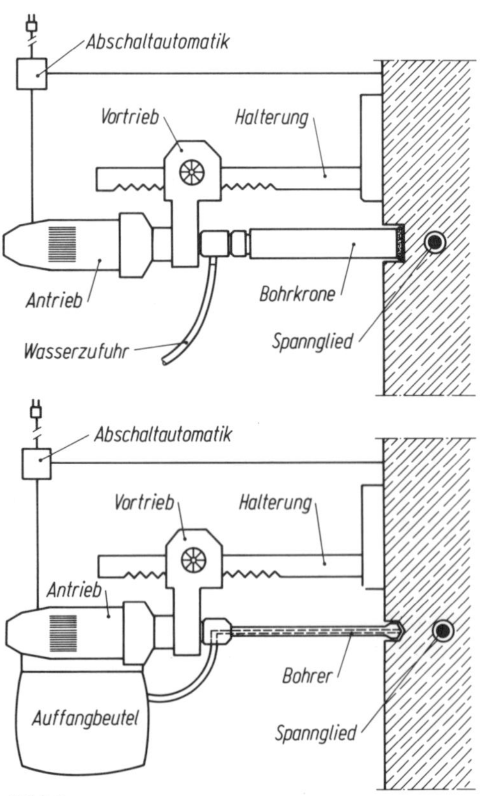

Bild 6
Anbohren von Spannkanälen

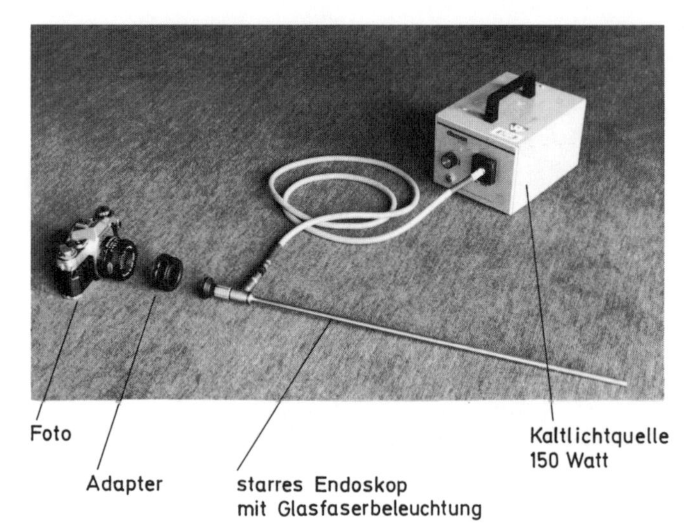

Bild 7
Endoskop

Geeignete Protokolle mit Schadensdifferenzierung liegen vor. Für die normalen Untersuchungen reicht ein Endoskop mit starrer Achse. Für größere Tiefen z. B. für Spannstahluntersuchung im Krümmungsbereich gibt es ein Endoskop mit flexibler Achse, dessen Kosten jedoch ein Vielfaches betragen. Die dabei verwendeten Glasfasern brechen leicht beim rauhen Baustelleneinsatz.

2.3.5 Hohlraummessung

Ein weiterer Vorteil des Anbohrens ist, den Hohlraum im Spannkanal im Anschluß an die Begutachtung des Spannstahles mit dem sogenannten Volumeßgerät auszulitern (Bild 8).
Dabei wird durch eine Vacuumpumpe der Spannkanal bis zu einem gewissen Unterdruck (z. B. halber Luftdruck) leergesaugt, dann durch eine Automatik der Meßvorgang eingeleitet, wobei die in den Kanal einströmende Luftmenge gemessen wird. Protokolle sind wiederum zu führen. Durch Versuche und Baustellenauswertungen wurde die Genauigkeit des Meßverfahrens ermittelt (Bild 9).
Meßgenauigkeit und Zuverlässigkeit des Meßvorgangs kann noch gesteigert werden. Die notwendigen Schritte sind mit den Herstellern abgesprochen und eingeleitet.
Haupteinsatz der Vacuumtechnik wird (Abschn. 4.4.1) bei der Schadensbehebung von mangelhaft verpreßten Kanälen liegen.

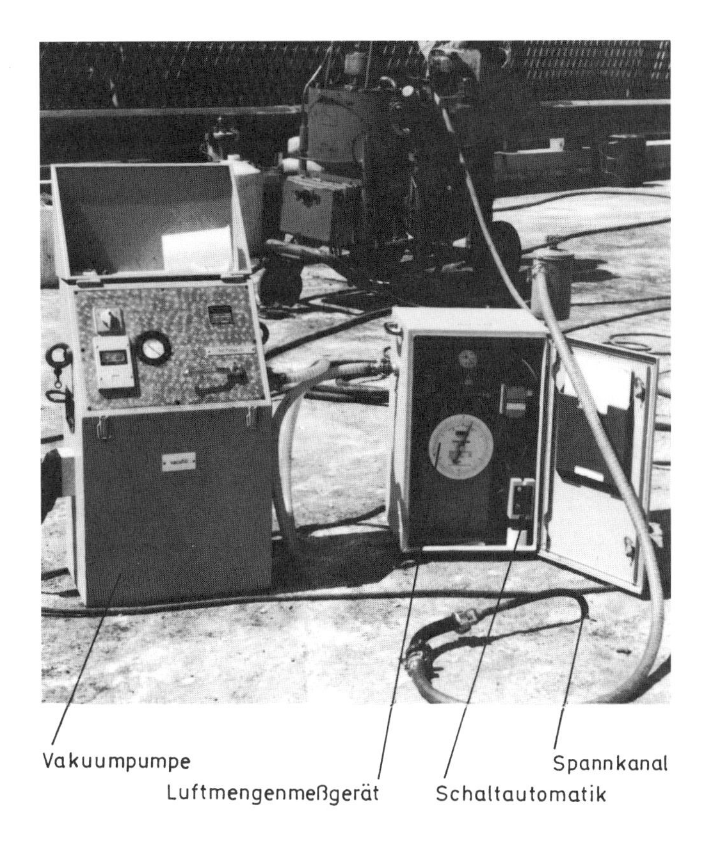

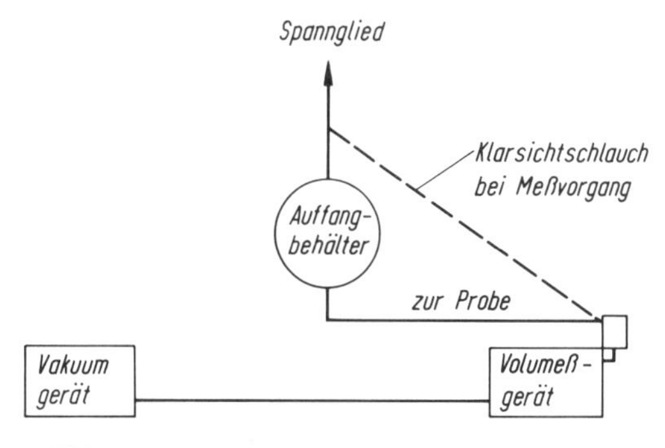

Bild 8
Hohlraummessung mit Vacuumtechnik

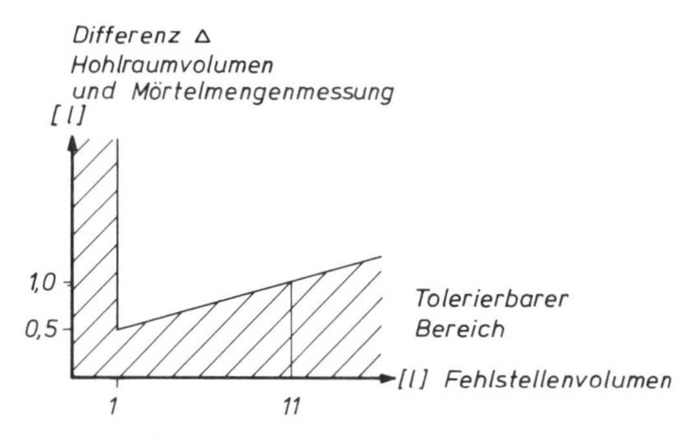

Bild 9
Genauigkeit der Vacuum-Volumen-Meßmethode

3 Beurteilung der Bewehrung [2]

3.1 Ursprüngliche Eigenschaften

Durch die Versuchsergebnisse in Verbindung mit der bauaufsichtlichen Zulassung und durch die Eigen- und Fremdüberwachung sind in etwa die Ausgangseigenschaften der Stähle bekannt. Es ist bisher noch nicht beobachtet worden, daß diese sich zeitabhängig verändern, es sei denn durch falsche Behandlung beim Einbau oder durch Korrosion während der Nutzung.
Leider entsprechen die Anforderungen bei der Zulassung nicht denen der späteren Beanspruchung; einige sind überbestimmt [10]. Die in Praxis zugelassenen Biegeradien, ganz zu schweigen von der Rückbiegung bei falscher Handhabung, sind dagegen geringer als die bei den Zulassungsversuchen. Häufig in Praxis vorkommende Verletzungen (Kerben) werden in den Eignungsversuchen nicht im erforderlichen Umfang getestet. Hier wären im Hinblick auf eine praxisfreundlichere Handhabung die Zulassungsbedingungen zu überdenken.

3.2 Beurteilung des Ist-Zustandes geschädigter Stähle

Um eine Aussage über die Standsicherheit und das Gebrauchsverhalten vornehmen zu können, muß der Ist-Zustand der geschädigten Stähle beurteilt werden.

3.2.1 Optische Beurteilung – Probeentnahme

Der Fachmann kann eine grobe Einstufung der mechanischen Eigenschaften von Beton- und Spannstahl aufgrund des optischen Bildes vornehmen. In den Erläuterungen zu DIN 4227 Teil 1 werden sogar Kerbtiefen benannt, bei denen noch keine Herabsetzung der Festigkeitseigenschaften zu erwarten ist (siehe auch [11]). Wie sich dabei die Stöße und Verbindungen verhalten, die von vorneherein gegenüber dem durchgehenden Stahl bis auf $\frac{1}{3}$ geminderte Schwingfestigkeiten aufweisen, ist weniger bekannt.
Werden am Stahl vereinzelt tiefere Kerben beobachtet, wäre eine Probeentnahme durch Herausschneiden eines Stahles angebracht. Sie kann natürlich als Einzelwert keine statistisch gesicherte Aussage liefern; die bei dieser Gelegenheit durch Verkürzungsmessung mit beobachtete vorhandene Spannkraft ist ein nützliches Nebenprodukt. Gesicherte Aussagen liefert das folgende Kapitel.

3.2.2 *Einfluß von Korrosionsschäden auf die Spannstahleigenschaften*

In einer umfassenden Arbeit hat REHM [2], [12] – im Zuge eines vom BMFT mitfinanzierten Forschungsvorhabens – im Auftrag von DYWIDAG die Auswirkung verschiedener Kerbformen auf die mechanischen Eigenschaften naturharter, vergüteter und gezogener Spannstähle untersucht. Das umfangreiche Schrifttum wurde gesichtet und Versuchsserien gefahren.
Es zeigte sich, daß die Zugfestigkeit nur unwesentlich von Korrosionsnarben beeinflußt wird. Gezogene Drähte erweisen sich hier als etwas empfindlicher. Je nach Narbenform stellt sich jedoch ein deutlicher Abfall im Biegeversuch und im Ermüdungsversuch ein. Der Biegeversuch hat dabei die Funktion, rasch, ggf. bereits vor Ort, eine Schnellaussage treffen zu können, bevor die Ergebnisse langwieriger Ermüdungsversuche vorliegen.
Die folgenden Bilder 10 bis 14 sollen der Praxis helfen, anhand der Kerbform und -tiefe eine rasche Beurtei-

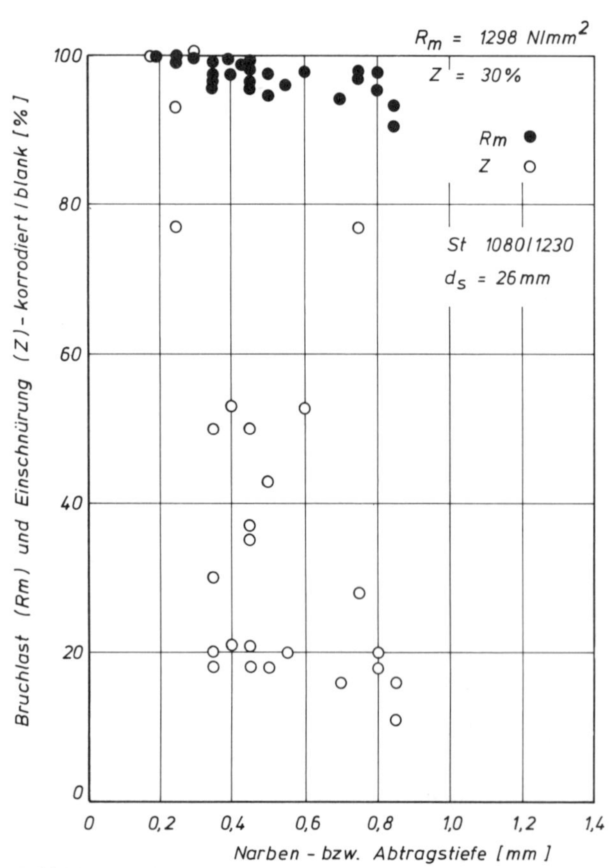

Bild 10
Einfluß der Korrosionsnarbe auf die Bruchlast und Einschnürung des naturharten Spannstahles St 1080/1230, $d_s = 26$ mm

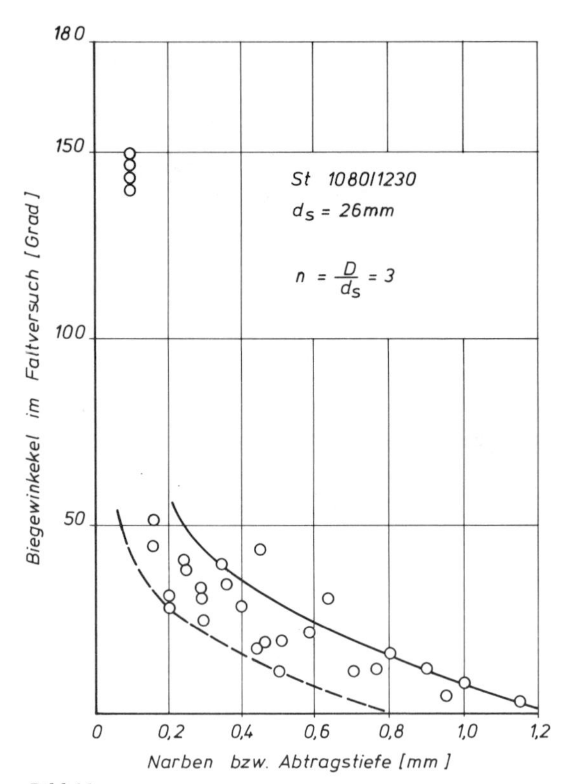

Bild 11
Einfluß der Korrosionsnarbe auf den Faltversuch des naturharten Spannstahles St 1080/1230, $d_s = 26$ mm

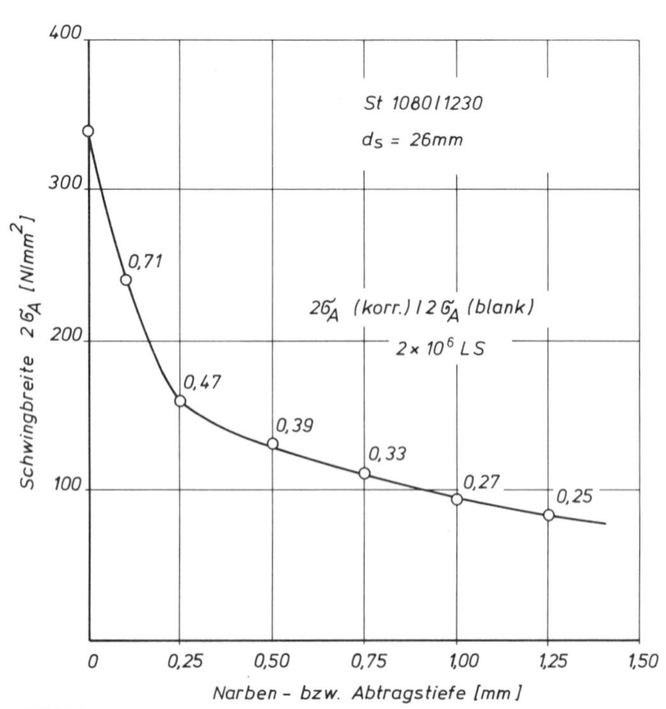

Bild 12
Einfluß der Korrosionsnarbe auf die Schwingfestigkeit des naturharten Spannstahles St 1080/1230, $d_s = 26$ mm

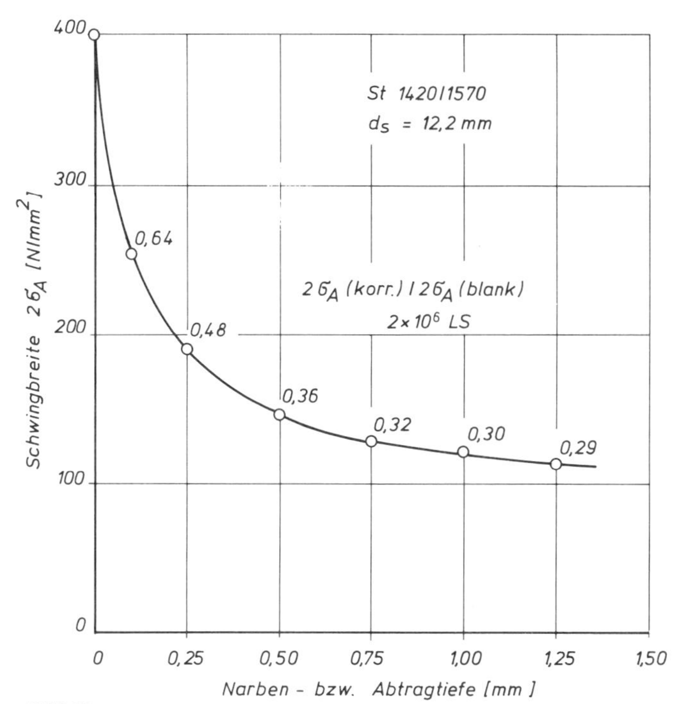

Bild 13
Einfluß der Korrosionsnarbe auf die Schwingfestigkeit des vergüteten Spannstahles St 1420/1570, d_s = 12,2 mm

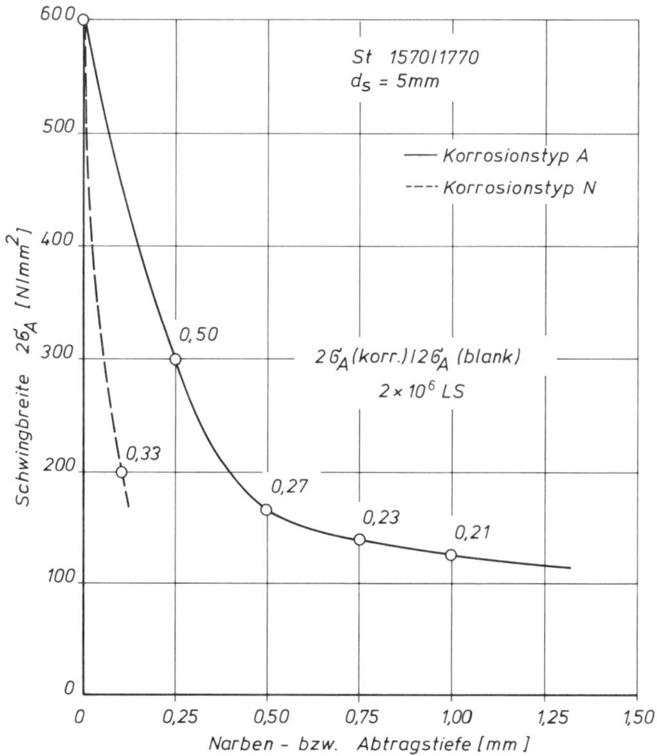

Bild 14
Einfluß der Korrosionsnarbe auf die Schwingfestigkeit des gezogenen Drahtes St 1570/1770, d_s = 5 mm

lung des statischen Tragverhaltens, des Biege- und des Ermüdungsverhaltens je nach Spannstahlsorte vornehmen zu können. Tabelle 2 und 3 faßt die Ergebnisse zusammen. Weitere Daten sind in [2] zu finden.
Nicht in die Betrachtung mit einbezogen ist lochfraßinduzierte Wasserstoffrißkorrosion. Hier genügen unter Sauerstoffarmut bereits geringe Narben, um den Bruch einzuleiten.

3.2.3 Einfluß von Korrosionsschäden auf die Betonstahleigenschaften

Ebenfalls aus [2], aber auch aus [13] kann für Betonstahl ähnliches Verhalten wie beim Spannstahl entnommen werden. Die Gefahr der Rißkorrosion entfällt natürlich.
Auch hier sollen die Bilder 15 und 16 Entscheidungshilfen für die Praxis liefern, um den Kerbeinfluß auf statische und dynamische Eigenschaften abschätzen zu können.

3.2.4 Extrapolation

Soll eine Aussage über das Langzeitverhalten geschädigter Konstruktionen gemacht werden, um ggf. rechtzeitig verstärken zu können, bräuchte man Daten über den zukünftigen zeitlichen Verlauf der Schädigung. Zunächst müssen die bevorstehenden Randbedingungen für Korrosion abgeschätzt werden (Feuchte, Sauerstoff, Chlorid usw.). Für diesen probabilistischen Vorgang fehlen noch die Kenntnisse, so daß sich keine Aussagen über den zukünftigen Verlauf des Korrosionskerbes machen lassen. Kann nicht der Grenzfall unterstellt werden, daß z. B. durch Trockenhalten jegliche weitere Korrosion gestoppt wird, bleibt für Voraussagen von Versagenswahrscheinlichkeiten (Abschn. 4.6) daher zunächst nur der Weg der Schätzung oberer und unterer wahrscheinlicher Werte.

4 Sanierungsmaßnahmen

4.1 Mangelhafte Betondeckung

Nicht ausreichend dichte und dicke Betondeckung bietet wegen rascher Karbonatisierung keinen aktiven Korrosionsschutz. Abplatzungen durch Volumenvergrößerung sind als Folge die am häufigsten auftretenden Bauschäden.

Tabelle 2
Abminderung der mechanischen Kennwerte von Spannstählen infolge Korrosion [12] (gemittelte Werte)

Stahlsorte	Meß-größe	Ausgangs-zustand	ND/NT ≦ 4		Ausgangs-zustand	ND/NT = 4 ÷ 10	
			NT = 0,4 mm	NT = 1,0 mm		NT = 0,4 mm	NT = 1,0 mm
St 1080/1230	R_m	1298	1270 (0,98)	1170 (0,90)			
d_s = 26 mm	A_{10}	7,7	4,6 (0,60)	1,5 (0,20)			
St 1420/1570	R_m	1640	1560 (0,95)	1510 (0,92)			
d_s = 12 mm	A_{10}	8,3	5,0 (0,60)	2,5 (0,30)			
St 1570/1770	R_m	1776	1600 (0,90)	1155 (0,65)	1810	1630 (0,90)	1450 (0,80)
d_s = 5 mm	A_{10}	8,7	3,1 (0,35)	0,9 (0,10)	8,8	4,0 (0,45)	1,8 (0,20)

R_m (N/mm²), A_{10} (%), in ( ): auf den Ausgangszustand bezogen, ND = Narbendurchmesser
NT = Narbentiefe

Tabelle 3
Abminderung der Dauerschwingfestigkeit von Spannstählen infolge Korrosion [12] (gemittelte Werte)

Stahlsorte	Meßgröße	Ausgangs-zustand	ND/NT ≦ 4		ND/NT = 4 ÷ 10	
			NT = 0,25 mm	NT = 1,0 mm	NT = 0,4 mm	NT = 1,0 mm
St 1080/1230 d_s = 26 mm	DSF	340	160 (0,47)	90 (0,27)		
St 1420/1570 d_s = 12,2 mm	DSF	400	190 (0,48)	120 (0,30)		
St 1570/1770 d_s = 5 mm	DSF	600	100 (0,17)		300 (0,50)	130 (0,21)

DSF = Dauerschwingfestigkeit (2×10^6 LW) (N/mm²), ND = Narbendurchmesser
in ( ): auf den Ausgangszustand bezogen NT = Narbentiefe

Als Sanierung bleibt nur die Möglichkeit, lose Stellen abzuklopfen, diese nach Abschn. 4.2 auszubessern und Korrosion, die noch nicht zu Abplatzungen führte, zum Stoppen zu bringen. Dazu bieten sich Versiegelungen und Beschichtungen an. Als Versiegelung kommen verdünnte Epoxydharzanstriche oder Akrylharze in Frage, als Beschichtung Epoxydharzspachtel. Bei dikkeren Schichten eignet sich Mörtel oder Spritzbeton. Die alte Betonoberfläche muß vorher gereinigt werden, z. B. mit Dampfstrahl, und für eine Haftbrücke gesorgt werden, siehe auch [14] bis [17].

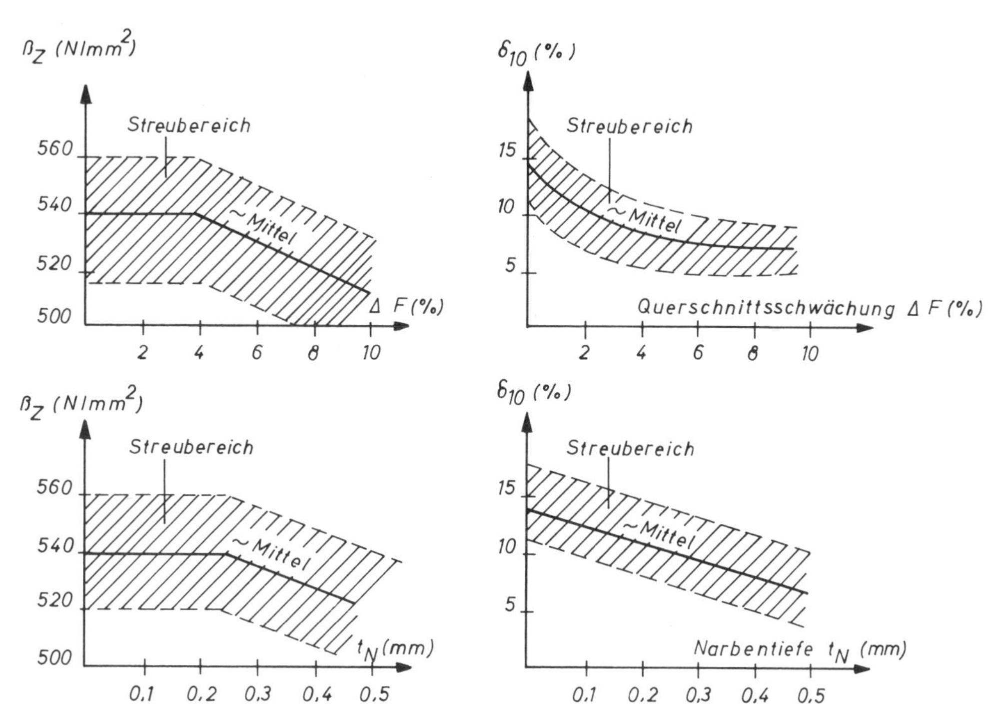

Bild 15
Auswirkung des Korrosionskerbes auf die Bruchfestigkeit und die Bruchdehnung des Rippentorstahles BSt 420/500, d_e = 8 mm

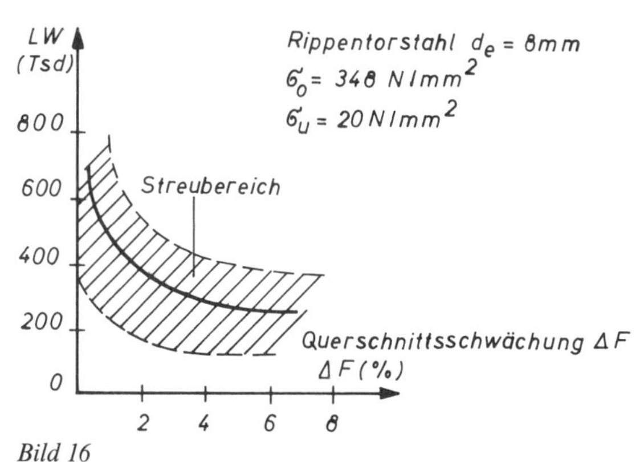

Bild 16
Auswirkung des Korrosionskerbes auf die Schwingfestigkeit des Rippenstahles BSt 420/500

Bild 17
Ausbesserung von Korrosionsabplatzungen

4.2 Ausbesserungen von Korrosionsabplatzungen

In Bild 17 wird gezeigt, wie Betonabplatzungen, verursacht durch Stahlkorrosion, zu beheben sind. Auf die Gefahr von neuen Potentialen wird hingewiesen.
Ist der umgebende Beton stark durch Chloridionen aus Tausalz verseucht, bleibt in der Regel nur das Trockenlegen der Zonen. Abbruch ist häufig nicht möglich, und Austreiben der Ionen z. B. über elektrische Felder ist für die Praxis wenig geeignet.

4.3 Rißverpressung

Breite Risse > 0,2 mm sind zur Vermeidung von Korrosion mit Epoxydharz zu verpressen. Geeignete Verfahren befinden sich am Markt und haben zum Großteil die Eignungsnachweise von BMV durchlaufen [18]. Schmälere Risse sind bei aggressiver Umgebung kapillar mit Harz zu schließen. Zur Vereinheitlichung der Sichtflächen ist die Gesamtfläche zu versiegeln, möglichst versehen mit Farbpigmenten.

Statisch wirksam sollte man solche Rißverklebungen nicht verwenden.

4.4 Verfüllen von nicht sachgemäß injizierten Spannkanälen

4.4.1 Vacuumverfahren (Bild 18)

Nicht nur zur Messung des Hohlraumes (Abschn. 2.3.5), sondern auch zu dessen Verfüllen hat sich das Vacuumverfahren vorzüglich geeignet. Es ist nur eine Bohrung an beliebiger Stelle erforderlich. Eine Kontrolle zwischen Hohlraummessung und Mörtelmessung gibt Aufschluß über den Erfolg der Maßnahme. Bei Unstimmigkeiten sind weitere Kontrollbohrungen erforderlich.

Nach dem Verpressen ist Druckaufbringung erforderlich, um die Restluft in die Poren zu verdrängen. Bei größeren Luftpolstern besteht die Gefahr, daß Absetzwasser zu Fehlstellen hin verdrängt wird und so Kanäle entstehen, die den Korrosionsschutz gefährden. Es ist daher Mörtel mit geringer Absetzneigung zu verwenden; zugelassene Sonderzemente existieren. Einzelheiten zu diesem Verfahren sind in eine Richtlinie in Zusammenarbeit mit Anwendern und BMV festgelegt.

In Sonderfällen kann störendes Wasser im Kanal unter Vacuum verdunstet und damit Trocknung erzielt werden. Spezielle Geräte und Kenntnisse sind dann erforderlich.

4.4.2 Verguß der Kanäle mit Spezialharz

Bei wassergefüllten Kanälen, bei denen durch Anbohren nicht entwässert werden kann und Trocknung nicht möglich ist (senkrechte, im Grundwasser stehende Kanäle), kann durch Einsatz viskoser Epoxydharze mit langer Verarbeitungszeit und hohem spezifischem Gewicht das Wasser verdrängt werden. Auf dem Markt sind geeignete Materialien vorhanden, die außerdem gutes Verbundtragverhalten und Aktivpigmente aufweisen (Tabelle 4).

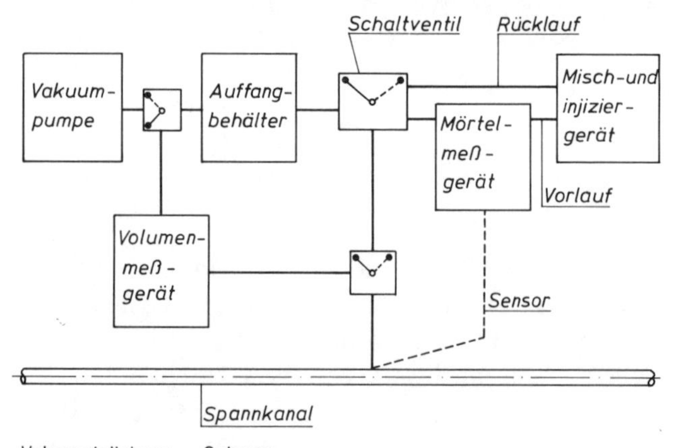

Bild 18
Hüllrohrverpressung mit dem Vacuumverfahren

4.5 Herausschießen gebrochener Spannglieder

Die Gefahr des Herausschießens nicht sachgemäß verpreßter bzw. sanierter Spannglieder im Laufe der Zeit ist zu überdenken. In der Regel verhindern dies Gesimskappen u.ä. Andernfalls ist die Flugbahn zu bestimmen. Ist mit einer Gefährdung zu rechnen, sind Sicherungsmaßnahmen vorzusehen, die in der Lage sind, die freiwerdende Energie in Verformungsenergie umzuwandeln. Ansätze hierzu sind in Bild 19 und 20 zu finden.

4.6 Lebenserwartung geschädigter Bauwerke; Versagenswahrscheinlichkeiten

Nach der Schadenserhebung und Behebung im ersten Schritt ist für die Beurteilung des zukünftigen rechnerischen Bruchzustandes der Zustand der Bewehrung durch die Wahl oberer und unterer Grenzen festzulegen:

	obere Grenze	untere Grenze
30% bis 40% der Spannglieder gut verpreßt	40	30
10% Spannglieder im nachhinein aktivierbar	10	10
25% Spannkanäle mit 100% bzw. 80% Erfolg wiederverpreßt	25	20
25% bzw. 35% nicht verpreßbar, Ausfall nach z. B. 50 Jahren halb bzw. ganz	12,5	0
Weitere Systemreserven 0 bzw. 5%	5	0
Verbliebene Spannkraft nach 50 Jahren	92,5	60
erf. Verstärkung für Bruchzustand	7,5	40
	100%	100%

Tabelle 4
Kennwerte eines Spezialvergußharzes

		DYWIPOX SPK
Topfzeit 100 g/23 °C		3 Stunden
Viskosität (mP.S)	23 °C	830
(Rotationsviskosi-	15 °C	1500
meter HAAKE)	10 °C	3380
	5 °C	6030
spez. Gewicht flüssig		1,75 g/ml
spez. Gewicht fest		1,81–1,87 g/ml
Shore-Härte		
A/24 h – trocken		60
7 d – trocken		98
7 d – naß		98
Glasübergangstemperatur (DIN 53445)		80 °C
kapillare Steighöhe		
∅ 0,02 mm		20 mm
∅ 0,083 mm		10 mm
∅ 0,84 mm		3 mm
pH-Wert-Bestimmung des Überstandwassers nach		
10 min		9,0
3 d		7,0–7,5
Haftverbund (Dauerlast 2 h)		
– naß		4,0 N/mm² – Schlupf 0,027–0,037
– trocken		4,0 N/mm² – Schlupf 0,025–0,035
Haftverband (Bruchlast)		
– naß		7,6 N/mm²
– trocken		11,4 N/mm²
Ausgießversuche 6 m Hüllrohr ∅ 36 mm, Stahl ∅ 26 mm, Wasserhöhe 1 m		gutes Kriechverhalten, keine Wassereinschlüsse, vollständige Verfüllung

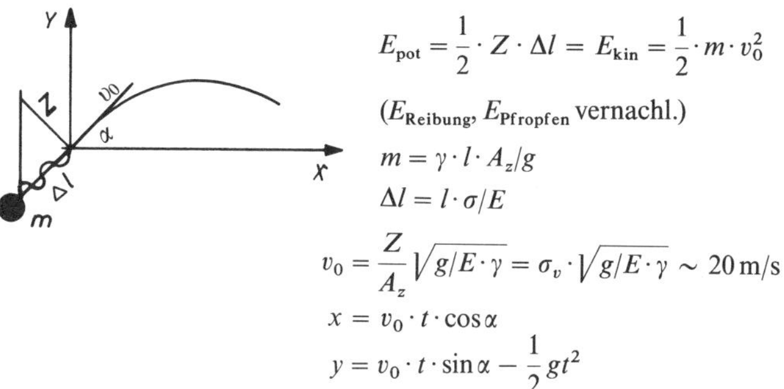

Bezeichnung
γ = spez. Gewicht kp/m³
E = E-Modul kp/m²
g = Erdbeschl. m/s²
σ_v = Vorspannung kp/m² = $0{,}55\beta_z$
β_z = Zugfestigkeit Spannstahl

Bild 19
Flugbahn gebrochener, gerader Spannglieder

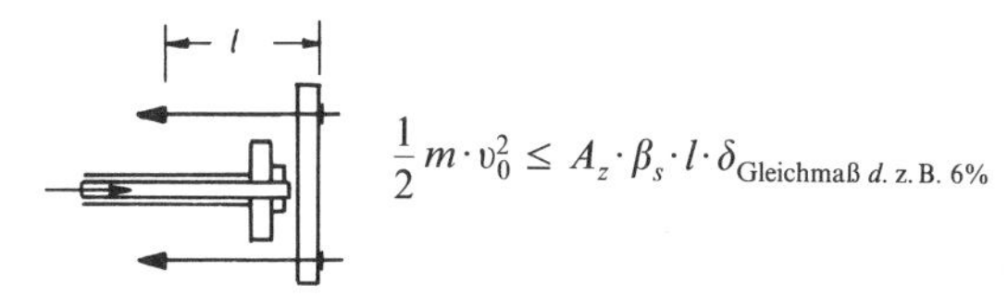

Bezeichnung
m = Masse
v_0 = Fluggeschwindigkeit
A_z = Stahlfläche Sicherung
β_s = Streckgrenze Sicherung

Bild 20
Sicherungsmaßnahmen zur Energieumwandlung gebrochener Spannglieder

Ähnliche Studien sind für den Gebrauchszustand (Rißbreitenberechnung) und für die Ermüdungsbeanspruchung durchzuführen. Daraus ergibt sich die Anforderung für eine ggf. erforderliche Verstärkung.
In Bild 21 sind die Zusammenhänge zeitabhängig ausgedrückt durch die Versagenswahrscheinlichkeiten für ein konkretes Bauwerk aufgetragen [2]. Deutlich läßt sich daraus der Einfluß streuender Größen erkennen. Mit dieser anschaulichen Darstellung kann der Bauherr entscheiden, ob eine einfache Sanierung genügt, ob eine Herabstufung der Brücke zweckmäßig ist, ob der eventuell bevorstehende Verkehrsausbau das Bauwerk erübrigen läßt oder ob verstärkt werden muß.

4.7 Verstärkungen

Die Bandbreite der Verstärkungen reicht von
Aufkleben von C-Faser
Aufkleben von Stahllaschen [19]

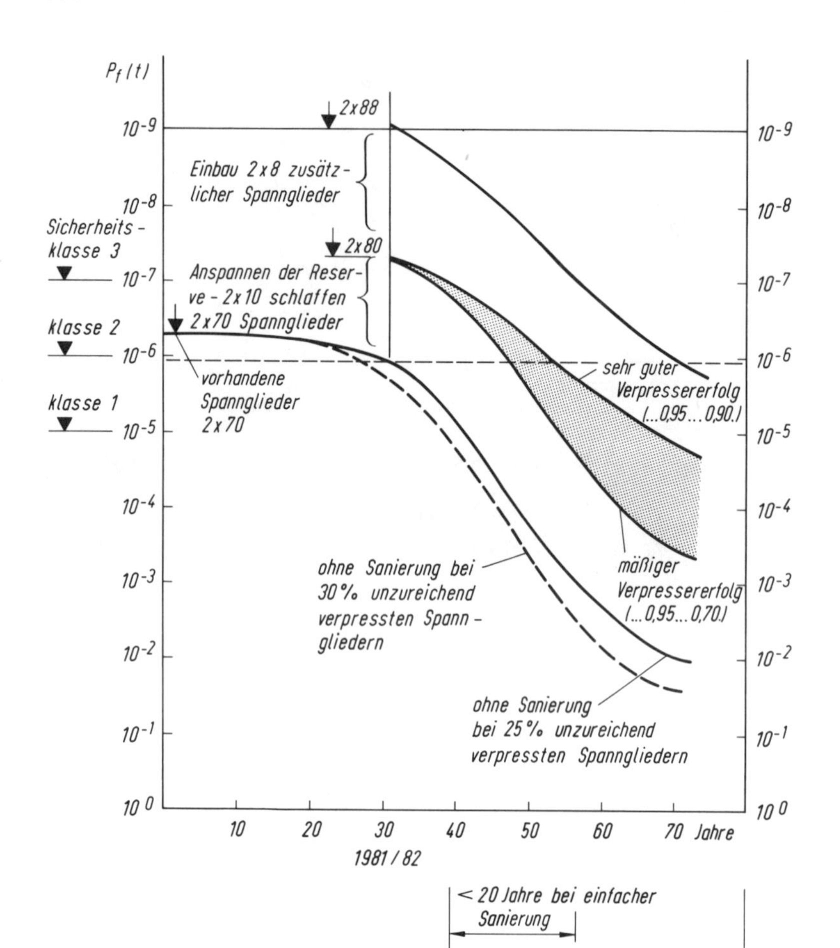

Bild 21
Verlauf der rechnerischen Versagenswahrscheinlichkeit ab Nutzungsbeginn

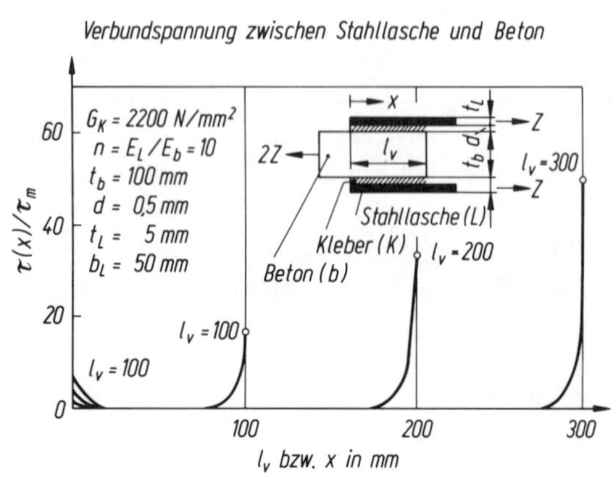

Bild 22
Aufkleben von Stahllaschen

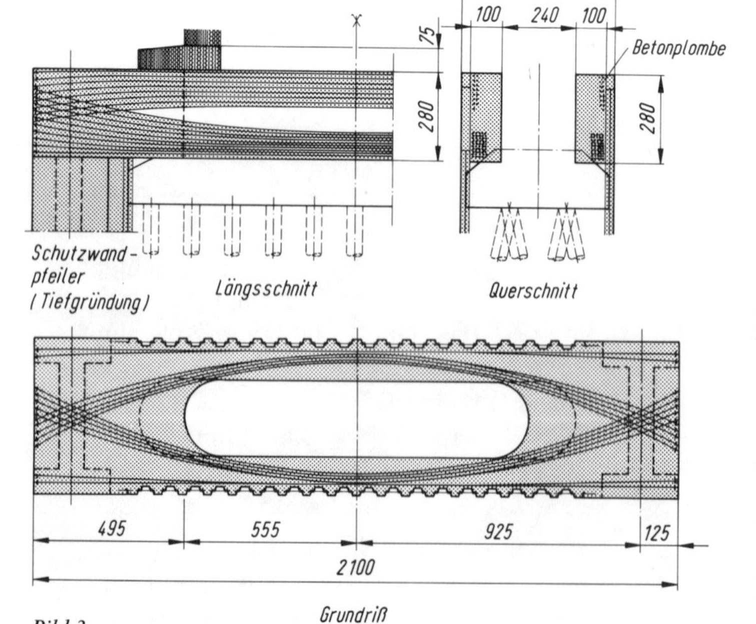

Bild 3
Beiblenden von Spannbetonbalken mit Verbindung des alten und neuen Teiles

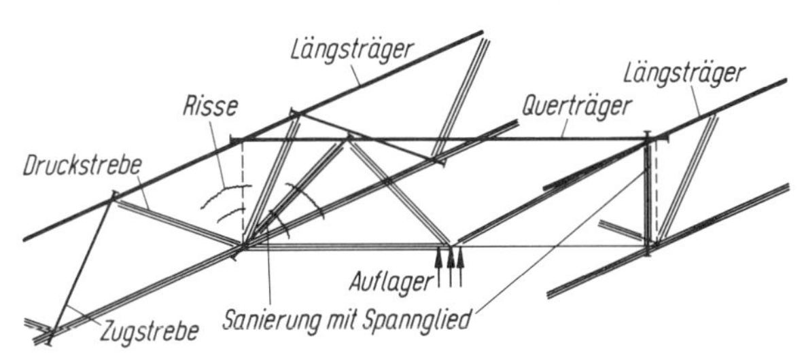

Bild 24
Einziehen von Schubnadel zur Herstellung des Schubtragverhaltens

Aufkleben von Betonfertigteilen [20]
Anbetonieren oder Aufspritzen von Betonverstärkungen [20]
Einziehen von Spannglieder [2], [8] bis zum
Anblenden von Spannbetonbalken [2], [8].
Die Bilder 22 bis 24 zeigen Anwendungsbeispiele, die zum Teil jetzt schon jahrelange Bewährung aufweisen. Das erstgenannte Beispiel wirkt erst bei Laststeigerung mit, wogegen die vorgespannten Lösungen Kraftumlagerungen beliebig steuern können.

5 Schlußbemerkung

Die Stahl- und Spannbetonbauweise hat sich bewährt. Vereinzelt aufgetretene Schäden lassen sich beheben und bei zukünftigem Bauen weitgehend vermeiden. Eine zentrale Stellung nimmt dabei der Baustoff Stahl ein. Er wurde in diesem Beitrag herausgestellt. Ziel dieses Beitrages war es, Stahlschäden zu erkennen, zu beurteilen und zu beheben, wobei bei der Ausführung von Sanierungen besonders hohe Anforderungen an die Ausführungsqualität zu richten sind. Selbstverständlich sollen die aus Schäden gewonnenen Erkenntnisse in zukünftiges Bauen einfließen. Dabei spielt die Sorgfalt der Ausführung, also der Mensch, eine bedeutende Rolle. Das damit verbundene Risiko muß über Qualitätssicherungssysteme [21], [22] weitgehend abgebaut werden.

6 Literatur

[1] Jungwirth, D.: Langzeitverhalten von Spannbetonkonstruktionen – Erfahrungen und Folgerungen; Betontag 1983, DBV Wiesbaden.
[2] BMFT-Abschlußbericht „Spätschäden" FKZ: Bau 7006. Spätschäden an Spannbetonbauteilen, Prophylaxe, Früherkennung, Behebung; DYWIDAG München, 1983.
[3] Grübl, P.: Früherkennung und Behebung von Spätschäden an Betonbauwerken; Tiefbau, Ingenieurbau, Straßenbau; Bertelsmann, Gütersloh 1983.
[4] Hillemeier, B. und Müller-Rau, U.: Bewehrungssuche mit der Thermographie; Beton- und Stahlbetonbau 4/1980.
[5] Stratfull, R.: The Corrosion of Steel in a Reinforced Concrete Bridge; Corrosion 13 (1957), No. 3.
[6] Geymayer, H.: Zur Auffindung aktiver Stahlkorrosion in Stahlbeton- und Spannbetonbauwerken, insbesondere Brükken; Österreichische Ingenieurzeitschrift 22 (1979), Heft 2.
[7] Ultraschalluntersuchung zur Auffindung von Injektionsfehlstellen und Spannstahlbrüchen in Spannbetonkonstruktionen; Dywidag-Versuchsbericht 14/76 vom 29. 2. 77; sowie Bestätigungsbericht No. 1585 vom Materialprüfungsamt München.
[8] Jungwirth, D. und Kern, G.: Langzeitverhalten von Spannbetonkonstruktionen; Beton- und Stahlbetonbau 11/1980.
[9] Thiele, H.: Durchstrahlungsprüfung an Spannbetonbauwerken; Aus dem Betriebsforschungsinstitut; Stahl und Eisen 98 (1978) No. 17.
[10] Jungwirth, D. und Kern, G.: Erfahrungen bei der Entwicklung und bei der Verarbeitung von Bewehrungsstählen im letzten Jahrzehnt; Bauingenieur 54/1979.
[11] Kern, G. und Jungwirth, D.: Betrachtungen zu neueren Entwicklungen von Spannverfahren am Beispiel des DYWIDAG-Verfahrens; Beton- und Stahlbetonbau 3/1974.
[12] Rehm, G.: Korrosionsschutz des Bewehrungsstahles im Beton; Tiefbau, Ingenieurbau, Straßenbau 5/1983.
[13] Schiessl, P.: Einfluß der Rißbreite auf die Rostbildung an der Bewehrung in Stahlbetonbauteilen; Lehrstuhl für Massivbau Nr. 1346 vom 14. 11. 79.
[14] Merkblatt Instandsetzen von Betonbauteilen. Deutscher Beton-Verein e. V. Fassung März 1982.
[15] Werse, H.-P.: Instandsetzung von Stahlbeton bei unzureichendem Korrosionsschutz der Bewehrung; Beton- und Stahlbetonbau (1983) Heft 2.
[16] Richtlinien für die Anwendung von Reaktionsharzen im Betonbau; Deutscher Betonverein, Teil 1–4.
[17] Richtlinien für die Ausbesserung und Verstärkung von Betonbauteilen mit Spritzbeton; Deutscher Ausschuß für Stahlbeton.
[18] Merkblatt für das Verpressen von Rissen mit Epoxidharzsystemen im Bereich von Spannglied-Koppelstellen (Fassung Mai 1980); Der Bundesminister für Verkehr, Abt. Straßenbau.
[19] Rostácy, F. S. und Ranisch, E. H.: Einseitig gerissene Koppelfugenbereiche durch angeklebte Stahllaschen; Forschung – Straßenbau und Straßenverkehrstechnik, Heft 378, Bonn, BMV 1983.
[20] König, G., Weigler, H. und A.: Nachträgliche Verstärkung von Stahlbeton- und Spannbetonquerschnitten. Prüfbericht Nr. 188.1.79 vom 2. 11. 79 d. Inst. f. Massivbau der TH Darmstadt mit Ergänzung v. 2. 6. 81 Nr. 153.1.81.
[21] Blaut, H.: Gedanken zum Sicherheitskonzept im Bauwesen; Beton- und Stahlbetonbau 9/82.
[22] Jungwirth, D.: Denkanstöße zur Qualitätsverbesserung im Spannbetonbau; FIP-Bericht, Stockholm 1982, DBV.

Entwicklung und Bewertung des Regelwerkes im Spannbetonbau aus korrosionstechnischer Sicht

Ulf Nürnberger und Bernd Neubert

1 Einführung

Im Rahmen des Wiederaufbaus seit Anfang der 50er Jahre und in engem Zusammenhang mit der Entwicklung der Bundesrepublik Deutschland und anderer Staaten zu modernen Industrieländern schaffte sich die Bauindustrie mit dem Spannbeton eine Bauweise, die heute einen festen und wichtigen Platz innerhalb der gesamten Bautätigkeit einnimmt. Anfänglich konnte man sich dabei hinsichtlich des Einsatzes der Baustoffe und bei der Bauausführung nur auf die Erfahrungen aus dem Stahlbetonbau stützen. Dies erwies sich jedoch insbesondere aus korrosionstechnischer Sicht sehr bald als nicht ausreichend, weshalb das entstandene Regelwerk ständig den gewonnenen Erfahrungen entsprechend verbessert werden mußte.
Unzureichende Angaben in den Normen und Vorschriften, aber auch die Tatsache, daß bestehende Regeln, die den Korrosionsschutz betreffen, aus Unkenntnis der Zusammenhänge nicht immer eingehalten wurden, führten im Spannbetonbau immer wieder zu Schäden. Der Ausfall einzelner Spannglieder, einzelner Bauteile und – in wenigen Fällen – sogar das Versagen ganzer Konstruktionen waren die Folge. Bis heute sind bei bestehenden Bauwerken immer wieder umfangreiche Sanierungsarbeiten notwendig geworden, um diese „Sünden" der Vergangenheit zu beheben und um die Ausführung dem gegenwärtigen Standard anzunähern.
Im vorliegenden Beitrag soll die Entwicklung des Regelwerkes im Spannbetonbau aus korrosionstechnischer Sicht von den Anfängen bis heute kurz aufgezeigt werden. Weiterhin erfolgt die Bewertung einiger wesentlicher, den Baupraktiker interessierender Regelungen, und es werden Ansatzpunkte für weitere Verbesserungen genannt. Anhand von Spannstahlschäden aus der jüngeren Zeit wird jedoch zunächst aufgezeigt, daß die Entwicklung eines Regelwerkes alleine nicht ausreichend ist, um Mängel am Bau gänzlich auszuschalten und daß es notwendig ist, stärker als bisher dem Anwender die korrosionstechnischen Regeln des Spannbetons vertraut und bewußt zu machen.

2 Analyse von Spannstahlschäden

Untersuchungen von Schadensfällen dienen nicht nur als Grundlage für juristische Entscheidungen, sondern helfen vor allem durch die Aufklärung ihrer Ursachen, künftig Schäden zu vermeiden. Für die Forschung ergaben sich aus Schadensfällen viele wertvolle Anregungen; dabei gewonnene Ergebnisse führten zu einer ständigen Verbesserung des Regelwerkes.
Bild 1 zeigt eine Übersicht der in [1] beschriebenen in- und ausländischen Schadensfälle mit korrosionsbedingten Spannstahlbrüchen bei vorgespannten Konstruktionen und bei Transport und Lagerung von Spannstählen. Für den Zeitraum 1950 bis 1981 werden die Schäden unterteilt nach Anzahl und Ausmaß sowie Schadensursachen aus bautechnischer Sicht. Darüber hinaus wird unterschieden, ob die Ursachen auf das Einhalten (Ziffern 3 bis 6 in Bild 1) oder Nichteinhalten (Ziffern 1 und 2 in Bild 1) der bei der Bauwerkserstellung gültigen (deutschen) Normen bzw. Richtlinien zurückzuführen sind (Abschnitt 3). In den meisten Fällen lagen mehrere Schadensursachen zugrunde, wobei Mängel am Beton oder Stahl (herstellungs- und verarbeitungsbedingt) in Kombination mit einer feuchten und/oder aggressiven Umgebung am häufigsten vorkamen. Berücksichtigt man, daß gegenwärtig auf der Welt pro Jahr etwa 1 Million Tonnen Spannstahl verarbeitet werden, sind die Schadensfälle durch Spannstahlversagen insgesamt jedoch als gering anzusehen. Dies gilt vor allem für jene Fälle mit katastrophalem Versagen der ganzen Konstruktion. Für Spannbetonbrücken beispielsweise sind Fälle mit totalem Versagen bisher nicht bekanntgeworden. Dies läßt insgesamt den Schluß zu, daß sich die Spannbetonbauweise auch aus korrosionstechnischer Sicht bewährt hat. Die Aus-

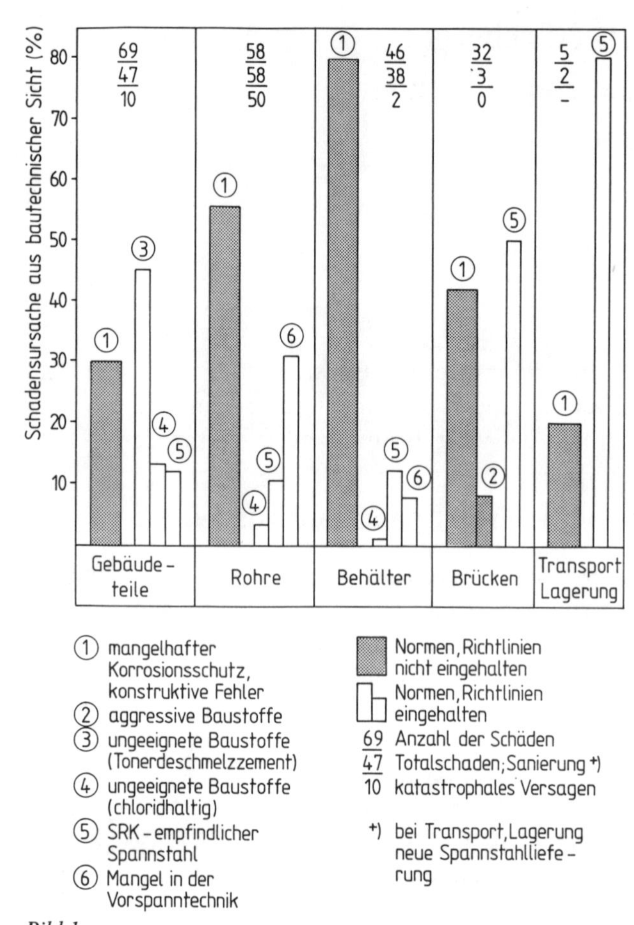

Bild 1
Unterteilung von Spannstahlschäden nach Ursache und Gewichtung

wertung zeigt aber auch deutlich, daß etwa die Hälfte aller Schäden auf Nichtbeachtung der bei der Bauwerkserstellung bestehenden Vorschriften, also auf menschliche Unzulänglichkeiten, zurückzuführen sind. Hierbei stehen mangelhafter Korrosionsschutz und konstruktive Fehler im Vordergrund. Auch in anderen Berichten zum Sicherheitsproblem von Bauwerken [2] wird herausgestellt, daß – wenn überhaupt – deren Dauerhaftigkeit vorwiegend durch Fehler der am Bau Beteiligten beeinträchtigt wird und nicht durch grundsätzliche Mängel der Bauweisen oder der bestehenden Richtlinien.

Zum gegenwärtigen Zeitpunkt haben die Baurichtlinien einen hohen Stand erreicht. Es ist davon auszugehen, daß zukünftig die meisten der festgestellten Schadensursachen auszuschließen sind, insbesondere dann, wenn auch stärker als bisher die korrosionstechnischen Regeln des Spannbetonbaus in das Bewußtsein der Konstrukteure Eingang finden.

Ordnet man die in Bild 1 aufgeführten Schäden nach dem Zeitpunkt der Bauteilerstellung und dem der Schadensfeststellung, so ergeben sich die in Bild 2 dargestellten Zusammenhänge. Auch hierbei wurde wiederum unterschieden, inwieweit zum Zeitpunkt der Bauteilerstellung gültige Normen beachtet wurden oder nicht. Im oberen Teilbild ist zu erkennen, daß die größte Schadenshäufigkeit bei den vor 1960 errichteten Bauwerken vorliegt; bei den späteren ist eine rückläufige Tendenz zu verzeichnen. Insbesondere nimmt jener Anteil von Schäden ab, dessen Ursache primär darin zu sehen ist, daß bestehende korrosionstechnische Regelungen mißachtet wurden.

Eine Unterteilung der Schäden nach dem Zeitpunkt ihrer Feststellung im unteren Teilbild zeigt einen starken Anstieg der Häufigkeit bis etwa zum Jahr 1975 mit im Mittel gleichen Anteilen von Fällen, bei denen Normen bzw. Richtlinien eingehalten bzw. nicht eingehalten wurden. Die hohe Zahl von Schadensfeststel-

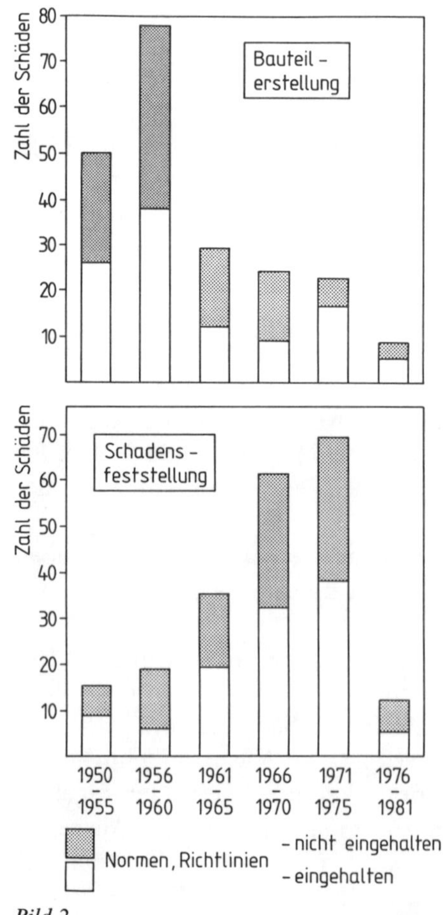

Bild 2
Zeitliche Reihenfolge von Schäden, getrennt nach Bauteilerstellung und Schadensfeststellung

lungen in den Jahren 1966 – 1975 ist darauf zurückzuführen, daß etwa bis 1960 eine größere Anzahl von Bauten korrosionstechnisch nicht dauerhaft ausgeführt wurde. D. h. aber wiederum, daß Mängel in der Ausführung i. M. erst nach 10 bis 15 Jahren als Schäden erkennbar wurden, was für die Anpassung der Regelwerke aus korrosionstechnischer Sicht an die tatsächlichen Erfordernisse hemmend war. Der starke Abfall der Schadensfeststellungen ab 1975 scheint darauf hinzudeuten, daß die Korrosionsprobleme der Vergangenheit weitgehend ausgestanden sind.

3 Schadensbeispiele

Aus den in Abschnitt 2 dargestellten Zusammenhängen wurde deutlich, daß etwa die Hälfte der ausgewerteten Schäden auf Unzulänglichkeiten der Bauausführenden, d. h. Nichtbeachtung der Normen bzw. Richtlinien zurückzuführen ist. Mit zu solchen Schäden beigetragen hat sicherlich auch die Tatsache, daß früher wie heute die Regelungen nicht immer eindeutig interpretierbare Formulierungen enthalten (Abschnitt 4) und ein gewisser Mangel in der Ausbildung vieler Bauingenieure hinsichtlich werkstofftechnischer Probleme zu beklagen ist. Ein Verständnis für die Notwendigkeit dauerhafter Stahlbeton- und Spannbetonbauwerke ist erst aufgrund einzelner spektakulärer Schadensfälle mit kostspieligen Sanierungen in das Bewußtsein gerückt.
Da auf die korrosionstechnischen Regelungen und deren Entwicklung in Abschnitt 4 noch näher eingegangen wird, seien an dieser Stelle einige jüngere Schadensbeispiele skizziert, die bei Beachtung des Regelwerkes prinzipiell vermeidbar gewesen wären.

Fall 1 [3]

Bei den Zugstreben einer 23 Jahre alten Halle wurden örtlich konzentriert, unverpreßte Bündelspannglieder vorgefunden. Die Hüllrohre waren von außen und/ oder innen durchkorrodiert, und die Drähte zeigten stellenweise Querschnittsminderungen von 10–70% (i. M. 30%). Über den Bereichen mit korrodierten Hüllrohren war die Betondeckung unzureichend (< 10 mm); der Beton zeigte auch Entmischungen. Die Korrosion hatte stellenweise zu einem Absprengen des Hüllbetons geführt. Bild 3 zeigt solche Stellen nach dem Sandstrahlen.

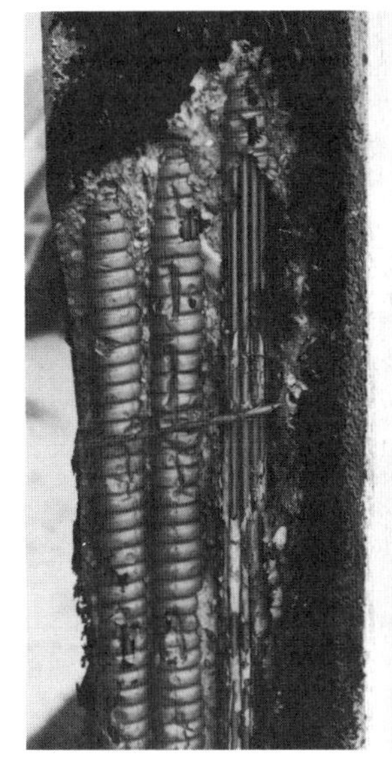

Bild 3
Unverpreßte Spannglieder von Zugstreben einer Halle nach 23 Jahren

Fall 2 [3]

Einige Längsspannglieder aus Litzenbündeln einer Straßenbrücke zeigten beim Abbruch nach 33 Jahren in den Hochpunkten auf mehreren Metern Länge unverpreßte Bereiche. Z. T. waren bis zu 90% der Einzellitzen wegen abtragender Korrosion ganz oder teilweise gebrochen (Bild 4). Die Korrosion war durch eindringendes Wasser und Chloride begünstigt worden, verursacht durch Mängel in der Fahrbahnabdichtung.

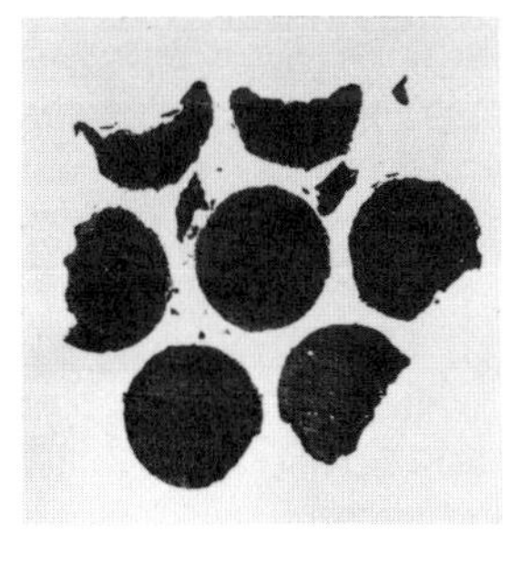

Bild 4
Durchkorrodierte Litzen in unzureichend verpreßten Spanngliedern einer 33 Jahre alten Straßenbrücke

Fall 3 [4]

Der Teileinsturz der 23 Jahre alten Berliner Kongreßhalle ist auf statische, konstruktive und Ausführungsmängel im Bereich der Anschlüsse Dachplatten/Ringbalken zurückzuführen. Dort führte die mangelhafte bzw. fehlende Betondeckung der Spannglieder mit Karbonatisierung und Zutritt von Feuchtigkeit zur Korrosion der Hüllrohre und Brüchen der Spanndrähte (Bild 5).

Fall 4 [5]

Bei einer vorgespannten Eisenbahnbrücke kam es zu Brüchen der Spannstähle noch vor dem Verpressen der Spannglieder. Die Untersuchungen ergaben lochfraßartige Korrosionsnarben auf den Stahloberflächen durch in die Hüllrohre eingedrungenes Betonrestwasser mit hohen Chlorid- und Sulfatgehalten. Brüche an Korrosionsnarben wurden durch die Verwendung eines für Spannbeton nicht zugelassenen, rhodanidhaltigen Betonverflüssigers ausgelöst (Bild 6).

Fall 5 [1]

Ein langzeitig gespanntes, noch unverpreßtes Spannglied einer im Bau befindlichen Brücke wurde zwecks Entfernung der Rostprodukte vor dem Verpressen mit einem „Rostlöser" gespült. Dieser verursachte innerhalb weniger Stunden Brüche der Drähte (Bild 7), da er zwar Rost löste, jedoch gleichzeitig im hohen Maße Wasserstoffrißkorrosion förderte.

Bild 5
Fehlende bzw. unzureichende Betondeckung bei Spanngliedern des Daches der Kongreßhalle, Berlin, nach 29 Jahren

Bild 6
Korrosion und lochfraßinduzierte Wasserstoffrißkorrosion durch Betonrestwasser bei Verwendung eines für Spannbeton nicht zugelassenen Betonverflüssigers

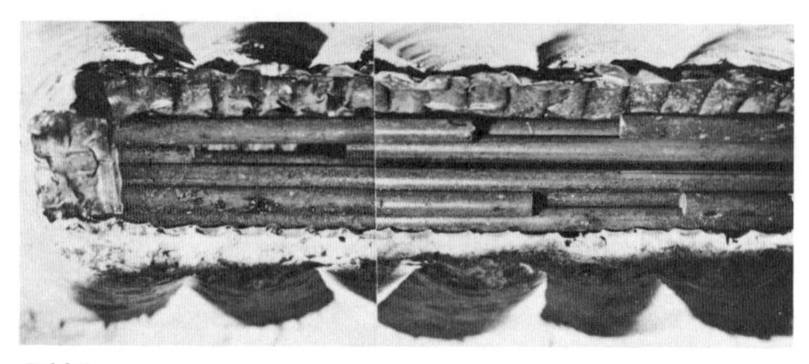

Bild 7
Wasserstoffrißkorrosion nach Spülen unverpreßter Spannkanäle mit einem stark sauren, sog. „Rostlöser"

Fall 6 [3]

Bei Verpreßankern (Erdankern) für vorübergehende Zwecke wurde abweichend von der vorgeschriebenen Ausführung im unverpreßten Bereich hinter dem Ankerkopf die Verbindung Ankerkopf/Hüllrohr nicht besonders abgedichtet. Das Hinzutreten von aggressivem Grundwasser führte bei dem relativ empfindlichen Spannstahl nach wenigen Monaten zu Brüchen (Bild 8).

4 Entwicklung der Vorschriften aus korrosionstechnischer Sicht

In Bild 2 wurde herausgestellt, daß der größte Teil der in der Vergangenheit aufgetretenen ernsthaften Schä-

Bild 8
Bruch eines Erdankers außerhalb der Verpreßstrecke wegen Abweichung vom vorgeschriebenen Korrosionsschutzsystem

den die vor 1960 erstellten Bauwerke betrifft. Der Rückgang der Schäden ab 1960 ist zu einem erheblichen Teil auf eine Verbesserung des Regelwerkes und dessen stärkere Beachtung zurückzuführen. Die Verbesserungen der Normen und Richtlinien betreffen sowohl die Baustoffe als auch die Bauausführung. Tabelle 1 stellt die wesentlichsten Änderungen bei den Baustoffen, Tabelle 2 diejenigen der Bauausführungen bei den deutschen Regeln, jeweils aus korrosionstechnischer Sicht gegenüber. Wiedergegeben ist der heutige Stand sowie jener von 1950 bis 1958. Im folgenden sollen die wesentlichsten Änderungen genannt und begründet werden; eine kritische Wertung der heutigen Regeln erfolgt in Abschnitt 5. Zunächst wird auf die Entwicklung bei den Baustoffen eingegangen (Tab. 1).

Spannstahl

Bei den warmgewalzten Stählen wurde die Festigkeit durch legierungs- und verfahrenstechnische Maßnahmen im Laufe der Jahre stetig erhöht. Eine Steigerung der Festigkeit bis auf Werte entsprechend einem St 1080/1320 konnte sich wegen der damit zusammenhängenden Erhöhung der Wasserstoffrißkorrosionsempfindlichkeit und der Schadenshäufigkeit [1] nicht durchsetzen.

Die vergüteten Spannstähle erfuhren durch legierungstechnische Maßnahmen (Herabsetzung des Kohlenstoffgehaltes, Zulegierung von Chrom) gegenüber früher eine erhebliche Verbesserung ihres Rißkorrosionsverhaltens [6]. Bei dieser Stahlsorte wurde die Gruppe mit der niedrigsten Festigkeit (St 125/140) aufgegeben.

Bei den gezogenen Drähten und Litzen wurden die Festigkeiten und damit die Empfindlichkeit gegenüber Korrosion herabgesetzt.

Bezüglich des Durchmessers von Einzeldrähten, Stäben und Litzen ist ein ständiger Trend zu größeren Durchmessern zu verzeichnen. Zusätzlich wurden die Mindestdurchmesser erhöht [7]. Neben wirtschaftlichen Gesichtspunkten wurde unterstellt, daß größere Mindestdurchmesser das Versagensrisiko bei einem Korrosionsangriff herabsetzen. Diese Entwicklung ging auf Kosten der guten Verbundeigenschaften dünnerer Stähle und machte im Bereich der gezogenen und vergüteten Sorten Neuentwicklungen notwendig.

Nachweise der Unempfindlichkeit gegenüber Spannungsrißkorrosion, die früher in der Eigenverantwortung der Spannstahlhersteller lagen, werden heute durch Richtlinien geregelt.

Tabelle 1
Regelungen bezüglich der Baustoffe für Spannbeton aus korrosionstechnischer Sicht

		Stand 1984		Stand 1950–1958	
	Baustoffe	Vorschrift	Regelung	Vorschrift	Regelung
1	Spann-stahl	Zulassung	R_m^{Nenn} – warmgewalzt 1080–1230 N/mm² vergütet (neu) 1470–1570 N/mm² gez., Litze (7) 1570–1770 N/mm²	Zulassung	R_m^{Nenn} – warmgewalzt 833–1030 N/mm² vergütet (alt) 1375–1570 N/mm² gez., Litze 1670–1960 N/mm² (2, 3, 7)
		Richtlinie	SRK-Verhalten ist nachzuweisen		SRK-Nachweis nicht gefordert
		DIN 4227 (1)	d_s-Draht: ≧ 5 mm (≧ 30 mm², oval) d_s-Litze: 7 × ≧ 3 mm		d_s-Draht: ≧ 1,5 mm (≧ 20 mm², oval) d_s-Litze: ≧ 2 × ≧ 1,5 mm

Fortsetzung Seite 116

Tabelle 1 (Fortsetzung von Seite 115)
Regelungen bezüglich der Baustoffe für Spannbeton aus korrosionstechnischer Sicht

		Stand 1984		Stand 1950–1958	
	Baustoffe	Vorschrift	Regelung	Vorschrift	Regelung
2	Zement-güte (Beton)	DIN 4227 (1) DIN 1164	PZ, EPZ: Z 35 F, Z 45, Z 55 HOZ, TrZ: Z 45, Z 55 TSZ seit 1962 verboten	DIN 1045 DIN 1164 DIN 1167	PZ, EPZ, HOZ: Z 225, Z 325, Z 425 TrZ TSZ erlaubt
			Cl^--Gehalt: ≦ 0,1 Gew.-% (seit 1958)		Cl^--Gehalt: keine Angabe
3	Zement-güte (Einpreß-mörtel)	DIN 4227 (5)	PZ 35 F, PZ 45 F, PZ 55	Richtlinie 1957	PZ
		DIN 1164	Cl^--Gehalt: ≦ 0,1 Gew.-%		Cl^--Gehalt: keine Chorid-zugaben erlaubt
4	Beton-zuschläge	DIN 4227 (1)	Cl^--Gehalt: nachtr. Verbund: keine Angaben, sofort. Verbund: ≦ 0,02 Gew.-% löslich		Cl^--Gehalt: keine Angaben
5	Anmach-wasser	DIN 4227 (1)	Cl^--Gehalt: ≦ 600 mg/Liter	DIN 1045 Richtlinie 1957	Beton: alle natürlichen Wässer geeignet Einpreßmörtel: keine Angaben
6	Zusatz-mittel	Zulassung, Prüfbescheid	Cl^--Gehalt: < 0,002% (bezog. Zement) SCN^-: Zugabe nicht erlaubt	Richtlinie 1957	Cl^--Gehalt: nicht geregelt, z. T. zugegeben Einpreßmörtel: Chloridzugabe nicht erlaubt SCN^-: keine Zugabe erfolgt
7	Beton-zusatz-stoffe	Zulassung, Prüfbescheid	nachträglicher Verbund: erlaubt sofortiger Verbund: nicht erlaubt		nicht geregelt
8	Beton-güte	DIN 4227 (1)	nachträglicher Verbund: ≧ B 25, sofortiger Verbund: ≧ B 35	DIN 4227	nachträglicher Verbund, sofortiger Verbund: ≧ B 25 (B 300)
		Richtlinie 1983	Zementgehalt: ≧ 300 kg/m³, w/z: ≦ 0,6	DIN 1045	Zementgehalt: ≧ 300 kg/m³, w/z: nicht geregelt
9	Einpreß-mörtel-güte	DIN 4227 (5)	$\beta_{c,28}$: ≧ 30 N/mm², w/z ≦ 0,44	Richtlinie 1957	$\beta_{c,28}$: ≧ 30 N/mm², w/z ≦ 0,44
10	Hüll-rohre	DIN 18 553	Dichtigkeit der Nähte und Stöße, keine Knicke, keine Beschädigungen		keine Angaben zur Hüllrohrausführung

Tabelle 2
Regelungen der Bauausführung für Spannbeton aus korrosionstechnischer Sicht

		Stand 1984		Stand 1950–1958	
	Bauausführung	Vorschrift	Regelung	Vorschrift	Regelung
1	Transport, Lagerung der Spannglieder bzw. der Spannstähle bis zur Spanngliedherstellung	DIN 4227 (1)	Leichter Flugrost erlaubt, frei von schädigendem Rost, keine Berührung mit schädigenden Stoffen, Trockenlagerung, Fertigspanngliedherstellung in geschlossenen Räumen	DIN 4227	Leichter Flugrost erlaubt, rostfreier Zustand
		Zulassung	Trocken, geschützt, beschädigungsfrei befördern und lagern, keine Berührung mit schädigenden Stoffen (Chloride, Nitrate, Säuren)	Zulassung	Trocken transportieren, einwandfrei lagern, Transportmittel frei von Rückständen, die Korrosion hervorrufen
2	Einbau der Spannstähle bzw. Spannglieder	DIN 4227 (1)	Zeitbegrenzung: max. 12 Wochen zwischen Herstellung und Verpressen, davon max. 4 Wochen frei in der Schalung und max. 2 Wochen gespannt	Zulassung	Zeitbegrenzung: Verarbeitung innerhalb von 3 Monaten nach Lieferung
3	Besonderer Korrosionsschutz		erforderlich, falls Bedingungen der Zeilen 1 und 2 nicht erfüllt		Keine Anforderungen
4	Betondeckung	DIN 4227 (1)	nachträglicher Verbund: Hüllrohre $\geqq$ 3 cm, sofortiger Verbund: nach DIN 1045, Tabelle 10, ($\geqq$ 2–4 cm), > Zuschlagkorn	DIN 1045	generell: $\geqq$ 1,5 bis 2 cm, große Abmessungen, aggressive Umgebungsverhältnisse: bis 4 cm
5	lichter Spanngliedabstand	DIN 4227 (1)	nachträglicher Verbund: $\geqq$ 2,5 cm, $\geqq 0{,}8\,d_i$ Hüllrohr, sofortiger Verbund: > Zuschlagkorn	DIN 4227	generell: ausreichende Umhüllung der Spannglieder muß gewährleistet sein
6	Abstand Spannglieder – verzinkte Teile	DIN 4227 (1)	$\geqq$ 2 cm, keine metallische Verbindung Spannglied – verzinkte Teile		keine Angaben

Fortsetzung Seite 118

Tabelle 2 (Fortsetzung von Seite 117)
Regelungen der Bauausführung für Spannbeton aus korrosionstechnischer Sicht

		Stand 1984		Stand 1950–1958	
	Bauausführung	Vorschrift	Regelung	Vorschrift	Regelung
7	Stahlspannung	DIN 4227 (1)	$\leqq 0{,}55\,R_m$, $\leqq 0{,}75\,R_{p,0,2}$	DIN 4227	$\leqq 0{,}55\,R_m$, $\leqq 0{,}75\,R_{p,0,2}$
8	Einpreß-vorgang	DIN 4227 (5)	geregelt zwecks Erzielung einer vollständigen Verpressung, Wasserspülung erlaubt	Richtlinie 1957	geregelt zwecks Erzielung einer vollständigen Verpressung, Wasserspülung der Spannkanäle erforderlich
9	Rißbreiten		nur Zustand I erlaubt		nur Zustand I erlaubt

Zementgüte

In der Entwicklung der Zemente für Spannbeton stellt das Verbot des Tonerdeschmelzzementes im Jahre 1962 den wesentlichsten Einschnitt dar. Die Vielzahl der zuvor zu verzeichnenden Schadensfälle beim TSZ-Beton nach Einwirken von Feuchtigkeit und Wärme machte die Verwendung von Tonerdeschmelzzement untragbar. Wegen der zeitabhängigen Porositätszunahme wurde die Karbonatisierung des ohnehin $Ca(OH)_2$-armen Zementsteins erheblich beschleunigt, was im Zusammenhang mit dem erhöhten Sulfidschwefelgehalt (deutscher Zemente) eine Wasserstoffrißkorrosion der Spannstähle begünstigte [8].
Seit 1958 dürfen nur noch Zemente mit Chloridgehalten $\leqq$ 0,1 Gew.-% verwendet werden. Diese Maßnahme wurde notwendig, nachdem in erhöhtem Maße chloridbedingte Spannstahlschäden (vor allem bei chloridhaltigen Zusatzmitteln) aufgetreten waren [1].

Betonzuschläge, Anmachwasser, Zusatzmittel

In den Anfängen des Spannbetons waren die Chloridgehalte dieser Betonbestandteile nicht geregelt. Als Anmachwasser war z. B. die Verwendung von Meerwasser nicht ausgeschlossen. Nachdem bei Spannbeton ein Chloridgehalt von 0,2 Gew.-%, bezogen auf den Zement, als ein für die Korrosion kritischer Grenzwert festgelegt wurde, mußten die maximalen Cl^--Gehalte aller Betonbestandteile entsprechend begrenzt werden.

Die seinerzeitige Vorschrift, Rhodanide (SCN^-) in Zusatzmitteln für Spannbeton nicht zu erlauben, hat sich nachträglich als richtig erwiesen, da diese Bestandteile sich in nicht abgebundenen Betonwässern aufkonzentrieren und Wasserstoffrißkorrosion hervorrufen können [5]. Zwischenzeitlich wurden rhodanidhaltige Betonzusatzmittel auch für Stahlbeton verboten [9], um eine Verwechslungsgefahr (versehentliche Verwendung im Spannbetonbau) von vornherein auszuschließen.

Betongüte, Hüllrohre

Bei Beton mit sofortigem Verbund wurde gegenüber früher die Anforderung an die Betongüte erhöht. Die im Vergleich zu Bauteilen mit nachträglichem Verbund möglichen geringeren Betondeckungen und das Fehlen zusätzlich schützender Hüllrohre ließen es ratsam erscheinen, einen hochfesteren und damit dichteren Beton zu verwenden, der zusätzlich auch einen besseren Verbund liefert. Bei Konstruktionen mit nachträglichem Verbund wurden die erforderlichen Festigkeiten gegenüber früher nicht angehoben, da die höheren Anforderungen an die Zementgüten und die zwischenzeitlich erfolgte Qualitätsverbesserung der Zemente ohnehin zu höheren Festigkeiten führen, was in Verbindung mit der Forderung nach dichten Hüllrohren einen erhöhten Korrosionsschutz bietet. In die gleiche Richtung zielt die neuerliche Forderung der Begrenzung des w/z-Wertes für Spannbetonbauwerke auf $\leqq 0{,}6$.

Im folgenden wird die Entwicklung bei der Bauausführung (Tabelle 2) kurz abgehandelt:

Transport, Lagerung der Spannstähle

Die Vorschriften für Transport und Lagerung der Spannstähle bis zur Spanngliedherstellung sind in den letzten 30 Jahren nicht grundsätzlich geändert bzw. verbessert worden.

Einbau der Spannstähle bzw. Spannglieder

Die Zeitbegrenzung für die Verarbeitung von Spannstählen (3 Monate) wurde im Grundsatz beibehalten. Der wachsende Umfang der Spannbetonbauweise mit nachträglichem Verbund und die dabei aufgetretenen Spannstahlbrüche im gespannten, aber noch nicht verpreßten Zustand [1], machten es notwendig, zusätzliche Grenzen für die freie Lagerung der Spannglieder in der Schalung und im gespannten, nicht verpreßten Zustand einzuführen.

Besonderer Korrosionsschutz

Die vorgenannten Zeitbegrenzungsvorschriften für die Verarbeitung der Spannstähle lassen sich in der Praxis vielfach nicht einhalten. Gründe sind beispielsweise Verzögerungen im Bauablauf, Frostperioden und spezielle Bau- oder Spannverfahren. Deshalb und auch wegen der von Fall zu Fall gegenüber früher gestiegenen Aggressivität der Atmosphäre müssen Spannstähle in Ausnahmefällen besonders geschützt werden.

Betondeckung

Im Hinblick auf die Verbesserung der Dauerhaftigkeit wurden die Betondeckungen gegenüber früher generell heraufgesetzt. Die in den ersten Spannbetonvorschriften noch möglichen geringen Betondeckungen ($\geqq$ 1,5 cm) ließen sich nicht immer mit der gewünschten Zuverlässigkeit einhalten. Bei Schadensuntersuchungen wurden gelegentlich Betondeckungen von nur wenigen Millimetern festgestellt [1], [3], [4].

Lichter Spanngliedabstand

Bei älteren Bauwerken wurde insbesondere bei dünnen Stegen und schmalen Balken eine zu dichte Anordnung der Spannglieder festgestellt, die häufig zu Betonierfehlern führte [10]. Auch waren dadurch vielfach Abweichungen von der planmäßigen Spanngliedlage bedingt [3]. Dieser Sachverhalt machte es erforderlich, insbesondere bei der Vorspannung mit nachträglichem Verbund, den lichten Spanngliedabstand besonders zu regeln.

Abstand Spannglieder – verzinkte Teile

In den Anfängen der Spannbetonpraxis sind bei empfindlichen Spannstählen sehr vereinzelt kurz nach dem Betonieren Spannstahlbrüche dort aufgetreten, wo die Stähle Kontakt mit verzinkten oder aluminierten Teilen hatten [1]. Um eine solche Elementbildung in Zukunft auszuschließen, wurde ein Mindestabstand gefordert und eine metallisch leitende Verbindung der Spannglieder mit verzinkten Teilen untersagt.

Einpreßvorgang

Zur Überprüfung oder Schaffung der Durchgängigkeit der Spannkanäle für das Einpressen kann eine Spülung der Hüllrohre mit Wasser erfolgen. Die diesbezüglich frühere „Mußbestimmung" ist einer „Kannbestimmung" gewichen.

5 Bewertung der derzeitigen Spannbetonregeln aus korrosionstechnischer Sicht

Neuere Forschungsergebnisse [11], [12] auf dem Gebiet der Spannstahlkorrosion sowie die Baupraxis haben gezeigt, daß die derzeitigen korrosionstechnischen Vorschriften für den Spannbeton in einzelnen Punkten schwer interpretierbar sind und auch nicht immer den Kern der Probleme treffen. Der Baupraktiker ist darüber hinaus bezüglich der Einhaltung der Vorschriften in vielen Fällen überfordert.

Spannstahl, Stahlspannung

Zwischen der bereits in Abschnitt 4 erläuterten Tendenz der Stahlhersteller, Spannstähle höherer Festigkeit anzubieten und gewissen Vorbehalten der Anwender in der Bundesrepublik Deutschland, diese Stähle auch höher als bisher auszunutzen, besteht ein gewisser Widerspruch. Während im westlichen Ausland unabhängig von der Stahlgüte vielfach zulässige Vorspannungen bis zu 0,8 R_m erlaubt sind, glaubt man bei uns aufgrund einer damit erhöhten Korrosionsgefährdung, diesen Schritt nicht mitvollziehen zu können. Dazu ist folgendes anzumerken:

Sowohl aus Laboruntersuchungen als auch aus praktischen Erhebungen ist bekannt, daß höhere Festigkeiten und höhere Gebrauchsspannungen die Spannungsrißkorrosions-Sicherheit herabsetzen. Gemäß der Beziehung [13]

$$L = \frac{C}{\sigma^3 \cdot R_m^9}$$

nimmt die Lebensdauer L mit der 3. Potenz der Spannung σ und der 9. Potenz der Festigkeit R_m ab (der Faktor C ist ein Kennwert für die Werkstoffbeständigkeit im betreffenden Medium). Bild 9 verdeutlicht am Beispiel eines warmgewalzten Spannstahls die Zusammenhänge zwischen Festigkeit (1070 – 1320 N/mm^2), Spannung (0,45 – 0,85 R_m) und Standzeit bei Wasserstoffrißkorrosion. Am tabellarischen Beispiel im Bild ist erkennbar, daß zur Erzielung einer Vorspannung von z. B. 800 N/mm^2 die höhere Ausnutzung eines niedrigfesteren Stahls eine höhere Sicherheit gegenüber Wasserstoffrißkorrosion bietet als die niedrigere Ausnutzung eines höherfesten Stahls. Beispielsweise würde demnach die höhere Ausnutzung (0,75 R_m) eines St 835/1030 eine 6,5fach höhere Standzeit erbringen als die geringere Ausnutzung (0,60 R_m) eines St 1080/1230. Diese Zusammenhänge zeigen, daß die Entwicklung immer höherfesterer Spannstähle aus korrosionstechnischer Sicht nicht sinnvoll erscheint. Andererseits spricht auch nichts dagegen, niedrigfestere Stähle höherer Beständigkeit auf Dauer auch höher auszunutzen, wenn dadurch nicht andere Eigenschaften der Stähle beeinträchtigt werden.

Stahldurchmesser

Nach den gegenwärtigen Regelungen dürfen Mindestwerte der Stahldurchmesser primär aus korrosionstechnischen Gründen nicht unterschritten werden, was jedoch unbegründet erscheint. In der Praxis sind nämlich Langzeitschäden infolge anodischer Korrosion der Stähle äußerst selten, und nur extreme Korrosionsangriffe sind in der Lage, die Dauerhaftigkeit der Stähle zu beeinträchtigen [14]. Das Versagen von Spannstählen ist in den allermeisten Fällen auf eine Form von Rißkorrosion zurückzuführen. Sowohl die Schadenserhebungen [1] als auch Laborversuche [15] haben gezeigt, daß die Durchmesser der Stähle nur von untergeordnetem Einfluß auf die Rißkorrosionsbeständigkeit sind, sofern sich nicht mit der Veränderung des Durchmessers die Stahleigenschaften ändern.

Rißkorrosionsverhalten

Zum gegenwärtigen Zeitpunkt muß das Spannungsrißkorrosions-Verhalten von Spannstählen nach vorläufigen Richtlinien [16], die sich an die FIP-Richtlinien anlehnen, nachgewiesen werden. Die darin aufgeführten Prüfungen in chemisch konzentrierten Lösungen sind jedoch weder in der Lage, zwischen den einzelnen Stahlsorten ausreichend zu klassifizieren, noch deren Verhalten in der Praxis ausreichend wiederzugeben. Neuere Untersuchungen [11], [17] haben ergeben, daß dieser Mangel z.B. durch Prüfung in verdünnten, wäßrigen Lösungen behoben wird. Nachdem hierbei bereits in großem Umfang reproduzierbare und dem praktischen Verhalten der Stähle konforme Er-

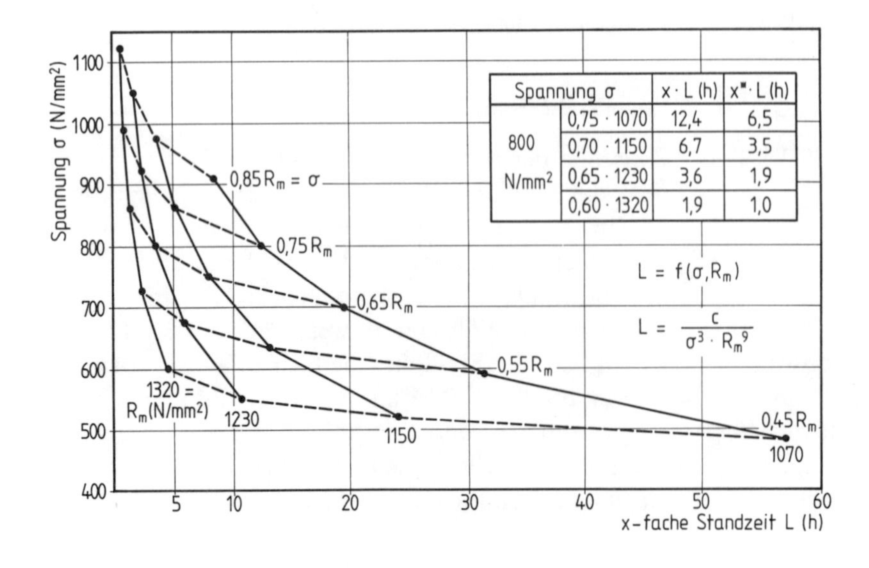

Spannung σ		x · L (h)	x* · L (h)
800 N/mm²	0,75 · 1070	12,4	6,5
	0,70 · 1150	6,7	3,5
	0,65 · 1230	3,6	1,9
	0,60 · 1320	1,9	1,0

Bild 9
Zusammenhang zwischen Festigkeit, Spannung und Standzeit bei Wasserstoffrißkorrosion (Beispiel warmgewalzter Stahl)

gebnisse vorgelegt werden konnten [18], wäre es empfehlenswert, diese Prüfmethode in das zukünftige Regelwerk aufzunehmen.

Chloridgehalte der Betonbestandteile

Die Begrenzung der Chloridgehalte der einzelnen Betonbestandteile auf Obergrenzen, die die Gewähr bieten, daß der Gesamtgehalt an Cl^--Ionen im Beton 0,2 Gew.-%, bezogen auf den Zement, nicht übersteigt, ist als eine grundsätzlich sinnvolle Maßnahme anzusehen. Dem Bauausführenden sollte dabei jedoch stets klar sein, daß es sich hier lediglich um Richtwerte handelt, deren fallweises Überschreiten nicht immer gleich als Alarmsignal zu bewerten ist. Der für die Korrosion kritische Grenzgehalt an Chloriden hängt nämlich ganz entscheidend von den Belüftungsverhältnissen des Betons am Stahl (w/z-Wert, Betondeckung) ab. Bei den für Spannbeton in der Regel hohen Betongüten und Betondeckungen sind bei fachgerechter Ausführung weit höhere Chloridgehalte noch unschädlich [19], da der für die Korrosion notwendige Sauerstoff nicht nachdiffundieren kann. Bild 10 zeigt in diesem Zusammenhang Ergebnisse zum Einfluß von Chloridgehalt, w/z-Wert und Betondeckung auf die Korrosion auf. Im Mittel aller Chloridgehalte wird die Korrosion um den Faktor 10 verringert, wenn z. B. der w/z-Wert von 0,60 auf 0,45 herabgesetzt und gleichzeitig die Betondeckung von 1,5 cm auf 3,0 cm erhöht wird.

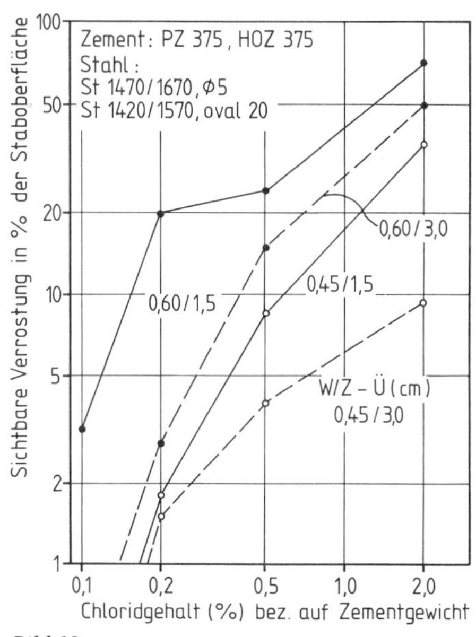

Bild 10
Einfluß von Chloridgehalt, w/z-Wert und Betondeckung auf die Korrosion nach 6,5 Jahren (nach MÜLLER, RAUEN)

Dichtigkeit der Hüllrohre, Einpreßvorgang

Nach den bisherigen Anforderungen müssen Hüllrohre dicht sein, damit beim Betoniervorgang kein sich absetzendes Wasser eindringen kann. Vor dem Verpressen kann mit Wasser gespült werden, um den freien Durchgang sicherzustellen. Zur Vermeidung von Hohlstellen im Einpreßmörtel ist das verbleibende Wasser mit Druckluft auszublasen. Aufgrund jüngerer Schadensfälle [5] und umfangreicher Untersuchungen [11] ist diese Vorgehensweise in einem völlig neuen Licht zu sehen. Tatsächlich sind die verwendeten Hüllrohre herstellungs- oder verarbeitungsbedingt nicht immer dicht, so daß Betonwässer mit einer Aufkonzentrierung von Schadstoffen in das Hüllrohrinnere gelangen können. Das Eindringen ist auch bei einigen Spannverfahren systembedingt möglich. Deshalb sollten prinzipiell alle Spannglieder unmittelbar nach dem Abbinden des Betons mit Leitungswasser gespült werden. Hinsichtlich einer Korrosion ist das trotz Ausblasen mit Druckluft noch verbleibende Restwasser einschließlich der möglichen Kondenswasserbildung weit weniger kritisch als im Hüllrohr bis zum Verpressen verbleibendes Betonrestwasser.

Transport, Lagerung der Spannglieder

Rostbildung auf den Spannstahloberflächen und augenscheinlich erkennbare Korrosionseffekte sind kein ausreichender Maßstab für die Beurteilung einer Gefährdung. Es existieren nämlich Spannstähle, bei denen bereits kleinste Korrosionsangriffe in der Lage sind, Wasserstoffrißkorrosion auszulösen [1], [15], während die überwiegende Zahl der Stähle starke Korrosionseffekte erträgt, ohne daß die Verwendbarkeit eingeschränkt wird. Im Hinblick auf diesen Sachverhalt erscheint die Definition des sog. schädigenden Rostes („Flugrost“) in der DIN 4227 äußerst fragwürdig. Es kommt hinzu, daß für den Anwender die Grenze zwischen Flugrost und Narbenrost nicht zu erkennen ist. Rostbildung auf Spannstählen läßt sich nur dann vermeiden, wenn klare Anweisungen für Transport, Lagerung und Einbau vorliegen. Die jetzigen Formulierungen der DIN 4227 und der Zulassungen (Tabelle 2) können diesbezüglich als noch nicht ausreichend erachtet werden. So wird z. B. die geforderte trockene und geschützte Lagerung von vielen Bauingenieuren noch so verstanden, daß Spannstähle oder Spannglieder mit oder ohne zusätzlichen Schutz längerzeitig unter einer Abdeckplane gelagert werden.

Tatsächlich kann unter solchen Bedingungen durch Kondenswasserbildung bewirkte Narbenkorrosion auftreten [11]. Der ideale Schutz für Spannstähle sind gut belüftete oder temperierte Räume.

Einbau der Spannglieder

Die gegenwärtigen Regelungen gehen davon aus, daß in „lufttrockenen", noch nicht verpreßten Hüllrohren korrosionsbedingte Spannstahlschädigungen durch Kondenswasserbildung am wahrscheinlichsten sind. Da frei in der Schalung liegende Spannglieder diesbezüglich gefährdeter sind als zusätzlich einbetonierte, wurden in der DIN 4227 die maximalen Zeitbegrenzungen von 4 Wochen frei in der Schalung und 2 Wochen in unverpreßtem Zustand gespannt aufgenommen.
Bei der Festschreibung dieser Zeitspannen wurde noch nicht berücksichtigt, daß das sich beim Betonieren absetzende und in das Hüllrohrsystem eindringende Wasser aufgrund seiner Aufkonzentration mit aggressiven Stoffen für den Spannstahl eine erheblich höhere Korrosionsgefährdung darstellt als das bloße Einwirken von Kondenswasser [5]. Selbst im noch nicht gespannten Zustand ist innerhalb weniger Tage erhebliche Narbenkorrosion und lokale Wasserstoffbildung und -aufnahme durch den Stahl möglich. Daraus resultierten in der Vergangenheit Schadensfälle mit Spannstahlbrüchen unmittelbar beim Vorspannen.
Aus den vorgenannten Gründen sind Zeitbegrenzungen beim Einbau der Spannstähle zwar sinnvoll, sie sollten jedoch nicht durch zu starre Auslegung die Gegebenheiten des Einzelfalles unberücksichtigt lassen. Auch sollte, wie bereits erwähnt, das Spülen der Hüllrohre unmittelbar nach dem Betonieren in das Regelwerk mit aufgenommen werden.

Temporärer Korrosionsschutz

Art und notwendige Wirkungsweise temporärer Korrosionsschutzmaßnahmen für Spannstähle sind nicht im einzelnen geregelt. Sie sollten jedoch eine Schutzwirkung erzielen und die Verarbeitbarkeit der Stähle nicht nachträglich beeinflussen [20], [21]. Allen bisher entwickelten besonderen Korrosionsschutzmaßnahmen ist gemeinsam, daß sie auf eine Verhinderung der anodisch abtragenden Korrosion (Narbenbildung, Rostbildung) abzielen.
In der Praxis hat sich allerdings gezeigt, daß der damit verbundene, vielfach hohe Aufwand nicht immer zu dem erwarteten Schutz geführt hat. Obwohl die Maßnahmen zu einer Begrenzung der korrodierten Oberfläche geführt haben, konnte in manchen Anwendungsfällen Rißkorrosion nicht verhindert werden [1]. Es fehlt bisher der Nachweis, ob die gängigen Maßnahmen aufgrund ihrer physikalisch-chemischen Wirkungsweise überhaupt in der Lage sind, Rißkorrosion auszuschließen. Hier wäre ein wichtiger Ansatzpunkt für zukünftige Forschungsaktivitäten.

Abstand Spannglieder – verzinkte Teile

Der in der DIN 4227 geforderte Mindestabstand zwischen Spanngliedern und verzinkten Einbauteilen und das Verbot einer leitenden Verbindung zwischen diesen Bauteilen beruht im wesentlichen auf folgenden Annahmen, die vor allem auf Laborversuche zurückzuführen sind:

- zusätzliche Gefährdung der Spannstähle durch Wasserstoffrißkorrosion,
- zusätzliche Gefährdung der Spannstähle durch abtragende Korrosion.

Wie aus Bild 11 durch Aufzeichnung der Korrosionspotentiale beispielhaft hervorgeht, kann sich Zink in frischem Beton vorübergehend erheblich unedler als Stahl verhalten. Stehen Zink und unverzinkter Spannstahl in leitender Verbindung, so findet eine Elementbildung statt. Bei unmittelbarem Kontakt zwischen Zink und Stahl stellt sich ein Mischpotential an der Kontaktstelle ein, das im wesentlichen jenem des Zinks entspricht. Durch den fließenden Elementstrom wird Zink im Kontaktbereich verstärkt aufgelöst und der Stahl kathodisch geschützt (polarisiert), wobei eine Wasserstoffabscheidung am unverzinkten Stahl erfolgen kann.
Bei zunehmender Betonerhärtung nimmt die Gefahr der Wasserstoffabscheidung bei unmittelbarem Kontakt zwischen Zink und Stahl bereits nach kurzer Zeit erheblich ab, da sich auf dem Zink Deckschichten bilden. Bild 11 verdeutlicht dies anhand des zeitlichen Anstiegs des Zink-Potentials und des Abfalls des Elementstroms.
Der erläuterte Vorgang ist jedoch noch an weitere Voraussetzungen gebunden, um auch unter praktischen Verhältnissen einen tatsächlichen Schaden beim Spannstahl entstehen zu lassen. Die Unkenntnis dieser weiteren Voraussetzungen führte häufig zu Verunsicherungen, und durch Begutachtungen im Einzelfall

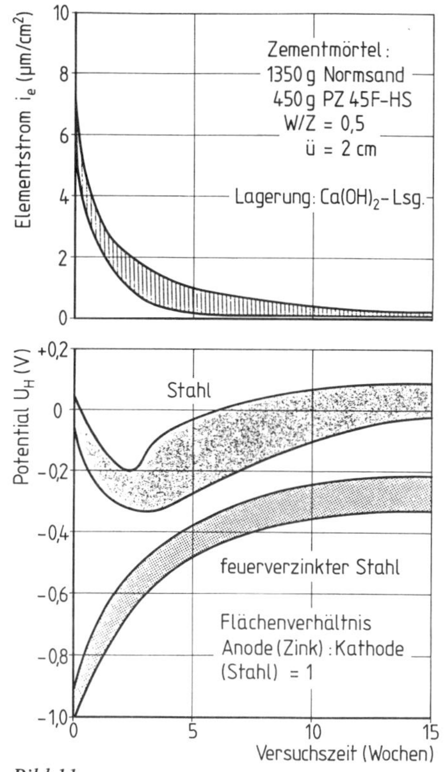

Bild 11
Zeitlicher Verlauf der freien Ruhepotentiale und des Elementstroms von Stahl und feuerverzinktem Stahl in Zementmörtel (nach RÜCKERT, NEUBAUER)

wurde mehrfach nachgewiesen, daß die Forderungen der DIN 4227 in begründeten Einzelfällen nicht gerechtfertigt waren.
Seitens der konstruktiven Ausbildung gilt nämlich, daß die kathodische Polarisation des Stahls in Beton mit zunehmender Entfernung von der unmittelbaren Kontaktstelle Stahl/Zink stark abnimmt und höchstens wenige Millimeter erreicht. Zeitlich erhöhte Elementströme fließen demzufolge nur im unmittelbaren Kontaktbereich. Spannglieder, die über eine metallisch leitende Stahlverbindung (z. B. Betonstahl) über mehr als einige Millimeter Entfernung mit einem Zinkteil verbunden sind, sind daher bezüglich Wasserstoffabscheidung nicht mehr gefährdet.
Die Polarisierbarkeit von Spannstahl (Kathode) durch Zink (Anode) ist zudem vom Flächenverhältnis Anode/Kathode abhängig. Nur hohe Verhältniszahlen, die in der Baupraxis kaum vorliegen, können sich ungünstig auswirken.
Schließlich setzt eine Wasserstoffrißkorrosion eine besondere Empfindlichkeit der Spannstähle voraus. Der im frischen Beton kurzzeitig entstehende Wasserstoff reicht jedoch bei den heute zugelassenen Spannstählen nicht aus, um bleibende Schäden herbeizuführen.
Aus dem vorgenannten Zusammenhang wird deutlich, daß das Verbot eines unmittelbaren Kontaktes Spannstahl/Zink als vorbeugende Maßnahme noch sinnvoll erscheint. Das Verbot jeglicher leitender Verbindung von Spanngliedern mit verzinkten Teilen ist jedoch sachlich ungerechtfertigt.
Daß ein Kontakt von Spannstahl mit Zink nicht zu einer verstärkten anodischen Korrosion beim Spannstahl führen kann, versteht sich wegen der kurz dargelegten Zusammenhänge von selbst. Der Spannstahl ist höchstens in der Lage, die Korrosionsgeschwindigkeit des Zinks im Kontaktbereich kurzfristig zu erhöhen.
Ausgehend von den Verhältnissen bei Wasserinstallationen wurde zeitweise auch für den Betonbau angenommen, daß ab etwa 60° eine sogenannte Potentialumkehr stattfindet, wobei der Stahl an der Kontaktstelle bevorzugt anodisch aufgelöst wird. Durch neuere Untersuchungen [22] wurde zwischenzeitlich jedoch eindeutig gezeigt, daß diese Potentialumkehr bei erhöhten Temperaturen im Beton nicht stattfindet und daß demnach eine Verbindung der Bewehrung mit verzinkten Stahlteilen seitens einer abtragenden Korrosion bei der Bewehrung grundsätzlich unbedenklich ist.

6 Schlußfolgerung

Die vorstehenden Ausführungen zeigen, daß seit der Einführung der Spannbetonbauweise eine stetige Anpassung des Regelwerkes in Anlehnung an die in der Praxis gewonnenen Erfahrungen stattgefunden hat, um aus korrosionstechnischer Sicht zu einer Verbesserung der Konstruktionen zu gelangen. Die korrosionstechnische Bewertung des heutigen Vorschriftenstandes läßt jedoch auch erkennen, daß noch nicht alle Regelungen diesbezüglich so abgefaßt sind, um dem Konstrukteur als verständliche und eindeutig interpretierbare Hilfsmittel dienen zu können. Die aufgezeigten Ansatzpunkte zur weiteren Verbesserung sind als Beitrag letztlich im Hinblick auf eine Erhöhung der Dauerhaftigkeit der Spannbetonbauwerke und der Wirtschaftlichkeit dieser Bauweise anzusehen.

7 Literatur

[1] NÜRNBERGER, U.: Analyse und Auswertung von Schadensfällen an Spannstählen. Forschung, Straßenbau und Straßenverkehrstechnik 308 (1980), S. 1–195.

[2] MATOUSEK, M. und SCHNEIDER, J.: Untersuchungen zur Struktur des Sicherheitsproblems von Bauwerken. Bericht des Institutes für Baustatik und Konstruktion, ETH Zürich, Februar 1976.

[3] Interne Versuchsberichte der FMPA Baden-Württemberg, 1981–1983, unveröffentlicht.

[4] ISECKE, B.: Failure Analysis of the Collapse of the Berlin Congress Hall. In: [12].

[5] REHM, G., NÜRNBERGER, U. und FREY, R.: Zur Korrosion und Spannungsrißkorrosion von Spannstählen bei Bauwerken mit nachträglichem Verbund. Bauingenieur 56 (1981), S. 275–281.

[6] JÄNICHE, W., STOLTE, E. und LITZKE, H.: Sicherheit gegen Bruch von Bewehrungsstählen in Stahlbeton- und Spannbetonbauteilen. Materialprüfung 7 (1965), S. 449–458.

[7] Ministerialblatt für das Land Nordrhein-Westfalen Nr. 59 vom 5. Mai 1967.

[8] REHM, G.: Schäden an Spannbetonbauteilen, die mit Tonerdeschmelzzement hergestellt wurden. Betonstein-Zeitung 29 (1963), S. 651–661.

[9] MANNS, W. und EICHLER, W. R.: Zur korrosionsfördernden Wirkung von thiocyanathaltigen Betonzusatzmitteln. Betonwerk + Fertigteiltechnik 48 (1982), S. 154–162.

[10] STANDFUSS, F.: Schäden an Straßenbrücken – Ursache und Folgerungen. Straße und Autobahn 30 (1979), S. 417–426.

[11] Korrosionsverhalten von Spannstählen – Neue Forschungsergebnisse. Gemeinsame Vortragsveranstaltung des IfBt (Berlin) und des VDEh (Düsseldorf), Berlin 7./8. 6. 1983.

[12] Stress Corrosion of Prestressing Steel. Third FIP-Symposium, Madrid 22./23. 9. 1981.

[13] STOLTE, E.: Über die Spannungskorrosion an Spannstählen. Beton- und Stahlbetonbau (1968), S. 116–118.

[14] NEUBERT, B. und NÜRNBERGER, U.: Prophylaxe, Früherkennung, Behebung von Spätschäden an Spannbetonbauteilen. Forschungsbericht II.6–13675 der FMPA Baden-Württemberg, Stuttgart 31. 1. 1983.

[15] REHM, G., NÜRNBERGER, U. und FREY, R.: Zur Korrosion und Rißkorrosion bei Spannstählen. Werkstoffe und Korrosion 32 (1981), S. 211–221.

[16] Richtlinien für die Zulassungs- und Überwachungsprüfungen von Spannstählen. Vorläufige Fassung der Bestimmungen für die Durchführung von Spannungsrißkorrosionsversuchen an Spannstählen, Institut für Bautechnik, Berlin, Juli 1978.

[17] REHM, G. und NÜRNBERGER, U.: Neue Methoden zur Beurteilung des Spannungsrißkorrosionsverhaltens von Spannstählen. Betonwerk + Fertigteiltechnik 48 (1982), S. 287–294.

[18] NÜRNBERGER, U.: Versuche zur praxiskonformen Beurteilung des Spannungsrißkorrosionsverhaltens von Spannstählen. Forschungsbericht II.6–14055 der FMPA Baden-Württemberg 31. 3. 1984.

[19] NÜRNBERGER, U.: Chloridkorrosion von Stahl in Beton. 1984 in Betonwerk + Fertigteiltechnik.

[20] REHM, G., RIECHE, G. und DELILLE, J.: Erfahrungen bei der Prüfung von temporären Korrosionsschutzmitteln für Spannstähle. Schriftenreihe des DAfStb 298 (1978), S. 3–19.

[21] JUNGWIRTH, D. und KERN, G.: Baupraktische Maßnahmen zum temporären Korrosionsschutz am Beispiel der Dywidag-Spannglieder. Techn. Beitrag 8. FIP-Kongreß, London 1978.

[22] RÜCKERT, J. und NEUBAUER, F.: Zum Kontaktverhalten von feuerverzinktem und unverzinktem Bewehrungsstahl in Beton bei erhöhter Temperatur. Werkstoffe und Korrosion 34 (1983), S. 295–299.

Die Entwicklung des Dauerankers – Eine Bilanz nach 25 Jahren

Manfred Stocker

1 Einführung

Der Daueranker im Lockergestein ist in diesem Jahr 25 Jahre alt geworden. Bereits ein Jahr nach der Entwicklung des Verpreßankers im Lockergestein auf der Baustelle Bayerischer Rundfunk in München im Jahre 1958 durch die Firma Bauer wurde der Anker bei zwei Projekten als Daueranker mit insgesamt 280 lfdm eingesetzt: beim Einlaßbauwerk Leitzachkraftwerk in Seeham und bei einer Rohrhängebrücke in Mühldorf. Im Jahre 1961 waren es schon 13400 lfdm. Heute liegt die jährliche Gesamtmenge an Dauerankern in Deutschland etwa bei 60000 lfdm, was einem Umsatz von ca. 9,5 Mio. DM entsprechen dürfte.
Der Daueranker hat dank seiner vielseitigen Anwendbarkeit (Verankerung von Stützbauwerken, Krag- und Seilkonstruktionen, Hangstabilisierungen, Auftriebssicherungen) eine sehr rasche Verbreitung gefunden und dabei eine große Entwicklung durchgemacht.
Die Ankerentwicklung wurde entscheidend durch die Entwicklung des Spannstahles beeinflußt. Die Probleme lagen weniger in der Erhöhung bzw. im Erhalt der Tragfähigkeit aus bodenmechanischer Sicht als vielmehr in der Korrosionssicherheit, setzt man für den Daueranker doch eine Lebensdauer von mindestens 50 bis 100 Jahren voraus.

2 Entwicklung des Spannstahles

2.1 Einstabanker

Die Entwicklung des Ankers begann mit einem Einstabanker, Durchmesser 26 mm. Auch heute ist der Einstabanker noch der am häufigsten verwendete Ankertyp in Deutschland. Dies hängt zusammen mit der Einfachheit der Handhabung, dem problemlosen Spannverfahren und dem gegenüber Bündelankern einfacheren Korrosionsschutz.

In den Jahren 1959 bis 1966 gab es Spannstähle, die sich als Ankerzugglieder nur bedingt eigneten. Es handelte sich um warm gewalzte, gereckte und angelassene, glatte Stähle mit Durchmessern von 26 und 32 mm und einer Güte St 785/1030 (siehe Tab. 1). Zwar waren damit bereits Gebrauchslasten bis zu 360 kN möglich, doch bereitete der glatte Stahl Schwierigkeiten bezüglich der Verankerung sowohl im Verpreßkörperbereich als auch am Kopf. Zur Erhöhung des Verbundes wurde am erdseitigen Ende des Zuggliedes ein 50 cm langes Gewinde aufgerollt. Zusätzlich mußte der Ankerstab mit der „verlorenen Spitze" der Bohrrohre verschraubt werden.
Am luftseitigen Ende des Ankerstahles wurde ein 25 cm langes Gewinde aufgerollt, um den Stab mit einer Gewindemutter verankern zu können. Das wegen der hohen Kosten relativ kurze Gewinde erforderte eine sehr hohe Genauigkeit bezüglich der Bohrlochtiefe, da sonst der Ankerstahl nicht mehr ordnungsgemäß verankert werden konnte. Das aufgerollte Gewinde war zudem gegen mechanische Verletzungen sehr empfindlich.
Im Jahr 1966 kam ein anfangs halbseitig, später beidseitig durchgehend mit Gewinderippen versehener Spannstahl auf den Markt mit Durchmesser 26 mm, im Jahre 1969 mit Durchmesser 32 mm, Stahlgüte St 785/1030. Dadurch wurde die Verbundeigenschaft deutlich verbessert und das Aufschrauben der Verankerungsmutter an jeder beliebigen Stelle ermöglicht.
Im Jahre 1969 bzw. 1971 wurde die Spannstahlgüte auf St 830/1030 angehoben. Zugleich wurde 1969 ein glatter Spannstahl, Durchmesser 36 mm, angeboten, der Gebrauchslasten von 485 kN ermöglichte. Als Druckrohranker konnte dieser Stahl voll ausgenützt werden und bedeutete einen wesentlichen Fortschritt. Als gerippter Stahl wurde der Durchmesser 36 mm ab 1972 angeboten.
Ein weiterer Fortschritt wurde erzielt, als im Jahre 1973 die Stahlgüte des warm gewalzten, gereckten und

Tabelle 1
Entwicklung der Zugglieder für Daueranker

Jahr	Stahl ∅ (mm)	Stahlgüte (N/mm²)	Stahlform	max. zul. Gebrauchslast (kN)	Stahl-anzahl (Stück)	Ankertyp
1959	26/32	758/1030	glatt	360	1	Einstab-anker
1966	26	785/1030	gerippt	238	1	
1969	32	785/1030	gerippt	360	1	
	26/32/36	830/1030	glatt	485	1	
1971	26,5/32	830/1030	gerippt	384	1	
1972	36	830/1030	gerippt	485	1	
1973	26/32/36	1080/1325	gerippt	628	1	
1974	26,5/32/36	1080/1230	gerippt	628	1	
1972	45/54/65	St 520	glatt mit aufgesch. Gew.	637	1	
1975	16	1325/1470	gerippt	1060	7	Bündel-anker
1976	12	1325/1470	gerippt	600	7	
	0,5″	1570/1770	Litze	1169	14	
	0,6″	1570/1770	Litze	377	3	
1978	12	1420/1570	gerippt	642	7	
1983	0,5″	1570/1770	Litze	807	9	
1984	0,6″	1570/1770	Litze	754	5/6	

angelassenen Stabstahles auf St 1080/1325 angehoben wurde. Damit wurden Gebrauchslasten von 630 kN möglich. Dieser hochwertige Stahl mußte nach kurzer Zeit wieder aus dem Markt gezogen werden – die Ankerzulassungen wurden per Fernschreiben über Nacht zurückgezogen –, da sich herausgestellt hatte, daß unter bestimmten Bedingungen einzelne Brüche nach wenigen Wochen Einsatz aufgetreten waren. Auf zwei Dauerankerbaustellen mußten die Ankerköpfe vorsorglich gegen ein mögliches Herausschnellen bei Bruch gesichert werden. Es zeigte sich später, daß diese Bedenken unnötig gewesen waren, da bei den voll korrosionsgeschützten Zuggliedern kein Bruch auftrat.
Im Jahre 1974 kam ein Stabstahl St 1080/1230 auf den Markt mit Durchmessern 26,5, 32 und 36 mm, der annähernd die gleichen Eigenschaften besaß wie der Stahl 830/1030. Heute nach einem Gebrauch von über 10 Jahren kann bestätigt werden, daß dieser Stahl im praktischen Gebrauch in bezug auf Kerbschlagzähigkeit, Temperatur und Korrosion nicht empfindlicher reagiert als der Stahl St 830/1030.
Neben den hochwertigen Spannstählen kamen zu Beginn der siebziger Jahre auch Ankerstähle der Güte St 520 zur Anwendung. Wegen der niedrigen Festigkeiten waren die erforderlichen Querschnitte sehr groß. Es wurden mit Gewinde bzw. Profilierungen versehene Ankerstäbe mit Durchmessern von 45 bis 65 mm verwendet. Die Verankerung am Kopf erfolgte mit Gewindemuttern.

2.2 Bündelanker

Neben den Stabstählen kamen ab 1975 bzw. 1976 auch vereinzelt gerippte Stähle dünneren Durchmessers (12 und 16 mm) für Daueranker zur Anwendung. Die Stahlgüte bestand aus St 1325/1470 bzw. 1420/1570.
Ab 1976 wurden erstmals Spannstahllitzen 0,5 und 0,6″ der Güte St 1570/1770 für Daueranker eingesetzt.

3 Beurteilung der Ankerstähle für Daueranker

Die Eignung verschiedener Spannstahlsorten für Daueranker wurde im Laufe der Entwicklungen unter-

schiedlich beurteilt. Generell ging man davon aus, daß der Stahl um so empfindlicher gegen Korrosionsangriffe (Spannungsrißkorrosion, Wasserstoffversprödung, Lochfraß) reagiert, je höher die Stahlgüte ist. Dies konnte auch in Laborversuchen nachgewiesen werden. Zugleich beurteilte man – mehr gefühls- als ingenieurmäßig – ein Zugglied mit großem Durchmesser günstiger als ein Zugglied mit kleinem Durchmesser. Aus diesem Grunde wurden in der ersten Hälfte der siebziger Jahre Einstabanker mit Durchmessern 26,5, 32 und 36 mm Bündelankern mit mehreren Stählen kleineren Durchmessers vorgezogen.
Diese Meinung wurde im Laufe der Zeit geändert. Man ging davon aus, daß bei vorhandener Korrosionsmöglichkeit sowohl der dünne hochfeste als auch der dicke niederfeste Spannstahl im Laufe der Zeit zu Bruch gehen könne. Also müsse der Korrosionsschutz so sicher gestaltet sein, daß eine Korrosion des Stahles gar nicht möglich sei.
Zudem wurde argumentiert, daß ein Bündelanker aus mehreren Stählen oder eine Litze aus mehreren Drähten zugleich eine sogenannte systemimmanente Sicherheit aufweise. Sollte nämlich trotz aller Vorsichtsmaßnahmen ein Stahl eines z. B. 7gliedrigen Bündels oder ein Draht einer 7drähtigen Litze korrodieren, so würden immer noch 86% des Stahlzuggliedes tragen; bei einer angenommenen ursprünglichen Sicherheit von 1,75 wäre bei Bruch eines Stahlgliedes immer noch eine 1,5fache Sicherheit gegen totalen Bruch vorhanden.
Somit können heute für Daueranker praktisch sämtliche zugelassenen Spannstähle verwendet werden, mit Ausnahme von Stählen mit Querschnitten von weniger als 100 mm². Diese Einschränkung beruht auf konstruktiven Gründen. Voraussetzung ist ein für Spannbeton bzw. Daueranker zugelassenes Spannsystem und ein einwandfreier Korrosionsschutz.

4 Entwicklung des Korrosionsschutzes

4.1 Stahlzugglied

Der Korrosionsschutz hat im Laufe der Zeit eine noch deutlichere Entwicklung durchgemacht als der Spannstahl selbst.
In den sechziger Jahren war die Konstruktion der Daueranker eine jeweils projektabhängige Angelegenheit zwischen wenigen Spezialisten und dem Bauherren. Anfänglich bestand der Korrosionsschutz lediglich aus einem PE-Rohr-Überzug in der freien Ankerlänge. Im Bereich der Verankerungslänge wurde der Stahl durch den Zementverpreßkörper geschützt. Die meisten der vor 20 bis 25 Jahren so hergestellten Anker sind heute noch in Betrieb ohne irgendwelche Schäden.
Als erste Verbesserung wurde der Ringraum zwischen PE-Rohr und Stahl mit Kunstharz ausgepreßt. Diese Lösung war jedoch, wie spätere Untersuchungen gezeigt haben, nicht vollständig ausreichend. Durch das Spannen des Stahles bekommt die Kunstharzverfüllung Risse und kann sich infolge örtlicher Spannungen vom Stahl lösen. Dadurch kann die durch das nicht ausreichend diffusionsdichte PE-Rohr eindringende Feuchtigkeit eine Rostunterwanderung hervorrufen.
Es wurde versucht, das Stahlzugglied auch im Bereich des Verpreßkörpers durch eine Kunstharzbeschichtung oder einen Anstrich gegen Korrosion zu schützen. Es zeigte sich aber, daß damit kein ausreichender Verbund mit dem Zementverpreßkörper erzielt werden konnte und es zu Abplatzungen und Längsrissen kam. Eine Besandung der frischen Harzbeschichtung führte ebenfalls nicht zum Erfolg, da die Besandung die Wasserdurchlässigkeit förderte.
Eine sprunghafte Verbesserung des Korrosionsschutzes brachte der Bau des Olympia-Stadions in München, bei dem das Zeltdach durch Daueranker gespannt werden sollte (Bild 1). Es fanden zum erstenmal

Bild 1
Verankertes Olympia-Zeltdach, München

Grundsatzversuche statt, in denen Anker hergestellt, geprüft, anschließend vollständig freigegraben und kontrolliert wurden. Wegen der Größe und Brisanz des Projektes ging man von einem neuen Konzept aus: Der Korrosionsschutz sollte an jeder Stelle doppelt sein, bestehend aus einem geeigneten Korrosionsschutz und einem möglichst diffusionsdichten mechanischen Schutz.
Damit wurde der Olympia-Anker, allgemein der sogenannte Druckrohranker, geschaffen (Bild 2, 3). Das Stahlzugglied, am untersten Ende in ein Druckrohr eingeschraubt, wurde auf die ganze Länge mit einem Korrosionsschutzanstrich aus Inertol-Poxitar dreimal beschichtet. Die Lückenlosigkeit wurde mit Hilfe eines Hochspannungsprüfgerätes mit einer Spannung von 1000–3000 V überprüft. Dieses so beschichtete Zugglied wurde im unteren Teil von einem 12 mm dicken Stahlrohr St 520, im oberen Teil von einem 3 mm dikken PE-Rohr mechanisch geschützt.
Später wurde der Anstrich durch ein Verfüllen des Hohlraumes zwischen Stahl und Hüllrohr mit einer Öl-Bitumenmasse (Palesit 209) ersetzt. Das PE-Rohr wurde durch ein 4,5 mm dickes PVC-Rohr ersetzt. Damit war ein Anker geschaffen, der allen Anforderungen gerecht wurde. Das korrosionsgeschützte Stahlzugglied konnte vollständig im Werk zusammengebaut werden, es war widerstandsfähig gegen rauhe Transport- und Baustellenbehandlung und hatte einen nach menschlichem Ermessen ausreichenden Korrosionsschutz. Abgesehen davon war der Druckrohranker auch in weiterer Hinsicht ideal. Die Kraft wird vom

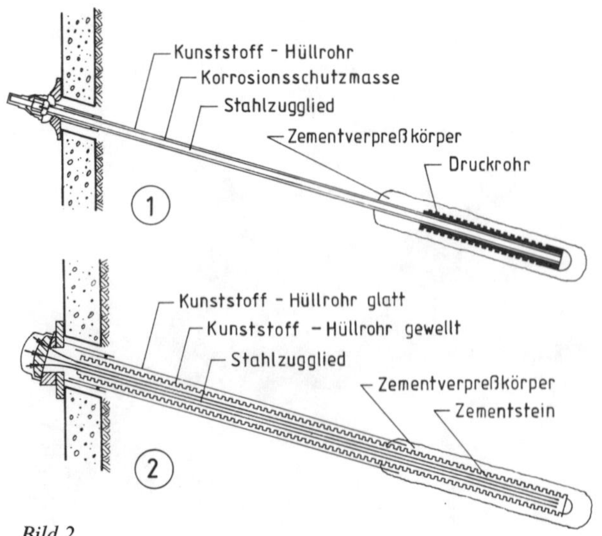

Bild 2
Konstruktionsprinzip für Druckrohranker [1] und Wellrohranker [2]

Bild 3
Druckrohranker-Querschnitt durch Verpreßkörperbereich

Ankerende her zur Luftseite hin in den Verpreßkörper eingeleitet. Dadurch steht der Zementverpreßkörper unter Druck und es können keine Zugrisse entstehen. Guter Korrosionsschutz ist teuer, zudem stand der Druckrohranker unter Patentschutz. Das System ist außerdem schlecht geeignet für Bündelanker. Es wurden deshalb andere Korrosionsschutzsysteme gesucht. Durchgesetzt hat sich das Prinzip des Wellrohrankers (Bild 2, 4). Bei diesem Ankertyp ist das Stahlzugglied im Bereich der Verpreßstrecke mit einem profilierten Kunststoff- oder beschichtetem Stahlwellrohr mit einer Wanddicke von nur 1 mm überzogen. Der mindestens 5 mm große Ringraum zwischen Wellrohr und Zugglied wird mit Zementsuspension ausgepreßt. Der Zementstein bildet den eigentlichen Korrosionsschutz, das Wellrohr die diffusionsdichte Hülle für den Fall, daß der Zementstein beim Spannen Querrisse erhält. Durch vielfache Versuche wurde gezeigt, daß ein Wellrohr aus PVC selbst bei Ankerkräften von 2000 kN die Verbundkraft auf den umgebenden Zementverpreßkörper übertragen kann, ohne dabei zerstört zu werden.
Im Bereich der freien Ankerlänge wird das Zugglied entweder durch eine Korrosionsschutzmasse oder ebenfalls durch Zementstein und ein doppeltes Kunststoff-Hüllrohr geschützt.
Der Wellrohrankertyp ist bezüglich des Korrosionsschutzsystems ein einwandfreier Anker. Die gute Korrosionsschutzwirkung des Zementsteines auf Dauer ist bekannt. Nachteilig ist die relativ große Empfindlichkeit bei Transport, Lagerung und Behandlung auf der Baustelle. Bisher wurden die Wellrohranker zumeist im

Werk vorgefertigt. Nach Aushärten der Zementfüllung zwischen Wellrohr und Stahlzugglied sind die Anker deshalb sehr unhandlich und steif und bereiten besonders bei großen Längen Schwierigkeiten beim Einführen in das Bohrloch.

Neuentwicklungen gehen deshalb dahin, das Verfüllen des Wellrohres mit Zementsuspension aus Flexibilitäts- und Gewichtsgründen auf der Baustelle, das heißt im Bohrloch vorzunehmen. Dies widerspricht jedoch der derzeitigen Forderung, daß der Korrosionsschutz vollständig außerhalb des Bohrloches aufgebracht werden muß, um diesen nochmals kontrollieren zu können. Die Praxis und weitere Versuche werden zeigen, ob mit diesem für die Praxis günstigeren System die gleiche Sicherheit erzielt werden kann.

Neben diesen beiden Ankergrundsystemen gibt es auch den Daueranker aus Stahl St 520. Wegen der geringen Festigkeit ist einerseits das Gewicht des Ankers relativ hoch, andererseits besteht bei diesem Stahl kaum die Gefahr einer Korrosion infolge Wasserstoffversprödung oder Spannungsrißkorrosion. Im Laufe der siebziger Jahre wurden noch mehrere Projekte mit St 520-Ankern ausgeführt mit Stahldurchmessern von 45 bis 65 mm und einer etwa 3 mm tiefen Profilierung. Die zusätzlichen Korrosionsschutzauflagen für diesen aus niederfestem Stahl bestehenden Anker waren jedoch – unverständlicherweise – so hoch, daß eine Wirtschaftlichkeit kaum gegeben war.

4.2 Ankerkopf

Eines der schwierigsten Korrosionsschutzprobleme stellt der Ankerkopf dar, einerseits durch die technischen Anforderungen, andererseits dadurch, daß der Korrosionsschutz nicht werkseitig, sondern auf der Baustelle aufgebracht werden muß und somit von allen Unbilden einer Baustelle betroffen ist.

In den Anfängen der Entwicklung wurden die Ankerköpfe – einer gesunden ingenieurmäßigen Überlegung folgend – nach dem Spannen und Festlegen zumeist fest einbetoniert.

Mit der Aufstellung der Norm (1972) bzw. der allgemeinen Zulassungsrichtlinien für Daueranker wurde gefordert, daß die Ankerköpfe zugänglich und nachspannbar bleiben müssen. Damit wurden die Ankerköpfe einem erhöhten Korrosionsrisiko unterworfen, da sie zumeist der Witterung und Temperatur sowie mechanischen Verletzungsmöglichkeiten ausgesetzt waren. Mit sehr aufwendigen Konstruktionen wurde

Bild 4
Wellrohranker-Querschnitt durch Verpreßkörperbereich

versucht, diesen Forderungen nach einem auf Dauer wartungsfreien und korrosionssicheren Ankerkopf gerecht zu werden (siehe Bild 5, 6).

5 Bilanz nach 25 Jahren

Es wird kaum ein Land geben, in dem der Daueranker im Lockergestein so häufig eingesetzt worden ist wie in Deutschland. Trotzdem sind keine nennenswerten Schadensfälle bekannt geworden. In einer Studie berichtet Nürnberger [11] 1978 von 6 Schadensfällen. Mag die Dunkelziffer doppelt oder dreifach sein, so ist dies doch ein verschwindend kleiner Anteil am gesamten Ankervolumen. Die Schadensfälle infolge Korrosion an Spannbetonbauwerken betragen prozentual ein Vielfaches davon.

Die Gründe für dieses hohe Qualitätsniveau sind vielseitig. Ein entscheidender Hauptgrund ist sicherlich der, daß vor Einsatz jedes Dauerankertyps Grundsatzprüfungen erforderlich sind, in denen mindestens drei Versuchsanker komplett ausgegraben werden und der gesamte Korrosionsschutz einer genauen Prüfung unterzogen wird. Durch diese Versuche werden viele Erfahrungen gesammelt und Details verbessert. Weitere Gründe mögen sein die hohen konstruktiven Anforderungen, die in den Vorschriften der Normen und Zulassungen erhoben werden und die gute Ausführungsqualität durch Spezialfirmen. Trotz dieser positiven Bilanz wird heute nach 25 Jahren den Dauerankern von vielen Bauherrn und Bauherrnvertretern noch häufig Mißtrauen entgegengebracht. Dieses Mißtrauen stammt

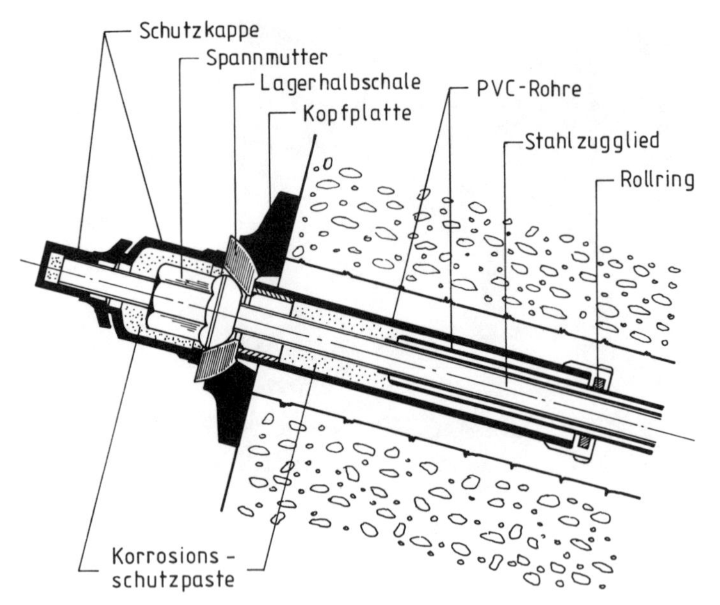

Bild 5
Konstruktionsprinzip des Ankerkopfes

aus den in Norm und Zulassung gemachten Auflagen, daß Nachprüfungen an den Ankern erforderlich sind. Der Bauherr hat das Gefühl, er bekommt weder ein wartungsfreies noch ein absolut sicheres Bauwerk. Aus diesem Grunde versucht er häufig, durch andere Konstruktionen die Daueranker gänzlich zu umgehen.

Woher rühren diese Auflagen? In den Anfängen der Dauerankerzulassungen wurde gefordert, daß 5 bis 10% aller Anker in Abständen von 2 Jahren einem Abhebeversuch oder einer Kraftmessung unterworfen werden. Diese Kontrolle sollte sicherstellen, daß zwischenzeitlich keine Korrosion an den Zuggliedern erfolgt ist. Die Korrosionssachverständigen waren bald der Meinung, daß diese Art der Prüfung keine Sicherheit biete. Sie forderten daher einen sichereren Korrosionsschutz und verzichteten ihrerseits auf diese Kontrollprüfungen. Übrig blieb die Forderung der Nachspannbarkeit und Nachprüfung aus bodenmechanischen Gründen. Diese Forderung erweckt jedoch wiederum Mißtrauen beim Anwender.

Es ist verständlich, daß es aufgrund von nur kurzdauernden Ankerprüfungen schwierig ist, auf das Langzeitverhalten zu schließen. Das bisherige Kriechkriterium nach DIN 4125/II hat sich aber in der Praxis sehr gut bewährt (Bild 7). Andererseits sind kaum Fälle bekannt, in denen Anker nachgespannt wurden, es sei denn bei Hangsicherungen. Also ist diese Forderung in der Praxis nicht nötig. Gerade aufgrund dieser Forderungen kam es aber häufig zu unsinnigen Konstruktionen, die im Laufe der Zeit zu Korrosionsschäden führen müssen. Zum Beispiel wurden Ankerköpfe offen in Nischen versetzt, in denen sich Wasser ansammeln konnte; oder Ankerköpfe lagen offen an Stützwänden unterhalb von Straßen, so daß Salzwasser über die Ankerköpfe rinnen konnte; oder die offenen Ankerköpfe wurden durch spätere Nebenarbeiten mechanisch beschädigt. In diesen Fällen wurde zwar die offizielle Forderung erfüllt, die Langzeitwirkung wurde aber dabei nicht bedacht.

Zieht man rückblickend Schlüsse, so gibt es nur eine brauchbare Lösung: Vollständiges Verschließen des Ankerkopfes – möglichst durch Einbetonieren – nach der Fertigstellung des Bauwerkes. Bestehen echte Bedenken gegen die Langzeittragfähigkeit aus bodenmechanischer Sicht, so können z. B. 3 bis 5% der Ankerköpfe offenbleiben, an denen in gewissen Zeitabständen Abhebeversuche oder Kraftmessungen durchgeführt werden können. Noch einfacher sind Verschiebungsmessungen an den Bauwerken.

Bild 6
Korrosionsgeschützter Kopf eines Einstabankers

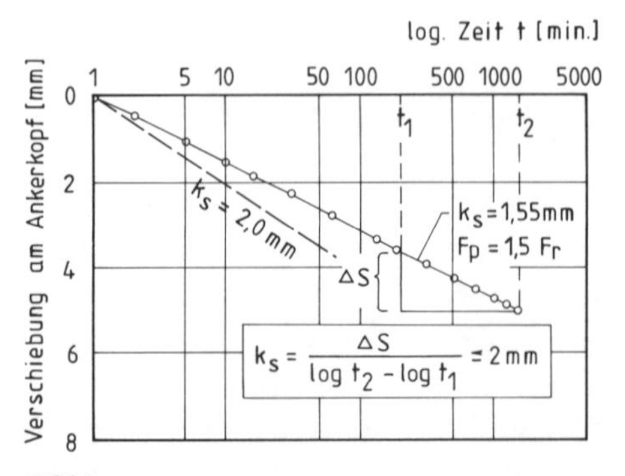

Bild 7
Zeit-Verschiebungskriterium für Daueranker

Nur in Ausnahmefällen, z. B. bei Hangsicherungen, kann das Offenhalten aller Ankerköpfe in Erwägung gezogen werden. Man sollte sich aber darüber bewußt werden, daß in den meisten Fällen ein Nachspannen der Anker nicht erforderlich ist, ein teilweises Entspannen aber nur eine vorübergehende Lösung des Problemes darstellt. Wurde der Erddruck zu niedrig angesetzt, so helfen in der Regel nur Zusatzanker. Zudem sollte berücksichtigt werden, daß bei manchen Ankersystemen ein Entspannen im Regelfall nicht möglich ist.

Stellt man heute nach 25 Jahren Erfahrung mit Dauerankern eine Bilanz auf, so kann man dieses Bauverfahren nur positiv beurteilen. Es ermöglicht sowohl technisch als auch wirtschaftlich optimale Lösungen bei vollkommen ausreichender Sicherheit. Mit den hohen Qualitätsanforderungen und den wenigen Schadensfällen an Dauerankern scheinen wir in Deutschland auf dem richtigen Weg zu sein. Dies wird auch durch das Nachziehen vieler ausländischer Vorschriften bestätigt. Wenn noch die Vorschriften dahingehend geändert werden, daß der Anker absolut wartungs- bzw. nachprüfungsfrei ist, wird das teilweise noch vorhandene Mißtrauen schwinden und der Daueranker in Zukunft einen weiteren Aufschwung nehmen.

6 Literatur

[1] Martin H. und R. Schneider: „Verbundverhalten und Korrosionsschutz des Spannstahles St 80/105, ∅ 26 mm, durch Beschichtung mit Epoxidharz“. Prüfbericht, Materialprüfungsamt für das Bauwesen der Techn. Hochschule München, 1966.

[2] Rehm, G.: „Zur Frage der Korrosionsbeständigkeit von Injektionszugankern, System Bauer.“ Gutachten, Braunschweig, 1969.

[3] Rehm, G.: „Zur Frage der möglichen Gefährdung von Injektionszugankern durch chloridhaltigen Einpreßmörtel.“ Gutachten, Braunschweig, 1969.

[4] Rehm, G.: „Beurteilung der Korrosionsschutzmaßnahmen der Erdanker System Duplex der Firmen Karl Bauer KG und Stump Bohr GmbH“. Gutachten, Braunschweig, 1970.

[5] Rehm, G.: „Beurteilung des Erdankers, System Duplex, hinsichtlich Korrosionsschutz.“ Gutachten, Braunschweig, 1970.

[6] Rehm, G. und Russwurm, D.: „Korrosionsschäden an Spannstählen St 80/105 ∅ 26 mm, glatt, die als Einstabanker (Erdanker) Verwendung fanden.“ Bericht, München, 1971.

[7] Rehm, G.: „Zu korrosionsgeschützten Dauerankern System Bauer der Typen Wellrohranker, Druckrohranker und St 52-3.“ Gutachten, Braunschweig, 1972.

[8] Rehm, G. und Russwurm, D.: „Versuche im Rahmen der Zulassungsbeurteilung für die Druckrohranker der Fa. Bauer KG, Schrobenhausen.“ Versuchsbericht, München, 1973.

[9] Rehm, G. und Rieche, G.: „Überprüfung der Eigenschaften von Inertol-Poxitar-Beschichtungen auf doppelseitig gerippten Spannstählen St 85/105, ∅ 32 mm, als dauerhaften Korrosionsschutz für Verpreßanker.“ Versuchsbericht, Braunschweig, 1974.

[10] Kern, G. und Jungwirth, D.: „Betrachtungen zu neueren Entwicklungen von Spannverfahren am Beispiel des Dywidag Verfahrens.“ Beton und Stahlbetonbau, 3, 1974.

[11] Nürnberger, U.: Analyse und Auswertung von Schadensfällen an Spannstählen. „Forschung, Straßenbau und Verkehrstechnik“, Heft 308, 1980, BMW, Bonn-Bad Godesberg.

[12] Rehm, G.: „Verwendung von Verpreßankern für vorübergehende Zwecke aus nur einer 7-drähtigen 0,5″-Litze.“ Gutachten, Stuttgart, 1983.

Zur Auswahl dauerhafter Kombinationen aus Korrosionsschutzanstrich und Brandschutzputz für den Stahlgeschoßbau*

Nils V. Waubke

1 Einleitung

Bauwerke sind nicht nur standfest und funktionsgerecht, sondern auch dauerhaft zu planen und auszuführen. Dazu bedarf es unter anderem der Beachtung ausreichender Korrosions- und ausreichender Feuerbeständigkeit. Im Falle von Stahlgeschoßbauten kommen zur Sicherstellung der notwendigen Feuerwiderstandsfähigkeit der Konstruktion vorzugsweise

- entweder flächige bzw. kastenförmige Verkleidungen der tragenden Bauteile
- oder profilfolgende Ummantelungen, insbesondere mit spritzfähigen Putzen,

zum Einsatz. Für profilfolgenden Feuerschutz haben sich dabei Putze auf der Basis von Zement als Bindemittel bewährt, welchen Vermiculite, Perlite, Mineralfasern, Hüttenwolle o. ä. als „Dämmstoff" beigemengt sind.

Soweit vom Inneren von Stahlgeschoßbauten die Rede ist, können solche Putze vielfach direkt auf die entrostete Oberfläche der Stahlbauteile aufgebracht werden, weil an den Korrosionsschutz der Bauteile – zumindest in beheizten und ausreichend trockenen (besonders: klimatisierten) Gebäuden – in weiten Bereichen nur sehr moderate, teilweise auch gar keine Anforderungen gestellt sind [1]. In solchen Fällen haben wir hinsichtlich der Dauerhaftigkeit des Brandschutzes auch keine Schwierigkeiten zu erwarten, nachdem die hier erwähnten Putze selbst mit dem Stahluntergrund verträglich sind und ihre langfristige Haftung am Bauteil

- entweder ausreichend nachgewiesen
- oder durch Hinzunahme von Putzträgern oder anderen „Hafthilfen" verbessert ist.

Es wird allerdings immer wieder Fälle geben, in denen mittels eines konventionellen Anstrichs, Überzuges oder einer Beschichtung korrosionsgeschützte Stahlbauteile auch mit Brandschutz zur Erzielung ausreichender Feuerwiderstandsdauer zu versehen sind und dafür – sei es aus Gründen der Wirtschaftlichkeit, aus konstruktiven Gründen oder aus Gründen des Bauablaufes – eine profilfolgende Ummantelung mit einem der hier behandelten Spritzputze bevorzugt wird. Aus langjähriger Erfahrung läßt sich überdies feststellen, daß die einer derartigen Maßnahme zugrundeliegende technisch-wirtschaftliche Entscheidung sehr häufig zu einem Zeitpunkt gefällt wird, zu dem die Stahlbauteile bereits gefertigt und mit Korrosionsschutz versehen, eventuell sogar schon an die Baustelle geliefert oder auch schon montiert sind. Eine Möglichkeit, auf Art und Applikation des Korrosionsschutzes (etwa auf Anregung des Spritzputzlieferanten) noch Einfluß zu nehmen, besteht in aller Regel jedenfalls dann nicht mehr.

Gerade eine solche Einflußnahme wäre in der Vergangenheit aber manchmal von Vorteil gewesen: Während viele heute in der Bundesrepublik zum Einsatz gelangende, insbesondere ‚airless' spritzbare Korrosionsschutzbeschichtungen für den Stahlbau eine ausgezeichnete Widerstandsfähigkeit gegen saure Medien – und damit gegen die üblicherweise aus der Umwelt zu erwartenden Angriffsformen – aufweisen, zeigen sie Schwächen bei stark alkalischem Angriff: Sieht man von Sonderbereichen (etwa dem Kessel- und Rohrleitungsbau) ab, gab es in der historischen Entwicklung von Korrosionsschutzanstrichen für den Stahlbau keinen Anlaß, möglichem alkalischem Angriff ein besonderes Augenmerk zu schenken; Regen und Atmosphäre in unseren Breiten – wie in allen typischen Industrieländern – gaben im Gegenteil zunehmend Anlaß zu der Annahme, daß die natürliche Gefahr für Anstrich und Bauteil laufend weiter zur ‚sauren Seite' – also zu niedrigeren pH-Werten – hin verschoben würde.

---

* Die hier berichteten Versuche wurden im Auftrage des Deutschen Ausschusses für Stahlbau durchgeführt und mit Mitteln des Bundesministeriums für Raumordnung, Bauwesen und Städtebau gefördert [2]. Hierfür ist der Autor zu Dank verpflichtet.

So war es natürlich und angemessen, daß sich für die Anwendung im Stahlhochbau z. B. die sog. Alkydharze wegen ihres günstigen Preises und ihrer problemlosen Verarbeitbarkeit (auch im Airless-Verfahren) ein breites Einsatzfeld sicherten [3]. Wo anders gelagerte, teilweise verschärfte Bedingungen erwartet wurden, vermochten sich ebenso die sog. Acrylharze oder auch acrylmodifizierte Alkydharze durchzusetzen. Demgegenüber vermochten sich die – allgemein als chemisch widerstandsfähiger akzeptierten – Reaktionsharze der PUR- oder EP-Klasse nicht in gleichem Maße für eine Massenanwendung zu etablieren, was vermutlich sowohl kosten- als auch applikationsbedingt war: Vor allem scheiterte der gelegentlich vorgetragene Wunsch auf verstärkten Einsatz von EP-Harz für bestimmte Anwendungen u. a. auch daran, daß diese Materialgruppe weniger gut für eine Verarbeitung durch Spritzen (und überhaupt in dünnen Schichten) geeignet ist. Soweit EP-Material verstärkt verwendet wurde, beschränkte sich dies vornehmlich auf einige spritzfähige Epoxidharzester, welche in der Tat verbesserte Korrosionsschutzeigenschaften – im Vergleich zu anderen konventionellen Anstrichen und, vor allem, wiederum im „herkömmlichen" Sinn gegen sauren Angriff – aufweisen; ihre Stabilität bei alkalischer Beanspruchung ist aber ähnlich gering wie jene anderer Ester.

Zementgebundene Brandschutzputze weisen – wie alle frischen Kalk- und Zementmörtel – einen pH-Wert von mehr als 12 auf und stellen damit, bei ausreichendem Feuchteangebot, eine bedeutende alkalische Belastung des Untergrundes dar, auf den sie aufgebracht wurden. Während diese Alkalität dem Stahl allein nicht schadet (wo sie ‚ununterbrochen' und langfristig erhalten bleibt, eher schützend wirkt), vermag sie nicht ausreichend alkalienbeständige Anstriche anzugreifen und damit

- sowohl die Basis für den diesen Anstrichen zugedachten Korrosionsschutz zu mindern
- als auch die Grundlage der Haftung des Brandschutzputzes am Bauteil zu zerstören:

Was mit der Oberfläche des Anstrichs, auf den der Putz aufgebracht wurde, im einzelnen geschieht, läßt sich am besten als Kombination oder Wechselwirkung von „Quell-" und „Verseifungs"-Reaktionen zusammenfassen; die äußeren Erscheinungsbilder (Glanzverlust, Aufweichung, Bläschenbildung, Trübung, Unterrostung, Auflösung) sind hinreichend bekannt und für die verschiedenen Kombinationen von Anstrichen und Putzen untersucht und beschrieben [2], [4].

Das Ergebnis dieser Untersuchung besteht im wesentlichen darin, daß

- alle untersuchten Spritzputze die beschriebene Verseifungs- und Quellwirkung ausüben und damit die Grenzfläche Anstrich/Putz zu verändern vermögen,
- diese Wirkung in ihrer Intensität und in ihrem Ergebnis vom Feuchteangebot an der Grenzfläche sowie von der Karbonatisierungsgeschwindigkeit des Putzes (also insbesondere von der Porosität und Wasserrückhaltefähigkeit des Putzes) abhängt und
- die Widerstandsfähigkeit der so beanspruchten Anstriche/Beschichtungen grob in drei deutlich voneinander unterscheidbare Klassen eingestuft werden kann:

a) Weitaus am besten geeignet für eine Überschichtung sowohl mit Blähstoff gefüllter als auch mit Fasern/Gewöllen gefüllter Zementputze haben sich – wie aufgrund ihrer bekannten Laugenbeständigkeit erwartet – die Epoxide erwiesen, soweit nicht Sekundäreinflüsse (z. B. Aluminiumstaubfüllung des Anstriches oder schlechte Putzausführung) das Ergebnis beeinflussen und die notwendigen Verarbeitungsgrenzen (Mindestanstrichdicke, Härter-Binder-Verhältnis) eingehalten werden. Die Epoxide besitzen hinsichtlich der Putzhaftung darüber hinaus einen Vorsprung gegenüber anderen Anstrichen bei höheren Temperaturen.
   Diese „höchstrangige" Kategorie zeichnet sich im übrigen durch Verseifungszahlen VZ = 0 und Säurezahlen SZ = 1 … 2 aus.
b) In die zweite Kategorie lassen sich zahlreiche andere Korrosionsschutzmittel (z. B. auf der Basis von Epoxiestern, Acrylharz und Alkylsilikat – aber auch von besonders „langöligen" Alkydharzen als Bindemittel) einordnen, welche sich teilweise nur als zur Überschichtung mit extrem leichten, rasch karbonatisierenden Faserputzen geeignet erweisen bzw. teilweise nur bei länger anhaltender Durchfeuchtung des Putzes versagen. Durch Zwischenschaltung einer haftvermittelnden Schicht auf PVA-Basis ist bei diesen Anstrichen regelmäßig bereits eine merkliche Verbesserung der Beständigkeit gegen den Putzeinfluß zu beobachten.
   Im übrigen sind bei den in diese Kategorie einzuordnenden Anstrichen Verseifungszahlen VZ = 50 … 250 und Säurezahlen SZ = 4 … 9 zu verzeichnen.
c) In eine dritte Kategorie lassen sich alle Anstriche einstufen, die ohne Haftzwischenschicht rasch verseifen und selbst mit einer „konventionellen" Haft-

schicht – z. B. auf Mowilithbasis – keine ausreichenden Ergebnisse liefern.

In dieser Gruppe wurden Verseifungszahlen VZ = 25... 125 und Säurezahlen SZ = 5... 12 festgestellt. Insgesamt haben die in [2], [4] beschriebenen Versuche jedoch gezeigt, daß eine Chance, die bewährten Alkyd-, Acryl- und Acryl-Alkyd-Harze als Bindemittel für Korrosionsschutzanstriche unter „zementgebundenem" Putz auch künftig einzusetzen, in der Verbesserung der bislang verwendeten Haftzwischenlagen bzw. in der Entwicklung/Erprobung neuer, besserer Haftlagen liegt.

2 Versuche mit Haftvermittlern

2.1 Versuchsprogramm und -durchführung

Nachdem die dominierende Rolle der Haftzone zwischen Anstrich und Putz für das Gesamtverhalten und die Dauerhaftigkeit eines derartigen „Korrosions- und Brandschutz"-Systems feststeht, liegt es nahe, diese Haftzone durch „Einfügen" einer (möglichst billigen) Zwischenlage beständig zu machen, die folgende Anforderungen wenigstens gleich gut oder sogar besser erfüllen muß als die bislang eingesetzten Mittel (z. B. auf PVA-Basis):

- Zuverlässige Haftung;
- Anwendbarkeit in Kombination mit den gängigen Korrosionsschutzanstrichen und Brandschutzspritzputzen;
- Temperaturstandvermögen bis ca. 300 °C;
- Zuverlässigkeit als zusätzliche Deckschicht;
- gute Verarbeitbarkeit.

Hinsichtlich der beiden ersten Forderungen bedarf es keiner Erläuterung. Die Temperatur von 300 °C wurde als Richtwert gewählt, weil viele der bisher untersuchten Korrosionsschutzanstriche bis zu dieser Temperatur – wenn auch unter mehr oder weniger starken Zersetzungserscheinungen – mechanisch genügend stabil bleiben und somit bis zu höheren Temperaturen ausgenützt werden könnten, wenn nicht die bislang eingesetzten Haftschichten zumeist bereits bei ca. 150 °C versagen würden. Die 4. Forderung betrifft vor allem die Anwendung auf metallstaubgefüllten Grundanstrichen oder auf Fertigungsbeschichtungen, die ihrerseits eines gewissen Schutzes bedürfen bzw. deren korrosionsschützende Wirkung ergänzt werden muß: Deshalb sollte die Haftschicht diffusionsdicht gegen korrosionsfördernde Ionen und gut deckend aufzutragen sein (z. B. auch nicht durchsichtig sein, um den lückenlosen Auftrag besser kontrollieren zu können). Schließlich umfaßt die letzte Forderung die Spritzbarkeit und einen für den Zweck günstigen Trocknungsverlauf, der nach dem Aufspritzen der Haftschicht genügend Zeit zum Auftragen des Putzes „naß in naß" läßt. Über erste „Tastversuche" mit einigen ausgewählten „Haftmitteln", von denen aus unterschiedlichen Gründen Eignung erwartet wurde – und die dabei erzielten Ergebnisse wird hier berichtet: Die Auswahl erstreckte sich

- sowohl auf im Stahlbau als Deckbeschichtungen hochwertiger Korrosionsschutzanstriche
- als auch auf als Betonanstriche (wegen ihrer Betonverträglichkeit)
- als auch auf bereits bislang als Haftvermittler eingesetzte Mittel (siehe Tabelle 1).

Da es sich – bis auf das gekennzeichnete „Gemisch" – um Handelsware handelte, ist Art und Menge der enthaltenen Füllstoffe nicht in allen Fällen bekannt:

Tabelle 1
Eingesetzte Haftschichten

Bindemittel	Füllstoff	Code Nr.
2 K Epoxid	Teer	1
2 K Epoxid-Dispersion	?	2
2 K Polyurethan	Eisen-glimmer	3
Acrylharz-Dispersion	?	4
Acrylharz-Dispersion	Mineral-fasern	5
Chlorkautschuk	?	6
Chlorsulfoniertes Polyäthylen	?	7
Asphalt-Standöl	Schuppen-pigmente	8
Anorg. Hochtemperatur-Mörtel auf Aluminat-Silikat-Basis	–	9
1 RT PVA + 1 RT Alkylsilikat + Verdünnung (Gemisch)	–	10

Die in Tabelle 1 aufgeführten Haftschichten wurden in den Versuchen jeweils mit den in Tabelle 2 zusammengestellten Korrosionsschutzanstrichen sowie mit einem typischen, zementgebundenen Vermiculiteputz kombiniert:

Tabelle 2
Eingesetzte Korrosionsschutzanstriche

Bindemittel	Füllstoff	Code	Bemerkung
2 K Epoxid	Zinkchromat	A	„Standard“
Alkydharz Alkydharz Acrylharz	Bleimennige Zinkchromat Blancfix, Zinkchromat, Titandioxid u. a.	B C D	ohne Zwischenschicht nicht geeignete Anstriche
Cumaronharz	Aluminium	E	hochtemperaturbeständiger Anstrich mit Metallstaubfüllung

Die Korrosionsschutzanstriche wurden mit einem Druckluft-Spritzgerät (7 bar) zweischichtig bis zu einer Dicke von insgesamt ca. 100 µm aufgetragen. Nach einer Trocknungszeit von 3 Wochen wurden die Bleche dann mit einer Lage Haftschicht versehen (wieder mittels Spritzgerät) und anschließend – ca. ½ bis 1 Std. nach dem Auftragen der Haftschichten – von Hand mit Vermiculite-Putz (6 RT Vermiculite + 0,8 RT Zement + 1 RT Kalkhydrat + 1 RT Wasser) überschichtet. Danach wurden die Proben weitere 3 Wochen bei Raumklima gelagert, bis der Putz vollständig erhärtet war.
Ein Satz Bleche wurde dann 8 Wochen im Schwitzwasser-Wechselklima beansprucht – und zwar jeweils 3 Tage lang bei 40 °C und 100% relativer Feuchte und 4 Tage lang in Raumklima. Die Verlängerung der Feuchtigkeits- und Trocknungs-Perioden erschien angebracht, da hierbei die unter praktischen Verhältnissen zu erwartende vollständige Durchfeuchtung bzw. Durchtrocknung des Putzes und die damit verbundene mechanische Beanspruchung des Systems infolge Quellung und Schrumpfung des Putzes besser simuliert wird als bei 1-Tages-Zyklen nach DIN 50017, bei denen der Putz nie richtig durchtrocknete [2]. Der Zustand der Proben wurde nach Entfernen des Putzes untersucht.

Bleche für Untersuchungen der Haftung bei höheren Temperaturen wurden noch weitere 5 Wochen bei Raumklima gelagert, so daß zum Zeitpunkt der Versuche der Korrosionsschutzanstrich 11 Wochen und Haftschicht + Putz 8 Wochen alt waren. Die Versuche wurden an waagerecht gelagerten Blechen mit einer Aufheizgeschwindigkeit von ca. 20 °C/min. ausgeführt, vgl. [4]. Die Temperatur wurde an drei verschiedenen Stellen mit Hilfe von Thermoelementen gemessen, deren Spitzen mittels „Schablone“ in der Mitte des Bleches (in 0,25 cm Tiefe) fixiert waren.
Zur Untersuchung der Hafteigenschaften der Haftschichten in Abhängigkeit von der Trocknungszeit wurden die Haftschichten ohne vorherige Grundierung mit Korrosionsschutzanstrich aufgetragen und dann sukzessive, in Zeitabständen von 15 bis 20 Minuten, mit Vermiculite-Putz überschichtet. Nach 3 Wochen Trocknungszeit wurden der Putz wieder abgenommen und die Veränderungen in der Haftfestigkeit qualitativ festgestellt.
Zur Prüfung auf Dichtigkeit wurde zunächst 1 Lage (ca. 50 µm) Haftschicht auf die ungrundierten Bleche aufgetragen und über Nacht trocknen gelassen, anschließend nochmals Haftschicht aufgespritzt und dann mit Vermiculite-Putz überschichtet. Eine Hälfte des Putzes war hierbei, statt mit reinem Wasser, mit einer 10%igen Natriumchlorid-Lösung angemacht; bei der anderen Hälfte wurde an Stelle von Zement und Kalkhydrat Maschinenputz-Gips als Bindemittel verwendet. Nach 3 Wochen Lagerung bei Raumklima wurden die Proben dann 6 Wochen im Schwitzwasser-Wechselklima (einwöchige Zyklen, siehe oben) beansprucht. Neben den bereits erwähnten, mit der Putztrocknung zusammenhängenden Nachteilen hätte eine Anwendung des genormten Schwitzwasserklimas (z. B. nach DIN 50018 mit Schwefeldioxid-Atmosphäre) auf mit Putz beschichtete Systeme weitere Nachteile, die mit dem „SO_2-Verbrauch“ des Putzes verknüpft sind [2]: Die durch das statt dessen gewählte Verfahren ermöglichte Anreicherung von korrosionsfördernden Ionen wie Chlorid bzw. Sulfat im Putz, in Verbindung mit der erhöhten Temperatur und Feuchtigkeit sowie der mechanischen Beanspruchung des Systems infolge des Quellens und Schrumpfens des Putzes, erschien dagegen als Weg, zu einer realistischen Einstufung der korrosionstechnischen Wirksamkeit der Systeme Korrosionsschutz – Haftschicht – Putz zu gelangen.

Tabelle 3
Ergebnisse der Schwitzwasserwechseltests (zusammengefaßt und codiert)

Haftlage	Korrosionsschutzmittel				
	A	B	C	D	E
1	intakt	intakt	intakt	leicht besch.	zerstört
2	intakt	fast intakt	intakt	leicht besch.	zerstört
3	intakt	intakt	intakt	intakt	zerstört
4	intakt	intakt	intakt	intakt	zerstört
5	intakt	intakt	fast intakt	intakt	zerstört
5*	intakt	intakt	intakt	intakt	zerstört
6	fast intakt	zerstört	leicht besch.	stark besch.	zerstört
7	intakt	stark besch.	leicht besch.	leicht besch.	zerstört
8	stark besch.	besch.	besch.	stark besch.	zerstört
9	zerstört	zerstört	zerstört	zerstört	zerstört
10	intakt	leicht besch.	intakt	leicht besch.	zerstört
Standard PVA-Disp.	leicht besch.	leicht besch.	leicht besch.	leicht besch.	zerstört

* Modifikation aus 1 RT Nr. 5 + 1 RT Sand + 0,1 RT Wasser

2.2 Ergebnisse

Die Ergebnisse der Versuche sind in den Tabellen 3 bis 6 zusammengefaßt. Die Beurteilung der Haftung bezieht sich dabei allein auf den Verbund unterste Schicht Putz – Haftschicht – Korrosionsschutzanstrich – Stahlblech. Diejenigen Haftschichten, deren Eignung nach der Beanspruchung im Schwitzwasserwechselklima zweifelhaft erschienen, wurden der Kürze halber bei den übrigen Ergebnissen (Tabellen 4 bis 6) nicht mehr mit aufgeführt.

Bei den Temperaturversuchen, bei denen mehr als ca. 250 °C erreicht wurden, wiesen Haftschichten und Korrosionsschutzanstriche ausnahmslos Zersetzungserscheinungen (Verfärbung nach braun bis schwarz) auf. Da sie überdies bei allen Produkten etwa gleich stark waren, wird auf eine Beschreibung verzichtet. Das Temperaturstandvermögen der Haftschichten wurde aus jenen Versuchen ermittelt, bei denen offensichtlich die Haftschicht (und nicht der Korrosionsschutzanstrich) versagt hatte.

2.3 Auswertung

Unter Berücksichtigung der eingangs formulierten Anforderungen an geeignete, die Chancen verseifungsanfälliger Anstriche verbessernde Haftlagen kann – aufgrund der in diesen Tabellen skizzierten Befunde – vermutet werden, daß sich folgende 5 der untersuchten Varianten auch in ausführlicheren und praxisnäheren Tests als deutliche Verbesserung gegenüber dem mituntersuchten (repräsentativen) Standard erweisen werden:

- Acrylharz-Dispersion (Code Nr. 4),
- PUR mit Eisenglimmerfüllung (Code Nr. 3),
- EP mit Teerfüllung (Code Nr. 1),
- Acrylharz-Dispersion mit MF-Füllung (Code Nr. 5),
- EP-Dispersion (Code Nr. 2),

Tabelle 4
Putzhaftung bei Temperaturen bis 340 °C; Aufheizgeschwindigkeit 20 K/min

Haftlage	Korrosionsschutzmittel			
	A	B	C	D
1	320 °C	280 °C	300 °C	140 °C
2	320 °C**	330 °C	320 °C	150 °C
3	290 °C**	240 °C**	240 °C**	170 °C
4	> 340 °C	> 340 °C	330 °C	160 °C
5	330 °C	320 °C	340 °C	200 °C
5*	> 340 °C	340 °C	330 °C	170 °C
10	330 °C**	> 340 °C	320 °C	150 °C
Standard	150 °C**	150 °C**	130 °C**	160 °C

* Modifikation aus 1 RT Nr. 5 + 1 RT Sand + 0,1 RT Wasser
** infolge Versagens der Haftschicht abgefallen

Tabelle 5
Putzhaftung in Abhängigkeit von der Zeitspanne Δt zwischen Auftragen der Haftlage und Putzaufbringung

Haftlage	1	2	3	4	5	10	Standard
Nachlassen der Haftung nach $\Delta t = \ldots$ min	150	210	>240	>240	>240	170	
Bemerkung	rostfrei	leicht unterrostet					stark unterrostet

Tabelle 6
Dichtigkeit der Haftlage gegenüber SO_4^{2-}- und Cl^--Ionen, beurteilt anhand des Unterrostungsgrades nach Überschichtung mit

Haftlage	Gipsputz	Zement-/Kalkhydratputz + 10% NaCl-Lösung
1	keine Rostbildung	
2	keine Rostbildung	
3	leichter Rost	kein Rost
4	kein Rost	leichter Rost
5	mittlerer Rost	leichter Rost
10	starke Unterrostung	
Standard	extrem starke Unterrostung	

und zwar in dieser Reihenfolge. Der Vorsprung dieser Haftschichten gründet sich vor allem auf deren wesentlich bessere Temperaturbeständigkeit und Dichtigkeit. Die Unterschiede in der Haftfestigkeit waren nicht ganz so groß, jedoch ließ sich auch hier deutlich ein Vorsprung feststellen. Die relativ geringfügigen Unterschiede zwischen den 5 Haftschichten selbst liegen dagegen in sehr engen Grenzen, so daß ihre Reihung noch nicht den Anspruch auf Endgültigkeit erheben kann. Zu erwägen ist lediglich eine etwas geringere Einstufung

- der 2-Komponenten-Haftschichten aufgrund des Mehraufwandes an Arbeitszeit (Mischen der Komponenten) und Kosten
- sowie der durchsichtigen bis milchigen Acrylharz-Dispersion Nr. 5, bei der ein gleichmäßig bedeckendes Auftragen schwierig zu kontrollieren ist.

Dagegen wurde die Acrylharz-Dispersion Nr. 5 auch mit Sand (Korngröße ca. 0,1 bis 0,3 mm) gefüllt (1:1 RT) eingesetzt: Zwar wurden nicht alle Tests mit dieser Variante durchgeführt – doch waren die erzielten Ergebnisse so günstig, daß – insbesondere aus Gründen der Wirtschaftlichkeit – auch mit Sand gefüllte Haftschichten in künftige Versuche einbezogen werden sollten.

3 Zusammenfassung

Aufgrund des Standes der Technik [4], der an „konventionellen" Kombinationen aus Anstrichsystem und Putz sowie an „vereinfachten" Kombinationen aus Fertigungsbeschichtung und Putz [2], [4] durchgeführten Untersuchungen und der hier vorgelegten, vorläufigen Ergebnisse mit zusätzlichen Haftlagen lassen sich folgende Hinweise für die Auswahl geeigneter, weil dauerhafte Wirksamkeit versprechender Materialkombinationen ableiten:

- Soweit dies fertigungstechnisch in Frage kommt und wirtschaftlich erscheint (z. B., weil an den Korrosionsschutz allein im Einzelfall bereits höchste Anforderungen zu stellen sind), erscheinen Kombinationen aus Korrosionsschutzbeschichtung auf EP-Basis und direkt aufgebrachtem Spritzputz als besonders günstige (langlebige) Lösungen.
  Allerdings ist auf eine sorgfältige und sachgerechte Verarbeitung des Korrosionsschutzes stets zu achten: Ungeeignete Härter-Binder-Verhältnisse sowie alkaliempfindliche Metallstaubfüllungen sind zu vermeiden.
- Korrosionsschutzsysteme (z. B. auf Acrylharz-, Alkydharz- oder Alkylsilikatbasis), welche nach dem Stand der Technik als „nur mäßig empfindlich gegenüber dem alkalischen Angriff des Zementes und/oder einer etwaigen korrosionsfördernden Wirkung des Putzes" oder als „nur zu geringer Putzhaftung führend" eingestuft und deshalb als „nur bedingt zur Überschichtung mit Putz geeignet" bezeichnet

werden [4], können mit einiger Sicherheit mit einem der beschriebenen, zementgebundenen Brandschutzputze überschichtet werden, wenn eine der hier als besonders günstig erkannten Haftlagen „zwischengeschaltet" wird. Die Temperaturfestigkeit des Gesamtsystems würde dabei nicht verschlechtert, weil diese Haftlagen – wie die meisten der fraglichen Korrosionsschutzanstriche – offenbar erst bei zwischen 250 °C und 350 °C ihre haftvermittelnde Wirkung verlieren.*

- Bei der Gruppe von Anstrichen, welche wegen zu schwacher korrosionsschützender Wirkung (Fertigungsbeschichtungen) und/oder besonderer Empfindlichkeit des Putzes (z. B. Leinölfirnis) als ungeeignet eingestuft wurden [4], müssen dagegen weiterhin Vorbehalte aufrechterhalten werden.
  Hinsichtlich der Fertigungsbeschichtungen ist immerhin denkbar, daß eine zusätzliche Schicht Haftvermittler zu ausreichendem Korrosionsschutz und somit zu einem befriedigenden System führt; von einer Überschichtung so hochempfindlicher Anstriche wie z. B. solcher auf der Basis von Leinölfirnis mit Putz wird jedoch nach wie vor abgeraten, da selbst kleinste Löcher (z. B. infolge mechanischer Verletzung in der Haft-/Deckschicht) zu Unterwanderung und großflächigen Abplatzungen führen.
- Die bereits ermutigenden Ergebnisse der hier dargestellten, eher noch tastenden (in der Auswahl der untersuchten Systeme unsystematischen) Versuche geben zu weiteren Bemühungen in dieser Richtung Anlaß und berechtigen zu der Hoffnung, daß mit noch besseren – eventuell von der chemischen Industrie gezielt entwickelten – Haftvermittlern das Spektrum der mit zementgebundenem Putz kombinierbaren Korrosionsschutzmittel wieder ähnlich breit wie jenes der bewährten Korrosionsschutzmittel an sich wird.

* Ungeachtet der Frage, ob bereits vorher Zersetzungserscheinungen einsetzen.

4 Literatur

[1] DIN 55928, Teil 1, Ausgabe November 1976, 3.1.2.
[2] WAUBKE, N. V., HERBELL, J.-D., und LUXENBERG, P.: Bericht II/80 DASt Nr. 7.2.03 zum FA BI 7-810705/247 an den BM Bau.
[3] Ergebnis einer Umfrage des Autors bei den Firmen MAN, Maurer und Krupp sowie beim BZA München und bei der SLV Duisburg in den Jahren 1977–1979.
[4] WAUBKE, N. V.: Profilfolgende Korrosionsschutz-Brandschutz-Systeme für den Stahlhochbau. Wilhelm Ernst & Sohn. Berlin (in Vorbereitung).
[5] WAUBKE, N. V., und LUXENBERG, P.: Bericht I/81 DASt Nr. 7.2.03 zum FA BI 7-810705/297 an den BM Bau.

Werkstoff-Forschung, Verbund

Das Verbundverhalten von Stahl und Beton unter besonderer Berücksichtigung der lokalen Stahlspannung

Josef Eibl
und Jörg Kobarg

1 Problemstellung

Trotz einer ganzen Reihe von Forschungsaktivitäten während der letzten 30 Jahre – hierzu hat zweifellos REHM [13] in Deutschland entscheidende Anstöße gegeben – wirft das Verbundverhalten von Beton und Bewehrungsstahl noch immer eine Reihe von Fragen auf. Deren Klärung ist notwendig, weil das sogenannte Verbundgesetz einerseits immer mehr zu einer wichtigen Ausgangsbasis für numerische Untersuchungen wird, andererseits weil eine Reihe von baupraktisch wichtigen Problemen unmittelbar davon abhängen. Dies sind insbesondere Fragen der Deformation und der Rißbildung unter Last bzw. Zwang und der daraus möglicherweise resultierenden Bauschäden.
Dazu soll über eine noch nicht abgeschlossene Untersuchung zu diesem Themenkreis, die im Rahmen eines Forschungsschwerpunktes „Stoffgesetze" der Deutschen Forschungsgemeinschaft gefördert wird, berichtet werden. Ziel dieser Arbeit ist es, den Einfluß der Parameter Querdruck- und Stahlzugspannung auf das Verbundverhalten zu studieren und experimentell gefundene Verbundgesetze in den Rahmen stoffgesetzlicher Beziehungen einzuordnen.

2 Verbund – Stoffgesetz

Derzeit wird die Beziehung zwischen einer fiktiven Schubspannung τ und der am gleichen Ort auftretenden Relativverschiebung von Beton und Stahl $\delta = u_S - u_B$ als Verbundgesetz bezeichnet. Daß es sich hierbei um kein echtes Stoffgesetz im Sinn der Kontinuumsmechanik handelt, sei einleitend an Hand einiger pauschaler Überlegungen gezeigt:
Bei einem echten Stoffgesetz können sich nicht, wie experimentell nachgewiesen, bei unterschiedlichen Prüfkörpern (Bild 1) verschiedene τ-δ-Zusammenhänge ergeben. Schon gar nicht dürfen sich bei ein und demselben Versuchskörper, wie erstmals von WATSTEIN [2] und NILSON [1] später u.a. auch von DÖRR [4] nachgewiesen, unterschiedliche τ-δ-Beziehungen an unterschiedlichen Stellen des Bewehrungsstabes (Bild 2) ergeben. Bevor weitere Feststellungen hierzu getroffen werden, seien zunächst nach Bild 3 die Verhältnisse z. B. im Innern eines Dehnkörpers näher erörtert.

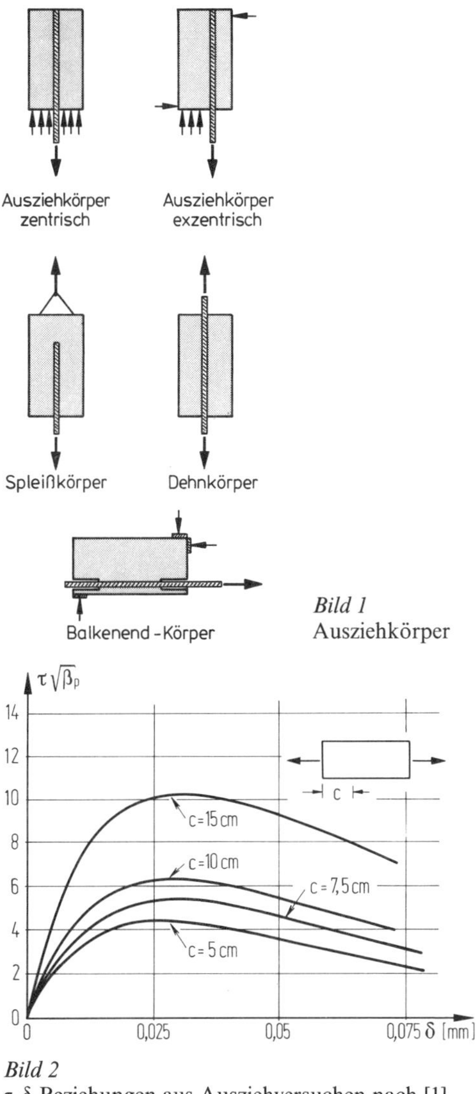

Bild 1
Ausziehkörper

Bild 2
τ-δ-Beziehungen aus Ausziehversuchen nach [1]

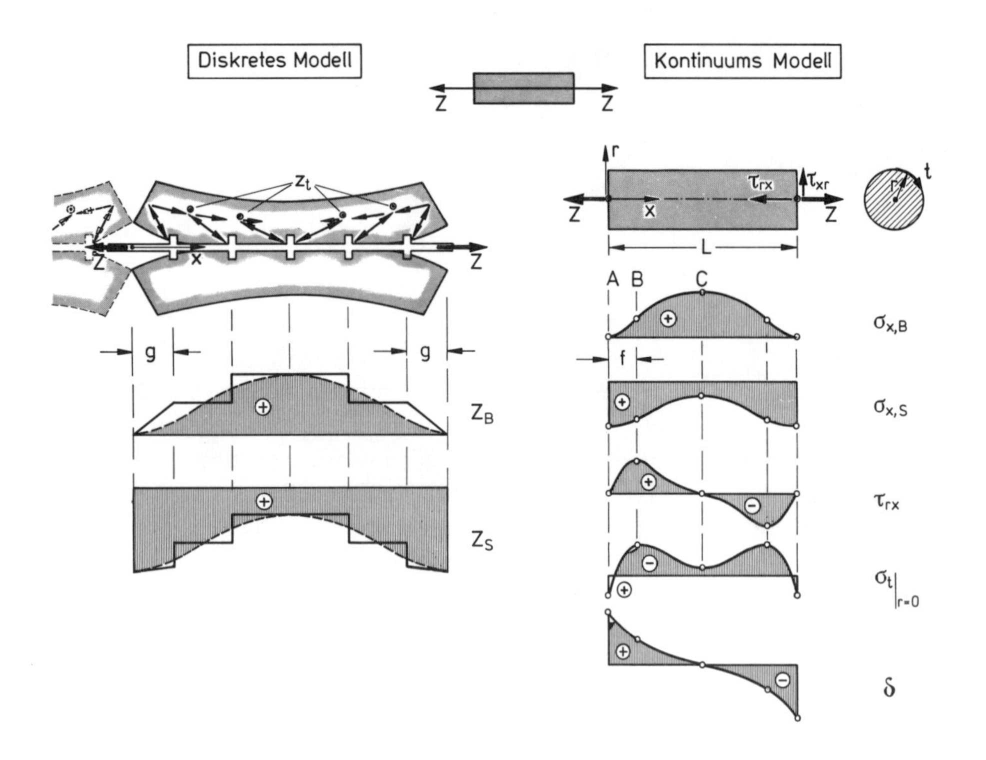

Bild 3
Verlauf der Kräfte bzw. Spannungen im Dehnkörper

Im linken Teil des Bildes 3 sieht man am diskreten Fachwerkmodell wie die schrägen Druckdiagonalen in Längsrichtung allmähliche Zugkräfte gegen das Innere des Betonkörpers hin aufbauen. An beiden Enden des Dehnkörpers muß dabei, ergänzend zum üblichen eindimensionalen Verbundmodell mit stetiger Schubübertragung, ein Störbereich „*g*", den wirklichen Verhältnissen entsprechend postuliert werden.

Am freien Rißufer $x = 0$; $x = L$ fehlen Schubspannungen τ_{xr} und damit aus Gleichgewichtsgründen auch Schubspannungen τ_{rx} in Längsrichtung. Erst allmählich mit zunehmendem Abstand vom freien Rand kann sich im Beton ein Widerstand gegen die Stahlverschiebung aufbauen. Dieser Störbereich „*g*" kann wirklichkeitsnäher nur durch ein zwei- bzw. dreidimensionales Modell abgebildet werden, wenn er eingehender untersucht werden soll.

Finite Elementstudien z. B. in [5] zeigen deutlich den dafür wichtigen Verlauf der „Querbeanspruchung" σ_t an der Stelle $r = 0$ (Bilder 3, 6) – zunächst ohne äußere Querbeanspruchung – und analoge Querverformungen u_r. Sie stehen mit der errechneten und beobachteten Rißbildung (Bild 4) in voller Übereinstimmung.

Außerhalb des Bereichs „*g*" kann mit dem üblichen Verbundmodell und den bekannten einfachen Zusammenhängen, die der Vollständigkeit halber im folgenden kurz wiederholt seien, durchaus gearbeitet werden.

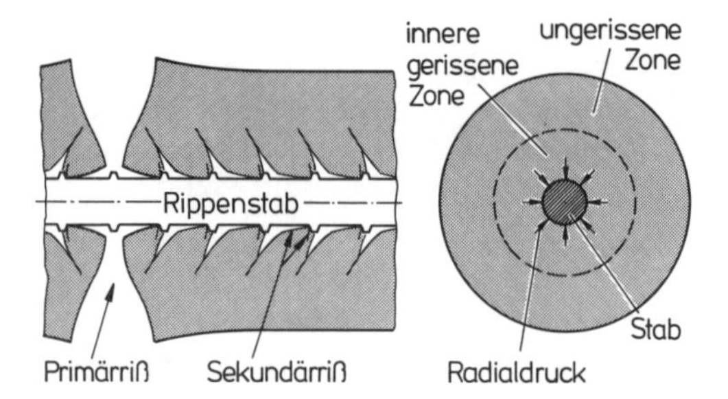

Bild 4
Rißbildung im Dehnkörper

Wegen der Gleichgewichtsbedingung

$$\tau = \frac{\varnothing}{4} \cdot \frac{d\sigma}{dx}$$

steigt die Schubspannung am Ende des Störbereichs vom Rißufer aus zunächst an, erreicht ihr Maximum dort, wo die Stahlspannung bzw. die Betonspannung einen Wendepunkt aufweist, und nimmt von hier ab wieder auf den Wert 0 in der Mitte des Prüfkörpers ab, wo

$$\frac{d\sigma_S}{dx} = 0 \quad \text{bzw.} \quad \frac{d\sigma_B}{dx} = 0$$

wird.

Später wird außerdem von Interesse sein, daß in Abhängigkeit von der Geometrie, d. h. bei zunehmender Rißbildung der größte Zug und Druckwert der σ_t-Verteilung mit kürzer werdendem Prüfkörper zunimmt (vgl. Bild 6).

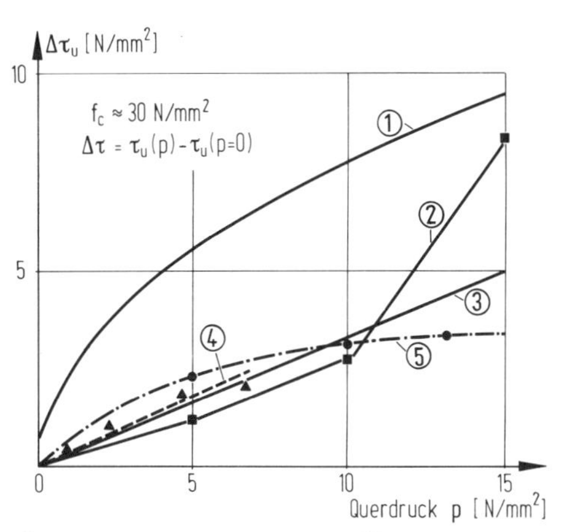

Bild 5
Einfluß eines äußeren Querdruckes auf die maximale Verbundspannung

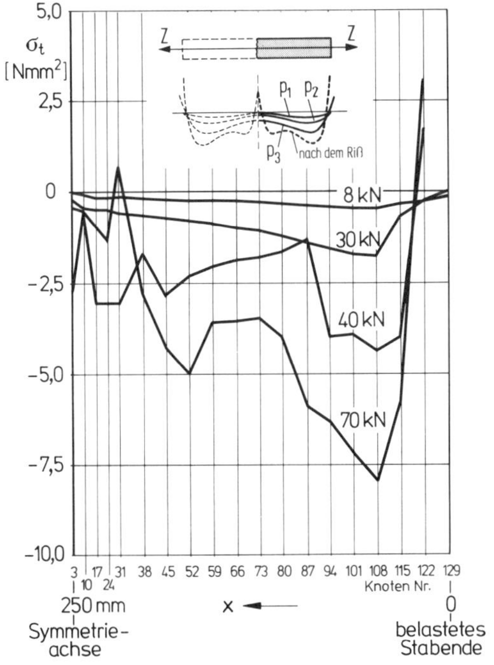

Bild 6
Berechnete Tangentialspannungen an der Stahloberfläche für verschiedene Beanspruchungsstufen [5]

Der Zusammenhang zwischen einer inkrementellen Erhöhung der Schubspannung $\Delta\tau$ und einer inkrementellen Veränderung der Relativbewegung $\Delta\delta$ zwischen Stahl und Beton während einer Belastung wird mithin *stoffgesetzlich* nur von der Spannungs-Dehnungs-Beziehung des Stahls bzw. des Betons bestimmt. Entscheidend sind jedoch außerdem Gleichgewichtsbedingungen, Kontinuitätsbedingungen und Randbedingungen des jeweiligen mechanischen Modells.
Deshalb muß zunächst festgestellt werden, daß die $\tau = f(\delta)$-Beziehung in üblicher Form kein originäres Stoffgesetz ist.
Nun werden aber auch die Spannungs-Dehnungs-Beziehungen des Betons bei mikroskopischer Betrachtung von komplexen Zusammenhängen zwischen Zuschlaggestein, Mörtelmatrix, Übergangsbedingungen, Porengehalt u. a. bestimmt, ohne daß hier die Existenz eines Stoffgesetzes in Frage gestellt würde.
Im Analogieschluß dazu kann wohl auch, weil es sich in der Praxis als zweckmäßig erweist, ein „technisches Stoffgesetz" für den Verbund

$$\Delta\tau = f(\Delta\delta \ldots); \quad \tau = f(\delta \ldots)$$

in inkrementeller oder finiter Form postuliert werden. Es muß jedoch dann sorgfältig zwischen Parametern, die konstant gehalten werden – z. B. Rippenform und bestimmten Randbedingungen der zu beschreibenden geometrischen Situation, d. h. des Prüfkörpers –, und Parametern mit wesentlichem Einfluß, die in einem solchen „Stoffgesetz" unmittelbar berücksichtigt werden sollen, unterschieden werden. Bei der Auswahl der letzteren ist naturgemäß eine Ermessensentscheidung zu treffen, da eine Verfolgung *aller* Einflüsse, wie sie die F-E-Methode prinzipiell ermöglicht, der Forderung nach einem „technischen Stoffgesetz" zum Zwecke der Vereinfachung widerspricht.
Auf der Suche nach einem solchen Stoffgesetz muß daher zunächst die Frage nach den signifikanten Parametern gestellt werden. Insbesondere ist zu prüfen, ob bisher wirklich alle relevanten Einflüsse überhaupt erfaßt wurden. Daß dies noch nicht generell der Fall gewesen sein kann (vgl. [6]), ergibt sich schon aus der erwähnten Feststellung, daß bislang selbst gleiche Versuchskörper zu lokal verschiedenen Verbundgesetzen führen.

3 Stoffgesetzliche Parameter – der Einfluß der lokalen Stahlspannung

Beschränkt man sich auf einen zylindrischen Dehnkörper etwa nach Bild 3, so ist bei gegebener Geometrie und gegebenen Werkstoffeigenschaften ein notwendiger Zusammenhang zwischen τ und δ unstrittig.
In zunehmendem Maße wird auch der Einfluß tangentialer bzw. radialer Spannungen σ_t, σ_r berücksichtigt.

Bei den experimentellen Untersuchungen nach Bild 5 wurde der Querdruck nicht in allen Fällen konstant über dem Umfang des Versuchskörpers aufgebracht. Entsprechende Versuche und daraus abgeleitete Ansätze liegen u.a. von DÖRR [4], UNTRAUER/HENRY [8], TASSIOS [9], CAIRNS/RAMAGE [11], VOS [7], ELIGEHAUSEN/POPOV/BERTERO [16] und nach [12] vor. Eigene Vorstellungen wurden in [5] entwickelt. Dabei muß – vgl. Abschnitt 4 – sorgfältig zwischen äußerer Beanspruchung p und durch den Längszug selbstinduzierten Spannungen σ_r bzw. σ_t im Bereich der Stahloberfläche (vgl. Bilder 3, 6) unterschieden werden, zumal sich die Maximalwerte der letzteren – positiv und negativ – mit abnehmendem Rißabstand vergrößern [5].
Bislang kaum beachtet wurde hingegen ein möglicher Einfluß der lokalen Stahldehnung bzw. Stahlspannung.
Bei gleicher aufzunehmender Gesamtschubkraft $T = (Z_1 - Z_2)/L$ (Bild 7) wird auch die „Basiszugkraft" Z_2 einen Einfluß haben, d.h. neben der erwähnten Querbeanspruchung σ_t wird u.U. auch die lokale Stahlspannung σ_S in einem Verbundgesetz

$$\tau = f(\delta, \sigma_t, \sigma_S)$$

zu berücksichtigen sein.
Man mag zunächst einwenden, die lokale Stahldehnung bewirke lediglich eine zusätzliche Verschiebung der Rippe A von ursprünglich

$$\delta = (A' - A)$$

auf

$$\delta = (A' - A) + \varepsilon_S \cdot c\,.$$

Und dieser Einfluß von $\delta_c \simeq \varepsilon_S \cdot c$ könne, obwohl bisher nicht berücksichtigt und obwohl noch in der Größenordnung von δ, nur von untergeordneter Bedeutung sein.

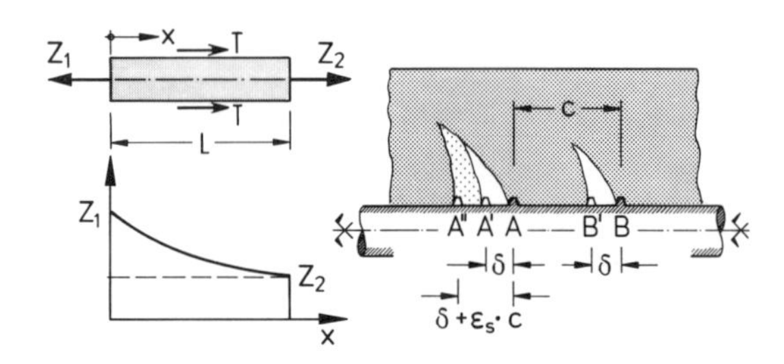

Bild 7
Beanspruchung des Dehnkörpers infolge der Stahlspannung

Zu einem anderen Schluß kommt man, wenn man berücksichtigt, daß der innere Zerstörungsgrad sicher maßgeblich durch die Zusatzverschiebung δ_c auf der kurzen Länge c, d.h.

$$\frac{\delta_c}{c} \simeq \frac{\mathrm{d}u_S}{\mathrm{d}x} = \varepsilon_S$$

mitbestimmt wird.

Anders ausgedrückt, die lokal ausgeprägte *Verschiebung* δ einer Rippe im Innern eines relativ langen Dehnkörpers bedingt u.U. eine geringere innere Zerstörung, als eine schnelle lokale *Änderung* der Verschiebung zwischen den Rippen.
Versuche von EIFLER [14] mit speziellen Testkörpern, bei denen der durchgehende Stahlstab vor den eigentlichen Ausziehversuchen bis weit in den plastischen Bereich auf Zug vorbelastet wurde, ergab in Übereinstimmung mit Experimenten von BENETT und SNOUNOU [15] einen deutlichen Einfluß einer solchen vorab aufgebrachten Stahlspannung.
Wenngleich eine derart hohe Stahlvorbelastung kaum praxisgerecht ist, so zeigt sie doch den prinzipiellen Einfluß einer unterschiedlichen lokalen Stahldehnung bei ansonsten gleichen Ausziehbedingungen und wird mit Rücksicht auf den jeweils resultierenden inneren Zerstörungsgrad verständlich.

4 Eigene Untersuchungen

Um eine mögliche Abhängigkeit der τ-δ-Beziehungen von der vorhandenen Stahlspannung σ_S zu prüfen, wurden zunächst die bekannten Versuche von DÖRR [4] neu ausgewertet. Dieser hatte zylindrische Dehnkörper ⌀ 15 cm, $L = 50$ cm mit Rippenstahl ⌀ 16 mm auf Zug und mit gleichzeitigem äußerem Querdruck p belastet und dabei $\varepsilon_S(x)$, $\varepsilon_B(x)$ sorgfältig registriert.
Mittels EDV wurden von den Verfassern aus diesen Versuchsreihen Werte $\Delta\tau/\Delta\delta = f(\delta,\ \sigma_S,\ p)$ an allen Meßstellen mit Ausnahme der unmittelbaren Rißumgebung (Bild 9) – Innenbereich – bestimmt. Studiert wurden im einzelnen für jeweils 3 Lastzyklen, 4 Stahldehnungen $\varepsilon_S = 0{,}25‰$; $\varepsilon_S = 0{,}5‰$; $\varepsilon_S = 1{,}0‰$; $\varepsilon_S = 1{,}5‰$; bei 4 äußeren Querdrücken $p = 0$ N/mm²; $p = 5$ N/mm²; $p = 10$ N/mm²; $p = 15$ N/mm² und in 48 Diagrammen zusammengestellt (vgl. [5]).

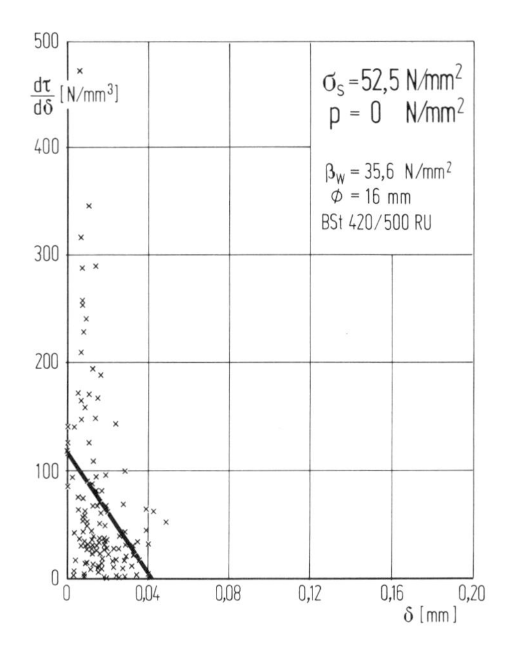

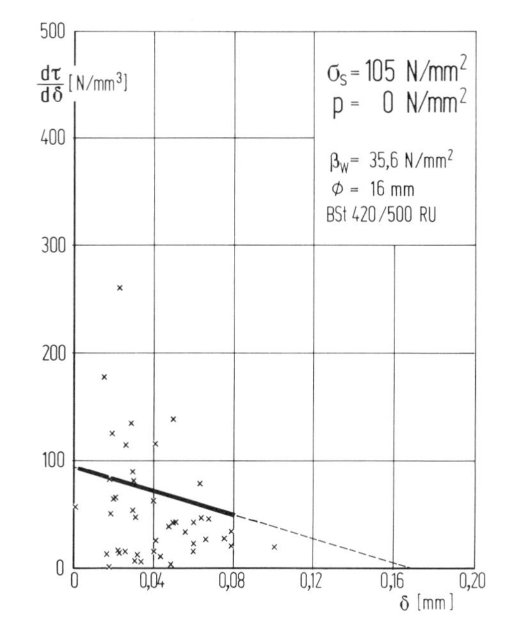

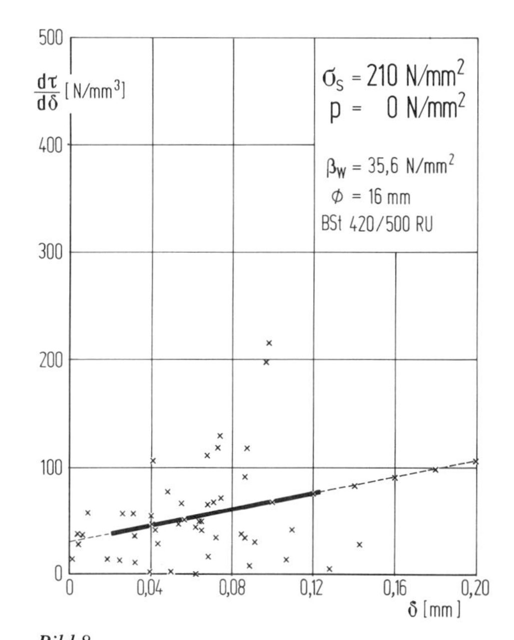

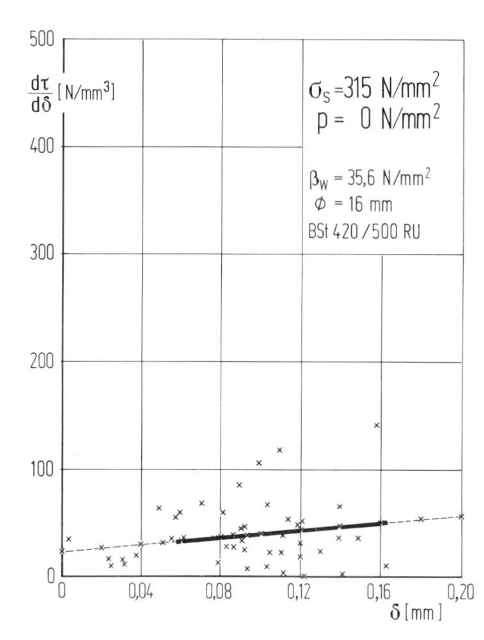

Bild 8
Steigung der τ-δ-Beziehung bei unterschiedlichen Stahlspannungen, bestimmt aus den Versuchen von DÖRR [4]

Bild 8 a–d zeigt eine Auswahl der so gefundenen Zusammenhänge anhand einer Serie von 5 Versuchen. Approximiert man in erster Näherung diese Ergebnisse abschnittsweise für konstantes $\sigma_{S,i}$ bei gleichbleibendem Querdruck $p = 0$ durch Gerade

$$\left.\frac{\mathrm{d}\tau}{\mathrm{d}\delta}\right|_{\sigma_{S,i}} = a_i(p = 0, \sigma_{S,i}) + b_i(p = 0, \sigma_{S,i}) \cdot \delta \quad (4.1)$$

– Bild 9 zeigt das Ergebnis bei Variation auch der p-Werte –, so folgt für Bild 8 a–d wegen

$$\left.\frac{\mathrm{d}\tau}{\mathrm{d}\delta}\right|_{\sigma_{S,i}} = a_i + b_i \cdot \delta \quad (4.2)$$

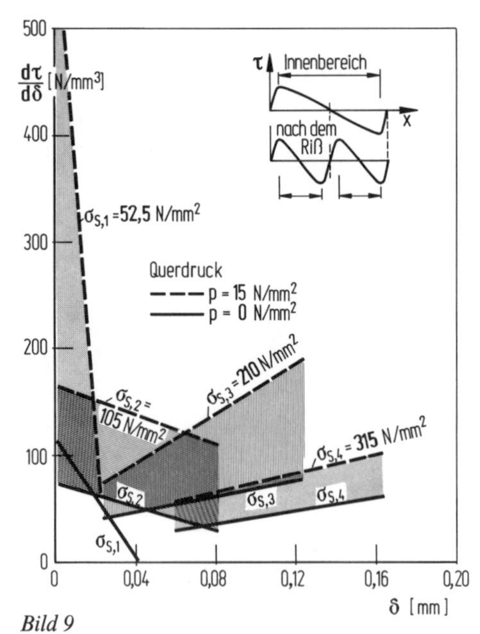

Bild 9
Steigung der τ-δ-Beziehung in Abhängigkeit von Stahlspannungen und Querdruck, bestimmt aus den Versuchen von DÖRR [4]

oder

$$\begin{aligned}
\sigma_{S,1} &= 52{,}5\,[\mathrm{N/mm^2}];\\
\frac{d\tau}{d\delta} &= 117 - 2846 \cdot \delta\,[\mathrm{N/mm^3}]\\
\sigma_{S,2} &= 105\,[\mathrm{N/mm^2}];\\
\frac{d\tau}{d\delta} &= 75 - 557 \cdot \delta\,[\mathrm{N/mm^3}]\\
\sigma_{S,3} &= 210\,[\mathrm{N/mm^2}];\\
\frac{d\tau}{d\delta} &= 31 + 376 \cdot \delta\,[\mathrm{N/mm^3}]\\
\sigma_{S,4} &= 315\,[\mathrm{N/mm^2}];\\
\frac{d\tau}{d\delta} &= 23 + 169 \cdot \delta\,[\mathrm{N/mm^3}]
\end{aligned} \tag{4.3}$$

Integriert man die Gleichungen (4.3), so gilt nach Bild 10 für $\sigma_{S,1} = 52{,}5$ N/mm² eine τ-δ-Beziehung etwa bis zum Punkt A. Eine Aussage über δ_A hinaus ist nicht möglich, da Werten $\delta > \delta_A$ in den Versuchen immer Stahlspannungen $\sigma_S > 52{,}5$ N/mm² zugeordnet waren. Für $\sigma_{S,2} = 105$ N/mm² läßt sich ein ebensolcher, durchaus plausibler Verlauf zwischen den Punkten B' und B gewinnen.

Bei $\sigma_{S,3} = 210$ N/mm² und $\sigma_{S,4} = 315$ N/mm² zwischen den Punkten $C'C$ bzw. $D'D$ fällt auf, daß sich die τ-δ-Kurven leicht nach oben krümmen entsprechend den positiven Steigungen in (4.3), (Bild 8c, d) bis zum maximalen τ-Wert bei etwa $\delta_D = 0{,}15$ mm. Bei beiden letztgenannten Stahlspannungen $\sigma_{S,3}$, $\sigma_{S,4}$ waren die Dehnkörper mindestens einmal durchgerissen, so daß die beobachteten Werte aus kürzeren Versuchskörpern resultieren.

Die so beobachtete Tendenz zu einer „vorübergehenden“ Versteifung bei relativ großen Verschiebungen und großen τ-Spannungen war auch in eigenen Versuchen [5], über die noch zu berichten sein wird, zu registrieren. Denkbar ist – es wurden nur randferne Punkte im Innenbereich (Bild 12) ausgewertet – daß

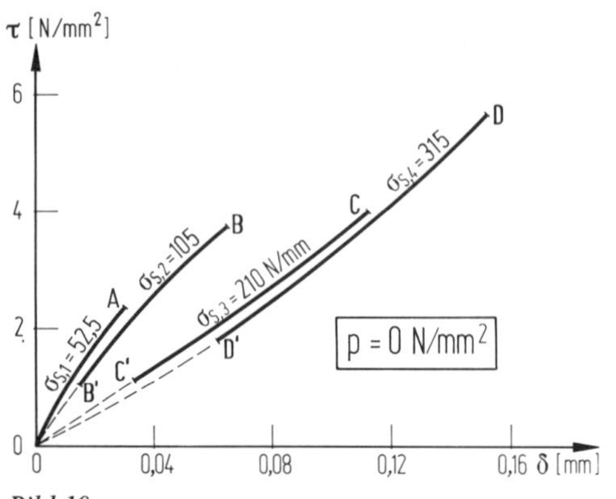

Bild 10
τ-δ-Beziehungen, bestimmt aus Bild 8

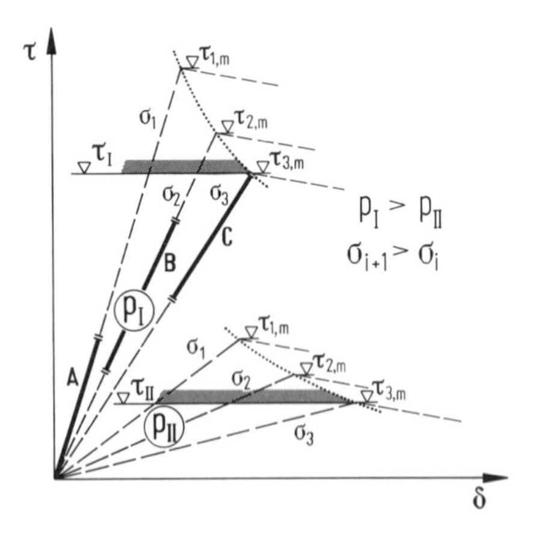

Bild 11
Schematische τ-δ-Beziehung für den Innenbereich

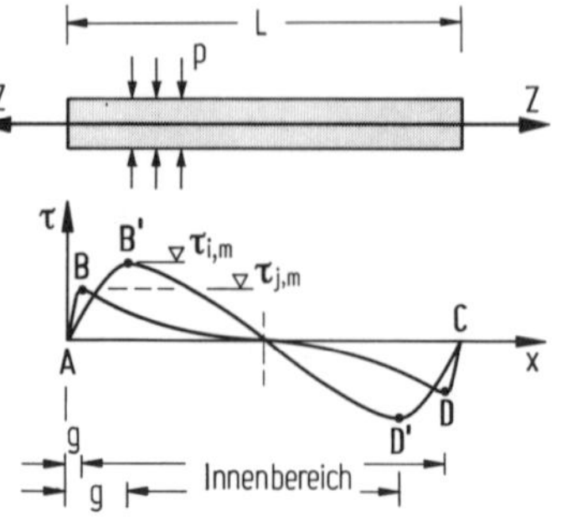

Bild 12
Schematischer τ-δ-Verlauf im Dehnkörper bei unterschiedlichen Beanspruchungen

sich nach erster innerer Rißbildung (Bild 3) die freigerissenen Druckdiagonalen zunächst noch komprimieren bis bei schnellem Fortschreiten der Radialrißbildung sich τ nicht mehr steigern läßt. Hinzu kommt, daß wie rechnerische Untersuchungen [5] zeigen, mit zunehmendem Zug bei weiterer Rißbildung außerhalb der Störzonen die selbstinduzierte Querdruckspannungen σ_t zunehmen und die Betonlängs-Zugspannungen zunächst abfallen, so daß der Beton auf den Bewehrungsstab „aufschrumpft". In Übereinstimmung mit der starken Stauchung der Druckdiagonalen beobachtete REHM [13] vor Kopf der Rippen eine pulverisierte, lokal stark verdichtete Betonzone bei hoher Beanspruchung.

Bedenkt man, daß sich bei der Ermittlung von τ an einer bestimmten Stelle x des Bewehrungsstabes im Dehnkörper neben δ auch σ_S mit der Belastung ständig ändert, so daß für die inkrementelle Zunahme von $\Delta\tau = f(\delta, \sigma_S, p)$ jeweils verschiedene Abschnitte der Kurven σ_i nacheinander maßgebend werden, so ergibt sich folgendes Bild:

Bei Beginn der Belastung erfolgt ein elastisches Eindrücken der Rippen mit nachfolgendem innerem Aufreißen und beginnendem Plastifizieren vor Kopf der Rippen, bestimmt durch die Verläufe von $\sigma_{S,1}$, $\sigma_{S,2}$. Daran schließt sich eine Verdichtungsphase mit steileren Abschnitten von $\sigma_{S,3}$, $\sigma_{S,4}$ an, bis schließlich τ_{max} bei schnellem Fortschreiten der Radialrisse erreicht ist.

5 Das Verbundgesetz – eine Hypothese

Da im Rahmen des eingangs erwähnten Forschungsvorhabens noch eine Reihe von Auswertungen und eigene Experimente durchgeführt werden, soll derzeit nicht versucht werden, ein endgültiges Stoffgesetz zu formulieren. Wenn dennoch Schlüsse aus den vorliegenden Untersuchungen gezogen werden, so müssen diese spekulativen Charakter haben und zur Diskussion herausfordern.

Betrachtet man zunächst den Bewehrungsstab in einem „unendlich" langen Dehnkörper mit einer begrenzten Anzahl von Rippen, weit ab von beiden Rißufern – Innenbereich in Bild 12 –, so kann, vernachlässigt man die Krümmungen der τ-δ-Beziehung in Bild 10 zugunsten einer bilinearen Darstellung, in Übereinstimmung mit dem Stand der Erkenntnis, ein prinzipielles Verhalten nach Bild 11 erwartet werden:

Zu unterschiedlichen äußeren Querdrücken z. B. p_I und p_{II} gehört – bei sonst gleichen geometrischen Bedingungen – jeweils ein Fächer von τ-δ-Geraden, deren Steigung mit zunehmendem $\sigma_{S,i}$ abnimmt. Entsprechend dem Grad der inneren Zerstörung nehmen auch die maximalen $\tau_{i,m}$-Werte zugehörig zur jeweiligen Stahlspannung $\sigma_{S,i}$ ab. Der daran anschließende Teil der τ-δ-Linien wird sich durch eine waagerechte, möglicherweise leicht abfallende Gerade approximieren lassen.

Der Verlauf der τ-Werte über die Länge eines Stabes im Dehnkörper wird an jeder Stelle x mit steigender Beanspruchung z. B. durch die Abschnitte A, B, C, bei veränderter Stahlspannung $\sigma_{S,i}$ bestimmt.

Werte $\tau > \tau_{II}$ bzw. $\tau > \tau_I$ entsprechend dem jeweiligen Querdruck können im Dehnkörper nie erreicht werden, wenn $\tau_{3,m}$ die der größten Stahlspannung – im τ-x-Verlauf am Rißufer – zugeordnete maximale Schubspannung bezeichnet.

Wie kommt nun der Abfall der τ-x-Werte (Bild 12) in der Nähe des Risses zustande?

Ein Abfall

$$\frac{\Delta\tau}{\Delta x} < 0$$

im Störbereich „g" kommt wegen der Zunahme von δ mit x nur zustande, wenn

$$\frac{\Delta\tau}{\Delta\delta} < 0$$

d. h. wenn im Stoffgesetz τ mit δ abfällt.

Die im Stoffgesetz eventuell leicht geneigte obere Begrenzung für den Innenbereich (Bild 11) bildet hierfür sicher keine ausreichende Basis. Vielmehr muß, wie eingangs betont, ein anderes mechanisches Modell, daß dem „Ausbrechen" bzw. „Weichwerden" der „Betonzähne" im Rißbereich, d. h. der besonderen Konstellation der Randzone – Störbereich g Rechnung trägt, betrachtet werden (Bild 13).

In Bild 13 wird zwischen einer gerissenen Zone, in der Umgebung des Stabes, die durch ein Vielfaches der Stabdurchmesser $\beta \cdot \varnothing$ – vgl. z. B. LEONHARDT [17] – begrenzt wird, und einem äußeren ringförmigen, nicht gerissenen Bereich unterschieden. Im randfernen Innenbereich nach Bild 12 bilden sich zwischen zwei Rissen sodann Druckstreben, wie etwas von A nach B aus, die im umgebenden Außenring Ringzugspannungen, an der Stahloberfläche Ringdruckspannungen erzeugen (vgl. auch Bild 3).

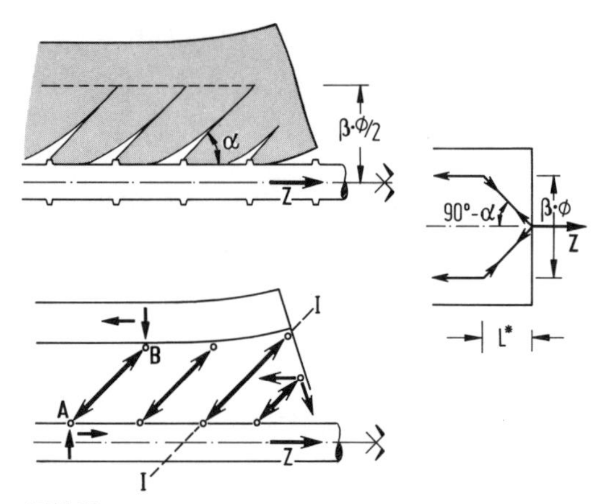

Bild 13
Mechanisches Modell für den Störbereich – I. Grenzfall (Aufbau der Zugkraft)

Am Rißufer findet jedoch die Druckdiagonale an ihrem oberen Ende wegen der Aufweitung des Zugringes und der Längsverformung des Rißufers nur ein sehr „weiches" Auflager bei gleichzeitig induzierten Radialzugspannungen σ_r. Das zunächst behandelte erste Modell für den Innenbereich ist deshalb erst ab etwa Schnitt I–I wirksam. Im hier zu betrachtenden Störbereich muß in einem zweiten Modell das Ausstrahlen der Betonzugkraft etwa über die Länge L^* in Rechnung gesellt werden. Eine solche „Ausbreitungs-Länge L^*" der Zugspannungen im Beton wird auch bei geringer Beanspruchung ohne innere Risse, d. h. bei weitgehend elastischem Verhalten existieren.
Mit β-Werten von $3 \div 5$ und einem Neigungswinkel $\alpha \sim 30°$ erhält man L^*-Werte von $0{,}8 \cdot \varnothing$ bis $1{,}5 \cdot \varnothing$ oder allgemein

$$L^* = \lambda \cdot \varnothing \tag{5.1}$$

wobei λ experimentell zu überprüfen ist.
Mit Rücksicht auf die Randbedingung $\tau_{xr} = \tau_{rx} = 0$ für $x = L$ (Bild 12), wonach die „Schubspannung" am Rißufer Null sein muß, erscheint es sinnvoll, bei einer τ-x-Berechnung die τ-Werte im Innenbereich mit den Stoffgesetzen nach Bild 11, 15 in inkrementeller Form bis zu einem Abstand L^* vom freien Rand zu bestimmen und von dort – mit Rücksicht auf die sehr komplexen Zusammenhänge und die geringe Größe von L^* – auf Null am Rißufer linear abfallen zu lassen. Mithin bestimmt ein *erster* Grenzwert L^* mit zugeordnetem Wert τ^* den Beginn des Abfalles von τ im Punkt D oder B (Bild 12) auf den Wert Null am Rißufer.
Beachtet man weiter, daß wegen des raschen Abfalles von τ auf der Strecke L^* sich σ_i nich mehr wesentlich ändern wird, so gilt für die Verschiebung des Rißufers relativ zur Stelle der größten Schubspannung τ^*:

$$\Delta\delta^* \cong \int_{L^*} \frac{\sigma_i}{E_S}\,\mathrm{d}x = \frac{\sigma_i}{E_S} \cdot L^* \tag{5.2}$$

Dieser *erste* Grenzwert L^* wird vor allem das Verhalten im Randbereich bei relativ *geringer* Längsbeanspruchung bestimmen.
Ein *zweiter* Grenzwert τ^{**} wird bei *hohen* τ-Beanspruchungen in Randnähe (Punkte B', D' in Bild 12) maßgebend, wenn ein „kegelförmiges" Ausbrechen eingeleitet wird.
Es erfolgt dabei ein rasches Abnehmen der Schubspannung bei B', D', weil der von den speziellen Randbedingungen bestimmte Maximalwert der Verbundspannung erreicht ist.
Für die stark simplifizierende Annahme einer Kegelfläche erhält man nach Bild 14 z. B. mit $\beta = 4$; $\alpha = 30°$ und

$$A_K \cong 6{,}2 \cdot \pi \cdot \bar{L} \cdot \varnothing$$

den Grenzwert τ^{**} aus der Gleichgewichtsbeziehung

$$\bar{L} \cdot U \cdot \tau^{**} = A_k \cdot \beta_z \cdot \sin\alpha$$
$$\bar{L} \cdot \varnothing \cdot p \cdot t^{**} = 6{,}2p \cdot \bar{L} \cdot \varnothing \cdot \beta_z \cdot 0{,}5$$
$$\tau^{**} \cong 3{,}0 \cdot \beta_z$$

kurz

$$\tau^{**} = \gamma \cdot \beta_z \tag{5.3}$$

wobei γ wieder anhand vorliegender Versuche zu überprüfen ist.
Für die Änderung der Verschiebung $\Delta\delta$ auf der Strecke $\bar{L}$ gilt aufgrund analoger Überlegungen ebenfalls Gl. (5.2), wobei jedoch an Stelle von L^*, $\bar{L}$ mit etwa $3 \cdot \varnothing \div 4 \cdot \varnothing$ zu setzen wäre.

$$\Delta\delta^{**} = \frac{\sigma_i}{E_S} \cdot \bar{L} \tag{5.4}$$

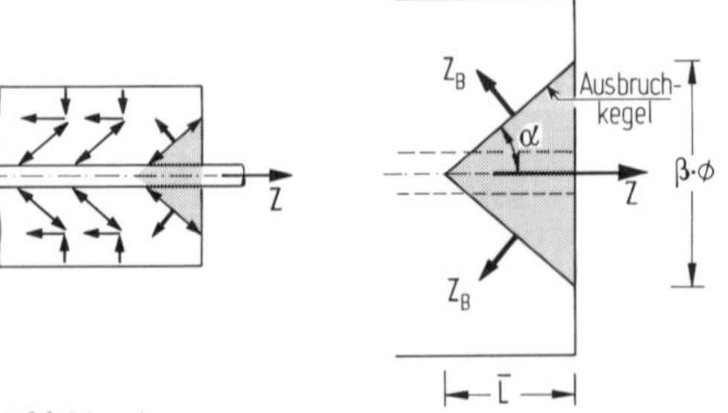

Bild 14
Mechanisches Modell für den Störbereich – II. Grenzfall (Ausbruchkegel)

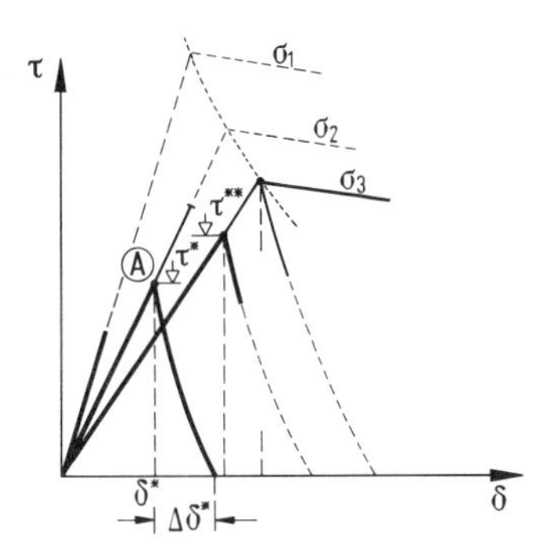

Bild 15
Technisches τ-δ-Stoffgesetz für den Dehnkörper

Der Gesamtzusammenhang der „Stoffgesetze für einen Dehnkörper" kann somit durch ein inkrementelles Stoffgesetz der folgenden Art erfaßt werden (Bild 15): Im Innenbereich des Dehnkörpers gilt für die ansteigenden Äste in Abhängigkeit vom Querdruck p und der Stahlspannung σ_S

$$\Delta\tau = [f(\sigma_S, p)] \cdot \Delta\delta \tag{5.5}$$

Wenn mit Annäherung an das Rißufer der Abstand L^* oder bei größerer Beanspruchung die Verbundspannung den Grenzwert $\tau = \tau^{**}$ erreicht, so wird vom jeweiligen Wert τ^* bzw. τ^{**} ein linearer Abfall bis auf den Wert Null am Rißufer angenommen.
Dieser lineare Abfall im τ-x-Verlauf tritt im τ-δ-Diagramm (Bild 15) sodann als gekrümmte Linie zwischen $\delta = \delta^*$ z. B. im Punkt A und $\delta = \delta^* + \Delta\delta^*$ (vgl. (5.2), (5.4)) bzw. $\delta = \delta^{**}$ und $\delta = \delta^{**} + \Delta\delta^{**}$ in Erscheinung.
Ein Abfall vom waagerechten Ast der τ-δ-Linie z. B. für $\sigma_{S,3}$ in Bild 15 ist wenig wahrscheinlich, da im Falle hoher Beanspruchung die bilineare τ-δ-Beziehung für den Innenbereich stets im waagerechten Teil einen Maximalwert aufweisen wird, der über dem Grenzwert τ^{**} am freien Rißufer liegen wird.
Selbstverständlich könnte im Rahmen einer solchen Approximation auch im τ-δ-Diagramm statt im τ-x-Verlauf ein linearer Abfall von τ^* bzw. τ^{**} auf den Wert Null an der Stelle $\delta^* + \Delta\delta^*$ bzw. $\delta^{**} + \Delta\delta^{**}$ postuliert werden.
Bei praktischen Berechnungen dürfte es jedoch einfacher sein, diesen direkt im τ-x-Verlauf vorzusehen, statt einen solchen im τ-δ-Diagramm vorauszusetzen und daraus den zugehörigen τ-x-Verlauf zu errechnen.
Derzeit sind die Verfasser dabei, einen entsprechenden Ansatz für Gleichung (5.5) an Hand der Versuche von DÖRR (z. B. Bild 8) zu optimieren und außerdem die beiden Grenzwerte

$$L^* = \lambda \cdot \varnothing \tag{5.6}$$

bzw.

$$\tau^{**} = \gamma \cdot \beta_z \tag{5.7}$$

festzulegen.

6 Zusammenfassung

Zunächst wird ein „technisches Verbundgesetz" für den randfernen Bereich des Dehnkörpers – es enthält echte stoffgesetzliche Zusammenhänge für Beton und Stahl, verknüpft mit Systemeigenschaften der jeweiligen Verbundkonstellation im Lichte neuerer Untersuchungen erörtert. Dabei wird der Parameter Querdruck und insbesondere der bislang kaum beachtete Einfluß der Stahlspannung auf die τ-δ-Beziehung an Hand eigener Untersuchungen diskutiert.
Ergänzt durch eine Beschreibung der Störzone unmittelbar am Riß wird sodann ein inkrementelles Stoffgesetz

$$\Delta\tau = [f(\sigma_S, p)] \cdot \Delta\delta$$

skizziert, das genügen sollte, den ganzen interessierenden Bereich hinreichend zu beschreiben.

7 Literatur

[1] NILSON, A. H.: Internal measurement of bond slip. ACI Journal, Proceedings Vo. 69, No. 7, Juli 1972, S. 439–441.
[2] WATSTEIN, D.: Distribution of bond stress in concrete pullout specimens. ACI Journal, Proceedings Vol. 43, No. 9, Mai 1947, S. 1041–1052.
[3] EIBL, J. und IVANY, G.: Studie zum Trag- und Verformungsverhalten von Stahlbeton. DAfStb, Heft 260, 1976.
[4] DÖRR, K.: Ein Beitrag zur Berechnung von Stahlbetonscheiben unter besonderer Berücksichtigung des Verbundverhaltens. Dissertation, Techn. Hochschule Darmstadt, 1980.
[5] KOBARG, J.: Veröffentlichung in Vorbereitung.
[6] KOBARG, J.: Verankerung druckbeanspruchter Bewehrungsstäbe unter Querdruck. Zwischenbericht DFG, Institut für Massivbau, Universität Karlsruhe, August 1983.
[7] VOS, E.: Influence of loading rate and radial pressure on bond in reinforced concrete. Diss. Delft, Delft University Press, 1983.
[8] UNTRAUER, R. E. und HENRY, R. L.: Influence of normal pressure on bond strength. ACI Journal, May 1965.
[9] TASSIOS, T.: Properties of bond between concrete and steel under loaded cycles idealizing seismic actions. AICAP-CEB Symposium, Structural Concrete under seimic actions, Volume 1 – State-of-the-art-report, Rome, 1979.
[10] MARTIN, H. und NOAKOWSKI, P.: Verbundverhalten von Betonstählen. Untersuchung auf der Grundlage von Ausziehversuchen in Heft 319 des Deutschen Ausschusses für Stahlbeton, 1981.
[11] CAIRNS/RAMAGE: In Comite Euro-International Du Beton. Bulletin D'Information No 151, Paris, 1982.

[12] Bond action and bond behaviour of reinforcement State-of-the-art-report. Comite Euro-International Du Beton. Bulletin D'Information No. 151, Paris, 1982.

[13] REHM, G.: Über die Grundlage des Verbunds zwischen Stahl und Beton. DAfStb, Heft 138, 1961.

[14] EIFLER, H.: Verbunduntersuchungen an gerippten Betonstählen als Grundlage für Formänderungsberechnungen. Forschungskoll. BAM, Berlin, 1981.

[15] BENETT, E.W. und SNOUNOU, J.G.: Bond-slip characteristics of plain reinforcing bars under varying stress in „Bond in Concrete" ed. by BARTOS, Appl. Science Publ., London, 1982.

[16] ELIGEHAUSEN, R., POPOV, E.P. und BERTERO, V.V.: Local bond stress-slip relationships of deformed bars under generalized excitations Earthquake Engineering Research Center, Report No. UCB/EERC – 83/23. University of California, Berkeley, 1983.

[17] LEONHARDT, F.: Vorlesungen über Massivbau, Dritter Teil. Grundlagen zum Bewehren im Stahlbetonbau. Springer-Verlag, Berlin, Heidelberg, New York, 1974.

Zur Rißbreitenbeschränkung im Massivbau*

Gert König
und Martin Krips

1 Vorbemerkung

Dauerhaftigkeit und Gebrauchsfähigkeit werden durch eine sinnvolle Steuerung der zu erwartenden Rißbreiten ganz wesentlich beeinflußt. Über diese grundlegende Aussage besteht weitgehend Einmütigkeit, so daß sich hieraus direkt die Forderung ableiten läßt, neben den Tragsicherheitsnachweisen eine zusätzliche Betrachtung des Zustandes vorzunehmen, in dem ein Bauteil sich planmäßig befindet. Die fortschreitende Entwicklung auf dem Gebiet der Standsicherheitsnachweise zeichnet sich durch eine stärkere Ausnutzung der beteiligten Baustoffe aus. Einerseits wird über wirklichkeitsnähere Belastungsannahmen ein Teil der ansonsten vorhandenen Gebrauchszustandsreserven aufgebraucht, andererseits erhöht sich das Beanspruchungsniveau des Gebrauchszustandes weiter über die Einführung höherwertiger Materialien, so daß sich die Notwendigkeit eines Rißbreitennachweises in Zukunft in verstärktem Maße stellen wird.
Nachdem man sich früher auf Konstruktionserfahrung verließ, ist heute in allen einschlägigen Entwurfsvorschriften dieser Forderung Rechnung getragen. Anspruch und Erfolg decken sich jedoch nicht in allen Fällen bzw. in allen Bereichen. Wenn auch weitgehende Übereinstimmung über die problembestimmenden Einflußgrößen vorhanden ist, so werden sie in den Theorien, die den einzelnen Nachweisverfahren zugrunde liegen, unterschiedlich gewichtet, so daß sich zwangsläufig Einschränkungen bezüglich Gültigkeitsbereich und Aussagekraft ergeben müssen.

* Die mitgeteilten Ergebnisse sind im Rahmen eines Forschungsvorhabens entstanden, das mit Förderung durch die Deutsche Forschungsgemeinschaft am Institut für Massivbau der Technischen Hochschule Darmstadt durchgeführt wurde.

2 Problemstellung, Zielsetzung

Die Verfahren zur Bestimmung von Rißbreiten lassen sich grob einteilen in empirische bzw. halb empirische Theorien und in Verfahren, die sich zur theoretischen Lösung der Methode der Finiten Elemente bedienen. Letztere werden vorwiegend dazu benutzt, Detailaspekte zu studieren. Die Entwurfsvorschriften werden jedoch von den halbempirischen Theorien beherrscht. Allen diesen Verfahren gemeinsam ist die Feststellung grundsätzlicher, theoretischer Abhängigkeiten und eine Anpassung an Versuchsergebnisse über additive und multiplikative Korrekturfaktoren.
Bei dieser Vorgehensweise wird naturgemäß der mathematische Grad der Abhängigkeit nicht gewonnen; er wird vielmehr als linear vorausgesetzt, so daß nur in einem begrenzten Bereich oder für eine beschränkte Anzahl von Parametern Übereinstimmung zwischen Theorie und Versuch herstellbar ist.
Beide beschriebenen Verfahren werden jedoch nur als Ausweg benutzt, da eine umfassende Lösung in geschlossener mathematischer Form nicht formulierbar erschien. Es soll deshalb einmal versucht werden, auf theoretischem Weg ein Verfahren zu entwickeln, das es erlaubt, die Einflußparameter in der Herleitung mitzuführen – ohne durch frühzeitige Vereinfachungen und empirische Angleichung an Versuche ihren Einfluß zu verwischen [19].

3 Grundlagen für Erstrißbildung

Steigt in einem bewehrten Betonbauteil die Beanspruchung über das Niveau der örtlichen Betonzugfestigkeit an, so erzwingt der damit einhergehende Riß eine Umverteilung der inneren Kräfte. Diese Rißproblematik läßt sich in der allgemein bekannten Art beschreiben, indem das Gleichgewicht der entlang des Bewehrungsstabes wirksamen Kräfte betrachtet wird:

$$\begin{aligned} d\sigma_s(x) \cdot A_s &= \tau_v(x) \cdot u \cdot dx \\ &= - d\sigma_b(x) \cdot A_b \end{aligned} \tag{1}$$

Als zweite Beziehung dient eine Verträglichkeitsbedingung zwischen Stahl- und Betonverformung, deren Differenz als unbekannte Verschiebung eingeführt wird.

$$\frac{dv}{dx} = \frac{1}{E_s} [\sigma_s(x) - n \cdot \sigma_b(x)] \tag{2a}$$

Da sich nach (1) Beton- und Stahlspannungen nur proportional verändern können, kann in (2a) die Betonspannung durch die Stahlspannung ausgedrückt werden. Als Proportionalitätsfaktor wird üblicherweise das Verhältnis zwischen Stahl- und Betonfläche herangezogen,

$$\frac{d\sigma_b(x)}{dx} = - \frac{A_s}{A_b} \cdot \frac{d\sigma_s(x)}{dx} = - \mu \cdot \frac{d\sigma_s(x)}{dx}$$

so daß sich die Beziehung (2a) auch schreiben läßt:

$$\frac{d^2v}{dx^2} = \frac{1 + \mu \cdot n}{E_s} \cdot \frac{d\sigma_s(x)}{dx} \tag{2b}$$

Allerdings besteht bei Verwendung des Bewehrungsprozentsatzes μ die Gefahr von Fehlinterpretationen, so daß es günstiger erscheint, hierfür nach Bild 1 das Verhältnis $\sigma_{SR}/\Delta\sigma_{SR}$ zu verwenden:

$$\frac{d^2v}{dx^2} = \frac{\sigma_{SR}}{\Delta\sigma_{SR}} \cdot \frac{1}{E_s} \cdot \frac{d\sigma_s(x)}{dx} \tag{2c}$$

Als dritte Bedingung wurde von Rehm in [2] ein Verbundgesetz eingeführt, das im Sinne eines Materialgesetzes die Verknüpfungseigenschaften zwischen Beton

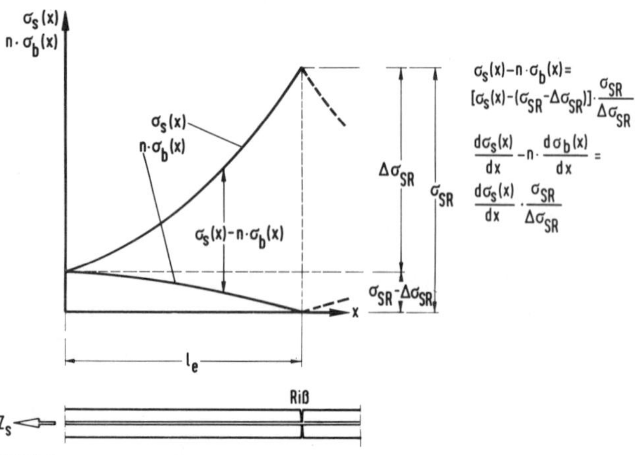

Bild 1
Substitution der Betonspannung durch die Stahlspannung

und Bewehrung beschreibt. Es wird dabei zwischen örtlicher Verschiebung und Verbundbeanspruchung ein Zusammenhang aufgestellt, der nach [5] in der Form (3) darstellbar ist.

$$\tau_v = \beta_w \cdot A \cdot v^N \tag{3}$$

Mit den Bedingungen (1) bis (3) läßt sich durch Substitution die Differentialgleichung des verschieblichen Verbundes (4) herleiten.

$$\frac{d^2v}{dx^2} = \frac{4}{d_S} \cdot \frac{\varepsilon_{SR}}{\Delta\sigma_{SR}} \cdot \beta_w \cdot A \cdot v^N \tag{4}$$

Es handelt sich hierbei um eine nichtlineare Differentialgleichung zweiter Ordnung, deren Lösung nach [1] problemlos angegeben werden kann:

$$v(x) = \left[\frac{2}{d_S} \cdot \frac{\varepsilon_{SR}}{\Delta\sigma_{SR}} \cdot \beta_w \cdot A \cdot \frac{(1 - N)^2}{1 + N} \cdot x^2 \right]^{\frac{1}{1-N}} \tag{5}$$

Die Anbindung dieser Verschiebungsbeziehung an die Rißstelle erfolgt über die beiden Randbedingungen, daß am Riß selbst die Betonspannung 0 sein muß, während sie am Ende des Einleitungsbereiches (Koordinatenursprung) gerade den Wert der Betonzugfestigkeit erreicht:

$$l_e = \frac{2}{1 - N} \cdot \left[\frac{\Delta\sigma_{SR}}{(\varepsilon_{SR})^N} \cdot \frac{d_S}{8} \cdot \frac{1 + N}{\beta_w \cdot A} \right]^{\frac{1}{1+N}} \tag{6}$$

Die Erstrißbreite w^1 kann dann als doppelter Wert der Verschiebung v am Endpunkt der Einleitungslänge angegeben werden:

$$w^1 = 2 \cdot \left[\Delta\sigma_{SR} \cdot \varepsilon_{SR} \cdot \frac{d_S}{8} \cdot \frac{1 + N}{\beta_w \cdot A} \right]^{\frac{1}{1+N}} \tag{7}$$

Stahl-, Beton- und Verbundspannungsverlauf lassen sich hieraus mit Hilfe der Eingangsbedingungen (1) bis (3) angeben. Zur Verdeutlichung des Ergebnisses sind sie in Bild 2 über der Einleitungslänge dargestellt.

4 Erweiterung der Erstrißtheorie

4.1 Verallgemeinerung der Lösung

In ähnlicher Form ist dieser Lösungsweg bereits in [2], [5] und [9] beschrieben. Allerdings handelt es sich bei dem so berechneten Verschiebungsverlauf um einen

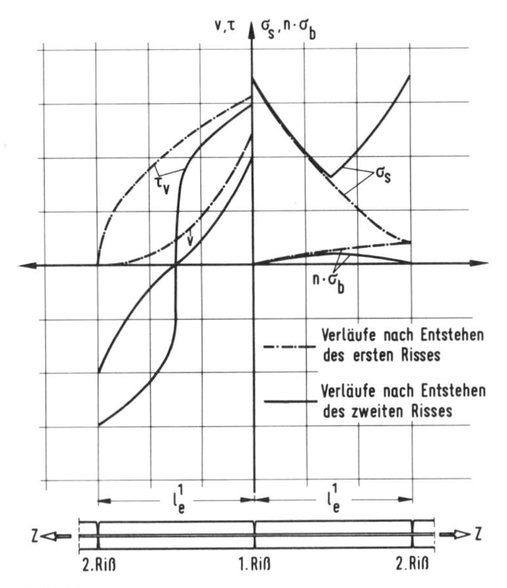

Bild 2
Verbundspannung und Verschiebung, Stahl- und Betonspannung bei Erst- und Zweitrißbildung

Spezialfall, der den wahren Sachverhalt nur im Anfangsstadium der Rißbildung treffend wiedergibt. Wie in den Randbedingungen zu (5) bereits zum Ausdruck kommt, sind am Endpunkt der Einleitungslänge dieselben Voraussetzungen vorhanden, wie sie zum ersten Riß geführt haben; ein zweiter Riß an dieser Stelle verändert den Verschiebungsverlauf aus der Erstrißbildung wesentlich.
Um auch diesen Fall einer analytischen Lösung zuzuführen, wird ein Ansatz bezüglich der erwarteten Gesamtverschiebung gemacht, die sich aus der bereits vorhandenen Verschiebung v_1 der Erstrißbildung nach (5) und einem zweiten, additiven Verschiebungsanteil v^* zusammensetzt.

$$v_{\text{ges}} = v_1 - v^* \tag{8}$$

Dieser Ansatz kann in (4) eingesetzt werden, da für die gesuchte Gesamtverschiebung unverändert die Aussage der Differentialgleichung, nämlich Kraftgleichgewicht, Verträglichkeit und Verknüpfungseigenschaften zutreffen muß. Man erhält dann eine neue Differentialgleichung für den unbekannten additiven Verschiebungsanteil.

$$\frac{d^2v^*}{dx^2} = \frac{2N \cdot (1+N)}{(1-N)^2} \cdot \frac{v^*}{x^2} \tag{9}$$

Die Lösung dieser Differentialgleichung hat eine ähnliche Form wie diejenige der Erstrißbildung, so daß sich die Gesamtverschiebung als Teil der Verschiebung aus der Erstrißbildung darstellen läßt.

$$v_{ges}(x) = v_1(x) \cdot \left(1 - \frac{c_1}{x} - \frac{c_2}{x^2}\right) \tag{10}$$

Die beiden Integrationskonstanten lassen eine weitere Verallgemeinerung dahingehend zu, daß die Lösung für beliebige Rißabstände und Beanspruchungszustände dargestellt werden kann: Als erste Bedingung ist einzuführen, daß die Stahlspannung am Riß unverändert den Wert σ_{SR} bzw. σ_S^{II} beibehält, während als zweite Randbedingung eingeführt werden kann, daß die Gesamtverschiebung in der Mitte zwischen zwei Rissen aus Antimetriegründen sich zu Null ergeben muß. Bei Verwendung bezogener Variablen kann die Lösung in allgemeiner Form wie folgt angegeben werden:

$$v(\xi) = v_1(\xi) \cdot \left[1 + \left(\frac{2\alpha - \eta}{\alpha}\right)^2 \cdot \frac{1}{1 - N\left(1 - \frac{\eta}{\alpha}\right)} \cdot \left(\frac{N}{2} \cdot \frac{1}{\xi} - \frac{1+N}{4} \cdot \frac{1}{\xi^2}\right)\right] \tag{11}$$

$$\alpha = \frac{l_e^{\text{II}}}{l_e^{\text{I}}} = \left(\frac{\sigma_S^{\text{II}}}{\sigma_{SR}}\right)^{\frac{1-N}{1+N}};$$

$$\eta = \frac{a}{l_e^{\text{I}}}; \quad \xi = \frac{x}{l_e^{\text{II}}}$$

Die Lösung hat einen stetigen Übergang zur ungestörten Erstrißbildung (5), was durch Einsetzen des Rißabstandes $a = 2 \cdot l_e$ leicht nachprüfbar ist.
In Bild 2 ist der hieraus resultierende Verformungs- bzw. Spannungszustand für den minimal möglichen Rißabstand dargestellt. Es zeigt sich, daß der additive Verschiebungsanteil bis zum benachbarten Rißufer durchschlägt, was letztlich eine Verkleinerung der Erstrißbreite durch Entstehen eines eng benachbarten Risses bedeutet – ein Verhalten, das vielfach in Versuchen zu beobachten ist.

4.2 Abgeschlossenes Erstrißbild, sukzessive Rißteilung

Wie aus Gleichung (10) und (11) ersichtlich, läßt sich die Gesamtverschiebung immer als Produkt der Verschiebung bei Erstrißbildung darstellen. Gleiches gilt für die Einleitungslänge; d.h. die Verschiebung und damit die Rißbreite, wie auch die Einleitungslänge bei Erstrißbildung stellen charakteristische Bezugsgrößen dar. Es liegt nun nahe, auch bezüglich des Rißbildes einen derartigen Bezugszustand zu definieren.

Während in den klassischen Rißtheorien das „abgeschlossene Rißbild“ verwendet wird – ein Begriff, der mehr versuchstechnisch definiert ist –, bietet es sich in einer analytischen Theorie an, aus der inneren Beanspruchung heraus einen Bezugszustand zu definieren. Hierfür wird der Zustand gewählt, der nach Stadium I im Beton einen charakteristischen Wert der Betonzugfestigkeit entstehen läßt. Für Lastbeanspruchungen ist es ausreichend genau, örtliche Streuungen zu vernachlässigen und mit dem Mittelwert der Betonzugfestigkeit zu rechnen bzw. bei Biegebeanspruchung den entsprechenden, nach [13] aus der zentrischen Betonzugfestigkeit abgeleiteten Wert der Biegezugfestigkeit zu verwenden.

In dem so beschriebenen Beanspruchungszustand können praktisch ohne Laststeigerung solange neue Risse entstehen, bis an keiner Stelle des betrachteten Bauteils mehr starrer Verbund vorliegt; denn starrer Verbund würde bedeuten, daß sich die Zone im Stadium I befindet, was nach obiger Definition einen weiteren Riß ermöglichen würde. Dieser Zustand wird im folgenden mit „abgeschlossener Erstrißbildung“ bezeichnet. Am Beispiel des zentrisch gezogenen Stahlbetonstabes (Bild 3) ist die abgeschlossene Erstrißbildung dadurch gekennzeichnet, daß die Betonzugspannung an keiner Stelle mehr den Wert der Betonzugfestigkeit

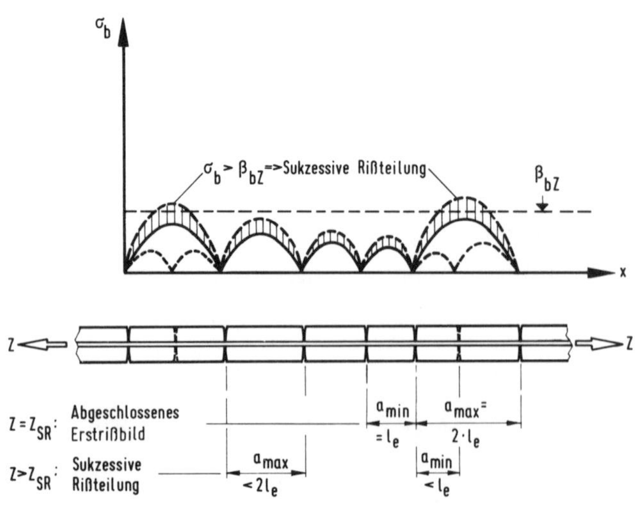

Bild 3
Erläuterung der Begriffe Abgeschlossenes Erstrißbild und Sukzessive Rißteilung

erreicht – was gleichbedeutend ist, daß alle vorhandenen Rißabstände zwischen einfacher und doppelter Einleitungslänge liegen.

Wächst die äußere Last über dieses Niveau an, so muß es logischerweise zu weiteren Relativverschiebungen zwischen Bewehrungsstahl und Beton kommen. Damit einher geht – wie in eigenen Versuchen auch festgestellt wurde – ein weiteres Anwachsen der Verbundspannungen. Mit höheren Verbundspannungen wird weitere Kraft vom Bewehrungsstab auf den Beton übertragen, so daß auch die Zugspannung im Beton weiter anwachsen muß. Speziell bei den großen Rißabständen des abgeschlossenen Erstrißbildes (siehe Bild 3) hat dies zur Folge, daß in Rißabstandsmitte wiederum die Betonzugfestigkeit erreicht werden kann, ohne daß an dieser Stelle starrer Verbund vorliegen müßte. Eine derartige Rißentstehung wird im Folgenden mit „sukzessiver Rißteilung“ bezeichnet.

Der Vorteil der analytischen Vorgehensweise besteht nun darin, daß ein derartiges Verhalten, wie die sukzessive Rißteilung, problemlos in die Theorie eingebaut werden kann. Hierzu ist die Bedingung zu formulieren, daß die Betonspannung in Rißabstandsmitte gerade die Betonzugfestigkeit erreicht.

$$\sigma_b\left(\xi = \frac{2\alpha - \eta}{2\alpha}\right) \stackrel{!}{=} \bar{\beta}_{bZ} \tag{12}$$

Es ist damit für jeden Rißabstand genau eine Laststufe definiert, für die das zutrifft. Allerdings führt diese Bedingung auf einen Ausdruck, der weder für die Laststufe noch für den Rißabstand explizit darstellbar ist, so daß auf eine Näherungslösung für den maximalen Rißabstand ausgewichen werden muß:

$$\begin{aligned} \eta_{\max}(a) &= 2 \cdot (\alpha^* - \beta^*)\,; \\ \alpha^* &= \left(\frac{\sigma_S^{\mathrm{II}}}{\sigma_{SR}}\right)^{\frac{1-N}{1+N} \cdot \frac{2+N}{2}}\,; \\ \beta^* &= \left(\frac{\sigma_S^{\mathrm{II}}}{\sigma_{SR}} - 1\right)^{\frac{1-N}{1+N} \cdot \frac{2+N}{2}} \end{aligned} \tag{13}$$

In Bild 4 ist der Funktionsverlauf zusammen mit der Näherungslösung (13) dargestellt. Es kommt hierin klar zum Ausdruck, daß nach Steigerung der Beanspruchung über das Niveau der abgeschlossenen Erstrißbildung hinaus der maximal mögliche Rißabstand anfangs schnell, bei höheren Laststufen jedoch nur noch langsam abnimmt.

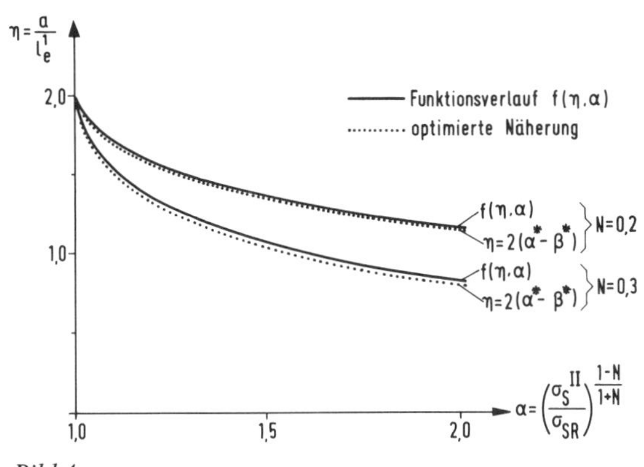

Bild 4
Maximaler Rißabstand bei sukzessiver Rißteilung

4.3 Verteilungsfunktion der Rißabstände

Mit den bisher entwickelten Beziehungen ist es möglich, für jeden konkreten Rißabstand und jede Laststufe den Zusammenhang zwischen Stahl-, Beton-, Verbundspannung und zugehörige Verschiebung anzugeben. Des weiteren ist der minimal – bzw. maximal mögliche Rißabstand für jede Laststufe bekannt. Zur Verallgemeinerung der Theorie fehlt noch eine Angabe über die Häufigkeit der in diesen Intervallgrenzen auftretenden Rißabstände.

Die vorherrschende Meinung besteht nun darin, daß die Rißabstände normal verteilt seien, so daß sich der mittlere Rißabstand aus dem arithmetischen Mittel zwischen maximal – und minimal möglichem Rißabstand ergibt. Neuere Arbeiten [8], [14] gehen aufgrund von Wahrscheinlichkeitsüberlegungen an Balken begrenzter Länge davon aus, daß der Mittelwert näher an dem minimal möglichen Rißabstand liegt (a_m = 1,33 bis 1,38 · l_e). Verallgemeinert man diese Überlegungen, so muß die Wahrscheinlichkeit eines bestimmten Rißabstandes ganz generell davon abhängen, wie oft er innerhalb einer vorgegebenen Länge überhaupt auftreten kann. Dies bedeutet, daß die Verteilungsdichte eines bestimmten Rißabstandes umgekehrt proportional zu seiner Größe sein muß. Ein kleiner Rißabstand hat also eine höhere Wahrscheinlichkeit als ein großer Rißabstand. Im Grenzfall bedeutet dies, daß der minimal mögliche Rißabstand doppelt so häufig vorkommen muß wie der maximal mögliche (vgl. Bild 3). Durch die sukzessive Rißteilung wird eine derartige Verteilungsfunktion unterstützt, da *ein* Rißabstand der Größe a_{max} *zwei,* im Idealfall halb so große, Rißabstände a_{min} entstehen läßt. Die Verteilungsdichtefunktion resultiert dann über die Bedingung, daß das Integral über minimalem und maximalem Rißabstand eins ergeben muß, zu

$$f(a) = \frac{1}{a \cdot \ln 2} \quad \text{bzw.} \quad f(\eta) = \frac{1}{\eta \cdot \ln 2} \tag{14}$$

5 Übergang auf Mittelwerte

Mit der Verteilungsdichtefunktion wird es möglich, die bisherigen Einzelergebnisse auf die wirklichkeitsnähere Basis von Mittelwerten zu bringen. Der mittlere Rißabstand verändert sich dabei nach (15) mit wachsender Beanspruchung stetig.

$$\begin{aligned} \eta_m &= \int_{\alpha^*-\beta^*}^{2\cdot(\alpha^*-\beta^*)} \frac{1}{\ln 2 \cdot \eta} \cdot \eta \cdot d\eta \\ &= \frac{1}{\ln 2} (\alpha^* - \beta^*) \end{aligned} \tag{15}$$

Nachdem NOAKOWSKI in [15] im Rahmen einer kontinuierlichen Rißtheorie bereits auf die Erfordernisse eines variablen Rißabstandes hingewiesen hat, gibt Gl. (15) eine theoretische Begründung für dieses Rißverhalten.

Es wird hiermit der Widerspruch der klassischen Rißtheorien auflösbar, daß Rißbreiten mit einem starren Rißabstand berechnet werden, auch wenn der betrachtete Beanspruchungszustand weit von jenem entfernt ist, in dem in Versuchen das „abgeschlossene Rißbild" festgestellt wird.

Berechnet man die mittlere Stahlspannung auf dem selben Wege,

$$\bar{\sigma}_{Sm} = \int_{\alpha^*-\beta^*}^{2\cdot(\alpha^*-\beta^*)} f(\eta) \int_{1-\eta/2}^{1} \sigma_S(\xi)\, d\eta\, d\xi \tag{16}$$

so resultieren aufgrund des Doppelintegrals für den praktischen Gebrauch schwer handhabbare Ausdrücke, die über eine Näherung auf die Form (17) gebracht werden können.

$$\begin{aligned} \bar{\sigma}_{Sm} = \sigma_S^{\mathrm{II}} - \Delta\sigma_{SR} \cdot \frac{\sigma_S^{\mathrm{II}}}{\sigma_{SR}} \cdot \frac{1}{4 \cdot \ln 2} \\ \cdot \frac{1+N}{1-N} \cdot \frac{\alpha - \beta}{\alpha} \left(1 - \frac{N}{2(1-N)} \cdot \frac{\alpha - \beta}{\alpha}\right) \end{aligned} \tag{17}$$

Die mittlere Stahlspannung setzt sich aus der Stahlspannung nach Stad. II und einem Abzugsglied zusammen, das vom Spannungssprung bei Erstrißentstehung $\Delta\sigma_{SR}$ abhängig ist und den Mitwirkungsanteil des Betons repräsentiert. Obwohl eine mit Laststeigerung progressive Betonmitwirkung angesetzt wurde, verkleinert sich der Betonmitwirkungsanteil aufgrund der sukzessiven Rißteilung. Nach Bild 5 strebt dieser Anteil asymptotisch einem Grenzwert zu, der nur noch vom Spannungssprung bei Erstrißbildung abhängig ist, was bereits in [16] auf experimentellem Wege festgestellt wurde.

$$\lim_{\alpha \to \infty} \bar{\sigma}_{Sm} = \sigma_S^{II} - \Delta\sigma_{SR} \cdot \frac{1}{4 \ln 2} \quad (18)$$

Die mittlere Rißbreite läßt sich für alle Beanspruchungszustände analog Gl. (11) in Abhängigkeit der Erstrißbreite darstellen. Trägt man diese Abhängigkeit über den Beanspruchungsparameter $\alpha = \left(\frac{\sigma_S^{II}}{\sigma_{SR}}\right)^{\frac{1-N}{1+N}}$ auf, so zeigt sich ein weitgehend linearer Zusammenhang, der über die Beziehung (19)

$$\frac{w^{II}}{w^{I}} = 1{,}6 \cdot \alpha - 0{,}7 \quad (19)$$

angenähert werden kann, so daß man für die mittlere Rißbreite eine einfach aufgebaute Gleichung erhält (20),

$$w_m^{II} = 2 \cdot \left[\varepsilon_{SR} \cdot \Delta\sigma_{SR} \cdot \frac{ds}{8} \cdot \frac{1+N}{\beta_W \cdot A}\right]^{\frac{1}{1+N}} \cdot (1{,}6 \cdot \alpha - 0{,}7) \quad (20)$$

die den direkten Nachweis der zu erwartenden Rißbreiten möglich macht – ohne auf Diagramme ausweichen zu müssen. Als wesentliche Einflußgrößen gehen in die Rechnung ein:

1. Stahlspannung σ_{SR} bzw. Stahlspannungssprung $\Delta\sigma_{SR}$ bei Erstrißentstehung. Hiermit wird ein Parameter in die Berechnung eingeführt, der alle geometrischen Eigenschaften über Beton- und Stahlquerschnittsfläche sowie Materialkennwerte über die mittlere Betonzugfestigkeit treffender erfaßt als der Ansatz des Bewehrungsprozentsatzes.
2. Stabdurchmesser d_S. Ersetzt man den Stabdurchmesser durch das Verhältnis zwischen Stabfläche und Stabumfang, so lassen sich auch gemischte Bewehrungen behandeln.
3. Verbundeigenschaften über die Verbundparameter A und N des angesetzten Verbundgesetzes sowie über die mittlere Würfelfestigkeit β_W des Betons.
4. Beanspruchungsniveau über die Stahlspannung σ_S^{II} nach Stadium II.

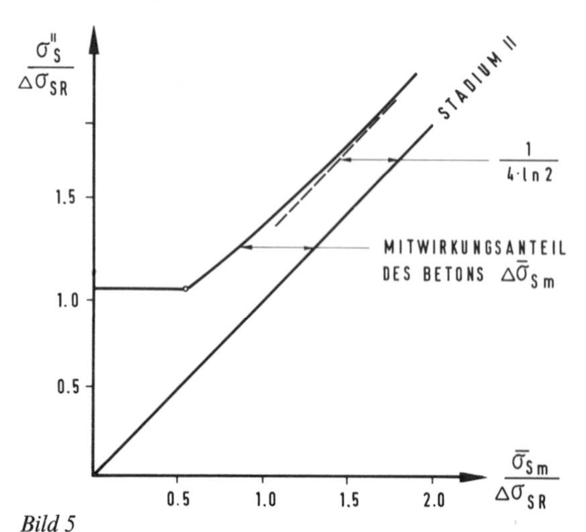

Bild 5
Mittlere Stahlspannung in Abhängigkeit der Stahlspannung am Riß

6 Verifizierung der Theorie

Nach rein analytischer Entwicklung der Bestimmungsgleichungen für den mittleren Rißabstand (15), die mittlere Stahlspannung (17) und die mittlere Rißbreite (20) ist zu überprüfen, ob der Realitätsbezug nicht durch die theoretische Betrachtungsweise verlorengegangen ist.

Bei dem Vergleich mit Versuchsergebnissen – es wurde die ausführlich dokumentierte Versuchsreihe nach [7] ausgewählt – zeigt sich in allen Laststufen eine sehr gute Übereinstimmung sowohl für den Rißabstand als auch für die Rißbreite. Zieht man den Quotienten aus Versuchswert/Rechenwert als Vergleichsgröße heran, so wurde im Mittel über alle Versuche und Laststufen (80 Einzelwerte) ein Wert von 1,07 bei einem Variationskoeffizienten von 16% erreicht.

Um den Vergleich zu halbempirischen Theorien zu ermöglichen, müssen die Werte $\Delta\sigma_{SR}$ und σ_{SR} wiederum rückwärts über den Bewehrungsprozentsatz ausgedrückt werden. Für zentrische Beanspruchung ist dies ohne Verlust der Allgemeingültigkeit möglich, so daß Gleichung (20) in der gebräuchlichen Art $d_S(\mu)$ dargestellt werden kann. In Bild 6 ist der Vergleich zu dem in [4] für Zwangsrißbildung angegebenen Diagrammen dargestellt. Für guten Verbund (obere Grenzlinie des schraffierten Bereiches) ergibt sich im Bereich der Zwangsrißbildung eine weitgehende Übereinstimmung. Auch für schlechte Verbundeigenschaften (un-

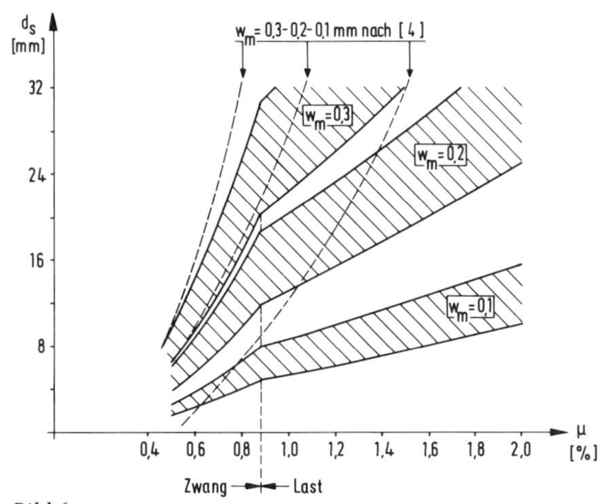

Bild 6
Vergleich der Ergebnisse mit Diagramm FALKNER

tere Begrenzungslinie des schraffierten Bereiches) ist eine Übereinstimmung vorhanden, wenn man die in [6] angegebene Empfehlung berücksichtigt, bei schlechtem Verbund den Bewehrungsprozentsatz um 30% zu erhöhen.
Der Vergleich zu [17] für biegebeanspruchte Bauteile erfordert zusätzliche Festlegungen, da der Spannungssprung in der Bewehrung nicht mehr allein durch den Bewehrungsprozentsatz μ ausgedrückt werden kann. Es ist zusätzlich noch für den Hebelarm der inneren Kräfte und die Bauteilhöhe eine zusätzliche Vorgabe zu treffen. Dies erfolgt in Bild 7 in der Form, daß ein Balken gebräuchlicher Höhe (40 cm) mit einer Bewehrungslage ($h/d = 0{,}93$) und variabler Betonausnutzung ($k_x = 0{,}35$ ./. 0,5) zugrunde gelegt wird. Für eine kritische Rißbreite von 0,2 bzw. 0,3 mm ergibt der Vergleich nur im Bereich höherer Bewehrungsprozentsätze (> 0,8%) eine befriedigende Übereinstimmung.

Abstriche in der Übereinstimmung sind für den Bereich kleiner Bewehrungsprozentsätze zu machen. Die Aussage, daß unterhalb von 0,3% Bewehrung keine rißbreitenbeschränkenden Maßnahmen ergriffen werden müssen, ist jedoch zu überdenken, da in ihr nur zum Ausdruck kommt, daß unter Lastbeanspruchung nicht mehr mit Rissen zu rechnen ist. Für eine Zwangsrißvorsorge sind jedoch wesentlich restriktivere Forderungen bezüglich des zu verwendenden Stabdurchmessers zu erheben.

7 Zusammenfassung

Es wurde versucht, die Vorgänge bei Rißbildung auf einem analytischen Weg nachzuvollziehen. Ausgangspunkt ist dabei ein Verbundgesetz der Form $\tau = \beta_W \cdot A \cdot v^N$, so daß die DGL des verschieblichen Verbundes einfach lösbar wird. Über einen zusammengesetzten Verschiebungsansatz wird die Lösung auf Fälle erweitert, in denen sich zwei benachbarte Risse gegenseitig beeinflussen.
Mit dem Begriff der „abgeschlossenen Erstrißbildung“ wird ein Bezugszustand definiert, der zusammen mit der Annahme fortschreitender Rißteilung zu einem kontinuierlich sich verändernden Rißmuster führt. Den hierin enthaltenen Rißabständen wird eine sinnvolle Häufigkeitsverteilung zugeschrieben, so daß sich für alle interessierenden Größen explizit Mittelwerte berechnen lassen. Es resultieren dabei einfache Gleichungen, die die wesentlichen Einflüsse ausreichend genau wiedergeben, was über entsprechende Vergleiche zu Versuchen und herkömmlichen Verfahren nachgewiesen wird.

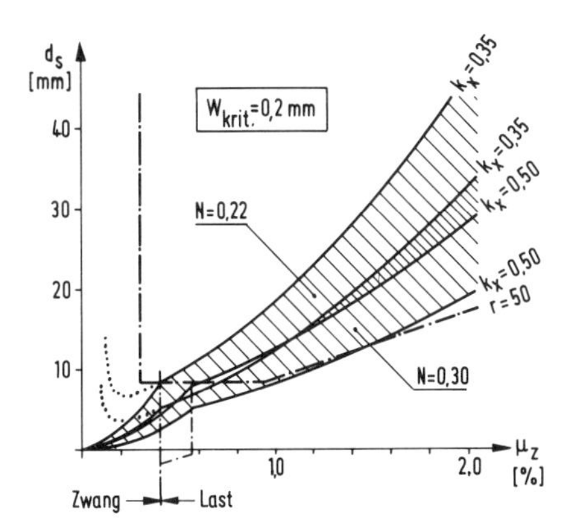

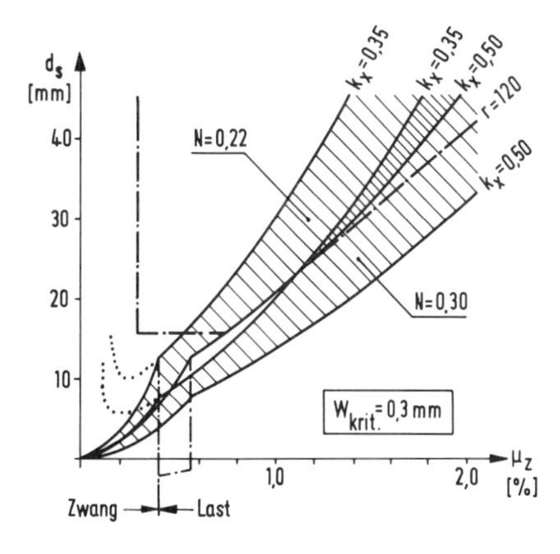

Bild 7
Ergebnisvergleich mit DIN 1045

8 Literatur

[1] KAMKE, E.: „Differentialgleichungen, Lösungsmethoden und Lösungen" Band I, Akademische Verlagsgesellschaft Becker u. Erler Kom.-Ges., Leipzig 1944.

[2] REHM, G.: „Über die Grundlagen des Verbundes zwischen Stahl und Beton". Deutscher Ausschuß für Stahlbeton, Heft 138, 1961.

[3] REHM, G. und MARTIN, H.: „Zur Frage der Rißbegrenzung im Stahlbetonbau". Beton- und Stahlbetonbau 8/1968.

[4] FALKNER, H.: „Zur Frage der Rißbildung durch Eigen- und Zwängspannungen infolge Temperatur in Stahlbetonbauteilen". Deutscher Ausschuß für Stahlbeton, Heft 208, 1969.

[5] MARTIN, H.: „Zusammenhang zwischen Oberflächenbeschaffenheit, Verbund und Sprengwirkung von Bewehrungsstählen unter Kurzzeitbelastung". Deutscher Ausschuß für Stahlbeton, Heft 228, 1973.

[6] LEONHARDT, F.: „Vorlesungen über Massivbau", vierter Teil. Springer-Verlag, Berlin, Heidelberg, New York, 1976.

[7] REHM, G., ELIGEHAUSEN, R. und MALLÉE, R.: „Rißverhalten von Stahlbetonkörpern bei Zugbeanspruchung". Lehrstuhl für Werkstoffe im Bauwesen, TU Stuttgart, Untersuchungsbericht Nr. 76/4 vom 2. 2. 76.

[8] BEEBY, A. W.: „Concrete in the Oceans, Cracking and Corrosion". Technical Report No. 1 (1978) CIRIA/UEG, Cement and Concrete Association Department of Energy.

[9] NOAKOWSKI, P.: „Die Bewehrung von Stahlbetonbauteilen bei Zwangsbeanspruchung infolge Temperatur". Deutscher Ausschuß für Stahlbeton, Heft 296, 1978.

[10] CEB/FIP – Mustervorschrift für Tragwerke aus Stahlbeton und Spannbeton, 3. Ausgabe 1978.

[11] MARTIN, H., SCHIESSL, P. und SCHWARZKOPF, M.: „Berechnungsverfahren für Rißbreiten aus Lastbeanspruchungen". Forschung, Straßenbau und Straßenverkehrstechnik, Heft 309, 1980.

[12] TROST, H., CORDES, H., THORMÄLEN, U. und HAGEN, H.: „Teilweise Vorspannung, Verbundfestigkeit von Spanngliedern und ihre Bedeutung für Rißbildung und Rißbreitenbeschränkung". Deutscher Ausschuß für Stahlbeton, Heft 310, 1980.

[13] JAHN, M.: „Zum Ansatz der Betonzugfestigkeit bei den Nachweisen zur Trag- und Gebrauchsfähigkeit von unbewehrten und bewehrten Betonbauteilen". Deutscher Ausschuß für Stahlbeton, Heft 341, 1983.

[14] MEIER, H.: „Berücksichtigung des Wirklichkeitsnahen Werkstoffverhaltens beim Standsicherheitsnachweis turmartiger Stahlbetonbauwerke". Dissertation TU Stuttgart 1983.

[15] NOAKOWSKI, P.: „Eintragungslänge, Verformung, Rißbreite, gleichzeitige Last- und Zwangsbeanspruchung im Stahlbetonbau". Vortrag an der Technischen Akademie Wuppertal 1983.

[16] HARTL, G.: „Die Arbeitslinie ‚Eingebetteter Stähle' unter Erst- und Kurzzeitbelastung". Beton- und Stahlbetonbau 8/1983.

[17] DIN 1045.

[18] DIN 4227.

[19] KRIPS, M.: „Rißbreitenbeschränkung im Stahlbeton und Spannbeton". Diss., Technische Hochschule Darmstadt 1984.

Einfluß der Betonzusammensetzung auf das Verbundverhalten von Bewehrungsstählen

Horst Martin

1 Problemstellung

Heute werden im Stahlbetonbau nahezu ausschließlich hochfeste Betonstähle der Güte BSt 420 und BSt 500 verwendet. Ermöglicht wurde dies durch die Entwicklung gerippter Betonstähle und die damit verbundene wesentliche Erhöhung des Verbundes zwischen Bewehrungsstahl und Beton. Die theoretischen und praktischen Grundlagen dafür stammen von G. REHM [2].
Allgemein bekannt ist, daß das Verbundverhalten im wesentlichen von der bezogenen Rippenfläche f_R, der Betondruckfestigkeit und der Lage der Stäbe beim Betonieren abhängt. In der Literatur gibt es allerdings Hinweise, daß auch die Betonzusammensetzung und die Konsistenz das Verbundverhalten in gleicher Größenordnung wie die bezogene Rippenfläche und die Lage der Stäbe beim Betonieren beeinflussen können [1], [2], [3], [4] und [8].
In den letzten Jahren hat sich die in der Praxis verwendete Konsistenz des Betons mehr in den flüssigen Bereich (K3) hin verschoben. Für die Zukunft muß auch davon ausgegangen werden, daß Betonzuschläge erster Qualität nicht mehr ausreichend zur Verfügung stehen werden.
Aber auch im Versuchswesen besitzen diese Einflußgrößen Betonzusammensetzung und Konsistenz eine nicht zu vernachlässigende Bedeutung. Vergleicht man z. B. die Ergebnisse von zeitlich verschiedenen Versuchsreihen oder solche unterschiedlicher Laboratorien untereinander, so findet man häufig sehr große Streuungen. Dies gilt vor allem für alle durch den Verbund beeinflußten Größen, wie z. B. Rißbildung, Schubtragverhalten und Verankerung. Eine Auswertung von mehr als 1000 Versuchen zur Rißbildung aus aller Welt hat die Problematik einer zusammenfassenden Auswertung aufgezeigt [7]. Es ist durchaus denkbar, daß zumindest ein größerer Anteil dieser Streuungen auf den Einfluß der Betonzusammensetzung zurückzuführen ist. Außerdem werden im Versuchswesen zur Abdeckung von möglichen Streuungen in der Praxis auch häufig ungewöhnliche Betonzusammensetzungen wie z. B. Wasser-Zementwerte bis zu 1,0 gewählt oder das Prüfalter herabgesetzt. Dies wäre sicherlich dann unproblematisch, wenn einfache, das Verbundverhalten des untersuchten Stahles beschreibende Verbundversuche mit durchgeführt werden würden. Leider ist dies fast nie der Fall.
Am Institut für Betonstahl und Stahlbetonbau e.V. München wurden in den letzten Jahren einige grundsätzliche Verbundversuche zum Studium des Einflusses von Betonzusammensetzung und Konsistenz durchgeführt; mit untersucht wurde dabei auch das wichtige Verhältnis Betonzugfestigkeit/Betondruckfestigkeit.

2 Untersuchungsprogramm

In der Versuchsreihe wurden folgende Varianten (siehe auch Tabelle 1) untersucht:

3 Wasser-Zementwerte:	0,5; 0,75 und 0,90
3 Konsistbereiche:	K1, K2 und K3
3 Sieblinien:	A16, B16 und C16 (DIN 1045)

und zusätzlich ein leichter Normalbeton, bei dem die Kornfraktion 4/8 aus Berwillit-Leichtzuschlag bestand – Sieblinie AB–L (siehe auch Bild 1).
Außerdem wurde bei zwei Varianten (Sieblinie B16, $w/z = 0{,}90$) ein Teil des Zementes durch Zugabe eines inerten Gesteinsmehls ersetzt.
Alle Versuche wurden mit dem gleichen Hochofenzement Z 25 NW ausgeführt; die Zuschläge stammten aus Münchner Vorkommen (Kalkstein) und standen in den drei Kornfraktionen 0/4, 4/8 und 8/16 zur Verfügung.
Die Betondruckfestigkeit wurde an 20-cm-Würfeln jeweils nach 3, 7, 28 und 90 Tagen und die Spaltzugfestigkeit an Zylindern 150/300 mm nach 7, 28 und 90 Tagen ermittelt.

Tabelle 1
Versuchsreihen – Betonzusammensetzung und Würfeldruckfestigkeit

w/z-Wert	Sieb-linie	Ver-dichtungs-maß v	Zement	Zu-schläge	$\beta_{w,28}$
			kg		N/mm²
0,50	B	1,39	344	1884	49,3
		1,15	378	1833	38,7
		1,07	424	1764	39,8
0,75	A	1,19	195	2059	20,7
		1,17	207	2038	19,5
		1,01	300	1800	26,6
	AB–L	1,16	240	1726[1])	24,1
	B	1,42	217	2077	28,0
		1,15	247	1968	26,8
		1,05	294	1886	23,8
	C	1,34	281	1908	20,6
		1,20	300	1875	20,6
		1,13	300	1875	27,9
0,90	B	1,37	178	2062	18,8
		1,17	193	2033	19,1
		1,10	218	1985	16,5
	B–G	1,24	150	2115[2])	16,8

[1]) Körnung 4/8 aus Leichtzuschlag
[2]) einschließlich 150 kg inertes Gesteinsmehl

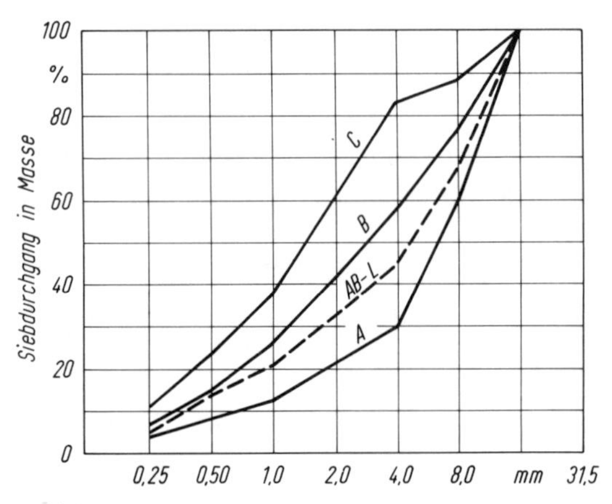

Bild 1
Verwendete Sieblinien (siehe auch DIN 1045 Bild 2)

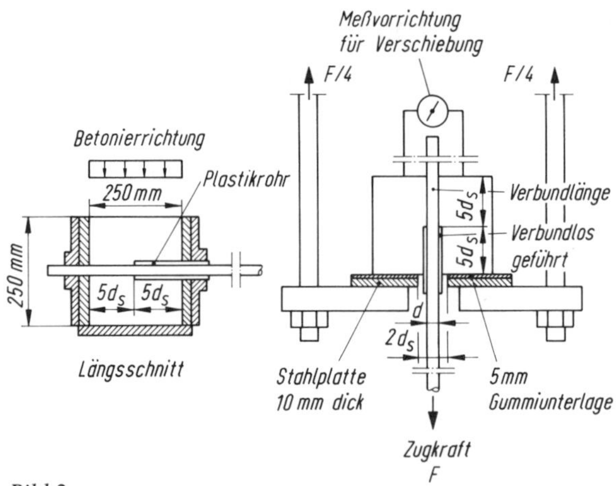

Bild 2
Ausziehversuchskörper und Versuchsanordnung

Das Verbundverhalten wurde in Ausziehversuchen mit mittig liegenden Prüfstäben nach Bild 2 mit einer wirksamen Verbundlänge von $l_v = 5d_s$ jeweils nach 28 Tagen bzw. bei einigen Varianten auch nach 7 Tagen geprüft. Um mögliche, sich allerdings unkontrolliert auswirkende, günstige Effekte auf das Verbundverhalten durch die Schalungsreibung und das Abziehen der Betonoberfläche (Nachverdichtungseffekt) auszuschalten, wurden die Ausziehversuchskörper mit einer gegenüber dem RILEM-Versuchskörper vergrößerten Kantenlänge von 250 mm ausgeführt.
Als Betonstahl wurde ein BSt 500 R, $d_s = 18$ mm, nach DIN 488 verwendet, der aus einer Schmelze stammte und im Anlieferungszustand eingebaut wurde:

$$R_{p,0,2} = 558\ \text{N/mm}^2;\ A_{10} = 18{,}0\%$$
$$R_m = 767\ \text{N/mm}^2;\ f_R = 0{,}075$$

Alle Prüfkörper wurden 7 Tage feucht abgedeckt und anschließend in der Prüfhalle ($\sim 20\,°C$ und $\sim 65\%$ relative Feuchte) gelagert. Die Durchführung der Ausziehversuche erfolgte in Anlehnung an RILEM Doc. 128.
Der in Ausziehversuchen mit kurzer Verbundlänge erhaltene Zusammenhang zwischen Verbundspannung und Verschiebung entspricht in guter Näherung dem von G. Rehm [2] eingeführten Grundgesetz des Verbundes.

3 Betondruck- und Spaltzugfestigkeit

Der Erhärtungsverlauf der Betondruckfestigkeit war praktisch unabhängig von der Konsistenz, der Sieblinie und dem Wasserzementwert; er entsprach im Mittel:

Tage	3	7	28	90
$\beta_{w,t}/\beta_{w28}$	0,34	0,55	1	1,24

Der Zusammenhang zwischen Spaltzugfestigkeit und Würfeldruckfestigkeit ist im Bild 3 dargestellt. Konsistenz, Sieblinie und Prüfalter waren ohne Einfluß.
Im Mittel ergab sich folgende in Einklang mit [6] stehende Beziehung:

$$\beta_{SpZ} = 0{,}222\ \beta_w^{2/3}$$

Die Abhängigkeit $\beta_{SpZ} \sim \beta_w^{2/3}$ hat zur Folge, daß Versuche, die im Alter von weniger als 28 Tagen geprüft werden, je nach Erhärtungsverlauf des verwendeten Zementes eine vergleichsweise höhere Zugfestigkeit als Druckfestigkeit aufweisen; im vorliegenden Fall sind dies z. B. im Betonalter von 7 Tagen – $(0{,}55)^{2/3}/0{,}55 \approx 1{,}20$ – ca. 20%.

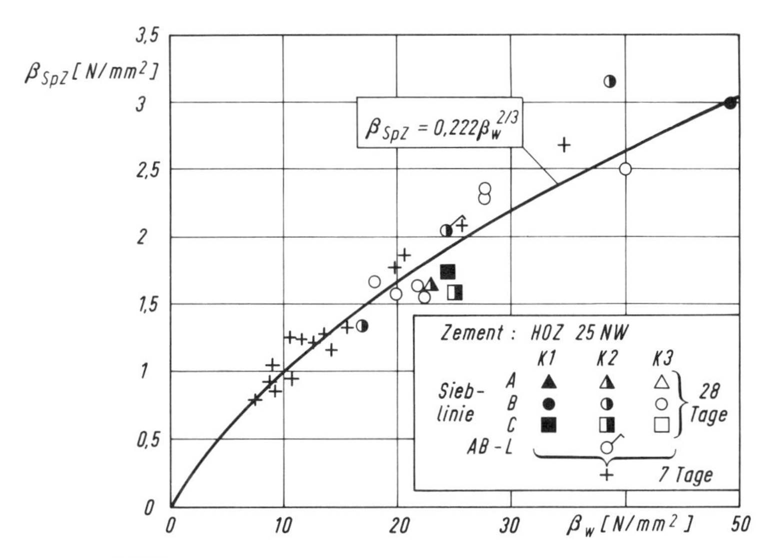

Bild 3
Zusammenhang zwischen Spaltzugfestigkeit β_{spz} und Würfeldruckfestigkeit

4 Verbundverhalten

Für die weiteren Betrachtungen werden die Mittelwerte der Verbundspannungen je Variante (bezogen auf die Verbundlänge $5\,d_s$) verwendet. Die Variationskoeffizienten lagen im für Verbundversuche üblichen Bereich, und zwar bei etwa 20% im Bereich sehr kleiner Verschiebungen und sonst bei etwa 10%.
Aus der Literatur [2], [5], [8], [9] ist bekannt, daß bei Rippenstählen im Bereich des reinen Scherverbundes (Verschiebungsbereich > 0,01 ÷ 1 mm) bei Ausziehversuchen mit Verbundlängen bis zu $10\,d_s$ Proportionalität zwischen Verbundspannung τ_v und Würfeldruckfestigkeit besteht:

$$\tau_v \sim \beta_w .$$

Lediglich im Bereich der reinen Haftung und beginnenden Verschiebungen (Bereich 0–0,01 mm) und naturgemäß im Bereich der Höchstlast (Verschiebungen > 1 mm) ist der Einfluß der Betondruckfestigkeit geringer, näherungsweise gilt:

$$\tau_v \sim \beta_w^{2/3}$$

Diese generellen Verhaltensweisen wurden durch die hier beschriebenen Versuche voll bestätigt. In der bautechnischen Anwendung ist vor allem der Verschiebungsbereich von 0,01 bis etwa 0,5 mm von Interesse. In diesem Bereich ist wegen der vorhandenen Proportionalität zwischen Verbundspannung und Betondruckfestigkeit eine Normierung möglich. Ohne wesentlichen Verlust an Genauigkeit kann die Normierung auch bis zur Höchstlast vorgenommen werden.
Die Auswertung für das Verhältnis τ_v/β_w (bezogene Verbundspannung) hat den Vorteil, daß alle Versuchsergebnisse zusammen betrachtet werden können und der Einfluß des Wasser-Zement-Wertes durch den Bezug auf die Betondruckfestigkeit entfällt

4.1 Einfluß der Betonzusammensetzung auf das Verbundverhalten

In Bild 4 ist beispielhaft der Einfluß der Konsistenzbereiche K1, K2 und K3 auf die Verbundspannungs-Verschiebungs-Charakteristik dargestellt. Konstant gehalten sind dabei die Sieblinie, der Wasser-Zement-Wert und das Prüfalter. Die höchsten bezogenen Verbundspannungen werden mit steifer Konsistenz K1, die niedrigsten mit weicher Konsistenz K3 erreicht. Der Unterschied im Verbundverhalten kann bis zu 100% betragen.
Der Einfluß der charakteristischen Sieblinien A, B und C nach DIN 1045 auf die bezogenen Verbundspannungen unter sonst gleichen Verhältnissen ist im Bild 5 wiedergegeben. Auch hier sind Unterschiede im Ver-

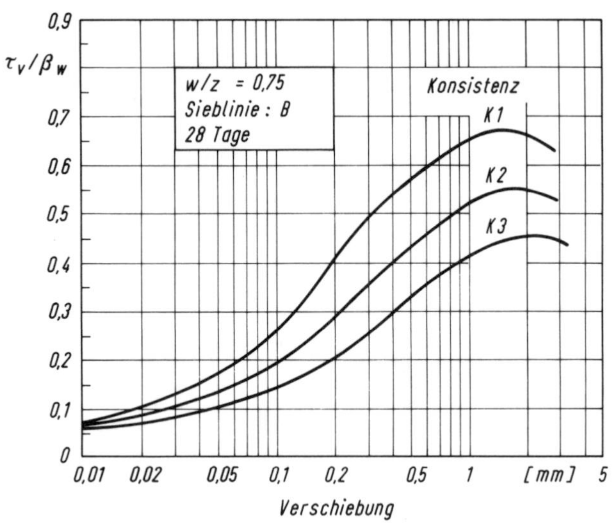

Bild 4
Bezogene Verbundspannungs-Verschiebungs-Charakteristik
Einfluß der Konsistenz (Verdichtungsmaß)

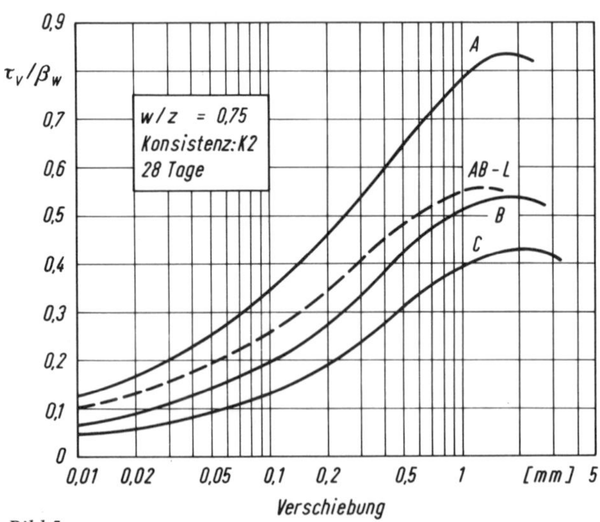

Bild 5
Bezogene Verbundspannungs-Verschiebungs-Charakteristik
Einfluß der Sieblinie

bundverhalten von 100% und mehr festzustellen, wobei das beste Verbundverhalten mit Sieblinie A und das niedrigste Verbundverhalten mit Sieblinie C erreicht wurde. Je flacher die Sieblinie insgesamt verläuft, um so besser ist das Verbundverhalten einbetonierter Rippenstähle. Ein Ersatz der Körnung 4/8 durch Leichtzuschlag (Sieblinie AB–L) ändert an diesem Verhalten nichts. Lediglich der Bruch erfolgt dann wegen der geringeren Kornfestigkeit des Leichtzuschlags etwas früher.

Die bezogenen Verbundspannungen τ_v/β_w, die bei einigen Varianten auch im Alter von 7 Tagen ermittelt wurden, lagen im Mittel bei kleinen Verschiebungen um 10% bis 15% und bei großen Verschiebungen um 20% bis 25% höher als die 28-Tage-Werte.

Der große Einfluß von Konsistenz und Sieblinie auf das Verbundverhalten erklärt sich sehr wahrscheinlich dadurch, daß im Bereich der Rippenkonsolen bei steifer Konsistenz bzw. bei feinteilarmer Sieblinie örtlich ein geschlossenes Betongefüge entsteht, das geringere Absetzerscheinungen und einen höheren Betonwiderstand aufweist.

4.2 Vergleich mit zulässigen Verbundspannungen nach DIN 1045

Um die untersuchten Einflußgrößen bewerten und deren Verbundverhalten mit den zulässigen Verbundspannungen der DIN 1045 vergleichen zu können, ist ein einfacher Vergleichsmaßstab zu finden. Am zweckmäßigsten eignet sich dafür die Festlegung einer charakteristischen Verbundspannung, mit der der wesentlichste Teil der Verbundspannungs-Verschiebungs-Charakteristik beschrieben werden kann.

Die zulässigen Verbundspannungen zul τ_1 der DIN 1045 sind für die Berechnung der Verankerungslängen geltende Mittelwerte; die Versuchswerte entsprechen dagegen dem örtlich wirksamen Verbundverhalten (Verbund-Grundgesetz). Mit Hilfe der Theorie des Scherverbundes [2], [5] läßt sich nun der Verhältniswert zwischen der mittleren Verbundspannung über größere Verbundlängen und der zu einer beliebigen Verschiebung gehörenden Verbundspannung des zugehörigen Verbundgrundgesetzes leicht herstellen. Wählt man als charakteristische Verbundspannung die zu 0,2-mm-Verschiebung gehörenden Werte $\tau_{0,2}$, so ergibt sich als Umrechnungsfaktor für die zul τ_1-Werte der DIN 1045 für mittlere Betongüten ein Wert von etwa 2. Die Wahl von $\tau_{0,2}$ hat gleichzeitig den Vorteil, daß diese Werte etwa dem mittleren Verbundverhalten unter rechnerischer Bruchlast entsprechen.

In Bild 6 sind alle Einflußgrößen im Zusammenhang dargestellt. Die Unterschiede im Verbundverhalten als Folge unterschiedlicher Betonzusammensetzungen und Konsistenzmaße sind beträchtlich. So verhalten sich z. B. die Verbundspannungen für Sieblinie C und Konsistenz K3 (v = 1,05) zu denen von Sieblinie A und Konsistenz K1 (v = 1,40) wie 1 : 5. Auch für die im Versuchswesen häufig benutzte plastische Konsistenz K2 (v = 1,15–1,20) betragen die Verbundspannungen für Sieblinie A noch das 2,2fache wie für Siebli-

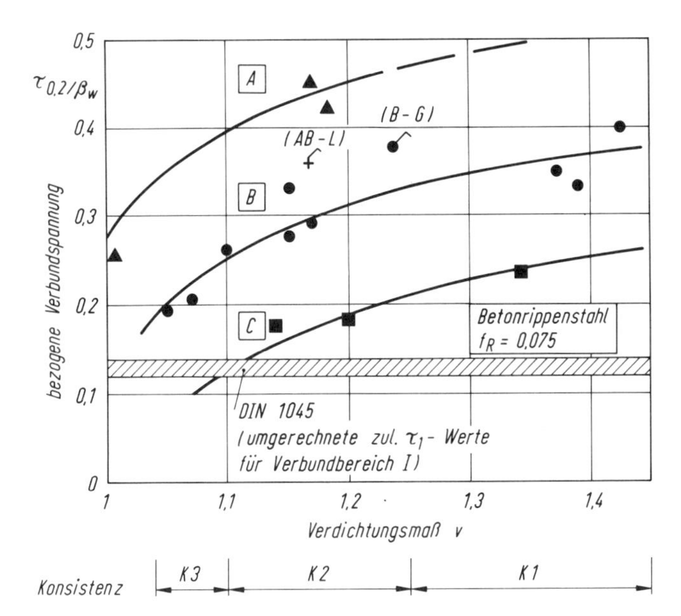

Bild 6
Bezogene Verbundspannung $\tau_{0,2}/\beta_w$ in Abhängigkeit von Verdichtungsmaß v und den Sieblinien A, B und C nach DIN 1045, Bild 2

nie C. Daraus folgt, daß z. B. die Verankerungslängen bei Beton mit Sieblinie A weniger als halb so groß wie bei Betonen mit Sieblinie C sind.
Das Verhalten der Betonmischung aus leichtem Normalbeton wurde bereits erläutert. Die Betonmischung mit teilweisem Ersatz des Zements durch den Zusatzstoff inertes Gesteinsmehl (B–G) ergab praktisch gleiches Verbundverhalten wie bei vergleichbaren Betonen ohne Zusatzstoffe. Inwieweit dieses Verhalten auch auf andere Zusatzstoffe übertragbar ist, muß durch ergänzende Untersuchungen geklärt werden.
Vergleicht man die Versuchswerte mit den ungerechneten zulässigen Verbundspannungen der DIN 1045, so zeigt sich, daß diese im Gegensatz zu seinerzeitigen anderen Bestrebungen richtigerweise sehr vorsichtig festgelegt wurden. Allerdings ist festzustellen, daß für Betone mit weicher Konsistenz oder geringer Verdichtung im Sieblinienbereich C der DIN 1045 keine zusätzlichen Sicherheiten für die im Bauteil gegebene Verbundwirkung mehr vorhanden sind.
Mögliche ungünstige Auswirkungen von Zusatzmitteln und Zusatzstoffen auf das Verbundverhalten sind dabei noch nicht berücksichtigt. Andererseits ist aber auch kritisch anzumerken, daß für Bauteilversuche vorzugsweise die Sieblinie A und plastische Konsistenz verwendet wird. Dies hat zur Folge, daß dann gegenüber den Gegebenheiten in der Praxis in der Regel viel zu günstige Verhältnisse für den Verbund, soweit es die Rißbildung, die Verankerung und das Verformungsverhalten betrifft, unter Umständen aber auch zu ungünstige Verhältnisse (Sprengwirkung) geprüft werden.

5 Schlußfolgerungen

Die Ergebnisse der vorgenannten Untersuchungen zum Einfluß der Betonzusammensetzung auf das Verbundverhalten von Betonrippenstählen können wie folgt zusammengefaßt werden:

- Durch die Variation von Sieblinie und Verdichtungsmaß im zulässigen Bereich der Betonzusammensetzung nach DIN 1045 kann das Verbundverhalten von Betonrippenstählen um mehr als 100% unterschiedlich ausfallen.
- Insbesondere Betone mit weicher Konsistenz, deren Kornaufbau im Sieblinienbereich C der DIN 1045 liegt, ergeben hinsichtlich des Verbundverhaltens geringe Werte.
- Bei der Durchführung von Bauteilversuchen aus Stahlbeton ist neben einer umfassenden Beschreibung des Betons und des Betonstahles daher auch die Verbundwirkung zu charakterisieren.
- Um die Übertragbarkeit von Versuchsergebnissen auf praktische Verhältnisse zu gewährleisten und eine bessere Vergleichsmöglichkeit unterschiedlicher Versuchsreihen untereinander zu ermöglichen, sollten daher in gleicher Weise wie die Betoneigenschaften (z. B. Druckfestigkeit) in einfachen Versuchen ermittelt werden, immer auch einfache Ausziehversuche zur Beschreibung des Verbundverhaltens mit einem einheitlich festgelegten Betonrippenstahl durchgeführt werden.
- Werden Bauteilversuche in einem jüngeren Alter als 28 Tage geprüft, so ist eine im Verhältnis zur Druckfestigkeit höhere Zugfestigkeit des Betons und ein besseres Verbundverhalten gegeben. Bei der Bewertung der Versuchsergebnisse ist dies zu berücksichtigen.
- Inwieweit durch die Verwendung der unterschiedlichsten Zusatzstoffe und insbesondere von Zusatzmitteln das Verbundverhalten über den hier ermittelten Rahmen hinaus noch verändert wird, muß durch entsprechende Untersuchungen noch geklärt werden. Ebenfalls zu überprüfen ist auch der Einfluß der Betonzusammensetzung für Betonstähle im Verbundbereich II.

6 Literatur

[1] RÜSCH, H., REHM, G. und MARTIN, H.: Berichte über Zulassungsversuche mit Rippenstählen, durchgeführt in den Jahren 1953 bis 1965, Institut für Massivbau, TU München – Berichte nich veröffentlicht.

[2] REHM, G.: Über die Grundlagen des Verbundes, Heft 138 der Schriftenreihe des Deutschen Ausschusses für Stahlbau, 1961.

[3] WALTER, R. und SORETZ, S.: Versuche über den Einfluß der Kornzusammensetzung des Betons auf den Verbund, Beton- und Stahlbetonbau 5/1967.

[4] MÜLLER, H. H.: Bericht über Verbundversuche an einer Probeschlitzwand, Materialprüfungsamt für das Bauwesen der TU München, Bericht Nr. 458/1969.

[5] MARTIN, H.: Zusammenhang zwischen Oberflächenbeschaffenheit, Verbund und Sprengwirkung von Bewehrungsstählen unter Kurzzeitbelastung, Heft 228 der Schriftenreihe des Deutschen Ausschusses für Stahlbetonbau, 1973.

[6] HEILMANN, H. G.: Beziehungen zwischen Zug- und Druckfestigkeit des Betons, beton, Herstellung–Verwendung 19 (1969) Heft 2, S. 68/70.

[7] MARTIN, H., SCHIESSL, P. und SCHWARZKOPF, M.: Berechnungsverfahren für Rißbreiten aus Lastbeanspruchung, Forschung Straßenbau und Straßenverkehrstechnik, Heft 309, 1980.

[8] MARTIN, H. und NOAKOWSKI, P.: Verbundverhalten von Baustählen – Untersuchung auf der Grundlage von Ausziehversuchen, Heft 319 der Schriftenreihe des Deutschen Ausschusses für Stahlbeton, 1981.

[9] MARTIN, H.: Bond Performance of Ribbed Bars (pull-out-tests) – Influence of Concrete Composition and Consistency –, International Conference on Bond in Concrete, Paisley College of Technologie, Scotland, 14 to 16 June 1982.

Werkstoff-Forschung, Beton und Stahl

Biegezugfestigkeit des dampfgehärteten Gasbetons

Dietmar Briesemann

1 Einleitung

Über die Biegezugfestigkeit des dampfgehärteten (= autoklavgehärteten) Gasbetons existieren in der Literatur nur wenige Angaben; systematische Arbeiten zu diesem Thema fehlen völlig.
Im Hinblick auf die weitere Erforschung des Tragverhaltens dieses Werkstoffes ist deren Kenntnis unumgänglich. So soll z. B. im Entwurf zur DIN 4223: Gasbeton, Bewehrte Bauteile, ein Verfahren zur Prüfung von Korrosionsschutzmitteln für den Stahl in Gasbeton genormt werden, das die Biegebelastung bewehrter Gasbetonplattenabschnitte vorsieht. Zur Abschätzung der Prüflasten fehlen bisher die Angaben zur Biegezugfestigkeit.
Im Werk Emmering der Hebel-Gruppe wurde unter anderem aus dem obengenannten Grunde eine umfangreiche Versuchsreihe*) durchgeführt, um die Biegezugfestigkeit des Gasbetons näher abzuklären. Dabei interessieren die die Biegezugfestigkeit beeinflussenden Parameter sowie ihr Verhältnis zur Druckfestigkeit.

2 Biegezugfestigkeit von Beton und Gasbeton in der Literatur

Die Biegezugfestigkeit des normalen Betons ist in der Literatur ausführlich abgehandelt worden. Zuletzt wurde in [1], [2] darüber berichtet. Danach wird sie durch:

- das Verhältnis Stützweite/Höhe (l/h) der Prüfkörper
- die Höhe der Prüfkörper (h)
- die Anzahl der Prüflasten (ein oder zwei Einzellasten) und
- den Feuchtigkeitsgehalt der Prüfkörper

beeinflußt. In [11] wird zwischen Biegezug- und Druckfestigkeit des Betons die Beziehung

$$\beta_{BZ} = c \cdot \beta_D^{2/3} \qquad (1)$$

angegeben. C beträgt danach im Mittel 1,07 für eine und 0,98 für zwei Prüflasten.
Die an dampfgehärtetem Gasbeton ermittelten Biegezugfestigkeiten aus der Literatur sind getrennt nach Prüfkörpergrößen in den Tabellen 1 (Prüfkörperhöhe ≥ 10 cm) und 2 (Prismen $4 \times 4 \times 16$ cm) dargestellt und ausgewertet.
Nach [3], [4], [5] und [8], siehe Tabelle 1, kann für das Verhältnis Biegezug- zu Druckfestigkeit des Gasbetons

$$\beta_{BZ}/\beta_D = 0{,}20 \text{ bis } 0{,}41 \qquad (2)$$

angegeben werden. Das Material in [7] ist nicht repräsentativ für dampfgehärteten Gasbeton; es wird bei der Auswertung daher nicht berücksichtigt. In [10] wurde nachgewiesen, daß die Anzahl der Prüflasten und der Feuchtigkeitsgehalt einen Einfluß auf die Biegezugfestigkeit des Gasbetons besitzen. Wenn zwei Einzellasten in den Drittelspunkten der Spannweiten statt einer mittigen Last angeordnet werden, fällt β_{BZ} im Mittel um 16% und bei Erhöhung des Feuchtigkeitsgehaltes von 1 auf 10 Gew.-% im Mittel um 40% ab, siehe Tabelle 2.

3 Versuche

Zur Bestimmung der Biegezugfestigkeit wurden die Prüfkörper aus zwei Formen entnommen, siehe Bild 1. Sie waren jeweils über die gesamte Form verteilt und besaßen die Solldruckfestigkeiten GB 3.3 und GB 4.4, siehe Tabelle 3. Das Versuchsprogramm wurde so ausgelegt, daß die von normalen Beton bekannten Einflüsse auf die Biegezugfestigkeit des Gasbetons untersucht werden konnten. Je Variante wurden zwischen 12 und 20 Biegezugprüfungen durchgeführt. Aus den zerbrochenen Hälften des so geprüften Gasbetons

*) Meinen Mitarbeitern H. Doblinger und Dipl.-Ing. (FH) G. Schönfeld danke ich herzlich für die Unterstützung bei diesen Arbeiten.

Tabelle 1
Angaben zur Biegezugfestigkeit von dampfgehärtetem Gasbeton aus der Literatur

Verfasser Jahr	Literatur	l/h	h (cm)	Last-anzahl	Feuchte-gehalt (Gew.-%)	β_{BZ}/β_D	Bemerkungen
GRAF 1949	[3]	16,2	15	–*)	–	0,21–0,41	Produktionsgasbeton
VINBERG 1953	[4]	–	–	–	–	0,22–0,27	zementhaltiger Produktionsgasbeton
VALORE 1954	[5]	–	–	–	–	0,20–0,33	keine Versuche, Literaturauswertung
SIMONSON u. a. 1954	[6]	1,8	10	2	10	$\beta_{BZ} = 36{,}3\,(\varrho - \frac{1}{3})$	ϱ = 460–950 kg/m³
CHAVKIN u. a. 1964	[7]	–	–	–	–	0,11–0,18	kalkgebundener Gasbeton aus UdSSR
GRIMER u. a. 1965	[8]	3	10,2	1	3,3–17,9	0,28–0,31	4 zementgebundene Produktionsgasbetone

*) keine Angaben

Tabelle 2
Angaben zur Biegezugfestigkeit von gehärtetem Gasbeton aus der Literatur, ermittelt an Prismen 4 × 4 × 16 cm (l/h = 2,5)

Verfasser Jahr	Literatur	Last-anzahl	Feuchtegehalt (Gew.-%)	β_{BZ} (N/mm²) bzw. β_{BZ}/β_D	Bemerkungen
GRAF 1949	[3]	1	–*)	$\beta_D = 0{,}75\,\beta_{BZ} + 0{,}07\,\beta_{BZ}$	„historischer" Gasbeton
SPIESS 1960	[9]	1	getrocknet bis Gew.-konstanz	$\beta_{BZ}/\beta_D = 0{,}30–0{,}39$	kalkgebundener Laborgasbeton
NERENST 1982	[10]	1	1	2,53	$\varrho = 591–597\,kg/m^3$
		1	10	1,48	
		2	1	2,09	
		2	10	1,27	

*) keine Angaben

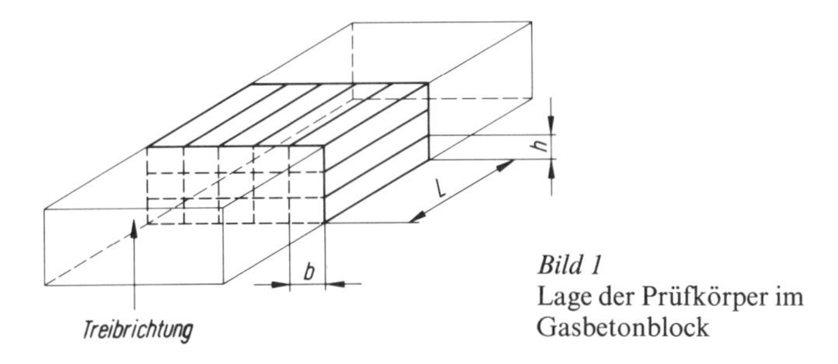

Bild 1
Lage der Prüfkörper im Gasbetonblock

wurden Würfel mit 10 cm Kantenlänge herausgeschnitten, die zur Ermittlung der Druckfestigkeit und der Rohdichte nach DIN 4223 dienten.
Der dampfgehärtete Gasbeton wurde unter Verwendung von Zement, Kalk, Wasser und gemahlenem Quarzsand aus dem Münchner Raum hergestellt. Als Treibmittel diente Aluminiumpulver.
Die Biegezugfestigkeit des Gasbetons wurde jeweils senkrecht zur Treibrichtung an frei drehbar gelagerten Balken auf zwei Stützen ermittelt. Die Belastungsgeschwindigkeit betrug ca. 0,01 N/mm^2s. Die Biegezugprüfkörper lagerten vor der Belastung ca. drei Wochen in einer temperierten Halle und besaßen mit Ausnahme von Versuchsreihe 3 zum Zeitpunkt der Prüfung einen Feuchtigkeitsgehalt zwischen 5 und 15 Vol.-%, siehe Tabelle 3. Die Versuchskörper der Reihe 3 wurden in Wasser gelagert, bis keine Gewichtszunahme mehr festgestellt wurde. Sie besaßen bei der Prüfung einen mittleren Feuchtigkeitsgehalt von 37,6 bzw. 43,6 Vol.-%.

4 Ergebnisse

Die Versuchsergebnisse sind in Tabelle 3 und den Bildern 2 bis 6 dargestellt. Mit Einzelwerten der Biegezugfestigkeiten, die aus dem Rahmen fielen, wurden statistische Ausreißertests durchgeführt. Auf diese Weise wurden 9 von insgesamt 284 Werten verworfen.

5 Diskussion der Ergebnisse

5.1 Streuung der Biegezugfestigkeit

Die errechneten Standardabweichungen der Biegezugfestigkeiten sind auf den Einfluß der Lage der Prüfkörper in der Gasbetonform und auf die Streuung bei der Prüfung zurückzuführen, siehe Bild 2. Gegenüber der Druckfestigkeit [12] und der Spaltzugfestigkeit [13] des Gasbetons zeigt die Biegezugfestigkeit größere Streuungen. Das ist im wesentlichen mit dem Einfluß der Prüfkörpergröße zu erklären, die hier ein bis zu 50fach größeres Volumen besaß als die in [12] und [13] untersuchten Prüfkörper.

5.2 Einfluß der Prüfkörperschlankheit *l/h*

Im Bild 3 sind die vergleichbaren Biegezugfestigkeiten (b = const., h = const.) gegen die Schlankheiten l/h der Prüfkörper aufgetragen. Als Parameter wurde die

Tabelle 3
Übersicht über die durchgeführten Versuche sowie mittlere Feuchtigkeitsgehalte zum Zeitpunkt der Biegezugfestigkeitsprüfung und mittlere Trockenrohdichten

Versuchs-reihe	Abmessungen (cm)			l/h	Prüf-körper je GB	GB 3.3		GB 4.4	
	b	h	l [1])			Feuchte (Vol.-%)	ϱ (kg/m^3)	Feuchte (Vol.-%)	ϱ (kg/m^3)
1	10	15	69	4,6	20	7,1	510	5,9	644
2	10	15	69 [2])	4,6	12	5,9	534	6,2	636
3	10	15	69	4,6	12	37,6	500	43,6	642
4	10	7,5	34,5	4,6	16	6,4	515	6,1	640
5	10	15	34,5	2,3	16	14,6	464	7,8	609
6	10	15	103,5	6,9	16	6,3	460	8,8	589
7	10	15	144	9,6	16	12,6	460	7,3	594
8	10	20	92	4,6	18	10,4	463	14,2	589
9	10	30	138	4,6	16	12,6	450	8,8	590

[1]) Stützweite; [2]) zwei Einzellasten

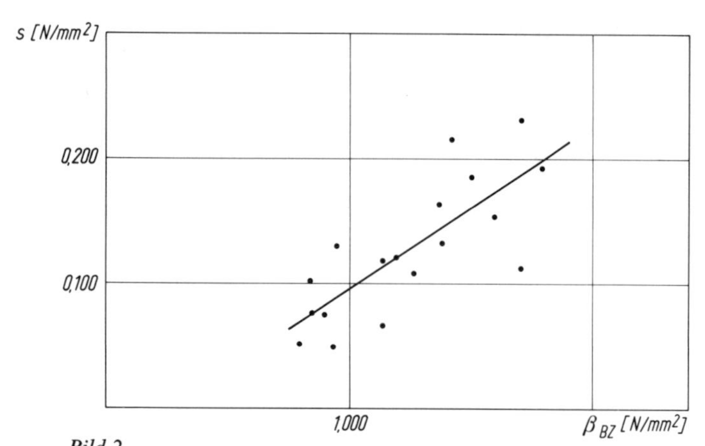

Bild 2
Abhängigkeit zwischen Standardabweichung und mittlerer Biegezugfestigkeit des Gasbetons

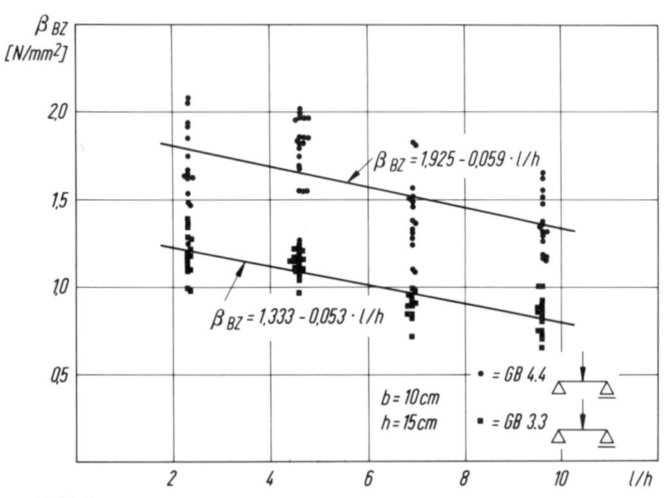

Bild 3
Abhängigkeit zwischen Biegezugfestigkeit des Gasbetons und der Prüfkörperschlankheit l/h

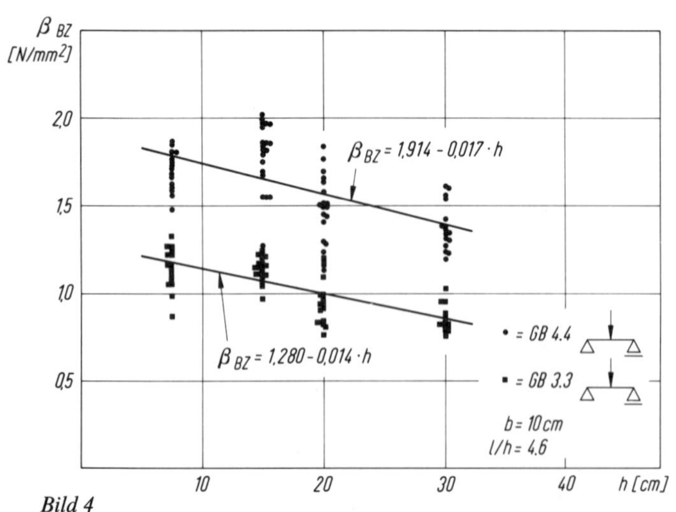

Bild 4
Abhängigkeit zwischen Biegezugfestigkeit des Gasbetons und Prüfkörperhöhe h

Solldruckfestigkeit gewählt. Die Biegezugfestigkeit des Gasbetons kann danach im Mittel für GB 4.4 (nach DIN 4223 E) mit

$$\beta_{BZ} = 1{,}925 - 0{,}059 \cdot l/h \tag{3}$$

und für GB 3.3 mit

$$\beta_{BZ} = 1{,}333 - 0{,}053 \cdot l/h \tag{4}$$

angegeben werden. Die Abnahme der Biegezugfestigkeit mit der Prüfkörperschlankheit stimmt mit elastizitätstheoretischen Ableitungen [14] überein.

5.3 Einfluß der Prüfkörperhöhe *h*

Die Biegezugfestigkeiten der Prüfkörper mit unterschiedlichen Höhen und b = const., l/h = const. sind in Bild 4 in Abhängigkeit von der Prüfkörperhöhe h dargestellt; Parameter ist ebenfalls die Solldruckfestigkeit des Gasbetons. Die Biegezugfestigkeit kann entsprechend Bild 4 für den Bereich GB 4.4 zu

$$\beta_{BZ} = 1{,}914 - 0{,}017 \cdot h \tag{5}$$

und für GB 3.3 zu

$$\beta_{BZ} = 1{,}280 - 0{,}014 \cdot h \tag{6}$$

ermittelt werden. Sie nimmt mit steigender Prüfkörperhöhe ab.
Dieser Tatbestand ist allgemein gültig und wird damit erklärt, daß in größeren Materialproben mehr Inhomogenitäten vorhanden sind als in kleinen.

5.4 Einfluß der Lastanordnung

In Bild 5 wurde untersucht, welchen Einfluß eine mittige Einzellast im Gegensatz zu zwei Lasten in den Drittelspunkten bei der Belastung der Prüfkörper hat (b = const., h = const., l = const.). Die Biegezugfestigkeit des Gasbetons im Mittel beträgt bei mittiger Einzellast

$$\beta_{BZ} = -\,0{,}0739 + 0{,}2885\,\beta_D \tag{7}$$

und bei zwei Lasten in den Drittelspunkten

$$\beta_{BZ} = -\,0{,}2718 + 0{,}2843\,\beta_D. \tag{8}$$

Der Abfall der Festigkeit bei zwei Einzellasten stimmt in der Tendenz mit [10] überein und ist dadurch zu erklären, daß die Wahrscheinlichkeit des Zusammentreffens von Inhomogenitäten im Material mit der Stelle des größten Moments bei dieser Belastung größer als bei mittiger Einzellast ist.

Der Unterschied der Gasbeton-Biegezugfestigkeit zwischen den beiden Belastungsarten beträgt im Mittel zwischen 20% und 12%. Aus Zweckmäßigkeitsgründen sollte bei der praktischen Prüfung stets die mittige Einzellast gewählt werden.

5.5 Einfluß der Prüfkörperfeuchtigkeit

Der Einfluß der Prüfkörperfeuchtigkeit zum Zeitpunkt der Prüfung wurde bei vergleichbaren Werten der Biegezugfestigkeit des Gasbetons (b = const., h = const., l = const.) in Bild 6 überprüft. Für die bei üblicher Prüffeuchte (Versuchsreihe 1 mit 7,1 bzw. 5,9 Vol.-%) ermittelte Biegezugfestigkeit gilt Gleichung (7) und für wassergesättigten Gasbeton (Versuchsreihe 3 mit 37,6 bzw. 43,6 Vol.-%) im Mittel die Beziehung

$$\beta_{BZ} = 0{,}0053 + 0{,}2316 \cdot \beta_D \tag{9}$$

An den wassergesättigten Proben wurden im Mittel um 15% geringere Biegezugfestigkeiten als an den lufttrockenen Proben ermittelt. Das steht im Einklang mit den Untersuchungen in [10] und im Gegensatz zu den Erkenntnissen bei normalen Beton [1] und [2]. Bei diesem erzielen wassergesättigte Prüfkörper in der Regel höhere Biegezugfestigkeiten als lufttrockene Proben, da sich bei diesen Schwindspannungen am Biegezugrand störend auswirken.

5.6 Beziehung zwischen Biegezug- und Druckfestigkeit des dampfgehärteten Gasbetons

Die Biegezugfestigkeit des Gasbetons nimmt mit steigender Druckfestigkeit zu, siehe Bild 7. Dort sind die mittleren Festigkeiten gegeneinander aufgetragen; jeder Punkt repräsentiert je nach Versuchsreihe 12 bis 20 Prüfungen. Für alle Einzelwerte wurde die Regressionsgerade sowie der zweiseitige Toleranzbereich angegeben, in dem 90% der Werte mit S = 95% zu erwarten sind. Danach beträgt die Biegezugfestigkeit des hier geprüften Gasbetons im Mittel

$$\beta_{BZ} = 0{,}3917 + 0{,}1821\,\beta_D. \tag{10}$$

Die in der Literatur angegebenen Werte der Biegezugfestigkeit [3], [4], [5] und [8] wurden in Bild 8 mit den Angaben des Verfassers (Bild 7) verglichen. Sie stimmen gut miteinander überein; die Biegezugfestigkeiten aus der Literatur liegen im 90%-Toleranzbereich der hier durchgeführten Versuche. Die Beziehung zwischen Druck- und Biegezugfestigkeit des Gasbe-

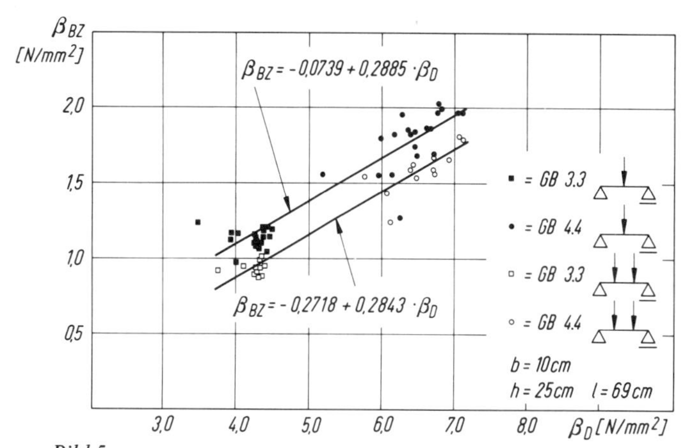

Bild 5
Einfluß der Lastanordnung bei der Prüfung auf die Biegezugfestigkeit des Gasbetons

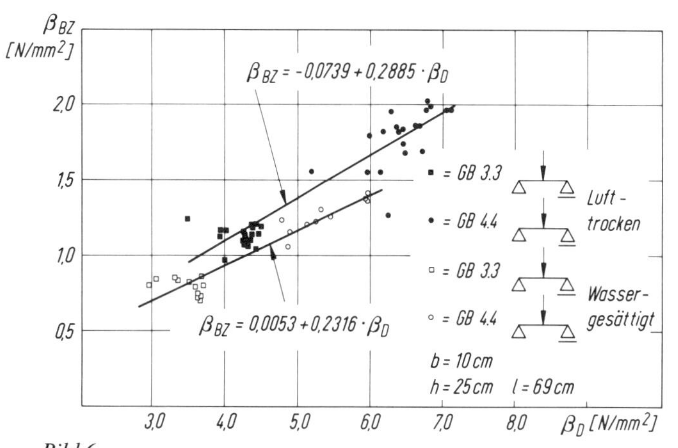

Bild 6
Einfluß der Prüfkörperfeuchtigkeit auf die Biegezugfestigkeit des Gasbetons

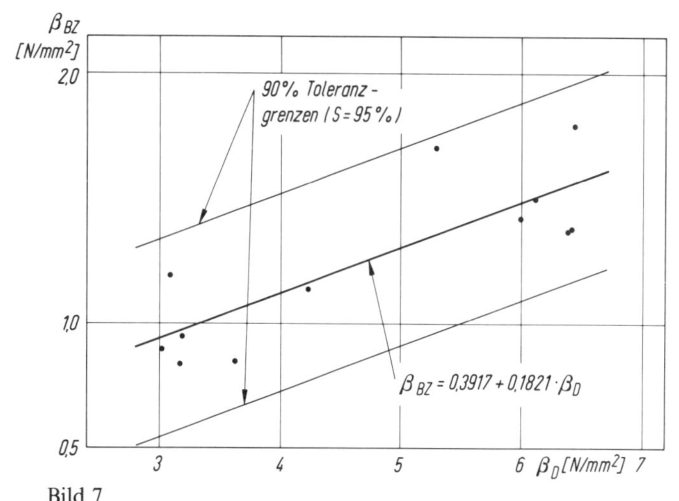

Bild 7
Abhängigkeit zwischen mittlerer Biegezug- und mittlerer Druckfestigkeit von Gasbeton bei den hier beschriebenen Versuchen

tons für die Versuche nach [3], [8] sowie des Verfassers lautet

$$\beta_{BZ} = 0{,}2666 + 0{,}2102 \cdot \beta_D \qquad (11)$$

In Bild 8 ist neben Gleichung (11) deren zweiseitiger 90%-Toleranzbereich angegeben. Die nach Bild 8 bestimmten Verhältnisse β_D/β_{BZ} liegen zwischen 2,58 und 4,75, im Mittel bei 3,67, so daß allgemein gilt

$$\beta_{BZ} = \frac{\beta_D}{3{,}67} \qquad (12)$$

Der lineare Zusammenhang zwischen Biegezug- und Druckfestigkeit des dampfgehärteten Gasbetons weicht von den vergleichbaren Vorstellungen bei normalem Beton ab. Das ist, wie in [13], u.a. mit dem geringen Festigkeitsbereich zu erklären.

6 Zusammenfassung

Zur Abklärung der Biegezugfestigkeit des dampfgehärteten Gasbetons wurde im Hebel-Werk Emmering ein umfangreiches Versuchsprogramm durchgeführt. Die Auswertungen zeigen, daß mit steigendem Verhältnis Stützweite/Prüfkörperhöhe und wachsender Prüfkörperhöhe die Gasbeton-Biegezugfestigkeit abnimmt. Die Belastung durch zwei Einzellasten in den Drittelspunkten der Stützweite führt gegenüber einer mittigen Einzellast zu einem Abfall der Biegezugfestigkeit. Das gleiche gilt für wassergesättigte Prüfkörper im Vergleich zu lufttrockenen Proben. Zwischen Biegezug- und Druckfestigkeit des dampfgehärteten Gasbetons gilt im Mittel die Beziehung $\beta_{BZ} = \beta_D/3{,}67$.

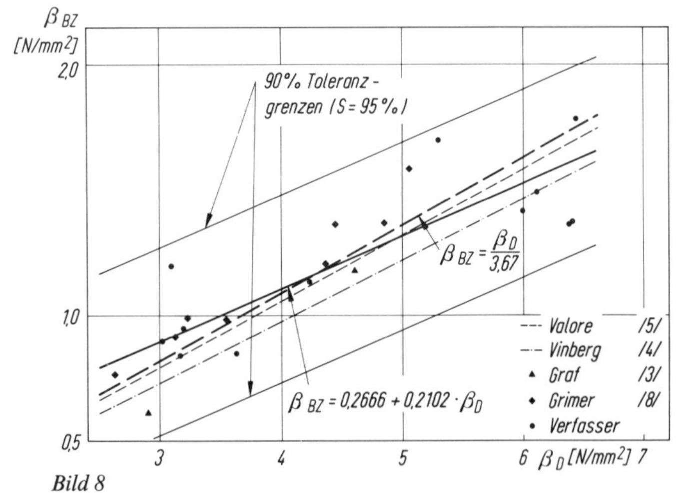

Bild 8
Allgemeiner Zusammenhang zwischen Biegezug- und Druckfestigkeit des dampfgehärteten Gasbetons

7 Literatur

[1] MEYER, A.: Die Biegezugfestigkeit als Gütemerkmal des Betons. Der Bauingenieur 38 (1963), S. 45–51.
[2] BONZEL, J.: Über die Biegezugfestigkeit des Betons. beton 13 (1963), S. 179–182 u. 227–232.
[3] GRAF, O.: Gasbeton, Schaumbeton, Leichtkalkbeton. Stuttgart: Verlag Konrad Wittwer, 1949, S. 40 u. S. 70.
[4] VINBERG, H.A.: Compressive Strength of Lightweight Concrete Brick Walls, Bulletin No. 13, Division of Bilding Statistics and Structural Engineering, Royal Institute of Technology, Stockholm, 1953, p. 95.
[5] VALORE, R.C.: Cellular Concretes. Journal of the American Concrete Institute 50 (1954), p. 773, 794–796, 817–836.
[6] SIMONSON, B. og WILCKEN, P.: Undersø gelse af letbetoners styrke – og elasticitets for hold. Beton og Jernbeton 4 (1954), S. 119–130.
[7] CHAVKIN, L.M. und KRYZANOVSKIJ, B.B.: Silicatbetonelemente für den Montage-Hausbau, zitiert nach Gundlach, H.: Dampfgehärtete Baustoffe, Wiesbaden und Berlin. Bauverlag GmbH, 1973, S. 212.
[8] GRIMER, F.J. and BREWER, R.S.: The within-cake variation of autoclaved aerated concrete. Int. Symp. on autoclaved calcium silicate building products. London 1965, p. 163–169.
[9] SPIESS, E.: Untersuchungen über die zweckmäßigste Art der Autoklavhärtung von Porenbetonen. Silikattechnik 11 (1960), S. 331–334.
[10] NERENST, P.: Determination of Modulus of Rupture of Gasconcrete, 1982, Mitteilung an Rilem 51 – ALC Committee.
[11] HEILMANN, H.G.: Beziehungen zwischen Zug- und Druckfestigkeit des Betons. beton 19 (1969), S. 68–70.
[12] BRIESEMANN, D.: Zur Ermittlung der Druckfestgkeit von Gasbeton. beton 22 (1972) S. 152–155.
[13] BRIESEMANN, D.: Spaltzugfestigkeit des Gasbetons. Die Bautechnik 51 (1974) S. 169–174.
[14] TIMOSHENKO, S. and GOODIER, J.N.: Theory of Elasticity, 2. Edit, New York/Toronto/London: McGraw – Hill Book Company, Inc. 1951.

The tensile properties of the Tempcore steels in two-peculiar applications

Jacques Defourny
und Adolphe Bragard

In this article, the tensile properties of the Tempcore steels are presented for two peculiar applications:
- in the case of high strain rates (dynamic loading)
- in the restraightened condition after bending.
In both here concerned applications, Tempcore reinforcements confirm their surprising resistance-ductility properties.

1 Introduction

While bend or rebend tests directly tell how a concrete reinforcing steel will behave during the shaping operations carried out on site for the framework preparation, the tensile test assesses the properties of the armature which govern the safety of the construction. The characteristics derived from the tensile test, specially the yield stress and the uniform elongation, are therefore of prime importance.
Many data have been published those last years about the tensile properties of the Tempcore steels, in the as received or welded conditions, and at ambient, low or elevated temperatures [1]–[5]. Those articles, among which those of Prof. Rehm, highlighted the optimum combination of strength and ductility built in those modern reinforcing bars.
The present article deals with the tensile properties of the Tempcore steels in two peculiar applications:
- in the case of high strain rates
- after bending and restraightening operations.

2 Influence of the strain rate on the tensile properties of Tempcore steels

For concrete constructions aimed to resist to impact loading (bunkers for instance), the tensile properties of the reinforcements at high strain rates are considered as important by several specialists.

By high strain rate, it is generally understood a deformation velocity ranging between 1 and 10 s^{-1} (or 100 and 1000%/s) while for simulating a quasi static loading, a strain rate equal to $5.10^{-5} s^{-1}$ is adopted. As a matter of fact, the latter strain rate corresponds, in the elastic part of the tensile diagramm, to the classical stress rate of 10 MPa/s ($\dot{\sigma} = \dot{\varepsilon}E$ with E = 205,000 MPa).
In order to assess the behaviour of Tempcore bars submitted to a dynamic loading, tensile tests were conducted at C.R.M. on a low inertia machine at three levels of strain rates.

2.1 Description of the tests

The tensile tests were performed on a fatigue testing machine. The deformation imposed to the reinforcement and the corresponding applied loads were measured by means of a transient recorder.
∅ 10 mm Fe B 400 (IIIS) and ∅ 12 mm Fe B 500 (IVS) Tempcore bars were tested. Their chemical composition is shown in Table 1.
Three levels of strain rates were applied:
- $5.10^{-5} s^{-1}$ (quasi static loading)
- $0.1 s^{-1}$
- $2 s^{-1}$ (dynamic loading).
The following characteristics were measured:
- yield stress (*Re*)
- ultimate tensile strength (*Rm*)
- elongation at maximum stress ($A_{\sigma_{max}}$).

Table 1
Chemical composition of the Tempcore steels examined in high strain rate tensile tests (on product 10^{-3}%)

	C	Mn	Si	P	S	N
∅10 Fe B 400	180	690	200	25	8	9.8
∅12 Fe B 500	150	840	250	18	15	12.0

Table 2
Results of tensile tests performed on Fe B 400 ∅ 10 Tempcore steel

Index	Strain Rate $\dot{\varepsilon}$ (s^{-1})	*Re* (MPa)	*Rm* (MPa)	*Rm/Re*	A5 (%)	A10 (%)	2 A10–A5 (%)	$A_{\sigma max}$ (%)
9-1	5.10^{-5}	551	648	1.18	23.8	16.5	9.2	9.3
10-2		511	605	1.18	26.2	19.1	12.0	13.1
11-3		528	621	1.18	25.3	19.5	13.7	12.5
12-4		547	637	1.16	23.5	16.6	9.7	10.8
		534	628	1.18	24.7	17.9	11.2	11.4
6-1	0.1	598	675	1.13	25.3	19.7	14.1	11.6
6-2		567	649	1.14	24.0	19.9	15.8	13.8
7-3		567	649	1.14	24.2	18.1	12.0	13.8
8-4		573	659	1.15	26.6	18.7	10.8	12.7
		576	658	1.14	25.0	19.1	13.2	13.0
1-1	2.0	605	681	1.13	26.0	20.3	14.6	12.9
2-2		592	662	1.12	26.8	20.6	14.4	15.7
3-3		586	649	1.11	28.0	21.5	15.0	13.7
4-4		618	681	1.10	22.8	17.4	12.0	12.5
		600	668	1.12	25.9	20.0	14.0	13.7

The uniform elongation was assessed through the formula 2 A10–A5, introduced and discussed in [6].

2.2 Results

The detailed results obtained on both steel grades are listed in Tables 2 and 3. The mean values are also calculated for each considered strain rate.
It will be first noticed that no significant difference appears between the elongation at maximum stress ($A_{\sigma max}$) and the uniform elongation calculated through the formula 2 A10–A5, the latter expression gives therefore a quite satisfying evaluation of the uniform elongation.
In Fig. 1 and 2, the mean values for *Re*-*Rm*–A5–A10 and Au (2 A10–A5) are plotted as a function of the strain rate, the latter parameter being reported in a logarithmic scale.
Examples of tensile diagrams are reproduced in Fig. 3.

2.3 Discussion

From the obtained results, it is quite clear that the higher the strain rate applied to Tempcore steels the

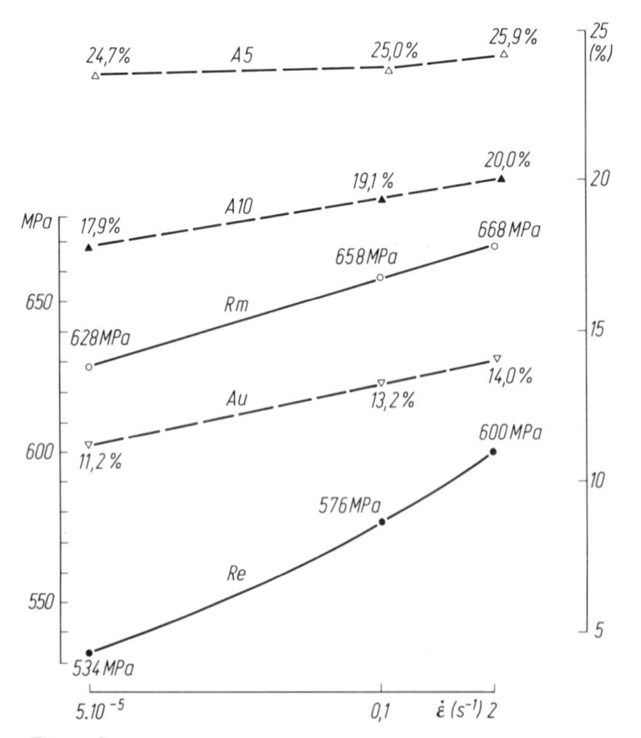

Figure 1
Influence of the strain rate on the tensile properties of Tempcore steel. Tempcore ∅ 10 Fe B 400

Table 3
Results of tensile tests performed on Fe B 500 ∅ 12 Tempcore steel

Index	Strain Rate $\dot{\varepsilon}$ (s^{-1})	*Re* (MPa)	*Rm* (MPa)	*Rm/Re*	A5 (%)	A10 (%)	2 A10–A5 (%)	$A_{\sigma max}$ (%)
13-1	5.10^{-5}	573	661	1.15	20.2	14.0	7.8	8.7
14-2		577	665	1.15	20.2	14.0	7.8	8.7
		575	663	1.15	20.2	14.0	7.8	8.7
23-1	0.1	599	685	1.14	23.3	16.7	10.1	10.1
24-1		592	676	1.14	22.2	16.3	10.4	10.8
20-2		606	690	1.14	21.7	16.1	10.5	10.1
21-2		606	694	1.15	21.7	16.7	11.7	10.4
22-2		608	694	1.14	Fracture near the grips			9.7
		602	688	1.14	22.2	16.5	10.7	10.2
15-1	2.0	615	707	1.15	23.2	16.8	10.4	11.5
18-1		623	710	1.14	24.0	17.9	11.8	11.5
19-1		628	710	1.13	25.7	18.8	11.9	11.5
16-2		632	714	1.13	23.3	16.7	10.1	11.1
17-2		628	710	1.13	24.2	17.9	11.6	11.4
		625	710	1.14	24.1	17.6	11.2	11.4

better the strength (*Re*-*Rm*) and the ductility (A5–A10–Au) of these rebars.
This result is not surprising: it has indeed been remarked on other types of rebars. Its extent depends mainly on the steel quality and is independent of the bar diameter [7].
As well known, the yield stress and the uniform elongation are important parameters for the safety of a construction.
Using a *Re*-Au diagram, those characteristics in either static ($5.10^{-5} s^{-1}$) or dynamic loadings ($2 s^{-1}$) are plotted in Fig. 4. Both Tempcore steels studied here have been reported as well as two similar grades of cold twisted steels, whose properties have been read in the litterature [7], [8]. Thus, the line joining in such a diagramm the static and the dynamic values is, for a given steel, a picture of its properties in the whole range of strain rates.
From Fig. 4, it is quite clear that Tempcore steels lie in the most favourable parts of this Re-Au diagram whatever the strain rate.

3 Tensile tests on bent and restraightened Tempcore steels

Rebending of reinforcing bars consists in straightening a reinforcement previously bent.
This operation is performed on site when the concreting of the construction is performed in several stages. In order to connect each other two concrete parts corresponding to two successive stages, prefabricated reinforcing elements are used. These elements are constituted by three-dimensional stirrups, the extremities of which are embedded, for instance in a box made with two encased steel sheets or in a plastic foam.
The elements are put at the right place before the first concreting stage. When the concrete has hardened, the extremities of each stirrup are straightened so as to allow a reinforcing connexion for the second concreting stage.

3.1 How to straighten?

Since it is performed on site, rebending has to be carried out manually. This implies of course that thin rebars (for instance ∅ 8 or 12 mm) are used.

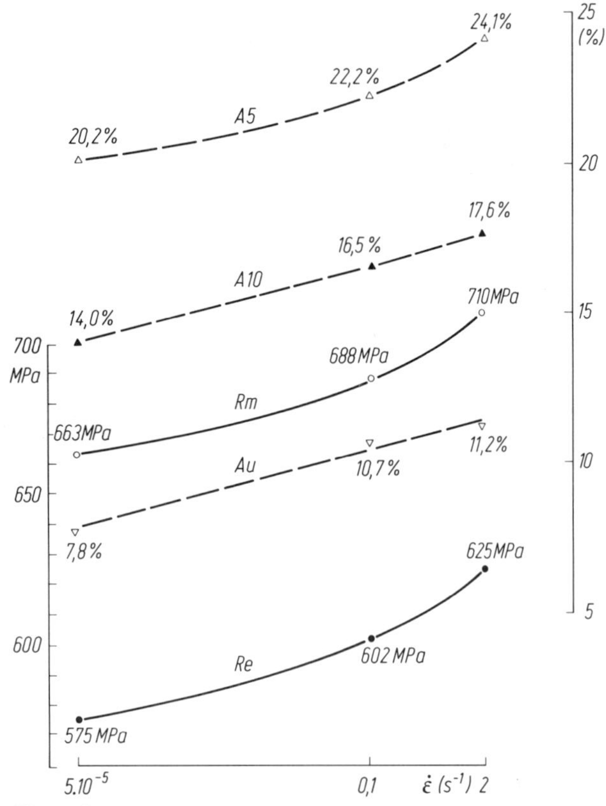

Figure 2
Influence of the strain rate on the tensile properties of Tempcore steel. Tempcore ∅ 12 Fe B 500

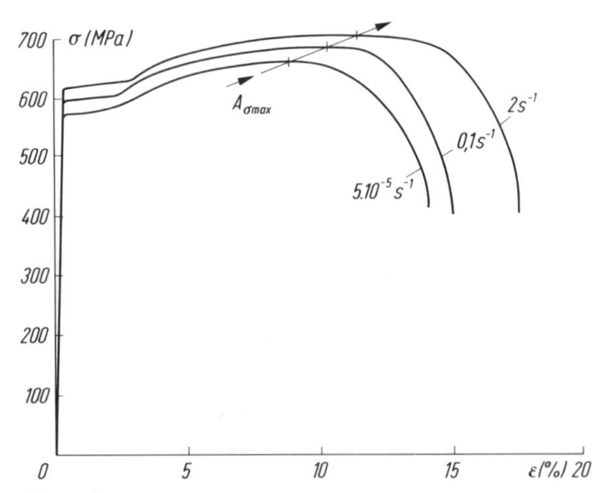

Figure 3
Tensile diagrams of Tempcore steel at different strain rates

In such a case, a tube is slipped around the bar the other extremity of which is embedded in concrete (Fig. 5). This system affords a sufficient straightness in usual cases, i.e. when the bar was bent on a mandrel whose diameter is equal to 4 or 5 d (d: diameter of the bar). For larger bending diameters (7 or 8 d for

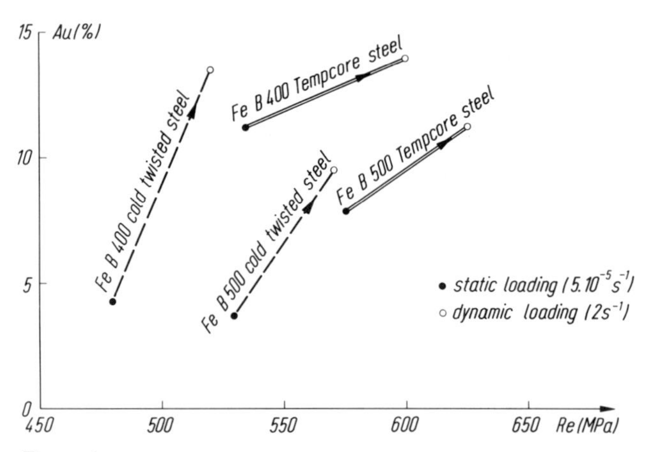

Figure 4
Comparison of Tempcore and cold twisted steels

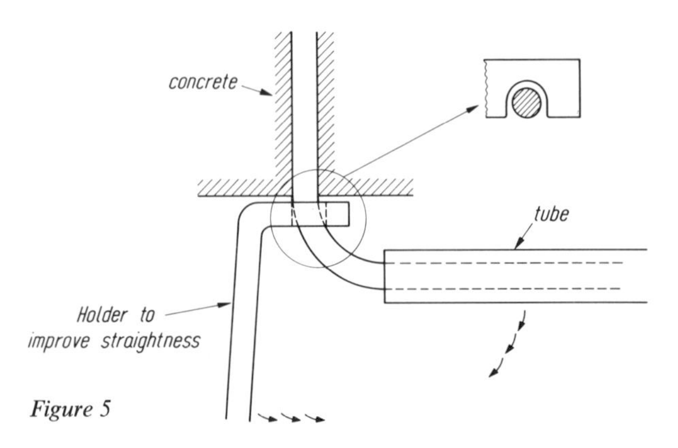

Figure 5

instance), which induce a longer curved area, a holder may be put on the bar in order to improve the straightness (Fig. 5).

3.2 Testing conditions

Tempcore steels of different chemical compositions and with diameters ranging from 10 to 14 mm were tested.

Table 4 a
Tempcore steels in tensile tests after straigthening
Chemical composition (on product 10^{-3}%)

Steel	C	Mn	Si	P	S	N
A-∅ 10	180	690	200	25	8	9.6
B-∅ 12	160	410	80	30	27	6.6
C-∅ 14	180	470	90	22	14	7.1
G-∅ 10	130	570	40	24	18	11.0
H–∅ 12	120	610	250	23	24	10.0*)
I-∅ 14	210	780	20	38	21	10.0

*) + Cu: 300-Cr: 120-Ni: 90 on steel H (electric steel)

Bars were bent to 90° on a mandrel diameter comprised between 4 and 5 d (d: bar diameter), artificially aged (100 °C – 1 hour), straightened and tensile tested.
Fracture elongation, elongation at maximum stress and total elongation were measured during these tests. Uniform elongation was calculated from fracture elongations through the formula: 2 A10–A5.
The chemical analysis and the mechanical properties of the studied Tempcore steels are listed in Tables 4 a and 4 b.

Table 4 b
Mechanical Properties

Steel	*Re* (MPa)	*Rm* (MPa)	A 5 (%)	A 10 (%)	2 A 10-A 5 (%)	$A_{\sigma max}$%	A_t^1(%)	A_t^2(%)
A-∅ 10	529	620	27.0	19.6	12.2	11.0	12.6	11.9
B-∅ 12	518	594	25.1	18.4	11.7	10.4	12.3	11.7
C-∅ 14	515	602	25.9	18.5	11.1	10.5	13.2	12.4
G-∅ 10	471	553	25.6	17.7	9.8	9.7	11.4	11.1
H-∅ 12	543	608	28.3	21.4	14.5	13.0	16.8	16.0
I-∅ 14	504	613	27.1	19.2	11.3	12.3	15.1	14.1

A 5: fracture elongation on a 5 d base length
A 10: fracture elongation on a 10 d base length
$A_{\sigma max}$: elongation measured on the tensile diagram at the maximum load
A_t^2: total elongation measured on the tensile diagram
A_t^2: total elongation measured on the broken specimen after tensile test
For $A_{\sigma max}$, A_t^1, A_t^2, the base length is 440 mm for steel index A, B, C; 340 mm for steel index G, H, I

Table 5 a
Tempcore steel after bending and straightening

Bending: at 90° on a bending machine
mandrel diameter: 40 mm for ∅ 10 mm bars (steel index: A)
50 mm for ∅ 12 mm bars (steel index: B)
50 mm for ∅ 14 mm bars (steel index: C)
70 mm for ∅ 14 mm bars (steel index: C)

Ageing: 100 °C – 1 hour
Sraightening: by hand
Tensile testing: (mean values – 5 specimens)

Steel	*Re* (MPa)	*Rm* (MPa)	A 5 (%)	A 10 (%)	2 A 10-A 5 (%)	$A_{\sigma max}$%	A_t^1(%)	A_t^2(%)
A-∅ 10	529	626	26.5	19.0	11.5	10.2	11.9	11.2
B-∅ 12	512	594	24.6	17.4	10.2	9.6	11.8	11.1
C-∅ 14 (50 mm)*	510	601	25.5	18.2	10.9	9.5	12.3	11.5
C-∅ 14 (70 mm)*	511	603	25.2	17.8	10.4	9.4	12.1	11.2

* 50 mm– 70 mm – mandrel diameter for bending.

Table 5 b
Tempcore steel after bending and straightening

Bending: at 90° on a press
mandrel diameter: 40 mm for ∅ 10 mm bars (steel index: G)
50 mm for ∅ 12 mm bars (steel index: H)
60 mm for ∅ 14 mm bars (steel index: I)

Ageing: 100 °C – 1 hour
Straightening: by hand
Tensile Testing: (mean values – 10 specimes)

Steel	*Re* (MPa)	*Rm* (MPa)	A 5 (%)	A 10 (%)	2 A 10-A 5 (%)	$A_{\sigma max}$%	A_t^1(%)	A_t^2(%)
G-∅ 10	465	550	26.6	18.2	9.8	9.1	11.1	10.7
H-∅ 12	531	598	27.0	19.3	11.6	10.9	13.6	12.9
I-∅ 14	504	610	25.8	17.9	10.0	11.1	13.6	12.6

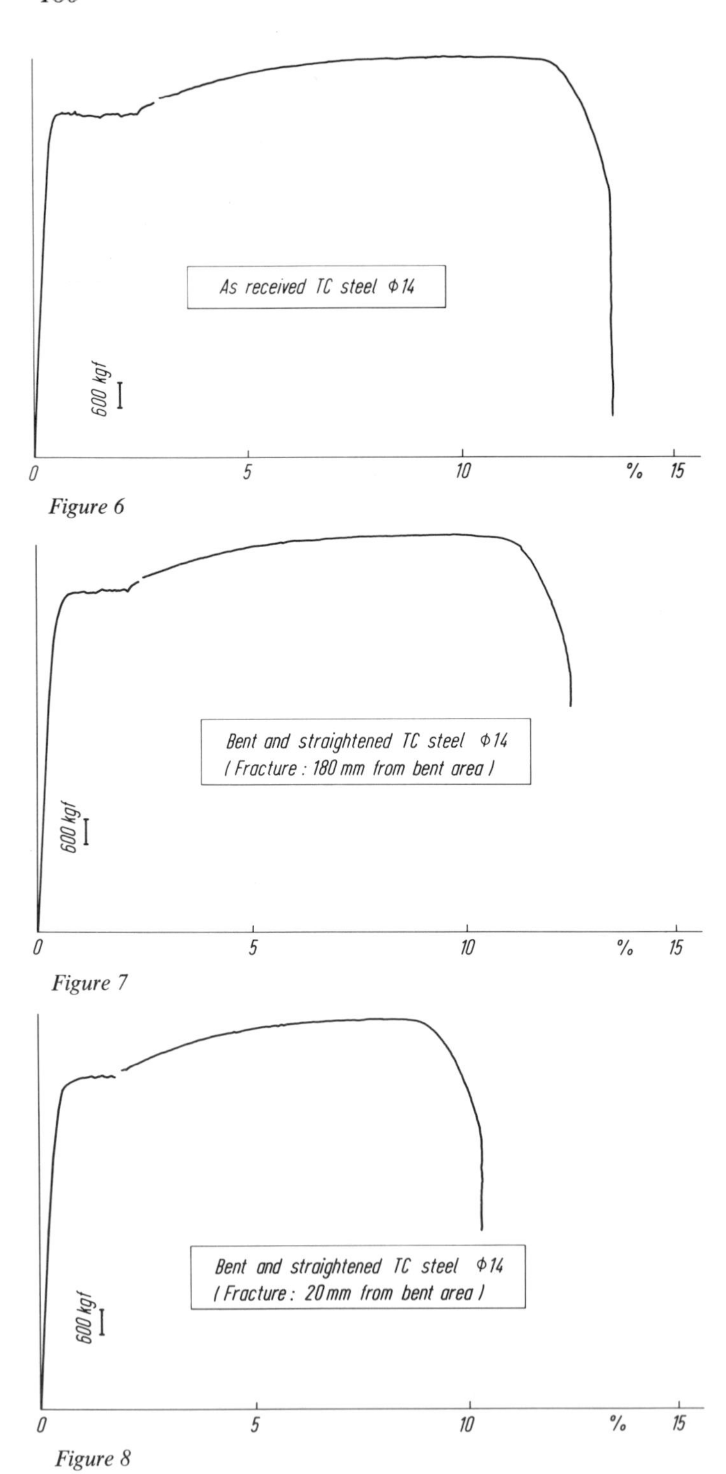

Figure 6

Figure 7

Figure 8

3.3 Results

The results of the tensile tests after rebending are listed in Tables 5a and 5b.

In comparison with the properties of the as received Tempcore bars, no significant reduction in strength or in ductility is to be noticed.

It is in particular worth pointing out the high values of uniform elongation reached by the Tempcore steels after rebending (over 8%). Here again, the uniform elongation assessed through the formula 2 A10–A5 is in good relationship with the elongation measured at the maximum stress.

In most cases, fracture at tensile test occurred far from the bent area. In two cases only, the distance was short: 20 and 30 mm, but the tensile properties were not significantly influenced by the distance of the fracture form the bent area.

Examples of tensile diagrams are shown in Fig. 6, 7 and 8. They relate respectively to an as received and to restraightened Tempcore steels which broke far from the bent area (usual case) or near to the center of the bent area (one of the two encountered cases). Those diagrams stress the excellent ductility shown by Tempcore steels after restraightening.

4 Conclusions

The tests presented in this paper demonstrate that even in peculiar applications such as in high strain rate loading or in the restraightened condition after bending, Tempcore steels display tensile properties, which correspond to the best resistance-ductility concept ever observed on reinforcements.

5 Literature

[1] Economopoulos, M., Respen, Y., Lessel, G.,and Steffes, G.: Application of the Tempcore process to the fabrication of high yield strength concrete reinforcing bars. CRM Metallurgical Reports, N° 45, December 1975.

[2] Defourny, J., and Bragard, A.: Tempcore – Process, the solution to Rebar Welding Problems, CRM Metallurgical Reports, N° 50, April 1977.

[3] Rehm, G., und Russwurm, D.: Beurteilung von Betonstählen hergestellt nach dem Tempcore-Verfahren. Betonwerk – Fertigteil – Technik, Heft 6/1977.

[4] Rehm, G., Russwurm, D., und Defourny, J.: Schweißen von Tempcore-Betonstahl. Betonwerk – Fertigteil – Technik, Heft 4/1979.

[5] Defourny, J., and Bragard, A.: Le diagramme de traction des armatures Tempcore, Rapport CRM, août 1978.

[6] Defourny, J., d'Haeyer, R., and van den Brink, S.H.: The weldability and the fatigue resistance of FeB 500 concrete reinforcing steels. Revue de la Soudure, N° 2, 1982.

[7] Berger, K.: Der Einfluß der Dehnungsgeschwindigkeit auf das mechanische Verhalten von Betonstählen. Report from BAM, Berlin.

[8] Ammann, W., Mühlematter, M., and Bachmann, H.: Stress-strain behaviour of non-prestressed and prestressed reinforcing steel at high strain rates. Report from Swiss Federal Institute of Technology Zürich.

Eine Anwendung des Reibschweißens im Bauwesen

Werner Fastenau

1 Allgemeines

Ein zumindest den älteren Bauingenieuren noch bekanntes Schweißverfahren ist das elektrische Abbrennstumpfschweißverfahren, mit dem man vor Aufkommen des Spannbetons Bewehrungsstäbe besonders im Stahlbetonbrückenbau zu größeren Längen zusammengeschweißt hat. Die dafür benötigten Geräte waren groß und unhandlich, sie erforderten so hohe elektrische Anschlußwerte, daß allein dadurch die Einsatzmöglichkeiten sehr beschränkt waren. Später wurde dann das Gaspreßschweißen ebenfalls zur Verbindung von Bewehrungsstäben entwickelt, bei dem mit verhältnismäßig leichten Maschinen die Stabenden mittels Gasbrennern aufgeschmolzen und dann mechanisch zusammengepreßt wurden.
An dieser Stelle soll über ein Schweißverfahren berichtet werden, das eine besondere Art des Stumpfpreßschweißens ist, nämlich das Reibschweißen. Dieses Verfahren wird seit etwa 20 Jahren in wachsendem Umfang im Maschinen- und Fahrzeugbau angewendet, ist aber bislang m. W. im Bauwesen unbekannt.

2 Beschreibung des Verfahrens

Voraussetzung für eine Reibschweißverbindung ist es, daß mindestens eines der zu verbindenden Teile rotatioinssymmetrisch ist.
Dieses Teil wird in ein Drehfutter (1) der Reibschweißmaschine eingespannt, wie es von Drehbänken bekannt ist. Das andere Teil wird auf einem Schlitten (2) festgespannt, der sich axial auf das im Drehfutter eingespannte Teil zu verschieben läßt. Bild 1 zeigt eine Reibschweißmaschine in schematischer Darstellung.
Zu Beginn des Schweißvorganges in der sog. Reibphase wird das feststehende Teil (4) mit relativ geringer Kraft (F_1) gegen das rotierende (3) gepreßt. Dabei kommt es infolge der Reibung zwischen den beiden

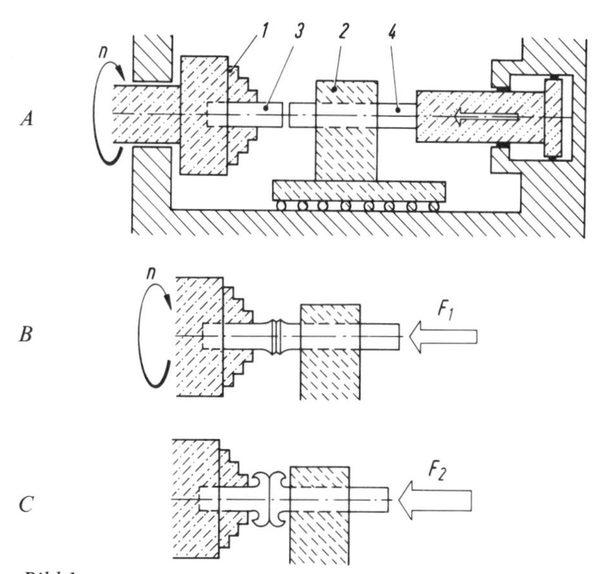

Bild 1
Reibschweißen in schematischer Darstellung

Teilen zu einer Erhitzung, die fast bis zur Schmelztemperatur ansteigen kann. Nunmehr wird die Rotation abgestellt und in der sog. Stauchphase beide Teile mit großer Kraft (F_2) gegeneinander gepreßt. Dabei wird der plastifizierte Werkstoff seitlich herausgedrückt und bildet einen Schweißwulst. In der Reibphase und insbesondere durch Bildung des Schweißwulstes tritt eine geringe Längenverkürzung auf. Die Schweißverbindung ist fertig. Der Schweißwulst kann, falls gewünscht, in einem weiteren Arbeitsgang, in dem die Schweißmaschine wie eine Drehbank arbeitet, abgedreht werden. Einen Makroschliff durch die Reibschweißverbindung von Rundstäben ∅ 100 mm aus CK 45 und 42 Cr Mo 4 zeigt Bild 2.

3 Vorteile des Reibschweißens

Eine Reibschweißverbindung kann in sehr kurzer Zeit hergestellt werden: Außer dem Ein- und Ausspannen

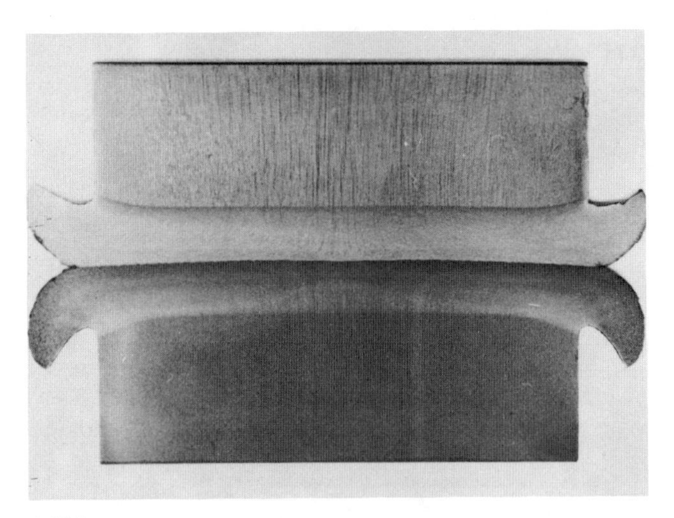

Bild 2
Makroschliff durch Reibschweiß-Verbindung

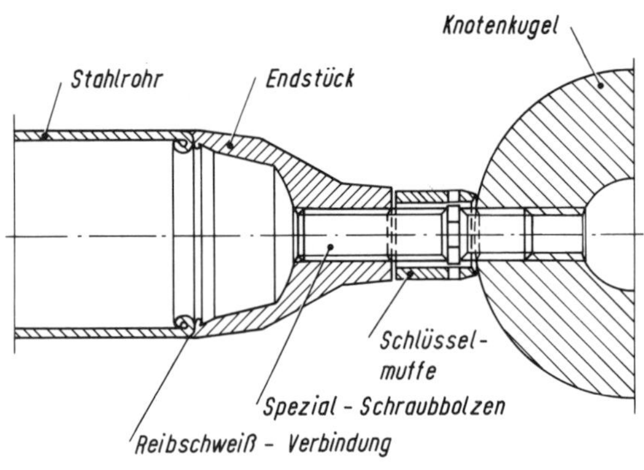

Bild 3
Anschluß eines Fachwerkstabes des RFW-Systems Züblin an die Knotenkugel

der Teile in die Maschine dauert der eigentliche Schweißvorgang nur 1 bis 100 s, abhängig vom Werkstoff und von der Größe des Querschnitts. Bei den weiter unten beschriebenen Anwendungen werden nur etwa 2 bis 15 s benötigt.
Der Bereich der thermischen Beeinflussung ist wegen der kurzen Dauer des Schweißvorganges sehr gering. Es können Verbindungen sehr verschiedener Werkstoffe hergestellt werden. Nicht nur unterschiedliche Stahlsorten, deren Verbindung mit den herkömmlichen Schweißverfahren problematisch wäre, lassen sich miteinander verschweißen, sondern auch beispielsweise Stahl und Aluminium oder Kupfer. Erfahrungstabellen geben Auskunft über eine erstaunliche Vielzahl von Werkstoffkombinationen. Reibschweißen ist unbeeinflußt von den Fähigkeiten des Maschinen-Bedienungspersonals. Die aufgrund von Erfahrungen und Vorversuchen der Maschine einprogrammierten Schweißparameter garantieren ein gleichmäßiges Qualitätsniveau.

4 Anwendungen

Schwierig geformte Guß- und Schmiedeteile können aus mehreren einfach herzustellenden Teilen zusammengefügt werden.
Drehteile mit unterschiedlichen Querschnitten können aus einzelnen vorgefertigten Teilen durch Reibschweißen verbunden werden, wodurch geringere Zerspanungsarbeit und einfachere Handhabung erreicht wird.
Teile, die nur in begrenzten Bereichen einen hochwertigen Werkstoff erfordern, müssen nicht vollständig aus diesem Werkstoff gefertigt werden, sondern können durch Reibschweißen von Teilen verschiedener Werkstoffe wirtschaftlich zusammengesetzt werden.

5 Fachwerkstäbe für Raumfachwerke

Im Bauwesen wurde m. W. bislang Reibschweißen nicht angewendet. Hierfür gibt es einleuchtende Gründe: Reibschweißen eignet sich nur für werksmäßige Fertigungen und setzt für eine wirtschaftliche Anwendung große Stückzahlen voraus, wie sie im Bauwesen nur selten auftreten.
Die Herstellung von Fachwerkstäben für Raumfachwerke jedoch bietet einen idealen Anwendungsfall für das Reibschweißen. Ein solcher Fachwerkstab besteht aus einem Stahlrohrstück, an dessen beiden Enden konische, geschmiedete Endstücke so anzuschweißen sind, daß bei einer Zugbeanspruchung die Bruchfestigkeit des Stahlrohres erreicht wird. Die Endstücke nehmen die Anschlußelemente, z. B. in Form von hochfesten Schraubenbolzen auf und ermöglichen durch ihre konische Form den Anschluß mehrerer Stäbe an einer relativ kleinen Knotenkugel. Bild 3 zeigt im Schnitt den Anschluß eines Fachwerkstabes an die Knotenkugel beim Züblin-Raumfachwerk-System, während Bild 4 einen Knoten zeigt, an dem Fachwerkstäbe verschiedener Größen angeschlossen sind.
In Tabelle 1 sind Fachwerkstäbe, bei denen die Endstücke durch Reibschweißen mit Stahlrohren verbunden werden, aufgelistet. Es werden so unterschiedliche Stabgrößen wie $\varnothing 44{,}5 \times 3{,}5$ mm, St 35 und $\varnothing$

Bild 4
Knoten des RFW-Systems Züblin

Bild 5
Reibschweißmaschine Typ LRS 30 von KUKA

127,0 × 5,0 mm St 52 mit der gleichen Maschine geschweißt, in die sich Teile bis ∅ 130 mm einspannen lassen. Die Bruchfestigkeit der größten auf Zug beanspruchten Stäbe liegt bei 1500 kN.
An die im Gesenk geschmiedeten Endstücke lassen sich Rohre mit einem auf sie abgestimmten Außendurchmesser anschweißen, deren Wandstärke jedoch in gewissen Grenzen variabel ist. Damit ist eine sehr wirtschaftliche Anpassung der Fachwerkstäbe an die elektronisch ermittelten Stabkräfte oder aber auch an die jeweiligen Liefermöglichkeiten der Stahlrohrhändler möglich.
Die Rohrwandstärke muß auf die Dicke des Endstückrandes abgestimmt sein. Sie kann maximal etwa so dick sein wie der Rand des Endstückes, darf aber nicht zu dünn werden, da dann das Rohrende beim Schweißen zu weich wird, wodurch die Gefahr besteht, daß es in der Stauchphase beult oder seitlich ausweicht, während der Rand des Endstückes nicht die zum Verschweißen erforderliche Erhitzung erreicht.
Die für eine bestimmte Kombination Endstück/Rohr erforderlichen Schweißparameter, Reibdruck–Reibzeit, Stauchdruck–Stauchzeit, werden durch Vorversuche ermittelt. Sie werden als Soll-Werte der Maschine einprogrammiert. Eine Kontrollvorrichtung gibt die beim Schweißvorgang erreichten Ist-Werte an. Für jeden Wert läßt sich noch eine obere und untere zulässige Abweichung eingeben, die wegen unvermeidlicher Maß- und Werkstofftoleranzen, z. B. der Rohre, erforderlich ist. Sobald ein Wert aus dem zulässigen Bereich

Tabelle 1
Standard-Fachwerkstäbe des Züblin-RFW-Systems mit reibgeschweißten Endstücken*)

Rohr		Bolzen/Güte		zul. Belastung Lastfall H (kN)	
∅ × t mm	Güte St	Zug	Druck**)	Zug	Druck, $\omega = 1$
44,5 × 2,9	37	M 14 – 8,8	M 14 – 8,8	60,8	53,2
60,3 × 2,9	37	M 20 – 8,8	M 14 – 8,8 D	84,6	74,0
76,1 × 3,2	37	M 20 – 10,9	M 14 – 8,8 D	118,2	103,5
88,9 × 4,0	37	M 27 – 10,9	M 14 – M 8,8 D	171,2	149,8
101,6 × 5,0	52	M 33 – 10,9	M 20 – 8,8 D	347,0	319,2
108,0 × 8,0	52	M 48 – 10,9	M 20 – 8,8 D	602,4	527,1
127,0 × 5,0	52	M 33 – 10,9	M 20 – 8,8 D	347,0	403,4

*) andere Wandstärken der Rohre sind möglich **) D = mit Druckmuffe

herausfällt, bleibt die Maschine stehen und zwingt so die Bedienung, dem Fehler nachzugehen und ihn zu beheben. Alle Ist- und Soll-Werte jeder Schweißung können außerdem ausgedruckt werden, so daß auch eine nachträgliche Kontrolle möglich ist. Eine Kontrollmöglichkeit ergibt sich nicht nur über das Einhalten der programmierten Schweißparameter, sondern auch durch Messung und Aufzeichnung der aufgetretenen Verkürzung.
Eine besondere Vorbereitung der Fügestellen ist nicht erforderlich; die im Gesenk geschmiedeten Endstücke sind sandgestrahlt, während bei den Rohren ein sauberer, winkelrechter Sägeschnitt genügt.
Die für das Schweißen der Fachwerkstäbe zur Verfügung stehende Maschine ist vom Typ LRS 30 der Firma KUKA, Augsburg. Bild 5 zeigt diese Maschine, die für Rohrlängen bis 5,0 m umgebaut wurde. Die Leistungsdaten dieser Maschine sind aus Tabelle 2 ersichtlich. Da Fachwerkstäbe Gewichte bis 150 kg haben können, ist die Maschine so eingerichtet, daß sie mit einem Kran von oben beschickt werden kann, was sehr wichtig ist, weil damit der Bedienungsmann in die Lage versetzt wird, allein die Maschine zu beschicken.
Die Schweißparameter werden von einem erfahrenen Fachmann durch Vorversuche für jede einzelne Stabtype festgelegt, sie können später von einer angelernten Kraft jederzeit wieder eingegeben werden. Eine Zerreißmaschine im Werk erlaubt Tragfestigkeitsproben zur Optimierung bei der Ermittlung der Schweißparameter, als auch Stichproben aus der laufenden Produktion.

Bild 6
Kontrolle der Schweißqualität durch Quetschen

Eine schnelle Kontrolle der Schweißqualität ergibt sich durch Quetschen des Stabendes in Nähe der Schweißnaht, Bild 6. Eine Schweißnaht, die eine solche Tortur ohne Schaden übersteht, wird mit großer Wahrscheinlichkeit auch im Zugversuch nicht versagen.
Typische Schweißparameter sind in Tabelle 3 angegeben für Stäbe ∅ 44,5 × 3,5 mm und ∅ 108 × 8 mm. Da es sich um tragende Bauteile handelt, sind die Fachwerkstäbe einer Güteüberwachung durch eine staatliche Materialprüfungsanstalt unterworfen.
Raumfachwerke sind Bauwerke, die nur für eine „vorwiegend ruhende" Belastung gem. DIN 1055, Blatt 3 zugelassen sind. Um nachzuweisen, daß die durch die Schweißwulste erzeugten Kerben bei wiederholter Belastung nicht zum Dauerbruch führen, wurde mit einem Stab ∅ 88,9 × 4,5 mm, St 37, ein Zugschwellversuch durchgeführt. Mit einer Unterlast von 95 kN

Tabelle 2
Leistungsdaten der Reibschweißmaschine Typ LRS 30 der Firma KUKA Schweißanlagen und Roboter GmbH, Augsburg

Schweißflächen bei Baustahl:	Stab	min.	170 mm^2 (14 mm ∅)
	Stab	max.	2500 mm^2 (55 mm ∅)
	Rohr	max.	3200 mm^2
Umgebaut für Fachwerkstäbe:	Durchmesser	max.	130 mm
	Länge	max.	5000 mm
Maschine:	Stauchkraft	max.	300 kN
	Spindeldrehzahl		1500/750 min^{-1}
	Schlittenhub	max.	300 mm
	Spindelantrieb		20/35 kW
	Anschlußleistung gesamt	ca.	50 kW

Tabelle 3
Parameter für das Reibschweißen von Züblin-Fachwerkstäben Maschinentyp: KUKA LRS 30, Drehzahl 1500 U/min

Rohr ∅ × t mm	Stahlgüte Rohr	 Endstück	Reibdruck N/mm²	Reibzeit s	Stauchdruck N/mm²	Verkürzung mm
60,3 × 2,9	St 37	St 37	5,1	2,6	8,1	9
108,0 × 8,0	St 52	St 52	8,0	15,0	15,0	13

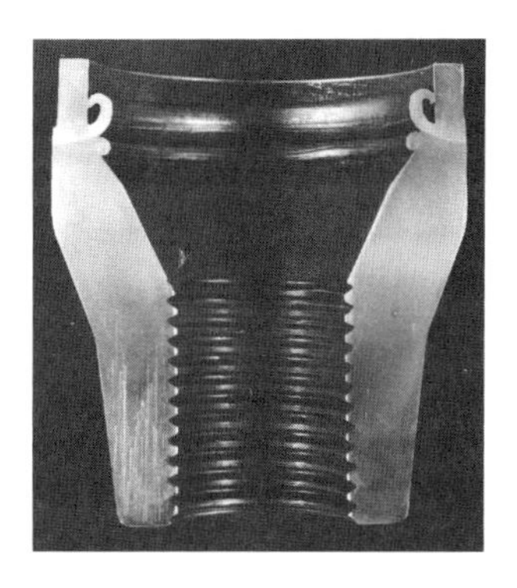

Bild 7
Aufgeschnittenes Stabende nach Zugversuch. Bruch im Rohr bei 1,14

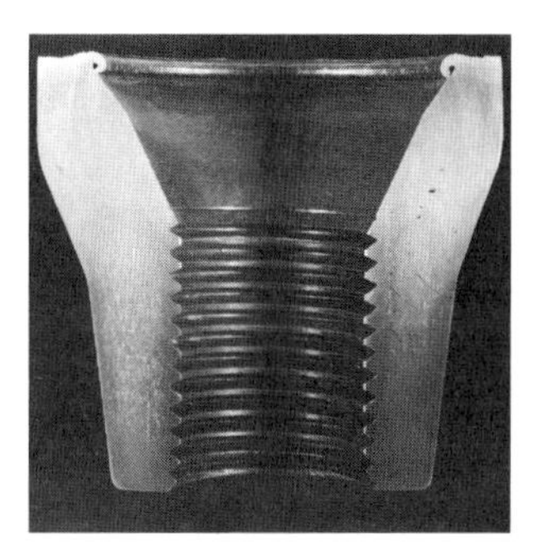

Bild 9
Aufgeschnittenes Stabende nach Zugversuch. Bruch der Schweißnaht bei 1,13

Bild 8
Makroschliff durch die Schweißnaht des Stabendes des Bildes 7

Bild 10
Ansicht der Schweißnaht von Bild 9 Bruchgefüge körnig und fehlerfrei

(80 N/mm²) und einer Oberlast 190 kN (160 N/mm²) wurden 502000 Lastspiele ohne Bruch ertragen. Danach wurde die Unterlast auf 19 kN (16 N/mm²) gesenkt, worauf nach weiteren 147000 Lastspielen der Bruch im Gewindebolzen eintrat, während die beiden Reibschweißungen noch keinen Schaden aufwiesen. Da für vorwiegend ruhende Belastung eine Grenzlastspielzahl von nur ca. 10^4 betrachtet wird, ist somit eine Gefährdung durch Dauerbruch der Reibschweißnähte nicht zu befürchten.
Bild 7 und 8 zeigen einige Aufnahmen von aufgeschnittenen, reibgeschweißten Stabenden nach Zugversuchen.
Bild 9 und 10 zeigen die Bruchfläche einer Reibschweißnaht nach einem bestandenen Zugversuch mit körniger und fehlerfreier Bruchfläche.

6 Zusammenfassung

Reibschweißen ist eine besondere Form des Stumpfpreßschweißens, bei der die zur Verschweißung benötigte Erhitzung als Reibwärme an der Fügestelle erzeugt wird.
Vorteile des Reibschweißens sind eine sehr rationelle Fertigung mit gleichmäßigem Qualitätsniveau. Reibschweißen gestattet die Verbindung sehr unterschiedlicher Werkstoffe.
Näher beschrieben wird die Anwendung des Reibschweißens zur Herstellung von Fachwerkstäben von Raumfachwerken, bei denen gesenkgeschmiedete, kegelförmige Endstücke mit Rundrohren verbunden werden.

Schadensakkumulation und Restfestigkeit im Licht der Bruchmechanik

Lutz Franke

1 Einleitung

Die Festigkeit von Werkstoffen wird bekanntlich durch langzeitig einwirkende konstante Belastungen oder durch dynamische Lasten beeinträchtigt. Die Ursache für die geringere Tragfähigkeit bei dynamischen Beanspruchungen und Zeitstandbeanspruchungen liegt nach einhelliger Auffassung in dem Anwachsen und der Entstehung von Fehlstellen bzw. Rissen, die zu einer entsprechenden Schwächung der Stoffstruktur führen. Um das Maß der Beeinträchtigung unter Betriebsbedingungen voraussagen zu können, bedarf es eines Beurteilungsverfahrens, das die Wirkung z. B. zeitlich veränderlicher Dauerstandbelastungen oder veränderlicher dynamischer Belastungen unter verschiedenen Bedingungen erfassen kann. Im folgenden wird nun auf der Basis des K_c-Konzeptes der Bruchmechanik unter Verwendung der Ergebnisse von Craze- und Rißfortschrittsmessungen ein Schadensakkumulationsgesetz unterbreitet, das die Wirkungen der verschiedenen Beanspruchungen erfaßt und als Hilfsmittel zur Vorhersage der Lebensdauer verschiedener Werkstoffe und Konstruktionen herangezogen werden kann. Unter anderem wird die bekannte empirisch begründete PALMGREN-MINER-Hypothese als Sonderfall des vorgeschlagenen Akkumulationsgesetzes ausgewiesen und gezeigt, daß diese Hypothese z. B. für Schweißverbindungen unter den getroffenen Voraussetzungen zu ungenauen Ergebnissen führen muß. Ferner wird die Frage nach der jeweiligen Restfestigkeit, also der Kurzzeitfestigkeit nach Betriebsbeanspruchungen behandelt.

2 Zeitstandbeanspruchung und Restfestigkeit

Nach dem K_c-Konzept der Bruchmechanik versagt ein Werkstoff dann, wenn die Spannungsintensität K in der Umgebung der maßgebenden Risse den Grenzwert bzw. kritischen Wert K_c erreicht, der als experimentell zu bestimmender Materialkennwert angesehen wird, vgl. z. B. HERTZBERG [16], KAUSCH [25], SCHWALBE [40], PARKER [49]. Der kritische Wert K_c an den Rißspitzen wird im Kurzzeitversuch bei monoton steigender Belastung in der Regel ohne wesentliche Rißverlängerung erreicht. Bei Zeitstandbelastungen mit konstanten Spannungen unterhalb der Bruchspannung wachsen dagegen die Rißlängen mit zunehmender Geschwindigkeit an, bis die kritische Spannungsintensität K_c und damit der Bruch des Stoffes eintritt. Berichte über entsprechende Rißfortschritts- und Crazefortschrittsmessungen liegen bereits für verschiedene Stoffe vor, vgl. WIEDERHORN [12], KAUSCH [17]. ARGON [20], AVESTON [56], SAKAGUCHI [58], PRIORI [61], KINLOCH [62].

Einige Ergebnisse dieser Messungen sind in Bild 1 zusammengestellt. Eine weitaus größere Zahl von Rißfortschrittsmessungen wurde für nicht ruhende Belastung vorgenommen. Diese Ergebnisse werden in Abschnitt 4 besprochen.

Die festgestellten Abhängigkeiten zwischen Rißgeschwindigkeit bzw. Crazegeschwindigkeit (bei Kunststoffen) und der jeweiligen Spannungsintensität zeigen den in Bild 2 schematisch dargestellten Verlauf, der in der Regel durch die Abhängigkeit (1) beschrieben werden kann.

$$\frac{\mathrm{d}a}{\mathrm{d}t} = A \cdot K^{n}_{1(t)} \tag{1}$$

Darin ist $\mathrm{d}a/\mathrm{d}t$ die Rißfortschrittsgeschwindigkeit, A ein vom Stoff und von den vorliegenden Bedingungen abhängiger Beiwert, n ein stoffabhängiger Exponent und $K_{1(t)}$ die jeweilige Spannungsintensität der maßgebenden Fehlstellen für den Beanspruchungsmodus 1. Nach dem K_c-Konzept kann weiterhin angeschrieben werden

$$K_{1(t)} = \sigma \cdot y \cdot \pi^{1/2} \cdot a_{(t)}^{1/2} \tag{2}$$

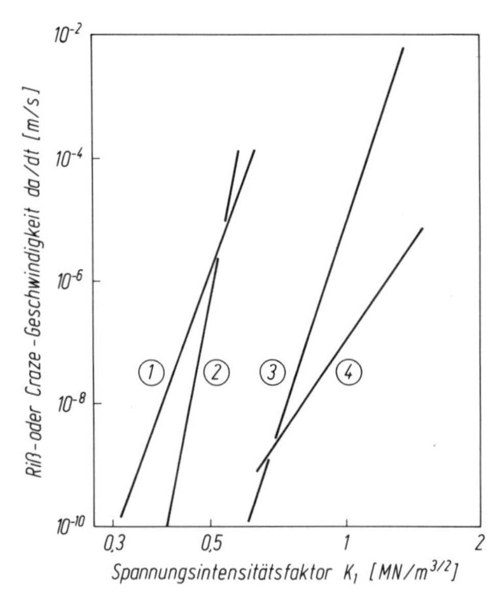

Bild 1
Ergebnisse von Riß- bzw. Craze-Geschwindigkeitsmessungen bei Zeitstandbelastung
① Natron-Kalk-Silikatglas (Fensterglas) bei 50% rel. F., nach WIEDERHORN [12]
② Quarz-Glas bei 50% rel. F. [12]
③ PMMA in Luft, nach KAUSCH [17]
④ Unidirekt. GFK (E-Glas/Polyesterharz) in 1 N – H_2SO_4, nach AVESTON/SILLWOOD [56]

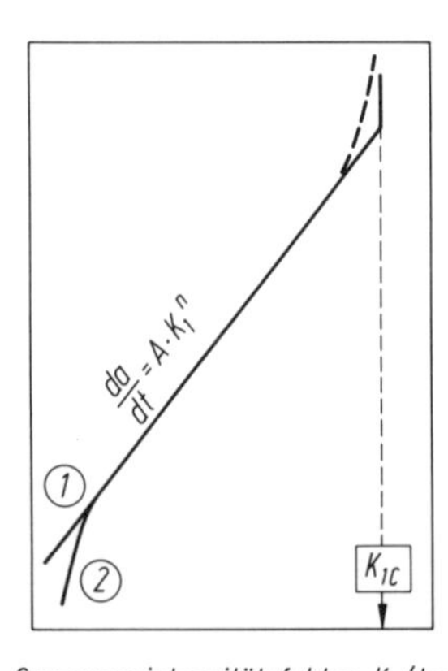

Bild 2
Schematische Darstellung des Verlaufs der Fehlstellenwachstumsgeschwindigkeit in Abhängigkeit von der Spannungsintensität im doppelt-logarithm. Maßstab;
je nach Stoff, Umgebungsbedingungen und Belastungsgeschichte kann bei niedrigen Beanspruchungen der Verlauf ① oder ② maßgebend sein

$$K_{1c} = \sigma_{\lim} \cdot y \cdot \pi^{1/2} \cdot a_0^{1/2} \tag{3}$$

σ ist darin eine außen angelegte (Zug-)Spannung, $\sigma_{\lim}$ die Materialfestigkeit, y eine von der Proben- bzw. Bauteilgeometrie abhängige Größe sowie a_0 und $a_{(t)}$ die wirksamen Rißlängen zu Beginn bzw. im Verlauf der Belastung.

Die zeitabhängigen Rißweiten unter der äußeren Belastung σ im Zeitstandversuch erhält man aus (1) und (2) nach Lösen der Dgl. zu

$$\frac{a_{(t)}}{a_0} = (1 - B_1 \cdot \alpha^n \cdot t)^{-\frac{2}{n-2}} \quad \text{für } n > 2 \tag{4}$$

Hierin ist α die bezogene äußere Belastung

$$\alpha = \frac{\sigma}{\sigma_{\lim}} \tag{5}$$

und B_1 ein Bauteil- bzw. Werkstoffkennwert

$$B_1 = \left(\frac{n-2}{2}\right) \cdot A \cdot (y \cdot \pi^{1/2})^n \cdot a_0^{\frac{n-2}{2}} \cdot \sigma_{\lim}^n \tag{6}$$

Die bis zum Erreichen eines bestimmten Verhältnisses $a_{(t)}/a_0$ benötigte Zeit beträgt dann

$$t_{(\alpha)} = \frac{1}{B_1 \cdot \alpha^n} \left[1 - \left(\frac{a_0}{a_{(t)}}\right)^{\frac{n-2}{2}}\right] \tag{7}$$

Vergleichbare Ansätze sind z. B. in [33], [56] und [62] enthalten. Benutzt man nun den über (2) und (3) erhältlichen Zusammenhang

$$\left(\frac{a_{(t)}}{a_0}\right)^{1/2} \cdot \frac{\sigma}{\sigma_{\lim}} = \frac{K_{1(t)}}{K_{1c}} \tag{8}$$

sowie die Bedingung

$$K_{1(t\max)} = K_{1c} \tag{9}$$

erhält man die Standzeit $t_{\max}$ in Abhängigkeit von der vorgegebenen äußeren Belastung α und dem Materialkennwert B_1 zu

$$t_{\max(\alpha)} = \frac{1}{B_1 \cdot \alpha^n} (1 - \alpha^{n-2}) \tag{10}$$

Der Materialkennwert B_1 nach Gl. (6) kann unter Verwendung von Gl. (10) ohne Kenntnis der enthaltenen Teilfaktoren direkt aus Zeitstandversuchen ermittelt werden. Für die praktische Handhabung des vorliegenden bruchmechanischen Konzeptes ist es nicht notwendig, die Anfangsgröße der Fehlstellen oder Risse zu kennen. Es können daher offenbar auch Werkstoffe mit einbezogen werden, bei denen Schädigungen erst nach einer gewissen Inkubationszeit oder kritischen Dehnung mikroskopisch bzw. visuell sichtbar werden, wenn man unterstellt, daß die submikroskopische Fehlstellenbildung ebenfalls mit der durch Gl. (1) vorgegebenen Geschwindigkeit abläuft. Dies würde z. B. einige Kunststoffe betreffen, vgl. KAUSCH [33], MENGES [35] oder KINLOCH/YOUNG [62].

Der Exponent n ist bei konstruktiv nutzbaren Stoffen in der Regel größer 20, meist in der Größenordnung von 30 bis 60, so daß der Faktor $(1 - \alpha^{n-2})$ zumindest für $\alpha \lesssim 0{,}9$ gleich 1 gesetzt werden kann. Statt (10) gilt dann

$$t_{\max(\alpha)} = \frac{1}{B_1 \cdot \alpha^n} \tag{11}$$

Die Umformung

$$\frac{\alpha_1}{\alpha_0} = \left(\frac{t_{\max(1)}}{t_{\max(0)}}\right)^{-\frac{1}{n}} \tag{12}$$

zeigt, daß es sich hierbei um eine Gerade im doppelt-logarithmischen Maßstab mit der Steigung $-1/n$ handelt.

Die Ergebnisse zahlreicher Zeitstandversuche an verschiedenen Werkstoffen bestätigen dieses Ergebnis. Andererseits weist die Gl. (10) darauf hin, daß bei Stoffen mit vergleichsweise niedriger Zeitstandfestigkeit bzw. kleinem Exponenten n im doppelt-logarithmischen Maßstab eine am Anfang gekrümmte Kurve erhalten wird. Bild 3 soll dies verdeutlichen. Bild 4 zeigt die zeitliche Rißentwicklung $a_{(t)}/a_0$ für einen Werkstoff mit $n = 35$ (z. B. GFK) bei einer Zeitstandbelastung $\alpha = 0{,}75$.

Für Sicherheitsbetrachtungen und für die Bewertung von Versuchen an bestimmten Werkstoffen, Verbindungen und Bauteilen ist die Frage der Restfestigkeit nach statischer und dynamischer Dauerbelastung von Belang. Hierzu gibt es eine Reihe von Überlegungen und Versuchsergebnissen, vgl. WITTMANN/ZAITSEV [10], RÜBBEN [43] oder MENGES/BIELING [50], die zeigen, daß an Zeitstandvorbelastungen anschließende Kurzzeitversuche in etwa die Ursprungsfestigkeit ergeben. Eine überzeugende Erklärung fehlte bisher. Weitgehend unbeantwortet ist weiterhin die Frage, welche Reststandzeit man erwarten kann, wenn man sich nach einer Anfangsstandzeit auf dem Beanspruchungsniveau α_1 auf ein Niveau α_2 z. B. oberhalb der Zeitstandfestigkeitskurve begibt, s. Bild 5. Diese letztgenannte Frage wird in Abschnitt 3 behandelt. Zur Ermittlung der Restfestigkeit σ_R nach Zeitstandvorbelastung muß von einem Stoffzustand ausgegangen werden, bei dem ein gegenüber dem Ausgangszustand vergrößerter Fehlstellen- bzw. Rißzustand $a_{(t)}$ entsprechend der Vorbelastungszeit und Vorbelastungshöhe vorliegt. Der Zusammenhang zwischen $a_{(t)}$ und σ_R wird über Gl. (2) und (3) erhalten

$$K_{1c} = \sigma_R \cdot y \cdot \pi^{1/2} \cdot a_{(t)}^{1/2} \tag{13}$$

Unter Bezugnahme auf den Ausgangszustand mit Rißlängen a_0 erhält man

$$\sigma_R = \sigma_{\lim} \cdot \left(\frac{a_0}{a_{(t)}}\right)^{1/2} \tag{14}$$

Durch Einsetzen von Gl. (4) in (14) und unter Verwendung von (10) ergibt sich schließlich der auf die Kurzzeitfestigkeit bezogene Verlauf der Restfestigkeit nach Vorbelastungszeiten $t \leq t_{\max}$ auf dem Niveau α zu

$$\frac{\sigma_R}{\sigma_{\lim}} = \left[1 - \frac{t}{t_{\max}}\left(1 - \alpha^{n-2}\right)\right]^{\frac{1}{n-2}} \tag{15}$$

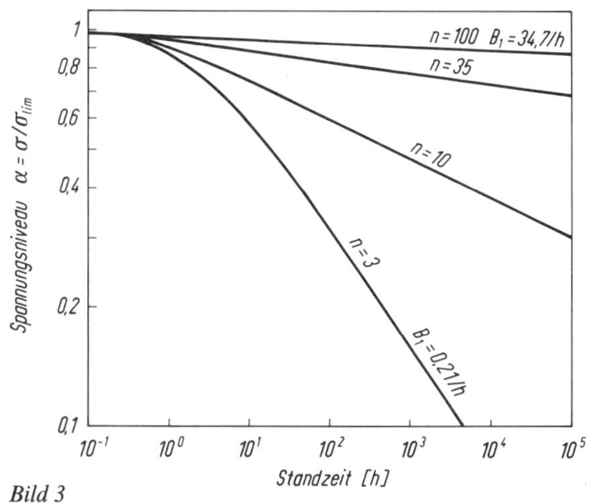

Bild 3
Verlauf der Zeitstandfestigkeit im doppelt-logarithm. Maßstab bei Zugrundelegung eines Fehlstellenwachstums entsprechend Bild 1 bzw. 2

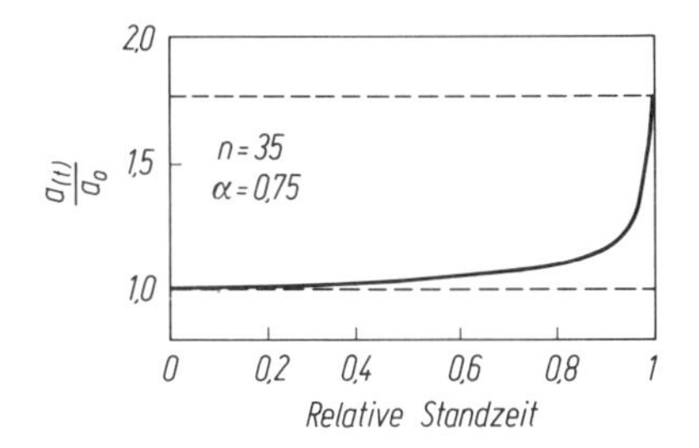

Bild 4
Beispiel für die Entwicklung der maßgebenden Rißlängen in Abhängigkeit von der relativen Belastungszeit, berechnet mit Gl. (4)

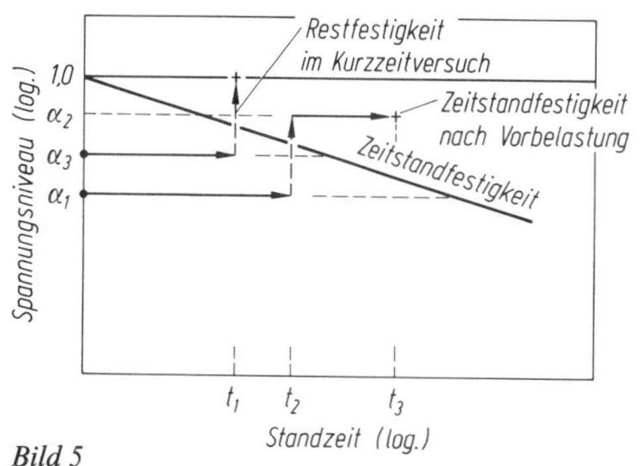

Bild 5
Schematische Darstellung der erwarteten Restfestigkeit nach Zeitstandvorbelastung sowie der erwarteten Zeitstandfestigkeit nach Zeitstandvorbelastung

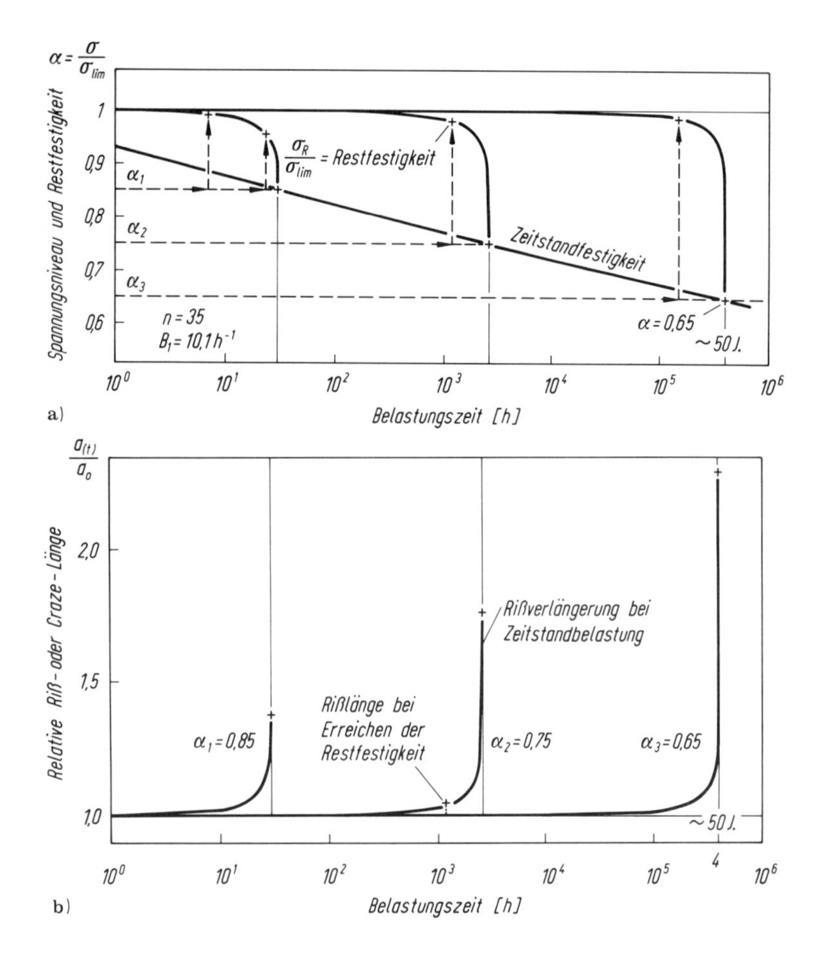

Bild 6
Beispiel für die Restfestigkeit nach unterschiedlichen Zeitstandvorbelastungen (Bild 6a) und zugehörige Rißlängen (Bild 6b), berechnet nach Abschnitt 3

Laut Beziehung (15) weist ein Stoff mit dem Exponenten $n = 40$ (z. B. Leichtbeton oder GFK-Spritzlaminat unter Zugbeanspruchung) auf dem Niveau $\alpha = 0{,}5$ bei 98% der maximalen theoretischen Standzeit noch eine Restfestigkeit von 90% des Ausgangswertes auf.
Bei $t = t_{max}$ entspricht erwartungsgemäß die Restfestigkeit der Höhe der Zeitstandbelastung.
Bild 6 zeigt die Verhältnisse für eine praxisnahe Zeitstandfestigkeitskurve und drei ausgewählte Zeitstandbelastungsniveaus in Relation zur zeitlichen Rißverlängerung im logarithmischen Maßstab. Zum Vergleich sei auf die Versuchsergebnisse in Bild 9 hingewiesen.

3 Schadensakkumulation bei Zeitstandbelastung

Bei Stoffen und Verbindungen mit ausgeprägter Abnahme der Tragfähigkeit bei Zeitstandbelastungen ist es erforderlich, die Auswirkungen von periodisch veränderlichen Zeitstandbelastungen auf die Tragfähigkeit zu kennen und qualitativ zu erfassen. Dies kann z. B. durch Angabe einer gleichwertigen als ständig wirkend angenommenen Ersatzlast geschehen. Von zusätzlichem Interesse können periodisch wirkende Temperatureinflüsse sein.
Im folgenden soll nun die Art und Weise der Akkumulation der Wirkungen auf die Materialfestigkeit gezeigt werden, wobei auf den bruchmechanischen Voraussetzungen des Abschnittes 2 aufgebaut wird und zunächst isotherme Bedingungen angenommen werden. Zunächst sei nach der restlichen Standzeit auf dem Lastniveau α_2 nach Zeitstandvorbelastung auf dem Niveau α_1 während der Zeitspanne $t_{(\alpha_1)}$ gefragt. Es ist laut Bild 7a

$$t_{R(\alpha_2)} = t_{\max(\alpha_2)} - t_{(\alpha_2)} \tag{16}$$

wobei $t_{(\alpha_2)}$ die Zeit ist, die auf dem Niveau α_2 benötigt wird, um dieselbe Rißverlängerung zu erzielen wie auf dem Niveau α_1 während $t_{(\alpha_1)}$.
Nach Einsetzen von Gl. (4) mit α_1 und $t_{(\alpha_1)}$ in (7) erhält man die $t_{(\alpha_1)}$ entsprechende Anfangszeit $t_{(\alpha_2)}$ zu

$$t_{(\alpha_2)} = \left(\frac{\alpha_1}{\alpha_2}\right)^n \cdot t_{(\alpha_1)} \tag{17}$$

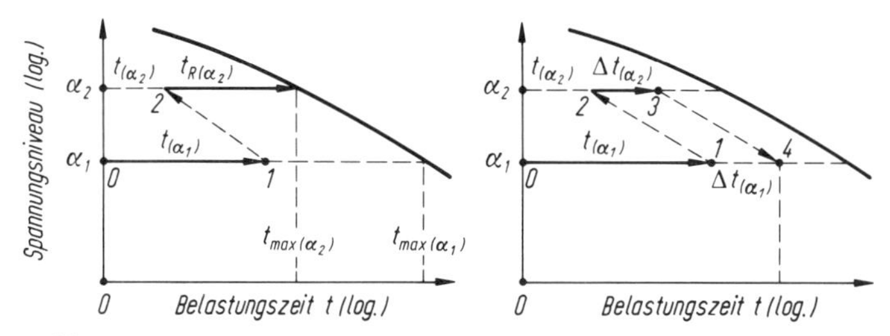

Bild 7
Schematische Darstellung für die Berechnung der äquivalenten Belastungszeiten beim Übergang von einem Lastniveau auf ein anderes, vgl. Erläuterungen im Text

Nach Kenntnis der Transformationsmöglichkeit für die „Anfangsstandzeiten" (17) wird untersucht, wie beliebige Zeitabschnitte Δt von einem Niveau auf ein anderes übertragen werden können, und zwar durch Vergleich der Wege 0–1–4 und 0–1–2–3–4 in Bild 7b. Unter Verwendung von (17) muß sein

$$(t_{(\alpha_2)} + \Delta t_{(\alpha_2)}) \cdot \left(\frac{\alpha_2}{\alpha_1}\right)^n = t_{(\alpha_1)} + \Delta t_{(\alpha_1)}$$

Nach Einsetzen von

$$t_{(\alpha_2)} = t_{(\alpha_1)} \cdot \left(\frac{\alpha_1}{\alpha_2}\right)^n$$

erhält man

$$\Delta t_{(\alpha_1)} = \left(\frac{\alpha_2}{\alpha_1}\right)^n \cdot \Delta t_{(\alpha_2)} \tag{18}$$

Dies bedeutet, daß unter den zuvor genannten Voraussetzungen beliebige Zeitstandabschnitte Δt wie Anfangsabschnitte von einem Niveau auf ein anderes umgerechnet werden können bzw. die einer beliebigen Verweilzeit auf dem Lastniveau α_i äquivalente Verweilzeit auf dem Niveau α_j angegeben werden kann.
Soll nun die Wirkung mehrerer Verweilzeiten auf beliebigen Niveaus α_i z. B. durch eine als ständig wirkend angenommene Ersatzlast ausgedrückt werden, so können folgende Beziehungen benutzt werden:

$$\sum_i \left(\frac{\alpha_i}{\bar{\alpha}}\right)^n \cdot t_{(\alpha_i)} + t_{R\,(\bar{\alpha})} = t_{\max\,(\bar{\alpha})} \tag{19}$$

hierin ist $t_{(\alpha_i)}$ die Gesamtverweilzeit auf dem Niveau α_i, ggf. aufaddiert aus nicht zusammenhängenden Teilverweilzeiten auf diesem Niveau, $t_{R\,(\bar{\alpha})}$ die nach Addition der Wirkungen noch auf dem Niveau $\bar{\alpha}$ ertragbare Restzeit bis zum Versagen zur Zeit $t_{\max\,(\bar{\alpha})}$.

Nach Division von (19) durch $t_{\max(\bar{\alpha})}$ und unter Verwendung der folgenden aus (10) abgeleiteten Beziehung

$$\frac{t_{\max\,(\bar{\alpha})}}{t_{\max\,(\alpha_i)}} = \left(\frac{\alpha_i}{\bar{\alpha}}\right)^n \cdot \left(\frac{1 - \bar{\alpha}^{n-2}}{1 - \alpha_i^{n-2}}\right) \tag{19a}$$

erhält man folgendes Akkumulationsgesetz für Zeitstandeinwirkungen

$$\sum_i \frac{t_{(\alpha_i)}}{t_{\max\,(\alpha_i)}} \cdot \left(\frac{1 - \alpha_i^{n-2}}{1 - \bar{\alpha}^{n-2}}\right) + \frac{t_{R\,(\bar{\alpha})}}{t_{\max\,(\bar{\alpha})}} = 1 \tag{20}$$

Für Stoffe, Verbindungen oder Konstruktionen, bei denen der Exponent n der Zeitstandfestigkeitskurve größer als ca. 20 ist und gleichzeitig die maximalen Beanspruchungen $\alpha \approx 0{,}7$ nicht überschreiten, was mit Ausnahme einiger Kunststoffanwendungen offenbar die Regel ist – ergibt sich aus (20) folgende Versagensbedingung für isotherme Verhältnisse.

$$\boxed{\sum_i \frac{t_{(\alpha_i)}}{t_{\max\,(\alpha_i)}} = 1} \tag{21}$$

Danach versagt das Bauteil dann, wenn die Summe der relativen Standzeiten auf den verschiedenen Beanspruchungsniveaus den Wert 1 erreicht. Diese Beziehung bestätigt z. B., daß Bauteile mit standzeitabhängigen Festigkeitswerten, die auf der Basis einer als ständig wirkend angenommenen Beanspruchung dimensioniert wurden, tatsächlich eine u. U. wesentlich längere Lebensdauer aufweisen, sofern die volle Dimensionierungslast nur periodisch eingewirkt hat.

4 Schadensakkumulationsgesetz für dynamische Beanspruchungen

Zur rechnerischen Voraussage der Lebensdauer bzw. der Betriebsfestigkeit von Bauteilen auf der Basis von einstufig ermittelten WÖHLERlinien und zur Erweiterung der Aussagefähigkeit von Betriebsfestigkeitsversuchen werden z. Z. vor allem im Metallbau zwei Wege beschritten, vgl. HAIBACH [18], KOSTEAS [54].
Beim ersten Weg wird der dynamische Beanspruchungsablauf mit Hilfe verschiedener Zählverfahren zu einem Amplitudenkollektiv zusammengefaßt und mit Hilfe einer Schadensakkumulationsrechnung auf der Basis der Wöhlerlinien, die verschiedenen Kerbfällen bzw. Bauteiltypen zugeordnet sind, die Lebens-

dauer bestimmt. Für die Schadensakkumulationsrechnung wird die seit Jahrzehnten bekannte empirisch begründete PALMGREN-MINER-Hypothese benutzt. Dieses Grundkonzept liegt z. B. DIN 15018 und DIN 4132 zugrunde, HAIBACH [29].

Der zweite in jüngerer Zeit, z. B. im Flugzeugbau [14a] und im Bereich der Forschung, beschrittene Weg besteht darin, unter Verwendung bruchmechanischer Ansätze die je Lastspiel auftretenden Rißverlängerungen aufzuintegrieren, um so die bei Erreichen der kritischen Rißlänge sich ergebende Grenzlastspielzahl zu erhalten [14], [54]. Das Verfahren benötigt die Anfangsrißlänge a_0.

Für die Ingenieurpraxis benötigt man allerdings ein Schadensakkumulationsgesetz, das möglichst entsprechend der PALMGREN-MINER-Hypothese handhabbar ist. Man hat daher auch für Beton- und Spannstähle, für ausgewählte Stahlbeton- und Spannbetonbauteile und für unbewehrte Betone die PALMGREN-MINER-Hypothese angewendet bzw. überprüft, s. z. B. HILSDORF/KESLER [2], TEPFERS [22], WEIGLER [52], KÖNIG/GERHARDT [57], MÜLLER u. Mitarb. [63] u. a.

Es wird festgestellt, daß die PALMGREN-MINER-Hypothese in vielen Fällen offenbar zufriedenstellende Ergebnisse liefert, in einer Reihe anderer aber nicht (SCHÜTZ [14], SCHIJVE [8], HAIBACH [53] oder [2], [18], [63]). Bei den Gründen ist man bisher auf Mutmaßungen angewiesen, da ein geschlossener, physikalisch begründeter Nachweis bekanntlich noch nicht vorlag.

Im folgenden soll nun allein auf der Basis bruchmechanischer Messungen und Überlegungen entsprechend Abschnitt 3 ein Schadensakkumulationsgesetz unterbreitet werden, das die PALMGREN-MINER-Hypothese als Sonderfall enthält und deren Grenzen aufzeigt. Ausgegangen wird entsprechend den bisherigen einschlägigen Arbeiten von Rißfortschrittsmessungen unter dynamischer Belastung (vgl. die Übersichten in [12], [14], [28], [40], [54], [62]).

In Bild 8 sind Ergebnisse solcher Messungen zusammengestellt.

Diese Rißfortschrittsdaten können entsprechend Gl. (1) für beliebig wechselnde Oberspannung und konstante Unterspannung oder näherungsweise für beliebige Spannungsamplituden bei einem vorgegebenen Mittelspannungsbereich durch die PARIS-Gleichung [1]

$$\frac{da}{dn} = C \cdot \Delta K_{(n)}^m \tag{22}$$

beschrieben werden.

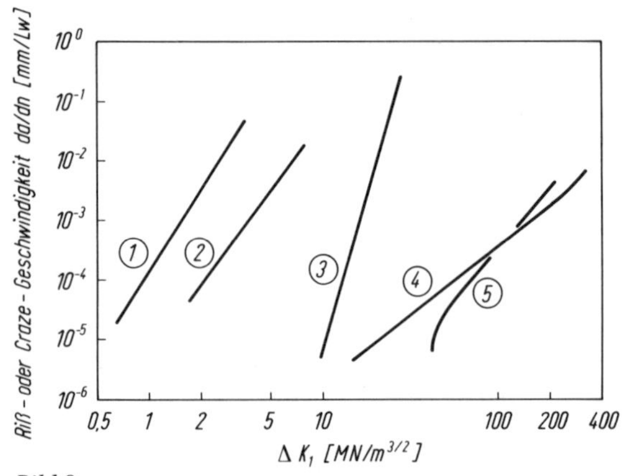

Bild 8
Ergebnisse von Rißfortschritts- bzw. Crazefortschrittsmessungen bei dynamischer Zugbelastung
① Polycarbonat [62]
② Polyamid [62]
③ GFK-Mattenlaminat [59]
④ Baustähle [54]
⑤ Baustähle [54]

n ist darin die jeweilige Lastspielzahl, m der stoffabhängige Exponent. Analog zu den Beziehungen (2) und (3) seien weiterhin auf der Basis des K_c-Konzepts die Abhängigkeiten

$$\Delta K_{(n)} = (\sigma - \sigma_u) \cdot y \cdot \pi^{1/2} \cdot a_{(n)}^{1/2} \tag{23}$$

und

$$\Delta K_c = (\sigma_{\lim} \cdot a_0^{1/2} - \sigma_u \cdot a_{(n)}^{1/2}) \cdot y \cdot \pi^{1/2} \tag{24}$$

gültig. Nach Einsetzen von Gl. (23) in (22) erhält man die Differentialgleichung

$$\frac{da}{dn} = C\,[(\sigma - \sigma_u) \cdot y \cdot \pi^{1/2}]^m \cdot a_{(n)}^{m/2} \tag{25}$$

Als Lösung von Gl. (25) kann entsprechend Gl. (4) die relative Rißverlängerung in Abhängigkeit von der Lastspielzahl angegeben werden:

$$\frac{a_{(n)}}{a_0} = (1 - D \cdot \beta^m \cdot n)^{-\frac{2}{m-2}} \tag{26}$$

für $m > 2$

Hierin ist

$$D = \left(\frac{m-2}{2}\right) \cdot C \cdot (y \cdot \pi^{1/2})^m \cdot a_0^{\frac{m-2}{2}} \cdot (\sigma_{\lim} - \sigma_u)^m, \tag{27}$$

$$\beta = \frac{\sigma - \sigma_u}{\sigma_{\lim} - \sigma_u}$$

Der Verlauf der Rißverlängerung entspricht bei großen m-Werten dem Verlauf in Bild 4. Meßwerte für die Schadensentwicklung bei Beton sind zum Vergleich in Bild 9 eingezeichnet.

Die Grenzlastspielzahl N in Abhängigkeit von der vorgegebenen relativen Schwingbreite β, von der relativen Oberspannung $\beta_o = \sigma/\sigma_{\lim}$ und von dem Werkstoff- bzw. Bauteilkennwert D, der aus Einstufenversuchen entnehmbar ist, erhält man über Gl. (26) zu

$$N_{(\beta)} = \frac{1}{D \cdot \beta^m} \left(1 - \beta_o^{m-2}\right) \tag{28}$$

Die Beziehung (28) stellt demnach die allgemeine Gleichung für einstufig ermittelte WÖHLERlinien dar für Stoffe, deren Rißausbreitung durch Gl. (22) beschrieben werden kann.

Entsprechend den Ausführungen in Abschnitt 3 ist (28) für kleinere m-Werte, wie sie z. B. für Schweißverbindungen vorkommen, im doppelt-logarithmischen Maßstab analog zu Bild 3 nicht mehr linear. Hierdurch läßt sich z. B. die von HAIBACH [18] angesprochene bisherige Diskrepanz zwischen den Steigungen der WÖHLERlinien für Schweißverbindungen und den zugehörigen Rißfortschrittskurven erklären. Die Schadensakkumulation infolge unterschiedlicher Schwingspiele n_i mit den zugehörigen (relativen) Schwingbreiten

$$\beta_i = \frac{\sigma_i - \sigma_u}{\sigma_{\lim} - \sigma_u} \tag{28.1}$$

kann entsprechend Gl. (19) auf eine als konstant angenommene Bezugsschwingbreite $\bar{\beta}$ umgerechnet werden.

Für diese Bezugsschwingbreite gilt dann

$$\sum_i \left(\frac{\beta_i}{\bar{\beta}}\right)^m \cdot n_{(\beta i)} + n_{R\,(\bar{\beta})} = N_{(\bar{\beta})} \tag{29}$$

hierin ist

$N_{(\bar{\beta})}$ die Grenzlastspielzahl für einstufige Beanspruchung mit der Schwingbreite $\bar{\beta}$,

$n_{R\,(\bar{\beta})}$ die ertragbare Restlastspielzahl.

Unter sinngemäßer Verwendung von Gl. (19a) erhält man für den Grenzfall $n_{R\,(\bar{\beta})} = 0$ und unter Benutzung von n_i statt $n_{(\beta i)}$ das folgende Schadensakkumulationsgesetz:

$$\boxed{\sum_i \frac{n_i}{N_i} \left(\frac{1 - \beta_{o\,i}^{m-2}}{1 - \bar{\beta}_o^{m-2}}\right) = 1} \tag{30}$$

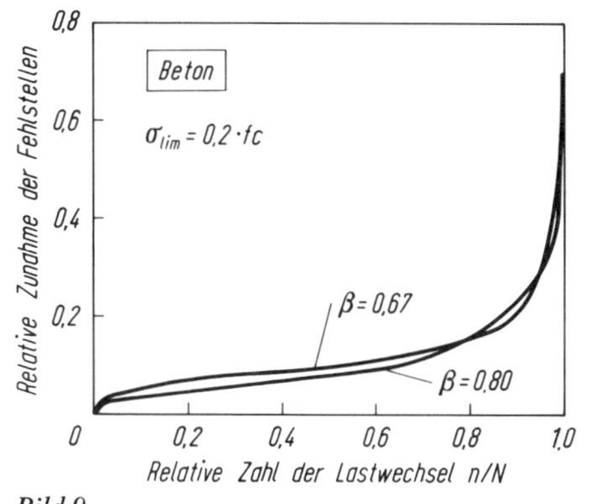

Bild 9
Fehlstellenentwicklung bei einstufiger dynamischer Druckbeanspruchung von Beton, Meßwerte von KLAUSEN [26]

Das auf den bruchmechanischen Voraussetzungen basierende Schadensakkumulationsgesetz (30) beinhaltet als Sonderfall für ausreichend große m die PALMGREN-MINER-Hypothese

$$\sum_i \frac{n_i}{N_i} = 1 \tag{31}$$

Auch in den Fällen, in denen die PALMGREN-MINER-Hypothese (31) zutreffend angewendet werden kann, ist die Schadensakkumulation – entgegen der Aussage in einer Reihe von Veröffentlichungen – nicht linear!

Betrachtet man die vorliegenden Ergebnisse von Dauerschwingversuchen für unbewehrten Leichtbeton und Normalbeton z. B. von KLAUSEN [26], TEPFERS [37] bzw. WEIGLER [52], so ergeben sich aus den WÖHLERlinien die Exponenten m zu mindestens ca. 20, so daß für diese Stoffe die vereinfachte Beziehung (31) anwendbar ist.

Die Exponenten m für bestimmte Natursteine (Granite) betragen nach KIM/MUBEEN [44] dagegen ca. 12; für GFK (Mattenlaminate) werden Werte zwischen ca. 9 und 13 angegeben [59], für Polyamid z. B. zwischen ca. 5 und 9 [51], für die Mehrzahl der Metalle zwischen ca. 2 und 10 [40], für geschweißte Baustähle bzw. Stahlkonstruktionen Werte zwischen ca. 3 und 5 [54]. Für die genannten Stoffe und zugehörigen Bauteilverbindungen mit relativ kleinen m-Werten muß demnach die PALMGREN-MINER-Hypothese zu abweichenden Schadenssummen führen.

Das Schadensakkumulationsgesetz (30) zeigt, daß die Wirkung der Schadensanteile in einem bestimmten Maß von der Bezugsschwingbreite bzw. der Reihenfolge der Schwingbreiten abhängt, vgl. Bild 10. In diesem Bild sind zum Vergleich die Schadenssummen dar-

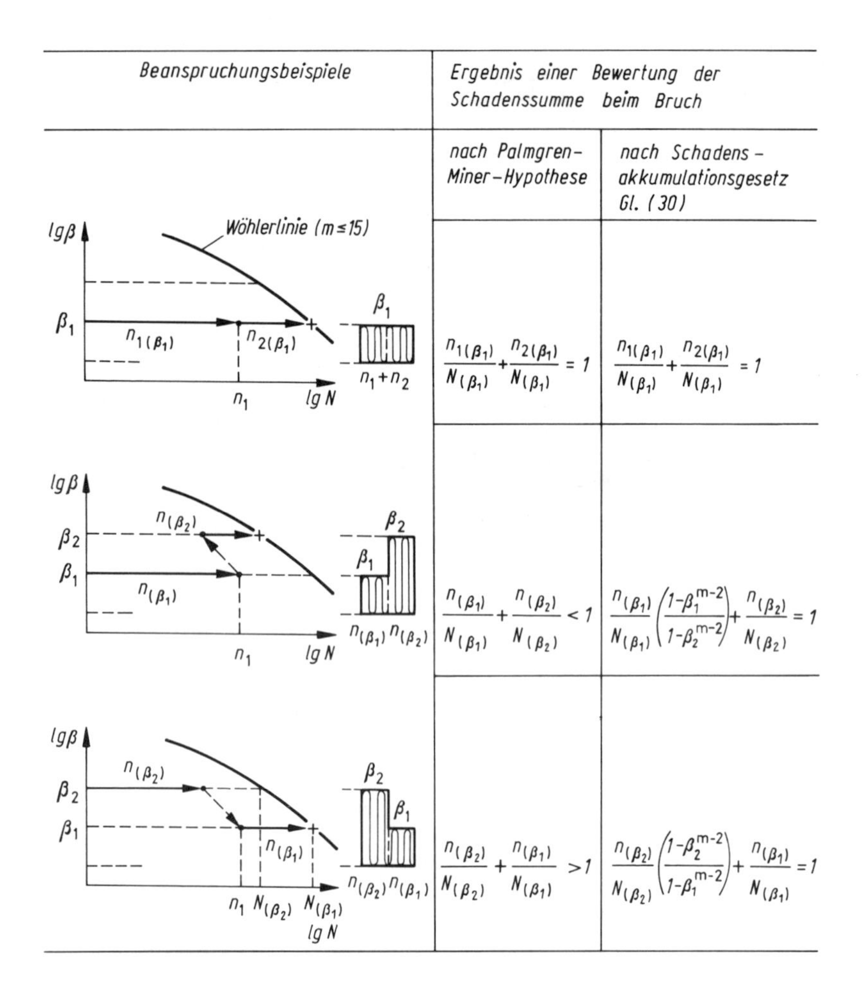

Bild 10
Vergleich der Aussagen der Palmgren-Miner-Hypothese und des Schadensakkumulationsgesetzes Gl. (30) zur Schadenssumme beim Bruch für ausgewählte Beanspruchungsbeispiele mit $m \leq 15$

gestellt, die sich für eine vorgegebene WÖHLERlinie und das dargestellte Kollektiv nach dem Schadensakkumulationsgesetz Gl. (30) und der PALMGREN-MINER-Hypothese ergeben.

Eine Schwingbeanspruchung aus mehreren Teilfolgen mit abnehmender Größe der Schwingbreiten muß danach zu einer größeren ertragbaren Schadenssumme führen als bei umgekehrter Reihenfolge der Teilfolgen. Die Versuchsergebnisse z. B. von SCHIJVE [8] an einem metallischen Werkstoff bestätigen dies deutlich.

Bei einer Betriebsbeanspruchung mit ausreichend häufig wiederkehrendem Kollektiv ist dagegen die Reihenfolge der Schwingbreiten von untergeordnetem Interesse. Man erhält hier die Schadenssumme ausreichend genau, wenn in Gl. (30) für $\bar{\beta}_o$ die maximale (relative) Oberspannung $\beta_{o\max}$ des Kollektivs eingesetzt wird (vgl. die Ausführungen zu Gl. (34)). Bei Anwendung des Schadensakkumulationsgesetzes Gl. (30) auf Betriebsbeanspruchungen findet daher eine höhere Wichtung der Schadensanteile der geringeren Schwingbreiten statt. Vorhersagen mit Hilfe der PALMGREN-MINER-Hypothese weichen daher um so mehr ab im Sinne einer Überbewertung der Lebensdauer, je mehr sich die Summenhäufigkeit des Kollektivs der Dreiecksform nähert.

Das grundsätzliche Problem der Bewertung der Kollektivanteile, die sich im Dauerfestigkeitsbereich der einstufigen WÖHLERlinie befinden, bleibt natürlich auch bei Verwendung von Gl. (30) erhalten. In den Schweizer und z. T. in den britischen Normvorschlägen wird eine nach unten unbegrenzte Zeitfestigkeits- bzw. WÖHLERlinie unterstellt (vgl. HIRT [19], HAIBACH [29]).

Zu diesem Fragenkomplex werden z. Z. an vielen Stellen Betriebsfestigkeits- und bruchmechanische Messungen durchgeführt.

Statt für vorgegebene Kollektive Lebensdauerlinien anzugeben, kann man versuchen, diese Kollektive durch eine äquivalente Beanspruchung konstanter Schwingbreite zu ersetzen und diese jeweils den einstufig ermittelten WÖHLERlinien gegenüberzustellen (HIRT [19], NIESER [46]). Auf der Basis des vorliegenden

bruchmechanischen Konzepts ergibt sich dabei folgendes:
Bei Vorgabe einer Bezugsschwingbreite $\bar{\beta}$ erhält man zunächst unter Benutzung von Gl. (29) aus der Betriebsbeanspruchung die zugehörige äquivalente Schwingspielzahl zu

$$n_{(\bar{\beta})} = \sum_i \left(\frac{\beta_i}{\bar{\beta}}\right)^m \cdot n_{(\beta i)} \tag{32}$$

Fordert man nun, daß die Schwingspielzahl $n_{(\bar{\beta})}$ der Summe der wirklichen Schwingspiele entspricht, also

$$n_{(\bar{\beta})} = \sum_i n_{(\beta i)}$$

so erhält man aus Gl. (32) die gesuchte äquivalente Schwingbreite zu

$$\bar{\beta} = \left(\frac{\sum_i \beta_i^m \cdot n_i}{\sum_i n_i}\right)^{\frac{1}{m}} \tag{33}$$

Setzt man darin statt $\bar{\beta}$ und βi die entsprechenden Spannungsdifferenzen (28.1) ein sowie $\Delta\bar{\sigma} = \bar{\sigma} - \sigma_u$ und $\Delta\sigma_i = \sigma_i - \sigma_u$, folgt

$$\Delta\bar{\sigma} = \left(\frac{\sum_i \Delta\sigma_i^m \cdot n_i}{\sum_i n_i}\right)^{\frac{1}{m}} \tag{33a}$$

Die so erhaltene Beziehung entspricht der in [24] oder z. B. von Fischer, Hirt [19] angegebenen und für die Schweizer Normung vorgeschlagenen Beziehung (Delta-Sigma-Konzept). Von Herzog [23] wird dieselbe Beziehung mit dem Exponenten $m = 4$ vorgeschlagen zur Berechnung einer Ersatzmomentenbeanspruchung für Stahlbeton- und Spannbetonträger, die einer Mehrstufenbeanspruchung ausgesetzt sind. Die für die äquivalente Schwingbreite $\Delta\bar{\sigma}$ bzw. für $\bar{\beta}$ über die zugehörige Wöhlerlinie berechnete Grenzlastspielzahl stimmt aber nur dann mit der Grenzlastspielzahl Der entsprechenden einstufigen Beanspruchung überein, wenn der Exponent m ausreichend groß ist, vgl. die Ausführungen zu Gl. (30). Für kleine m-Werte liegt man unter Verwendung von Gl. (33) zunächst auf der unsicheren Seite.
Die bei Einwirkung des tatsächlichen Kollektivs erreichbare Grenzlastspielzahl n_{Grenz} liegt zwischen folgenden Grenzen:

$$N_{\min(\Delta\bar{\sigma})} \leq n_{\text{Grenz}} \leq (N_{\min(\Delta\bar{\sigma})} + n_{\text{Koll}}),$$

$$N_{\min(\Delta\bar{\sigma})} = N_{(\Delta\bar{\sigma})} \cdot \left(\frac{1 - \beta_{o\max}^{m-2}}{1 - \bar{\beta}_o^{m-2}}\right) \tag{34}$$

hierin ist

$N_{(\Delta\bar{\sigma})}$ die Grenzlastspielzahl aus der zugehörigen Wöhlerlinie für einstufige Beanspruchung mit $\Delta\bar{\sigma}$ bzw. $\bar{\beta}$
n_{koll} die Lastspielzahl der Teilkollektive
$\beta_{o\max}$ die maximale (relative) Oberspannung des Kollektivs

Für ausreichend kleine Teilkollektive oder häufige Wiederholung der Spitzenspannungen kann angenommen werden:

$$n_{\text{Grenz}(\Delta\bar{\sigma})} = N_{(\Delta\bar{\sigma})} \cdot \left(\frac{1 - \beta_{o\max}^{m-2}}{1 - \bar{\beta}_o^{m-2}}\right) \tag{35}$$

Für ausreichend große m wird in Gl. (35)

$$n_{\text{Grenz}(\Delta\bar{\sigma})} = N_{(\Delta\bar{\sigma})}.$$

5 Schadensakkumulation bei kombinierter statischer und dynamischer Beanspruchung

Auf der Basis des vorliegenden Konzeptes erscheint es möglich, die akkumulative Wirkung von zusammen einwirkenden statischen und dynamischen Beanspruchungen auf die Tragfähigkeit von bestimmten Werkstoffen zu erfassen. Von Interesse wäre dies für Anwendungen von Kunststoffen, Verbundwerkstoffen, keramischen Stoffen, Gläsern und einigen mineralischen Stoffen. Hierüber soll an anderer Stelle berichtet werden.

6 Zusammenfassung

Es wird auf der Basis des K_c-Konzeptes der Bruchmechanik unter Verwendung der Ergebnisse von Craze- und Rißfortschrittsmessungen ein Schadensakkumulationsgesetz unterbreitet, das zur Vorhersage der Lebensdauer verschiedener Stoffe unter statischer und dynamischer Beanspruchung herangezogen werden kann. Unter anderem wird die bisher empirisch begründete Palmgren-Miner-Hypothese als Sonderfall des vorgeschlagenen Akkumulationsgesetzes ausgewiesen.

7 Literatur

[1] PARIS, P. C.: The fracture mechanics approach to fatigue. Proc. Tenth Sagamore Army Mat. Research Conf., Syracuse Univ. Press, 1964, 107.

[2] HILSDORF, H. K. und KESLER, C. E.: Fatigue strength of concrete under varying flexural stresses. ACI-J., 1966, 1059.

[3] BIERELT, G.: Über die Betriebsfestigkeit von geschweißten und genieteten Stahlverbindungen. Stahl u. Eisen 1967, 1465.

[4] HAIBACH, E.: Abhängigkeit der ertragbaren Spannungen schwingbeanspruchter Schweißverbindungen vom Beanspruchungskollektiv. Ber. FB-79 (1968) des Lab. f. Betriebsfestigkeit, Darmstadt.

[5] HAIBACH, E. und GASSNER, E.: Modifizierte lin. theor. Schadensakkumulations-Hypothese zur Berücksichtigung des Dauerfestigkeitsabfalls mit forschreitender Schädigung. TM Nr. 50/70, Lab. f. Betriebsfestigkeit, Darmstadt 1970.

[6] DUBUC, J., THANG, B. Q. und BAZERGIN, BIRON: Unified theory of cumulative damage in metal fatigue. WRC Bulletin 162, 1971.

[7] SCHÜTZ, W.: The fatigue life under three different load spectra – tests and calculation. Agard-CP-118, Symp. on random load fatigue, Danmark 1972.

[8] SCHIJVE, J.: Effect of load sequences on crack propagation under random and program loading. Eng. Fract. Mech. 1983, 269.

[9] GRIESE, F. W.: Über die Bedeutung der lebensdauerorientierten Dimensionierung von Bauteilen. VDI-Berichte Nr. 215, 1974.

[10] WITTMANN, F. und ZAITSEW, J.: Verformung und Bruchvorgang poröser Baustoffe bei kurzzeitiger Belastung und Dauerlast. Schriftenreihe des DAfStb, H. 232, 1974.

[11] Abeles Symposium „Fatigue of Concrete“, ACI-SP 41, Detroit 1974.

[12] WIEDERHORN, S. M.: Subcritical crack growth in ceramics, in „Fracture Mechan. of Ceramics“, New York 1974, Vol. 2.

[13] KÖRNER, CH.: Zur Betriebsfestigkeitsber. von Spannbetonquerschnitten. Bauplanung – Bautechnik, 1975, 293.

[14] SCHÜTZ, W.: Lebensdauer-Berechnung bei Beanspruchungen mit beliebigen Last-Zeit-Funktionen. VDI-Berichte Nr. 268, 1976.

[14a] SCHÜTZ, W.: Berechnung des Rißfortschritts bei schwingender Beanspruchung, in Angewandte Bruchmechanik. Ber. Symp. Bad Neuenahr, Verlag TÜV Rheinland, 1976.

[15] HAHN, H. G.: Bruchmechanik. Teubner 1976.

[16] HERTZBERG, R. W.: Deformation and fracture mechanics of engineering materials. 1976.

[17] KAUSCH, H. H.: Energetische Überlegungen zur Rißfortpflanzung in Thermoplasten. Kunststoffe, 1976, 538.

[18] HAIBACH, E.: Fragen der Schwingfestigkeit von Schweißverbindungen in herkömmlicher und in bruchmechanischer Betrachtungsweise. Schweißen und Schneiden, 1977, 140.

[19] HIRT, M. A.: Neue Erkenntnisse auf dem Gebiet der Ermüdung und deren Berücksichtigung bei der Bemessung von Eisenbahnbrücken. Bauingenieur, 1977, 255.

[20] ARGON, A. S. und SALAMA, M. M.: Growth of crazes in plassy polymers. Philosophical Magaz., 1977, 1217.

[21] HILLEMEIER, B. und HILSDORF, H. K.: Fracture mechanics studies on concrete compounds. Cem. + Concrete Research, 1977, 523.

[22] TEPFERS, R., FRIDEN, C. und GEORGSSON, L.: A study of the applicability to the fatigue of concrete of the Palmgren-Miner partial damage hypothesis. Magaz. of Concr. Res., 1977, 123.

[23] HERZOG, M.: Betriebsfestigkeit von Stahlbeton, Spannbeton und teilweise vorgespanntem Beton. Die Bautechnik, 1977, S. 73.

[24] SCHILLING, C. G., KLIPPSTEIN, K. H., BARSOM, J. M. und BLAKE, G. T.: Fatigue of welded steel bridge members under variable-amplitude loadings. Nat. Coop. High way Research Progr., Report 188, Washington, D. C., 1978.

[25] KAUSCH, H. H.: Polymer Fracture, Springer 1978.

[26] KLAUSEN, D.: Festigkeit und Schädigung von Beton bei häufig wiederholter Beanspruchung. Dissert. TH Darmstadt, 1978.

[27] SWARTZ, S. E., HU, K. K. und JONES, G. L.: Compliance monitoring of crack growth in concrete. ASCE Vol. 104, Aug. 1978, 789.

[28] RITTER, J. E. Jr.: Engineering design and fatigue failure of brittle materials, in „Fract. Mech. of Ceramics“, R. C. Brandt, 1978, Vol. 4.

[29] HAIBACH, E.: Grundlagen und Weiterentwicklung des Betriebsfestigkeitsnachweises für Schweißverbindungen im internationalen Regelwerk. DVS Düsseldorf, Vorträge Tagung Hamburg 1979.

[30] RUSSWURM, D. und REHM, G.: Dauerschwingfestigkeit (Betriebsfestigkeit) von Betonstahlmatten. Betonwerk + Fertigteiltechnik, 1979, 169.

[31] RADHAKRISHNAN, V. M.: Parameter representation of fatigue crack growth. Engin. Fract. Mech. 1979, 359.

[32] OKAMURA, H. und SAKAI, S.: Cumulative fatigue damage under random loads. Fatigue of Engin. Mat. + Sc., 1979, 409.

[33] KAUSCH, H. H.: Schädigungsprozesse in Kunststoffwerkstoffen. Material u. Technik, 1979, 111.

[34] BRÜLLER, O. S.: Theoretische Untersuchungen zum Kriechverhalten und Kriechversagen von Kunststoffen. Dissert. TH Aachen, Inst. f. Kunststoffverarb., 1979.

[35] MENGES, G.: Werkstoffkunde der Kunststoffe. Hanser 1979.

[36] HILSDORF, H. K.: Sinn und Grenzen der Anwendbarkeit der Bruchmechanik in der Betontechnologie. Kordina-Festschrift, Ernst + Sohn 1979.

[37] TEPFERS, T. und KUTTI, TH.: Fatigue strength of plain, ordinary and light-weight concrete, ACI-J., 1979, 635.

[38] SIEBKE, H.: Beschreibung einer Bezugsbasis zur Bemessung von Bauwerken auf Betriebsfestigkeit. Schweißen und Schneiden, 1980, 304.

[39] HARRE, W. und NÜRNBERGER, U.: Zum Schwingfestigkeitsverhalten von Betonstählen unter wirklichkeitsnahen Beanspruchungsbedingungen. Schriftenreihe des OGI – Stuttgart, 1980.

[40] SCHWALBE, K.-H.: Bruchmechanik metallischer Werkstoffe. C. Hanser, 1980.

[41] ASTM-745 Fracture mechanics for ceramics, rocks and concrete, 1980.

[42] HERTZBERG, R. W. und MANSON, J. A.: Fatigue of engineering plastics, 1980.

[43] RÜBBEN, A.: Versuchsmäßige Überprüfung der Begriffe Kriechlast und Restfestigkeit von Kunststoffkonstruktionen nach theoretischen Voraussagen. Bauingenieur (1980), 69.

[44] KIM, K. und MUBEEN, A.: Relationship between differential stress intensity factor and crack growth rate under cyclic tension in westerly granite. Fracture mechanics for ceramics, rocks and concrete, in ASTM-STP 745, 1980.

[45] GULATI, S. T.: Crack kinetics during static and dynamic loading. J. of Non-crystalline Solids, 1980, 475.

[46] NIESER, H.: Der Nachweis der Betriebsfestigkeit auf der Grundlage der Schadensakkumulation. Mitteilungen des IfBt, 1/1981.

[47] EGGERT, H. und SCHNEIDER, R.: Zum Nachweis der Betriebsfestigkeiten bei Bauwerken. Mitteilungen des IfBt, 3/1981.

[48] KOSTEAS, D. und GRAF, U.: Lebensdauervoraussage von Aluminiumkonstruktionen durch bruchmechanische Konzepte. 7. ILMT Leoben/Wien, 1981.

[49] PARKER, A. P.: The mechanics of fracture and fatigue, 1981.

[50] MENGES, G. und BIELING, U.: Untersuchung der Restfestigkeit von GF-UP-Mattenlaminaten unter Feuchte- und Temperatureinfluß. Bericht der AVK-Tagung 1981, Freudenstadt.

[51] SEFERIS, J.C. und NICOLAIS, L.: The role of the polymeric matrix in the processing and structural properties of composite materials. New York, 1981.

[52] WEIGLER, H.: Beton bei häufig wiederholter Beanspruchung. Betontechn. Berichte, Beton 1981, 189.

[53] HAIBACH, E.: Fatigue Data for Design Applications, in Materials, Experimentation and Design in Fatigue. Proc. of Fatigue '81, Soc. of Environmental Eng.

[54] KOSTEAS, D.: Grundlagen für Betriebsfestigkeitsnachweise. Stahlbau-Handbuch 1, 1982.

[55] IVBH-Kolloquium „Ermüdungsverhalten von Stahl- und Betonbauten". Vortragsband, Lausanne 1982.

[56] AVESTON, J. und SILLWOOD, J.M.: Long term strength of glass-reinforced plastics in dilute sulphuric acid. J. of Mat. Sc., 1982, 3491.

[57] KÖNIG, G. und GERHARDT, H.C.: Nachweis der Betriebsfestigkeit gemäß DIN 4212. Beton- und Stahlbetonbau 1982, 12.

[58] SAKAGUCHI, S. und SAWAKI, Y.: Delayed failure in silica glass. J. of Mater Sc., 1982, 2878.

[59] WANG, S.S., CHIM, E.S.M. und ZAHLAN, N.M.: Fatigue crack propagation in random short-fiber SMC-composites. J. of Comp. Mat., May 1983, 250.

[60] WANG, S.S. und CHIM, E.S.M.: Fatigue damage and degradation in random short-fiber SMC-composites. J. of Comp. Mat., March 1983, 114.

[61] PRIORI, A., NICOLAIS, L. und DIBENEDETTO, A.T.: The kinetics of surface craze growth in polycarbonate exposed to normal hydrocarbons. J. of Mat. Sc., 1983, 1466.

[62] KINLOCH, A.J. und YOUNG, R.J.: Fracture behaviour of polymers. London, 1983.

[63] MÜLLER, P., KEINTZEL, E. und CHARLIER, H.: Dynamische Probleme im Stahlbetonbau. Schriftenr. des DAfStb H. 342, 1983.

[64] HIBINO, Y., SAKAGUCHI, S. und TAJIMA, Y.: Crack growth in vitreous silica under dynamic loading. J. of Mat. Sc., 1983, 388.

Analyse des Betonstahlmarktes und Gedanken zur Gütesicherung in der Bundesrepublik Deutschland

Hans Peter Killing

Unabhängig von der allgemeinen weltweiten Stahlkrise hat kein anderes Stahlerzeugnis in den letzten Jahren so viele Schlagzeilen gemacht wie der Betonstahl. Er scheint ein gewisses Eigenleben zu führen. Die Gründe dafür sind vielfältig.
Im Folgenden wird versucht, anhand einer Analyse der Produzenten und ihrer Produktionsbedingungen sowie der Verhältnisse im Markt den Ursachen dieses Eigenlebens nachzugehen. Dabei wird sich die Untersuchung ausschließlich auf den Betonstabstahl beziehen. Andere Bewehrungsstähle werden nur vergleichsweise behandelt, d. h. sie werden nur in ihrem mengenmäßigen Verhältnis zum Betonstabstahl herangezogen.
Ebenso verzichtet wird auf eine eingehende Analyse der Nachfrageseite, d. h. Struktur und Bedarfslage der Bauindustrie. In den nachfolgenden Ausführungen werden die Bezüge der Bauindustrie als Faktum angesehen, und es wird nicht die Frage untersucht, ob das, was als Nachfrage im Markt anzutreffen ist, wirklich optimal ist und nicht vielleicht zu verbessern wäre.

1 Struktur der Produzenten

Traditionell wurde der Betonstahl wie jedes andere Walzstahlerzeugnis in integrierten Hüttenwerken produziert oder in selbständigen reinen Walzwerken, die Vormaterial von Hüttenwerken kauften oder Nutzeisen, wie z. B. ausrangierte Schienen, in Betonstahl umwandelten. Die Produktionslinie der Hüttenwerke umfaßt Hochofen (Roheisen) – Stahlwerk (Rohstahl) – Walzwerk (Betonstahl). Das Walzen des Betonstahls erfolgt entweder auf reinen Betonstahlstraßen oder Stabstahlstraßen; die dünnen Abmessungen (bis 10 mm) werden auch auf Drahtstraßen gewalzt. Mitte der 60er Jahre kamen alternativ zur herkömmlichen Produktionsweise sogenannte Mini-Stahlwerke in Betrieb, die sich im wesentlichen in zwei Positionen von den integrierten Werken unterscheiden:

- in der Produktionslinie,
- in der Beschränkung auf die Herstellung von ein oder zwei Walzerzeugnissen.

Die Stahlerzeugung dieser Werke erfolgt direkt in Elektro-Öfen unter Verzicht auf die Roheisen-Vorstufe. Statt Roheisen wird zur Stahlerzeugung Schrott und/oder Pellets in die Elektro-Öfen eingesetzt. Daraus ergibt sich ein nachhaltiger Kostenvorteil dieser Produktionsweise gegenüber der herkömmlichen: Schrott, den die Mini-Stahlwerke als Einsatzmaterial verwenden, ist in der Regel wesentlich billiger als Roheisen, erschmolzen aus Erz mit Koks im Hochofen, das die integrierten Werke zu Stahl umwandeln.
Dabei ist jedoch zu berücksichtigen, daß das Kostenverhältnis der beiden Verfahren nicht konstant ist. Schrott einerseits und Erz wie Koks andererseits haben unterschiedliche Preisbewegungen. Schrottpreise sind in Baissezeiten wegen der geringen Nachfrage niedrig im Verhältnis zum Erz-/Roheisenpreis. Sie haussieren, wenn der Stahlmarkt umschlägt, und können dann im Mini-Stahlwerkbetrieb zu höheren Kosten führen als im integrierten Hüttenwerk. Solche Haussezeiten mit internationaler Preisausgleichskasse für Schrott zur Herabschleusung des Schrottpreises hat es zwar vor vielen Jahren gegeben, die Prognosen für die Entwicklung des Stahlmarktes deuten jedoch nicht darauf hin, daß sich dieser Fall wiederholen wird.
Abgesehen davon gibt es für die Mini-Stahlwerke inzwischen eine Möglichkeit, bei Schrottknappheit auf Erz auszuweichen durch den Einsatz von Pellets zur Direktreduktion.
Die Mini-Stahlwerke hatten in den ersten Jahren ihres Entstehens einen weiteren Vorteil dadurch, daß der flüssige Stahl im Stranggußverfahren unmittelbar zu Vormaterial für die Betonstahlstraße vergossen wurde, während die integrierten Hüttenwerke noch im traditionellen Verfahren arbeiteten, d.h. der Rohstahl wurde zu Blöcken vergossen und diese mußten zunächst in einer Halbzeugstraße zu brauchbarem Vor-

material für die Betonstahlstraße umgewandelt werden. Der Nachteil gegenüber den Mini-Stahlwerken bestand nicht nur in der zusätzlichen Walzstufe, sondern auch in einem wesentlich geringeren Materialausbringen. Inzwischen arbeiten jedoch auch die integrierten Werke durchweg mit Strangguß, der direkt der Betonstahlstraße zugeführt wird, so daß dieser Nachteil mehr und mehr entfallen ist.
Ein weiterer Kostenvorteil der Mini-Stahlwerke besteht in der Beschränkung ihrer Produktionsprogramme. Im Normalfall stellen sie entweder nur Betonstahl her und bieten dazu Halbzeug ihrer eigenen Vorstufe an, oder sie produzieren neben Betonstahl noch Erzeugnisse, die man auf gleichen Typen von Walzstraßen herstellen kann, d.h. Stabstahl in kleineren Abmessungen oder Walzdraht, da die dünnen Betonstahlabmessungen meist ohnehin auf Drahtstraßen gewalzt werden. Eine solche Begrenzung des Produktionsprogramms hat niedrige Investitionskosten, d.h. einen niedrigeren Kapitaldienst zur Folge. Dagegen bieten integrierte Hüttenwerke normalerweise die gesamte Palette der Erzeugnisse an und müssen dadurch eine höhere Vorhaltekapazität finanzieren mit entsprechend höheren Kosten.
Es geht über den Rahmen dieser Studie hinaus, einen detaillierten Kostenvergleich anzustellen. Man kann aber ziemlich sicher davon ausgehen, daß der Kostenvorteil der Mini-Stahlwerke – gleiche Auslastung vorausgesetzt – gegenwärtig in der Größenordnung von 80 bis 100 DM/t Betonstahl liegt (ausgehend vom Schrottpreis von etwa 180 DM/t).
Dieser Kostenvorsprung hat dazu geführt, daß die Mini-Stahlwerke ihren Anteil an der Betonstahlproduktion in den letzten Jahren wesentlich erhöht haben: In der Europäischen Gemeinschaft lag ihr Anteil 1977 bei etwa 55% und stieg bis 1983 auf ca. 72%, in der BR Deutschland war der Anteil 1977 rd. 35%, dagegen 1983 bereits 60%.
Es ist aber durchaus denkbar, daß zwei Entwicklungen dem gegenwärtigen baissebedingten Schrottüberfluß ein Ende bereiten könnten:
- die weitere Zunahme des Stranggießens in den integrierten Hütten,
- eine Erhöhung des Schrotteinsatzes beim Rohstahlerschmelzen im Konverter.

Seit Einstellung der Roheisen-Schrottverfahren zur Stahlerzeugung (z.B. SM-Verfahren) ist die Schrottnachfrage der integrierten Hüttenwerke deutlich zurückgegangen, weil normalerweise der Schrottentfall im eigenen Werk in etwa ausreichte, um den verfahrensbedingten Schrottbedarf bei der Rohstahlerzeugung aus Roheisen im Konverter zu decken. Der Schrottentfall im eigenen Werk (Kreislaufschrott) vermindert sich aber und sinkt ggf. unter den verfahrensbedingten Schrottbedarf, wenn das Stranggußverfahren eingeführt wird. In diesem Fall muß Schrott zugekauft werden, um den Schrottbedarf zu decken. Bei einem im Verhältnis zum Roheisen sehr niedrigen Schrottpreis wie gegenwärtig versucht man möglichst viel Schrott einzusetzen, um dadurch die Kosten zu senken.
Für den Schrottverbrauch bei der Rohstahlherstellung aus Roheisen im Konverter gibt es bisher eine technische Grenze, deren nachhaltige Überschreitung nur durch zusätzlichen Investitionsaufwand möglich ist. Da aber seit Jahren die hohe Preisdifferenz zwischen Roheisen und Schrott anhält, besteht gegenwärtig ein erheblicher Anreiz, den Schrottverbrauch im Konverter weiter zu erhöhen. Wenn es den deutschen Werken gelingen sollte, den Schrottverbrauch im Konverter um etwa 50 kg/t Rohstahl, d.h. um 5% zu erhöhen, kann man davon ausgehen, daß es zu einer deutlichen Veränderung der Schrottmarktsituation in Deutschland kommt, d.h., daß der Schrottpreis steigen wird. Tritt ein solcher Zustand ein, bleibt den Mini-Stahlwerken im wesentlichen nur noch der Vorteil des niedrigen Kapitaldienstes aufgrund ihrer beschränkten Angebotspalette.

2 Herstellungsarten

Bei der Produktion von geripptem Betonstabstahl unterscheidet man zur Zeit zwischen
- unbehandelten, d.h. naturharten,
- kaltverformten
- und wasservergüteten Sorten.

Bei den *naturharten* Stählen wird die Festigkeit/Streckgrenze aufgrund der chemischen Zusammensetzung erreicht. Diese Stähle sind in der Regel nicht schweißbar; man kann aber durch Änderung der Gehalte der Begleitelemente eine Schweißeignung erreichen.
Kaltverformte Stähle wie z.B. der Torstahl werden nach dem Walzen einer Kaltverformung wie durch Verdrillen um die Längsachse (Torstahl), Ziehen oder Recken unterworfen. Dadurch erhält der Stahl die gewünschten Eigenschaften und ist ohne weitere Zusätze schweißgeeignet.

Wasservergüteter Stahl (Tempcore) wird seit 1976 hergestellt und hat in den letzten Jahren zunehmende Produktionsanteile zu verzeichnen. Es handelt sich dabei um ein Verfahren, bei dem der Stahl aus der Walzhitze gezielt abgeschreckt wird. Beim anschließenden Lagern auf dem Kühlbett wandert die im Stabinnern verbliebene Wärme in die Randschicht und bewirkt einen Anlaßeffekt. Dadurch bildet sich am Außenrand des Stabes ein verformungsfähiges und zugleich festes Vergütungsgefüge, während der Kern des Stabes weich, d.h. durch den Abschreckvorgang, unberührt bleibt. Derart hergestellter Betonstahl ist unbeschränkt schweißbar.

Alle Stähle werden beim Walzen mit Rippen versehen, durch die eine besondere Haftung im Beton erreicht wird. Die Anordnung der Rippen ist vorgeschrieben. Sie gibt Auskunft über die Gebrauchseigenschaften wie Festigkeit und Schweißeignung. Anhand eines Walzzeichens muß festgestellt werden können, in welchem Werk der Betonstahl gewalzt wurde.

Über die Anteile der einzelnen Verfahren an der deutschen Betonstahlproduktion liegen keine statistischen Angaben vor. Man kann jedoch für das Jahr 1983 etwa von folgenden Prozentanteilen ausgehen:

naturharter Stahl	36%
kaltverformter Stahl	29%
wasservergüteter Stahl	35%
	100%

Was die Zuordnung zu den Produzenten betrifft, so wird kaltverformter Stahl fast ausschließlich in integrierten Hüttenwerken hergestellt, während naturharter und wasservergüteter Stahl sowohl von integrierten als auch von Mini-Stahlwerken produziert wird.

3 Angebotene Sorten und Abmessungen

Ausgehend von den verschiedenen Herstellarten werden heute folgende Sorten angeboten:

- *unbehandelter, naturharter Stahl*
  III U bzw. BSt 420/500 RU
  III US bzw. BSt 420/500 RUS
  IV U bzw. BSt 500/550 RU
  IV US bzw. BSt 500/550 RUS
  d.i. gerippter Stahl (R)
  unbehandelt (U)
  schweißbar (S)
- *kaltverformter Stahl*
  III K bzw. BSt 420/500 RK (S)
  IV K bzw. BSt 500/550 RK (S)
  d.i. gerippter Stahl (R)
  kaltverformt (K)
  schweißbar (S)
- *wasservergüteter Stahl*
  III S bzw. BSt 420/500 RTS
  IV S bzw. BSt 500/550 RTS
  d.i. gerippter Stahl (R)
  wärmebehandelt – tempered (T)
  schweißbar (S)

Der Betonstahl IV hat, wie aus dem Kurznamen ersichtlich, eine wesentlich höhere Festigkeit als die Sorte III. Er wurde 1976 in Deutschland durch die Korf Stahl AG eingeführt. Bisher ist der Stahl IV von der Bauindustrie nicht in dem erwarteten Maße aufgenommen worden. Die Nachfrage hat sich nur auf bestimmte Objekte beschränkt. Bauplaner und Betonstahlverbraucher sind überwiegend auf den Stahl III eingestellt. Sie waren bisher nicht bereit, die Vorteile des Stahls IV zu nutzen, obwohl zum Beispiel der Preiszuschlag zum Stahl III geringer ist als der Preisvorteil infolge Gewichtseinsparung durch höhere Festigkeit.

Neben diesen Sorten findet man noch in geringem Umfang den Betonstahl I. Hierbei handelt es sich um einen glatten, also ungerippten Stahl geringer Festigkeit. 1960 hatte der Betonstahl I noch einen Marktanteil von einem Viertel, heute ist der Anteil fast Null.

Eine Weiterentwicklung des Betonstahls I war der Betonstahl II, ebenfalls ein glatter Stahl, der aber heute völlig vom Markt verschwunden ist.

Als Bewehrungsstähle sind in diesem Bereich noch zu erwähnen der Gewi-Stahl und der Spannstahl. Beides sind hochwertige Spezialerzeugnisse für bestimmte Verwendungszwecke, deren Marktanteil jedoch sehr klein ist.

Die Stähle III und IV umfassen rund 99% der deutschen Produktion an Betonstabstahl.

Was die Abmessungen betrifft, so werden derzeit bei Regellängen von 12 bis 14 m folgende Dicken angeboten:

6, 8, 10, 12, 14, 16, 18, 20, 22, 25, 28 mm.

Etwa die Hälfte der Produktion der deutschen Werke entfällt auf Stäbe bis 12 mm Durchmesser. Bei den dickeren Abmessungen (14–28 mm) hatten 1983 den höchsten Gewichtsanteil Stäbe mit 28 mm Durchmesser und den niedrigsten Stäbe mit 22 mm Durchmesser.

Wegen geringer Nachfrage werden aus Rationalisierungsgründen die Stäbe mit 18 mm Durchmesser und 22 mm Durchmesser in der neugefaßten DIN 488, Teil 2, nicht mehr enthalten sein.

4 Zugang zum Markt

Herstellung und Anwendung von Betonstahl unterliegen in Deutschland – wie auch in anderen Ländern – bestimmten öffentlich-rechtlichen Vorschriften.
Will ein Stahlproduzent Betonstahl im deutschen Markt anbieten, benötigt er eine amtliche Lieferberechtigung. Zuvor muß er nachweisen, daß er aufgrund seiner personellen und maschinellen Ausstattung in der Lage ist, bedingungsgemäßen Betonstahl zu produzieren. Nach erfolgreichem Nachweis wird ihm in einem Werkkennzeichenbescheid ein Werkzeichen zugeteilt. Die laufende Produktion des Betonstahls unterliegt einer Eigen- und Fremdüberwachung (DIN 488, Blatt 6, Ausgabe April 1972).
Während vor 1972 die Lieferberechtigung für den deutschen Markt über den relativ schwierigen Zulassungsweg geregelt werden mußte, brachte die Verabschiedung der DIN 488 danach für alle Lieferinteressenten leicht überschaubare Zugangswege zum deutschen Markt. Dabei erteilt Werkskennzeichen das Institut für Bautechnik (IfBt), Berlin, als Organ der obersten Baubehörden der Bundesländer, dem der Sachverständigenausschuß als beratendes Gremium zur Seite steht.
Im Laufe der Jahre hat die Zahl der Zulassungen zum Verkauf in Deutschland erheblich zugenommen, teils durch eigene Aktivität der ausländischen Produzenten, teils durch Aktivitäten deutscher Stahlhandlungen.
1976 gab es dreizehn inländische und fünfundzwanzig ausländische Betriebsstätten, deren Produktion zum Verkauf im deutschen Markt zugelassen waren. Heute haben nicht weniger als sechsundsechzig Produktionsstätten eine Zulassung, davon entfallen zweiunddreißig auf übrige EG-Länder und achtzehn auf Dritte Länder. Allein siebzehn italienische Werke sind daran beteiligt.
Zur Güteüberwachung sind bauaufsichtlich anerkannt sechzehn deutsche Prüfstellen, davon können elf Prüfstellen Erst- und Zulassungsprüfungen vornehmen.

5 Markt und Preise

Im Gegensatz zu den anderen Stahlerzeugnissen verläuft die Preisentwicklung beim Betonstahl weitaus hektischer. Das hat im wesentlichen drei Ursachen:
- die Vielzahl der Anbieter,
- die Vermarktung fast ausschließlich durch den Stahlhandel,
- die Preisbewegung beim Schrott.

Für viele ausländische Produzenten ist in den letzten

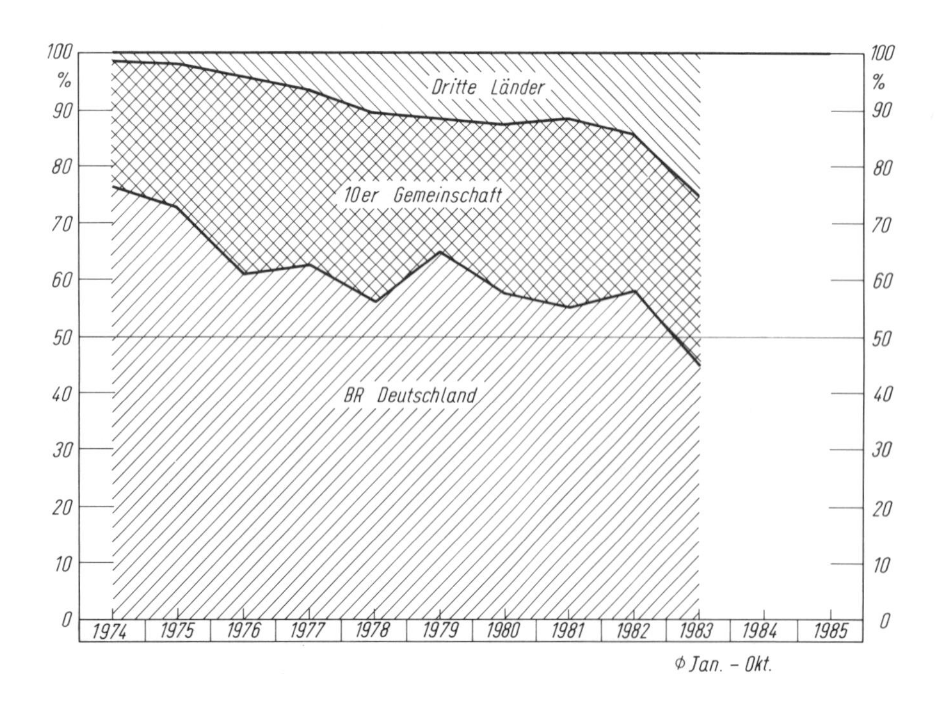

Bild 1
Marktversorgung der BR Deutschland mit Betonstabstahl in Prozentanteilen seit 1974
(Inland – EG-Importe – Drittlandimporte)

Jahren der deutsche Markt eine Art von Ausgleichsmarkt geworden: Wenn es im heimischen Markt Absatzschwierigkeiten gibt und die Preise im Drittländerexport unauskömmlich sind, liefert man nach Deutschland. Dort erlöst man immer noch einen besseren Deckungsbeitrag als im Export.
Die Veränderung der Marktanteile der Importe aus der übrigen EG und den Drittländern im Verhältnis zu den Lieferungen der heimischen Produzenten ist in Bild 1 dargestellt.
Seit 1974, dem Jahr der höchsten deutschen Rohstahlerzeugung, ist der Anteil der deutschen Werke an der Versorgung des deutschen Marktes rückläufig: 1974 lagen die Importe bei 23% und 1979 schon bei 35%. In 1983 lag der Marktanteil der Importe bei fast 55%, d. h. die deutschen Werke sind Außenseiter im eigenen Markt geworden.
Dabei hat sich bei den Importen die Struktur der Lieferländer erheblich verschoben. In der ersten Hälfte der siebziger Jahre wurde aus Belgien/Luxemburg mehr Betonstahl eingeliefert als von allen anderen ausländischen Lieferanten zusammen. Danach ist Italien der Hauptlieferant geworden durch das Vordringen der Mini-Stahlwerke. Eine überraschende Zunahme zeigt sich in den letzten Jahren bei den Drittlandeinfuhren: Bis 1975 gab es kaum Importe aus diesen Ländern, in 1983 dürfte ihr Anteil bei 25% liegen.
Die Vermarktung des Betonstahls, auch des deutschen, liegt fast restlos beim lagerhaltenden Stahlhandel. Dieser aber richtet sich seiner Natur und seiner Aufgabenstellung entsprechend ausschließlich nach Angebot und Nachfrage. Aber auch deutliche spekulative Elemente bestimmen sein Einkaufsverhalten. Das läßt sich anhand der Lagerbestandsstatistik leicht nachweisen.
Der dritte Faktor, der den Betonstahlpreis beeinflußt, ist der Schrottpreis. Man kann unterstellen, daß bei den Mini-Stahlwerken der Schrott mehr als ein Drittel der Gesamtkosten ausmacht.
Da die Mini-Stahlwerke in der EG inzwischen mit ca. 72% Produktionsanteil eine marktbeherrschende Stellung haben, gelingt es ihnen meistens, Schrottpreiserhöhungen gegen andere Wettbewerber in einem höheren Betonstahlpreis weiterzugeben, wie zum Beispiel im 2. und 3. Vierteljahr 1983.
Immer ist das jedoch nicht der Fall. So hatte die Europäische Kommission mit Wirkung vom 5. 5. 1977 einen Mindestpreis gemäß Artikel 61 EGKS-V von 550 DM/t festgesetzt, der zum 1. 7. 78 auf 525 DM/t gesenkt wurde. Damit sollte der ruinöse Preiswettbewerb unterbunden werden, der wegen der stark zurückgegangenen Nachfrage entstanden war. Das System erwies sich jedoch im Markt als wenig wirksam, weshalb es bald aufgehoben wurde.
Nach den Erfahrungen der letzten Jahre läßt sich sagen, daß der Preis im Bereich von etwa 100 DM/t, d. h. etwa zwischen 500 und 600 DM/t schwankt. Davon bleiben aber die Aufpreise für die Abmessungen unberührt.
Der Einfluß des Betonstahlpreises auf die Baupreise wird maßlos und manchmal auch gewollt überschätzt. Der Kostenanteil des Bewehrungsstahls (Betonstahl und andere Bewehrungsstähle) dürfte in der durchschnittlichen Größenordnung um 5% liegen, d. h. ein 100 DM/t, also um ca. 20% höherer Stahlpreis wirkt sich an den Baukosten mit 1% aus.
Bei einem Einfamilienhaus ist die Kostenerhöhung kaum meßbar, wenn man davon ausgeht, daß für ein solches Bauwerk etwa eine Tonne Bewehrungsstahl gebraucht wird. 100 DM/t Preiserhöhung wären dann auf einen Objektwert von 350 000 bis 450 000 DM zu beziehen. Schließlich sollte man bei der Kostenbetrachtung auch berücksichtigen, daß in die Baukalkulation Effektivpreise für die eingebaute Tonne, d. h. gebogen und verlegt, eingehen, die ohnehin eine Bandbreite von 1600 bis 2000 DM/t haben.
Es gibt längere Zeiträume, in denen sogar eine gegenläufige Bewegung zwischen den Betonstahlpreisen und den Baupreisen zu verzeichnen ist: Nach 1974 stand der Phase rückläufiger Betonstahlpreise ein steiler Anstieg der Baupreise gegenüber.

6 Bewehrungsstähle

Der Betonstabstahl nimmt seit einigen Jahren rund 60% der gesamten Bewehrungsstähle ein. Abgesehen von rund 2% Spannstahl entfällt der Rest auf Betonstahlmatten. Dieser Stand ist das Ergebnis einer langjährigen kontinuierlichen Entwicklung zu Lasten des Stabstahls (s. a. Bild 2). Vor etwa zwanzig Jahren lag der Mattenanteil noch bei 25%. Diese Strukturänderung ist weniger auf eine veränderte Relation der Betonstabstahlpreise zu den Mattenpreisen zurückzuführen, deren Preis vom Walzdrahtpreis abhängt. Vielmehr dürfte die Ursache im hohen Lohnniveau der Bauindustrie liegen. Der Einsatz einer in großer Stückzahl vorgefertigten Mattenbewehrung ist vielfach ko-

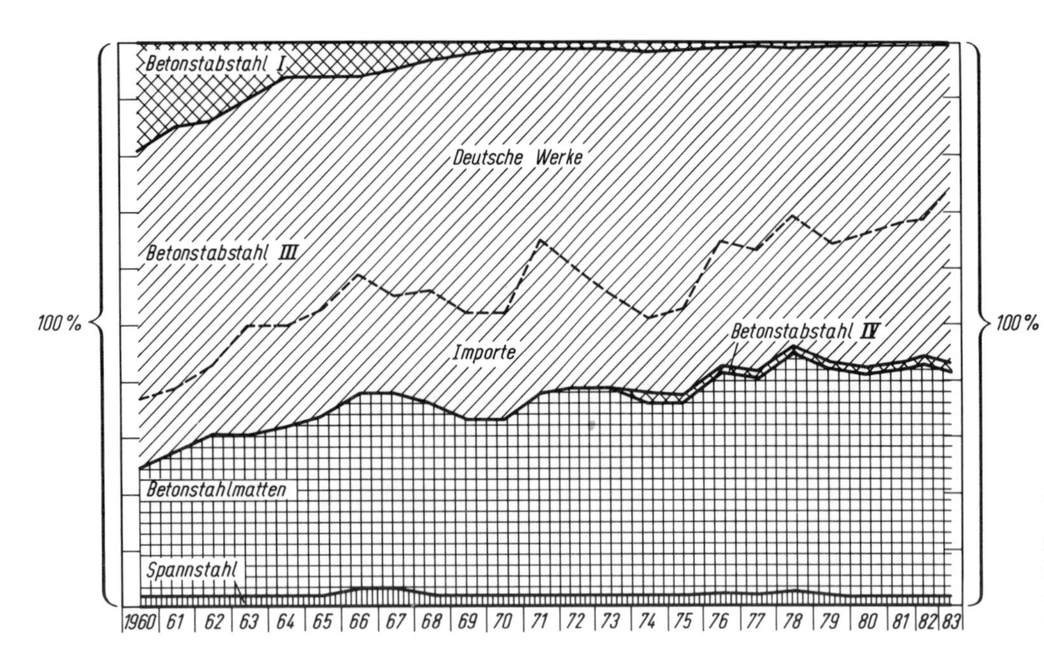

Bild 2
Marktversorgung der BR Deutschland mit Bewehrungsstahl in Prozentanteilen seit 1960 (Betonstabstahl, Betonstahlmatten, Spannstahl)

stengünstiger als das arbeitsintensive Schneiden, Biegen, Binden und Verlegen einzelner Stäbe. Man kann davon ausgehen, daß der Substitutionsvorgang weitgehend abgeschlossen ist, da die Verwendungsmöglichkeiten der Matte voll ausgeschöpft sind. Es kann jedoch zu temporären Verschiebungen kommen, wenn der mattenintensive Wohnungsbau seinen Anteil am gesamten Bauaufkommen deutlich erhöht.

Der Spannstahl hat seinen Anteil von rund 2% der Bewehrungsstähle in den vergangenen zwanzig Jahren fast unverändert beibehalten. Das bedeutet, daß eine weitere Substitution anderer Bewehrungen durch Spannstahl nicht wirtschaftlich ist, der Spannstahl jedoch seinen festen Platz im Bauwesen hat.

Die Darstellung im Bild 3 gibt bemerkenswerte Hinweise auf die Nachfrageschwankungen in den letzten zwanzig Jahren. Die Marktversorgung hatte eine eindeutige Spitze in den Jahren 1972/73 mit 300000 bis 350000 moto, davon entfielen 120000–130000 moto auf die Matten. Diese Tonnage wurde selbst in den Aufbaujahren nach dem Kriege nicht erreicht, wahrscheinlich weil in diesen Jahren in erster Linie bewehrungsarme Wohnbauten errichtet wurden.

Bemerkenswert ist, daß sich nach der Rezession 1974/75 der Bedarf an Bewehrungsstählen auf einem gegenüber den sechziger Jahren viel höheren Niveau eingependelt hat. Die Tonnage von rund 250000 moto bzw. rund 3 Mio jato dürfte, abgesehen von starken konjunkturellen Einflüssen, etwa dem Bedarf der nächsten Jahre entsprechen. Das verhältnismäßig hohe Verbrauchsniveau läßt sich dadurch erklären, daß heute mehr als früher beim öffentlichen Tief- und Verkehrswegebau aufwendige Gründungen eingesetzt werden müssen und der Schwerpunkt der Bautätigkeit mehr bei konstruktiven Ingenieurbauten für die gewerbliche Wirtschaft liegt.

7 Verarbeitung von Ringen

Händler und Verarbeiter, d. h. Biegereien, zeigen großes Interesse daran, warmgewalztes geripptes Material, das in Ringen aufgehaspelt angeliefert wird, maschinell zu Stäben und/oder zu Bügeln zu verarbeiten. Auch soll damit ausländischen Stablieferanten, besonders aus Drittländern, geholfen werden, da ihre Stäbe in dünnen Abmessungen durch mehrfaches Umladen oft die Geradheit verlieren.

Die maschinelle Herstellung der Bügel ist, entsprechende Stückzahl vorausgesetzt, billiger als heute im manuellen Biegebetrieb. Entsprechende Maschinen sind außerhalb der EG in verschiedenen Ländern in Betrieb, und die Produktion dieser sogenannten Bügelautomaten ist bauamtlich zugelassen.

Über die Auslastungsmöglichkeit solcher Bügelautomaten bestehen vielfach falsche Vorstellungen. Sachverständige schätzen, daß bei einem Betonstahlverbrauch von 120000 bis 130000 moto, wie er für 1983 anzusetzen ist, ca. 5% automatenfähig sind, also etwa 6000 moto. Hiervon dürfte etwa die Hälfte, das sind

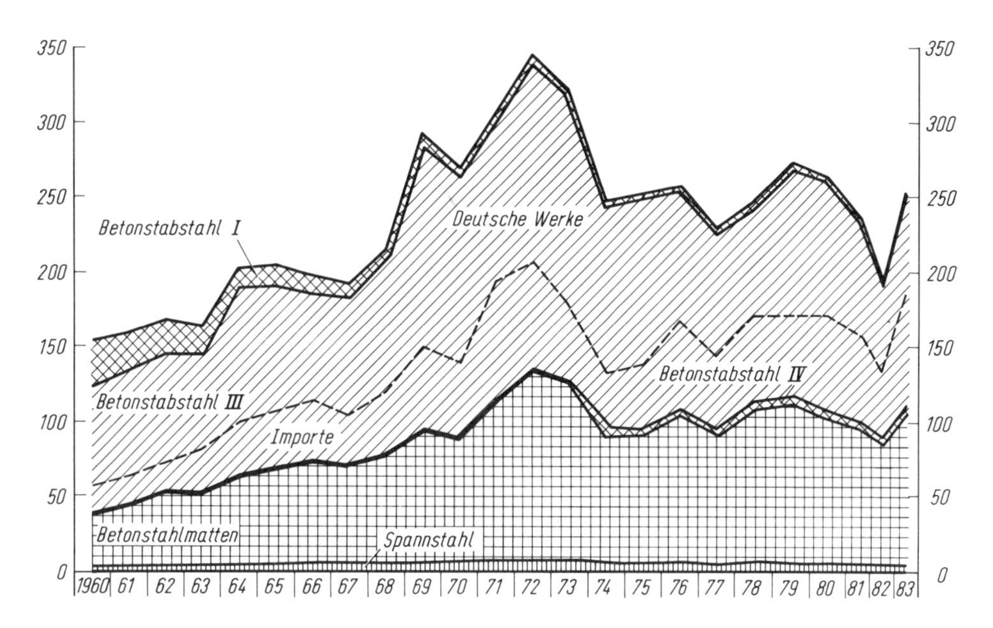

Bild 3
Marktversorgung der BR Deutschland mit Bewehrungsstahl in 1000 moto seit 1960
(Betonstabstahl, Betonstahlmatten, Spannstahl)

ca. 3000 moto, auf Großbetriebe entfallen. Bei dem derzeitigen Leistungsstand der Automaten und einem Verhältnis von Fixlängen zu Bügeln von etwa 60:40 dürften etwa 25 Automaten in Deutschland ausreichen, um den Bedarf zu decken.

In der BR Deutschland bestehen wie auch in anderen Ländern der EG erhebliche Bedenken gegen die Verarbeitung von Ringen. Untersuchungen haben ergeben, daß die Verformungsreserve des aufgehaspelten Materials nicht immer ausreicht, auch bei unsachgemäßer Verarbeitung die bedingungsgemäßen, technischen Anforderungen zu gewährleisten. Unbehandelte, naturharte Sorten nehmen an Festigkeit und an Verformungsvermögen ab. Außerdem werden beim Richten die Rippen beschädigt, wodurch sich die Verbundeigenschaften verschlechtern.

Bei Anwendung der maschinell gefertigten Erzeugnisse wäre daher ein besonderes Kennzeichnungs- und Überwachungssystem erforderlich, um bei Auftreten eines Bauschadens Hersteller und Verarbeiter identifizieren zu können.

Die Gewährleistung muß jedoch anteilig vom Hersteller wie vom Verarbeiter erbracht werden, da ein Hersteller nicht für Fehler haftbar gemacht werden kann, die bei der Verarbeitung unterlaufen. Das könnte dadurch erreicht werden, daß das Ringmaterial durch eine verstärkte Rippe gekennzeichnet wird auf der dem Werkskennzeichen gegenüberliegenden Stabseite und daß bei der Bügelfertigung ein weiteres Kennzeichen für den Verarbeiter aufgebracht wird. Dabei sollte die Herstellung von Stäben nur auf die Verarbeitung im eigenen Betrieb beschränkt werden. Inzwischen hat das IfBt durch die Herausgabe der „Grundsätze für die Verarbeitung von Betonstahl vom Ring" diesen Gedanken weitgehend Rechnung getragen.

Angesichts dieser Gütesicherungskosten sollte sich der Verarbeiter jedoch fragen, ob sich bei der Verarbeitung vom Ring noch ein Rationalisierungsgewinn errechnet. Sicherlich dürfte ihn ein hohes Investitionsrisiko erwarten.

8 Ausbau der Gütesicherung

Zahlreiche Anbieter und Sortenvielfalt sind für den deutschen Nationalökonomen der erfreuliche Beweis für eine rege Wirtschaftstätigkeit, besonders wenn sich unter den Anbietern viele Devisenausländer befinden. Die Händler sind daran interessiert, der deutschen Bauindustrie niedrige Einkaufspreise anzubieten. Sie verlassen sich voll auf das existierende Gütesicherungssystem.

Der Ingenieur sieht diese Situation wesentlich zurückhaltender, weil bei ihm die Qualität im Vordergrund stehen muß.

Qualität und Preis stehen aber, nicht nur beim Betonstahl, in einem engen Zusammenhang.

Die Fachöffentlichkeit weiß, daß hin und wieder Betonstähle auf dem Markt sind, die den Anforderungen nicht entsprechen. Der Händler hat dabei keine

Chance, qualitativ nicht bedingungsgemäße Ware zu erkennen. Jeder Biegebetrieb kann daher eine mehr oder minder lange Liste von Reklamationsfällen vorweisen. Man sollte sich hier nicht mit dem Gedanken trösten, daß spektakuläre Schadensfälle sehr selten sind.

Klar ist nur, daß bei der Komplexität von Marktversorgung und Technologie, verknüpft mit der Wirtschaftspolitik unterschiedlicher Länder, ein System sehr schnell Schwachstellen bekommen kann, wenn diese nicht durch rechtzeitige Vorsorge am Entstehen gehindert werden.

Der derzeitige Entwicklungsstand macht es nötig, darüber nachzudenken, welche Verbesserungen erreicht werden müßten.

Eine wesentliche Verbesserung könnte erzielt werden, wenn übergeordnete Aufgaben zentral koordiniert werden könnten, welche im derzeitigen System von keinem der Beteiligten wahrgenommen werden können.

Dazu gehören:

- Kontrolle der Betonstähle auf Baustellen, in Biegebetrieben und Händlerlagern. (Diese Prüfungen waren bisher in den Überwachungsverträgen als Vorbehalt der Bauaufsicht vorgesehen, sie wurden jedoch wegen der Kostenfragen nie durchgeführt.)
- Kontrolle der werkstoffgerechten Verarbeitung von Betonstahl auf den Baustellen,
- laufende Ergänzung der Gütesicherungsmaßnahmen entsprechend dem jeweiligen Erkenntnisstand.
- Begutachtung und Zustimmung zu werksinternen Prüfplänen,
- zentrale Auswertung der jährlichen Überwachungsberichte,
- Realisierung einer gleichmäßigen Strenge der Überwachung bei den verschiedenen Prüfstellen. Unterstützung der Prüfstellen bei der Durchsetzung von Ahndungsmaßnahmen bzw. bei der Anregung zur qualitativ bedingten Änderung der Technologie. Dieser Punkt ist insbesondere für die Zukunft von Bedeutung, da im Rahmen der europäischen Zusammenarbeit nationale Prüfinstitute die Überwachung der Exporte durchführen werden.
- zentrale Auswertung von Reklamationen,
- zentrale Sammlung und Auswertung der im Rahmen von DIN 4099 anfallenden Schweißuntersuchungen.

Diese vielschichtigen, übergeordneten Aufgaben machen deutlich, daß das Institut für Bautechnik und die einzelnen Prüfinstitute in ihrer Arbeit unterstützt werden müßten.

Welche Institution die genannten Aufgaben erfüllen kann, ist noch unklar. Sie muß auf jeden Fall in die baurechtlichen Gegebenheiten eingepaßt werden.

In den meisten europäischen Ländern sind Institutionen dieser Art schon vorhanden und haben sich dort bestens bewährt: z. B. Schweden, Niederlande, Belgien, Spanien, Frankreich und Österreich.

Rißwiderstand und Rißfortschritt bei Glasfaserbeton

Wolfgang Menz
und Jörg Schlaich

1 Einleitung

Faserbeton, ein *zug- und druck*fester Werkstoff, der von Hand auf der Baustelle hergestellt werden kann und in beliebig geformten Schalungen erstarrt – die Erfüllung des Traumes derer, denen Bauen Gestalten bedeutet?
In der Tat hat der Betonbau durch die Fortschritte bei der Entwicklung von Faserbetonen einen vielversprechenden Impuls erhalten. Die bisherige Einschränkung auf fabrikmäßig hergestellten Asbestzement konnte mit neuen Fasertypen, die vielfältige Fertigungsmethoden erlauben, überwunden werden [1], [2], [4].
Allerdings sind die heute verwendeten Fasern noch recht teuer; dort wo es konstruktiv möglich ist, ist es wirtschaftlicher, Zugkräfte durch Stabstahlbewehrung abzudecken. Zudem ist der Faseranteil fertigungstechnisch begrenzt, so daß mit Fasern die Zugtragfähigkeit von Beton nicht im gleichen Maße gesteigert werden kann, wie durch Einlegen von Stabstahl. Deshalb eignet sich Faserbeton vornehmlich für die Herstellung von dünnen Bauteilen einerseits und zur „hybriden" Verwendung andererseits, d. h. in Kombination mit Stahlbeton. Dabei ist vor allem an ein Einbetten von Stabstahlbewehrung in Faserbetonbauteile an Stellen mit Zugkraftkonzentrationen gedacht, aber auch an die Herstellung von Stahlbeton in verlorener Schalung aus Faserbeton.
Die Verfasser haben zu all den genannten Anwendungsfällen – allerdings ausschließlich mit Glasfasern – ausführliche Versuchsreihen durchgeführt [3], [5]. Dadurch wurde bestätigt, daß man von Faserbeton vor allem einen günstigen Einfluß auf die Rißerweiterung und das Rißbild erwarten kann.
Im Hinblick auf die Deutung und Übertragbarkeit dieser Versuchsergebnisse ist es von Interesse, sich theoretisch mit dem Einfluß einer Glasfaserbewehrung auf die Rißbildung von Beton auseinanderzusetzen. Dies wird in diesem Beitrag mit Hilfe der *Bruchmechanik* versucht.
Diese Arbeiten entstanden im Rahmen des Sonderforschungsbereiches 64 – Weitgespannte Flächentragwerke. Die Versuche wurden an der von Prof. Dr.-Ing. GALLUS REHM geleiteten FMPA Baden-Württemberg – OTTO-GRAF-Institut – durchgeführt. Wir widmen gerade diesen Beitrag Herrn REHM, um auch daran zu erinnern, daß die wissenschaftliche Auseinandersetzung mit der Rißbildung im Stahlbeton von ihm ausging.

2 Glasfaserbeton

Neben Asbestfasern werden in letzter Zeit vor allem Stahl- und Glasfasern für Faserbeton verwendet. Dem Einbetten von Glasfasern in eine Matrix, die Portland-Zement enthält, stand bisher der Festigkeitsverlust des Glases durch den Alkaliangriff entgegen. Als erfolgreiche Gegenmaßnahme erwies sich nach mancherlei fehlgeschlagenen Versuchen der Zusatz von Zirkonoxid ZrO_2 zur Glasschmelze. Heute werden solche Glasfasern für die technische Anwendung weltweit unter dem Namen CemFil-Glas vertrieben. Die Einzelfaser besitzt einen Durchmesser von 10–13 μm. Je 204 Fasern werden zu einem Bündel zusammengefaßt (Bild 1). Ein Überblick über die Beständigkeit dieser Fasern wird in [3], [6], [7] gegeben.
Bei Glasfaserbeton (GFB) besteht die Matrix aus einer Zement-Sand-Mischung im Verhältnis zwischen 1 : 1 und 1 : 2. Der Größtkorndurchmesser des Sandes beträgt 2 bis 3 mm, so daß man streng von Glasfasermör-

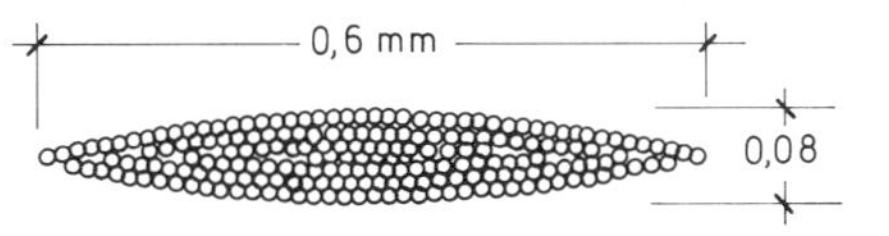

Bild 1
Faserbündel aus 204 Cemfil Filamenten

tel sprechen müßte. Im englischen Sprachraum sagt man meist glass fibre reinforced cement (GRC), also Glasfaserzement, wohl weil man dort sehr wenig Sand oder gar reinen Zementmörtel verwendet.

Für die *Herstellung* von GFB gibt es eine Reihe von Verfahren [4]. Besonders vorteilhaft ist das *Spritzen*. Dabei werden die Glasfasern als endlose Faserbündel in eine Spritzpistole geführt, dort auf eine Länge von ca. 50 mm geschnitten und zusammen mit der über eine Mörtelpumpe zugeführten Matrix mittels Druckluft ausgeblasen. Die Fasern verteilen sich zweidimensional auf der Fläche; der gewünschte Faseranteil wird sehr genau und gleichmäßig erreicht. Der Fasergehalt ist in Abhängigkeit von der Faserlänge auf etwa 6 Vol.-% begrenzt. Damit kann leicht ein dem Stahlbeton entsprechender echter Zustand II erreicht werden.

Solange die den Riß überbrückende Faserbewehrung die bei der Rißaufweitung freiwerdenden Kräfte aufnehmen kann, können sich parallel zum ersten Riß weitere Risse bilden. Der dafür erforderliche Fasergehalt wird überkritischer Fasergehalt $V_{f,cr}$ genannt. Das sich hierbei einstellende Rißbild *in Beanspruchungsrichtung* unterliegt den gleichen Gesetzmäßigkeiten wie bei Stahlbeton, das Rißbild des GFB ist aber äußerst fein verteilt (Bild 2) [3].

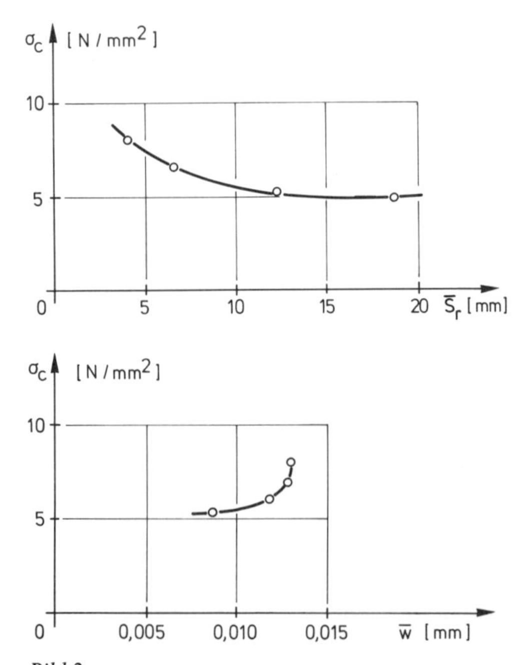

Bild 2
Rißabstand $\bar{s}_r$ und Rißbreiten $\bar{w}$ von GFB im Zustand II bei zentrischem Zug (V_f = 4,5 Vol.% und l_f = 50 mm, Alter 33 Tage) (σ_c: Composit-Spannung auf den Bruttoquerschnitt bezogen)

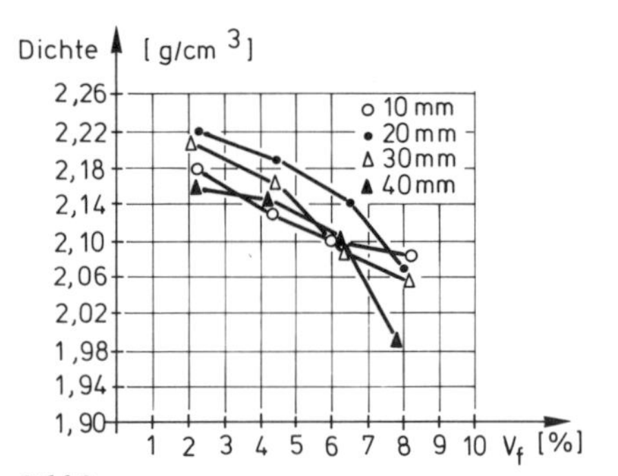

Bild 3
Der Einfluß des Glasfasergehaltes und der Faserlänge auf die Dichte von Glasfaserzement [8]

Die notwendige Nachverdichtung der aufgespritzten Mischung erfolgt mittels Handwalzen auf der Schalung. Fasergehalt und Faserlänge beeinflussen die Verarbeitbarkeit der Mischung. Eine schlechte Verarbeitung drückt sich in einer geringeren Dichte des Faserbetons aus (Bild 3). Mit abnehmender Dichte sinken die Druck- und Zugfestigkeiten. So ist es möglich, daß bei nachlässiger Verdichtung, die aber auch die Folge eines versehentlich zu hohen Fasergehalts sein kann, ein schlechteres Ergebnis erzielt wird als ohne Fasern.

3 Zum Bruchverhalten von Beton und Faserbeton

Das Versagen von Zementstein, Mörtel oder Beton ist bei Druck- und Zugbelastung auf das instabile Anwachsen von Mikrorissen zurückzuführen, die von lokalen Fehlstellen des Mehrphasen-Werkstoffes ausgehen. Bei Beton oder Mörtel treten diese Mikrorisse zunächst vornehmlich an der Kontaktzone zwischen Zuschlag und Zementstein auf (Verbundrisse). Mikrorisse im Zementstein selbst (Gefügerisse) folgen erst bei höherer Beanspruchung.

Bei *Zugbeanspruchung* wird mit dem Auftreten der Gefügerisse sehr rasch das Versagen eingeleitet. Die Richtung der Gefügerisse verläuft vorwiegend quer zur Belastungsrichtung. Die Nichtlinearität der Spannungs-Dehnungs-Linie im Zugversuch ist in erster Linie auf das Anwachsen der Rißflächen an den Kontaktzonen zurückzuführen und als Querschnittsverlust zu verstehen. Infolge des Querschnittsverlustes ist das Versagen wesentlich spröder als bei Druckbelastung, und die gemittelten Gesamtdehnungen von Zementstein und Zuschlagstoffen betragen nur $\varepsilon_{z,u}$ = 0,1–0,15‰.

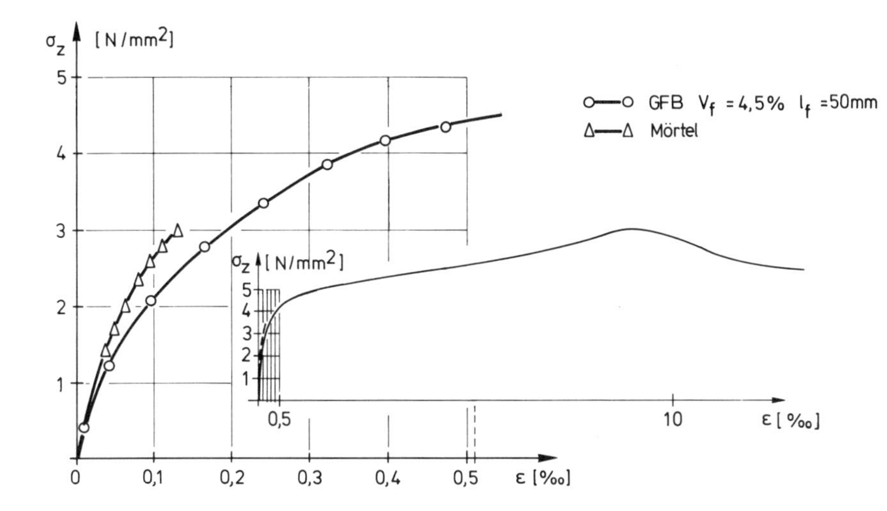

Bild 4
Die Zug-Arbeitslinie von Glasfaserbeton mit überkritischem Fasergehalt und von reinem Mörtel zum Vergleich

Bei *Druckbeanspruchung* pflanzen sich die Mikrorisse von den Kontaktflächen ausgehend, vorwiegend parallel zur Lastrichtung im Zementstein fort [9], ab 80 bis 90% der Bruchlast instabil. Die abfallenden Äste der Arbeitslinie jenseits der Bruchlast können im wesentlichen über die innere Reibung erklärt werden.
Mit *Fasern* können diese Vorgänge entscheidend beeinflußt werden. Bezogen auf die übertragenen Zugkräfte steht bei Faserbeton wegen des geringen Durchmessers der Fasern eine sehr viel größere Oberfläche für den Verbund zwischen der Faserbewehrung und der Matrix zur Verfügung als bei Stahlbeton. Deshalb kommen die Fasern nicht erst zur Wirkung, nachdem sich Trennrisse gebildet haben. Sie verzögern vielmehr bereits vorab das Entstehen durchgehender Risse dadurch, daß sie das Fortpflanzen der Mikrorisse von den Kontaktflächen in den Zementstein hinein, also das Entstehen der Gefügerisse hemmen und auch die innere Reibung erhöhen (Bilder 4 bis 6).
Im folgenden wollen wir uns auf das bruchmechanische Strukturmodell bei *Zugbeanspruchung* beschränken.

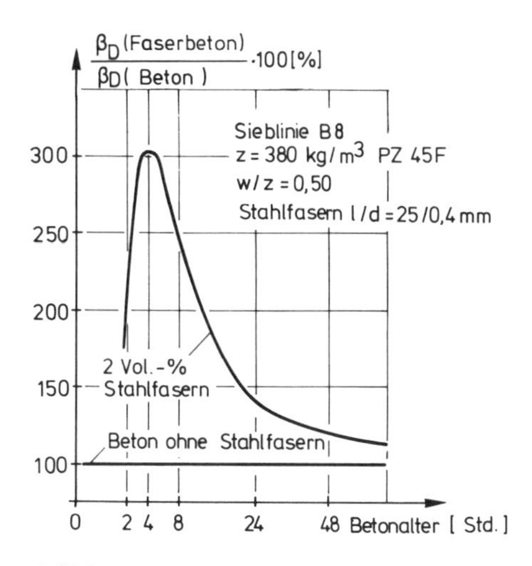

Bild 5
Relative Druckfestigkeit von jungem Beton ohne und mit Stahlfasern [10]

4 Der Rißvorgang bei Faserbeton

Obwohl die Ansätze der Bruchmechanik zunächst für homogene Werkstoffe gedacht waren, werden sie heute auch allgemein auf den heterogenen Beton angewandt. In diesem Beitrag wird unterstellt, daß dies zumindest näherungsweise zulässig ist.

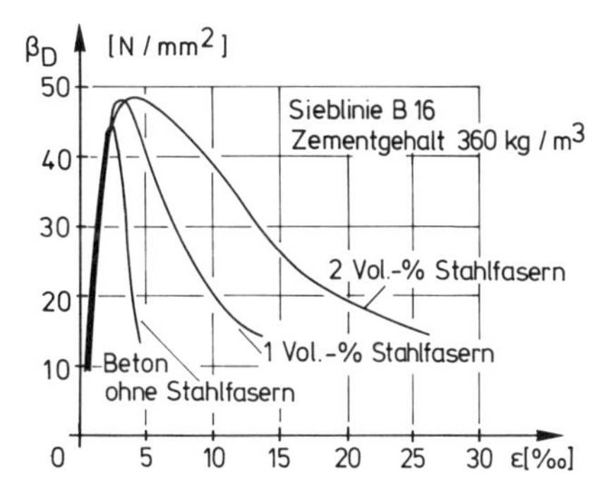

Bild 6
Druck-Arbeitslinien von Beton ohne und mit Stahlfasern [10]

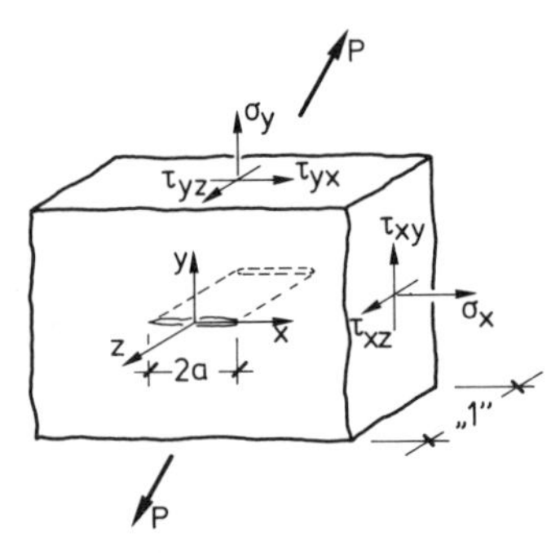

Bild 7
Einzelriß in unendlicher Scheibe mit der Rißoberfläche $\tilde{a} = 4a \cdot$ „1" unter einer beliebigen Belastung p

Bruchmechanische Untersuchungen über den Rißfortschritt in belasteten Körpern beruhen auf einer Betrachtung der beteiligten Energien. Bei einer Vergrößerung des Risses (Bild 7) muß die von einem äußeren Belastungsfeld geleistete Arbeit W neben der Formänderungsarbeit U auch die zur Erzeugung der erweiterten Rißoberfläche notwendige Arbeit T liefern. Inkrementelle Änderungen werden durch Änderungen der Belastung sowie ein Anwachsen der Rißoberfläche $\mathrm{d}\tilde{a}$ verursacht.

$$\mathrm{d}W = \mathrm{d}U + \mathrm{d}T \tag{4.1}$$

Wenn die Richtung der Belastungsänderung und des Rißfortschrittes festgelegt und die Verschiebungen klein sind, wird aus Gl. 4.1 nach [11]

$$\frac{\partial W}{\partial \tilde{a}} = \frac{\partial U}{\partial \tilde{a}} + \frac{\mathrm{d}T}{\mathrm{d}\tilde{a}} \tag{4.2}$$

$\mathrm{d}T/\mathrm{d}\tilde{a}$ enthält sowohl die für den Rißzuwachs benötigte spezifische Oberflächenenergie [12] als auch plastische Verformungsanteile in *unmittelbarer Nähe* der Rißwurzel [13].
Sollen auch *größere plastische Bereiche* berücksichtigt werden, teilt man die inkrementelle Formänderungsarbeit $\mathrm{d}U$ besser in einen elastischen Anteil $\mathrm{d}U_{el}$ und einen plastischen $\mathrm{d}U_{pl}$ auf. Faßt man die Anteile der äußeren Arbeit sowie die Formänderungsarbeit zusammen

$$\begin{aligned}\mathcal{G} &= \frac{\partial W}{\partial \tilde{a}} - \frac{\partial U}{\partial \tilde{a}} \\ &= \frac{\partial W}{\partial \tilde{a}} - \frac{\partial U_{el}}{\partial \tilde{a}} - \frac{\partial U_{pl}}{\partial \tilde{a}}\end{aligned} \tag{4.3}$$

dann läßt sich mit den Gl. 4.2 und 4.3 eine Funktion f für das *Rißwachstum* anschreiben [11]

$$f = \mathcal{G} - \frac{\mathrm{d}T}{\mathrm{d}\tilde{a}} \leq 0 \tag{4.4}$$

Solange die Funktion f negative Werte aufweist, ist die bei einer virtuellen Ausbreitung des Risses freiwerdende mechanische Energie kleiner als die für die Erzeugung neuer Rißoberflächen benötigte Arbeit und ein Fortschreiten des Risses ist nicht möglich.
Wird f gleich Null, so ist ein Rißwachstum möglich; die angelieferte Energie deckt den hierfür erforderlichen Energiebedarf. Die Rißausbreitung ist aber nur solange stabil, wie die Funktion f nicht zunimmt. Belastungsänderungen und Rißzuwachs müssen dann folgendermaßen verträglich sein:

$$\mathrm{d}f = \frac{\partial f}{\partial \tilde{a}}\mathrm{d}\tilde{a} + \frac{\partial f}{\partial \tilde{p}}\mathrm{d}\tilde{p} = 0 \tag{4.5}$$

Daraus ergibt sich die zu einem Rißzuwachs gehörige Belastungsänderung

$$\mathrm{d}\tilde{p} = -\frac{\partial f/\partial \tilde{a}}{\partial f/\partial \tilde{p}}\mathrm{d}\tilde{a} \tag{4.6}$$

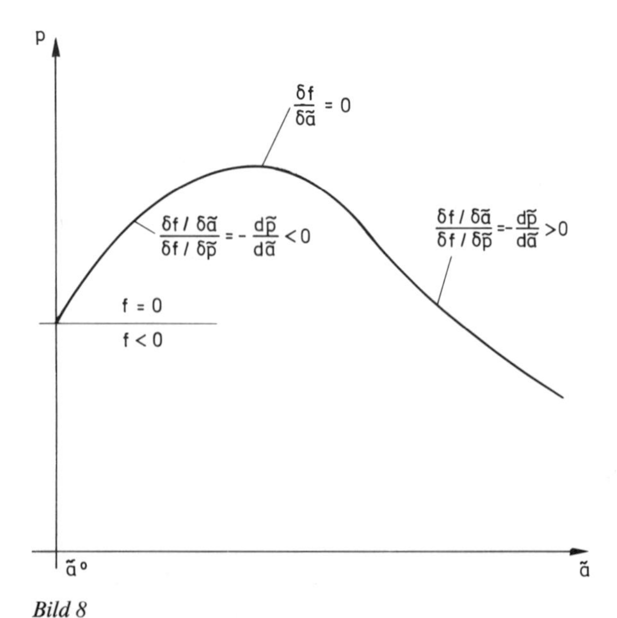

Bild 8
Zusammenhang zwischen Rißerweiterung und Belastung über die Funktion f nach Gl. 4.4 ausgehend von einer Anfangsrißfläche $\tilde{a}_0$ und stabilem Rißwachstum nach Gl. 4.5

Wenn $\partial f/\partial \tilde{a} = 0$ ist, kann sich der Riß ohne Lastanstieg ausbreiten. Bei einem kraftgesteuerten Versuch würde bis zu diesem Punkt ein allmähliches Rißwachstum, danach instabiles Rißausbreiten und Versagen eintreten. Bei einem dehnungsgesteuerten Versuch würde ab diesem Punkt die Kraft wieder zurückgehen (Bild 8).

Werkstoffe mit geringer Duktilität:

Für Werkstoffe mit weitgehend linear-elastischem Verhalten bzw. mit *kleinen plastischen Zonen* in unmittelbarer Umgebung der Rißwurzeln (Anteil $U_{pl} \ll U_{el}$) können die Gl. 4.3 und 4.4 auch über Spannungsintensitätsfaktoren K ausgedrückt werden [13]

$$\mathcal{G} = \frac{1-\nu^2}{E}(K_{\mathrm{I}}^2 + K_{\mathrm{II}}^2) + \frac{1}{2G}K_{\mathrm{III}}^2 \quad \text{ebener Dehnungszustand}$$

$$\mathcal{G} = \frac{1}{E}(K_{\mathrm{I}}^2 + K_{\mathrm{II}}^2) + \frac{1}{2G}K_{\mathrm{III}}^2 \quad \text{ebener Spannungszustand} \tag{4.7}$$

Darin sind 3 Faktoren K_i den 3 Beanspruchungsarten nach Bild 9 zugeordnet. Die Spannungsintensitätsfaktoren K beschreiben die Intensität des Spannungsfeldes in Rißnähe und können bei Einsetzen der instabilen Rißerweiterung als Werkstoffkonstante angesehen werden. Der Spannungs- bzw. Formänderungszustand in unmittelbarer Umgebung der Rißwurzel bleibt nämlich gleichartig, wenn die Geometrie und die Belastung der Körper verändert werden. Rißlänge, Last und Geometrie beeinflussen den Verlauf der Spannungen σ_y nur über die Scharparameter K (Bild 10). Der Bereich des stabilen Rißwachstums bei zunehmender Last entsprechend Bild 8 ist bei geringer Duktilität fast nicht vorhanden. Mit Beginn des instabilen Rißwachstums in lastabhängig gesteuerten Versuchen können über Gl. 4.3 ein kritisches $\mathcal{G}_{cr}$ und über Gleichung 4.7 für die jeweilige Beanspruchungsart die kritischen Spannungsintensitätsfaktoren $K_{i,cr}$ bestimmt werden. Experimentell ermittelte Spannungsintensitätsfaktoren $K_{\mathrm{I},cr}$ für einen Werkstoff mit geringer Duktilität, hier Zementstein, zeigt Bild 11.

Werkstoffe mit größerer Duktilität:

Bei Werkstoffen mit *großer Plastizität* kann eine stabile Rißerweiterung festgestellt werden. Gleichung 4.4 läßt sich in elastische und plastische Anteile aufspalten

$$f = \mathcal{G}_{el} - \mathcal{R} \tag{4.8}$$

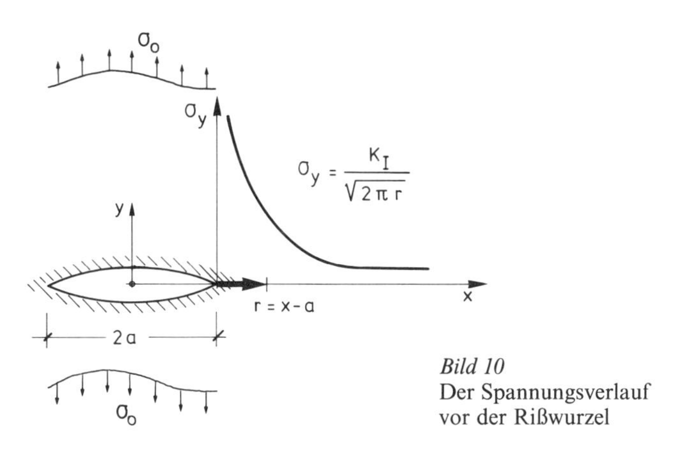

Bild 10
Der Spannungsverlauf vor der Rißwurzel

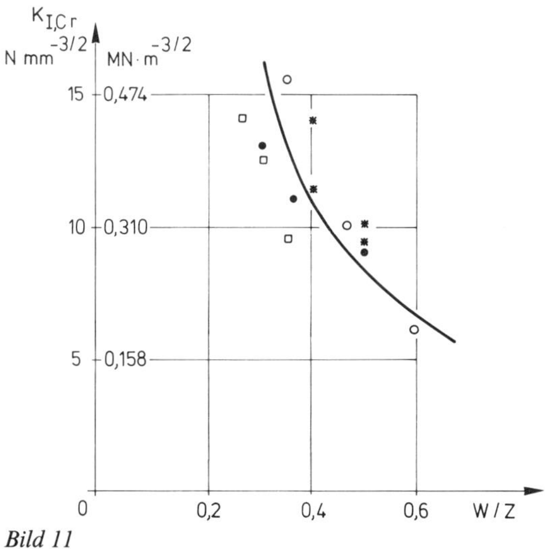

Bild 11
Werte des Spannungsintensitätsfaktors $K_{I,cr}$ (d.h. für Mode *I* nach Bild 9 für Zementstein in Abhängigkeit vom Wasser-Zement-Wert. Entnommen aus einer Zusammenstellung in [3].

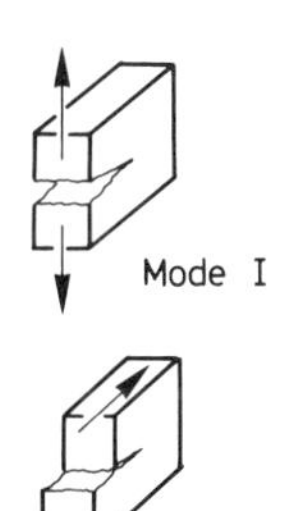

Mode II

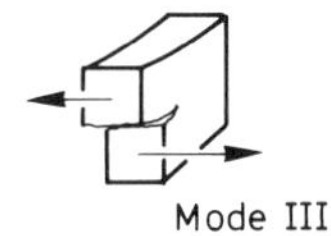

Bild 9
Die drei Beanspruchungsarten für einen Riß

mit

$$\mathcal{G}_{el} = \frac{\partial W}{\partial \tilde{a}} - \frac{\partial U_{el}}{\partial \tilde{a}}$$

$$\mathcal{R} = \frac{\partial U_{pl}}{\partial \tilde{a}} + \frac{dT}{d\tilde{a}}$$

Dabei beschreibt der Ausdruck $\mathcal{G}_{el}$ die Energieanteile, die die Rißoberfläche im Material vergrößern wollen, während $\mathcal{R}$ sämtliche Widerstände gegen die Rißausbreitung zusammenfaßt. Diese Vorgehensweise wird vielfach $\mathcal{R}$-Konzept genannt.
Instabile Rißerweiterung bei kraftgesteuerten Versuchen tritt auf, wenn

$$\frac{\partial f}{\partial \tilde{a}} = 0 = \frac{\partial \mathcal{G}_{el}}{\partial \tilde{a}} - \frac{\partial \mathcal{R}}{\partial \tilde{a}} \tag{4.9}$$

erfüllt ist.

Wegen des plastischen Anteils wird der Rißwiderstand $\mathcal{R}$ abhängig von der Probengeometrie, ist also keine Werkstoffkonstante. Eine oft benutzte Darstellung basiert auch hier auf Spannungsintensitätsfaktoren K. In Abhängigkeit von der stabilen Rißverlängerung Δa wird über Gl. 4.3 das zu $(a_0 + \Delta a)$ gehörende $\mathcal{G}_R$ bestimmt. Über Gl. 4.7 kann dann der dazugehörige scheinbare Spannungsintensitätsfaktor K_R ausgedrückt werden. Auch die K_R-Faktoren sind infolge der plastischen Anteile geometrieabhängige Werkstoffkennwerte. Den typischen Verlauf von K_R zeigt Bild 12. In Bild 13 sind an zementgebundenen Werkstoffen ermittelte K_R-Verläufe dargestellt.
Sind die plastischen Anteile in Gl. 4.8 sehr groß, so treten vor Beginn des Rißwachstums erhebliche Rißaufweitungen COD (crack opening displacements) an der Rißwurzel auf.

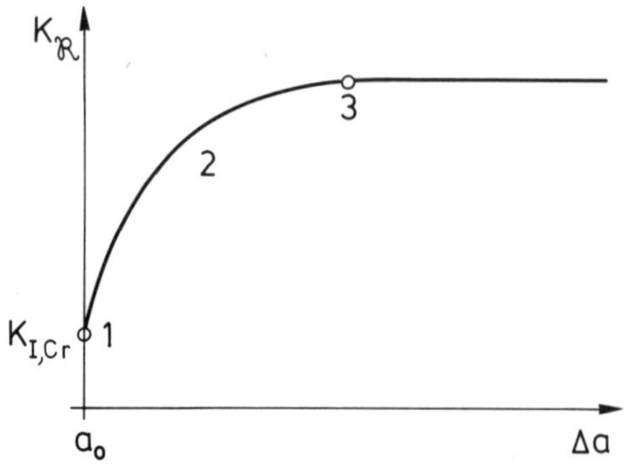

Bild 12
Typischer Verlauf einer K_R-Kurve

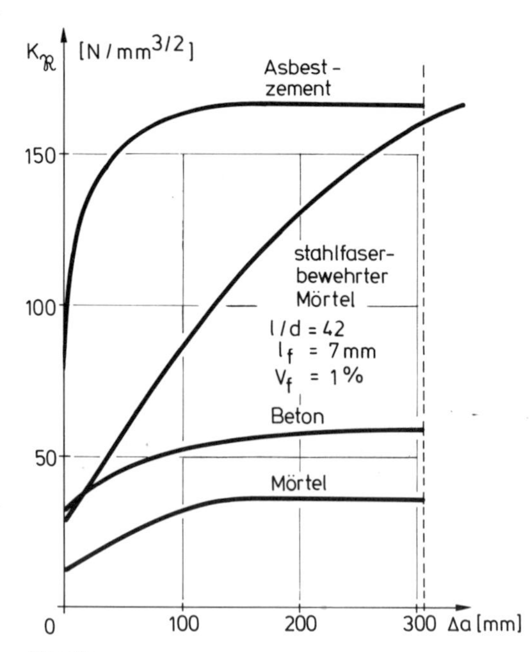

Bild 13
K_R-Kurven für verschiedene zementgebundene Werkstoffe [14]

Bei metallischen Werkstoffen wird bei einem bruchmechanischen Konzept davon ausgegangen, daß COD vor der Rißerweiterung eine werkstoffspezifische Kenngröße ist. Vielfach wird dabei von der vereinfachenden Annahme ausgegangen, daß sich die Rißwände geradlinig verschieben (Bild 14). Bei zementgebundenen Werkstoffen bildet sich infolge der vorhandenen Mikrorisse eine sogenannte Prozeßzone vor der Rißwurzel aus. Diese Prozeßzone ist von der Größe der Zuschlagstoffe abhängig. Es wird angenommen, daß sie bei Mörtel etwa 20 cm und bei Beton etwa 100 cm lang ist. Die stabilen Rißerweiterungen Δa können hierbei mehrere cm betragen (Bild 13).
Anstelle von COD wurde in [14] ein *konstanter Rißöffnungswinkel* COA (rack opening angle) an „double cantilever beams" nachgewiesen, der *genau dann erreicht* ist, *wenn stabile Rißerweiterung einsetzt* (Bild

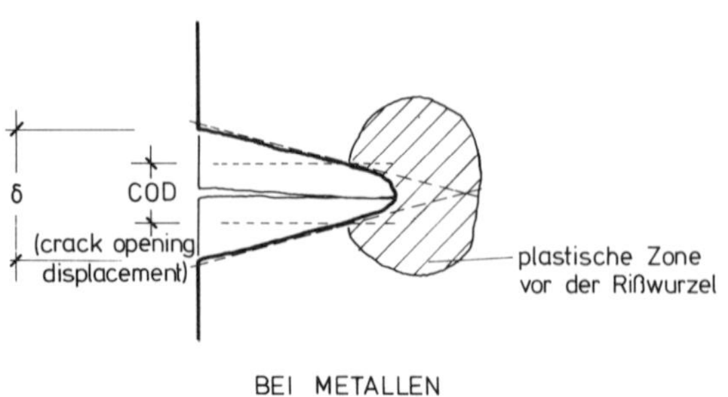

Bild 14
Die Rißuferschiebung bei Metallen

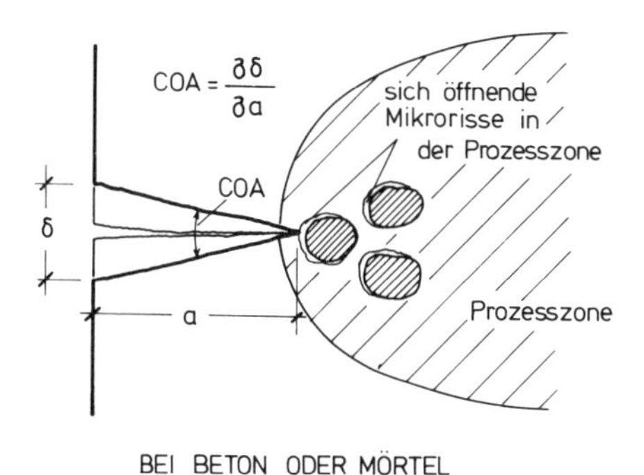

Bild 15
Die Rißöffnung bei zementgebundenen Werkstoffen

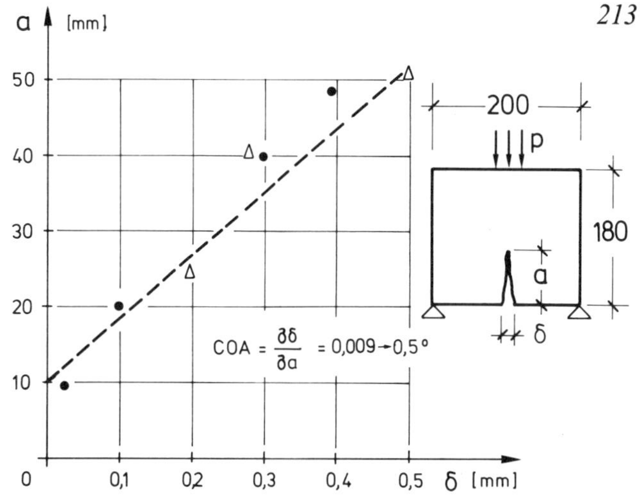

Bild 17
An hohen GFB-Biegeproben gemessene Rißöffnungswinkel bei Rißfortschritt

15). Ergebnisse derartiger Untersuchungen von [14] zeigt Bild 16. Man erkennt, daß der COA durch die Zugabe von Fasern erhöht, der Werkstoff aber zäher wird. Ergebnisse eigener Versuche an hohen Biegeproben aus GFB zeigt Bild 17. Der hierbei festgestellte COA beträgt 0,5° und erscheint recht hoch. Aber auch unsere Versuche an doppelt gekerbten Zugproben, auf die hier nicht näher eingegangen wird, zeigten dieselbe Größenordnung. Bei der folgenden Berechnung des Rißwiderstandes von GFB werden COA = 0,13° des Mörtels nach [14] gewissermaßen als untere Schranke, ein Zwischenwert COA = 0,3° und der eigene festgestellte Wert COA = 0,5° berücksichtigt. Für die Zukunft ist ein besseres Wissen über die Größe und die Streuung von COA nötig.

Formulierung der $\mathscr{R}$-Kurven für Glasfaserbeton

Der Anfangsriß mit der Länge a_0 am Rand einer unendlichen Halbscheibe beginnt sich unter einem Zugspannungsfeld zu öffnen. a_0 hat etwa die Größenordnung des Größtkorns oder des größten Faserabstandes.

Entlang der Rißachse kann man 4 Zonen unterscheiden (Bild 18):

Zone 1:

Die Anfangsrißlänge a_0, in der keine Fasern den Riß überbrücken.

Zone 2:

Hier überbrücken Fasern den sich öffnenden Riß. Diese Zone wird oft auch als pseudoplastischer Bereich bezeichnet. Ihre maximale Ausdehnung a_R ist erreicht, wenn die Fasern an der weitesten Rißöffnung herausgezogen sind oder zu reißen beginnen. Der Rißwiderstand $\mathscr{R}$ dieser Zone enthält sämtliche Anteile aus den Fasern, die die Rißweiterung behindern, im einzelnen: Die Kräfte in den Einzelfasern, das Lösen des Haftverbundes, das Herausziehen der Fasern aus der Matrix,

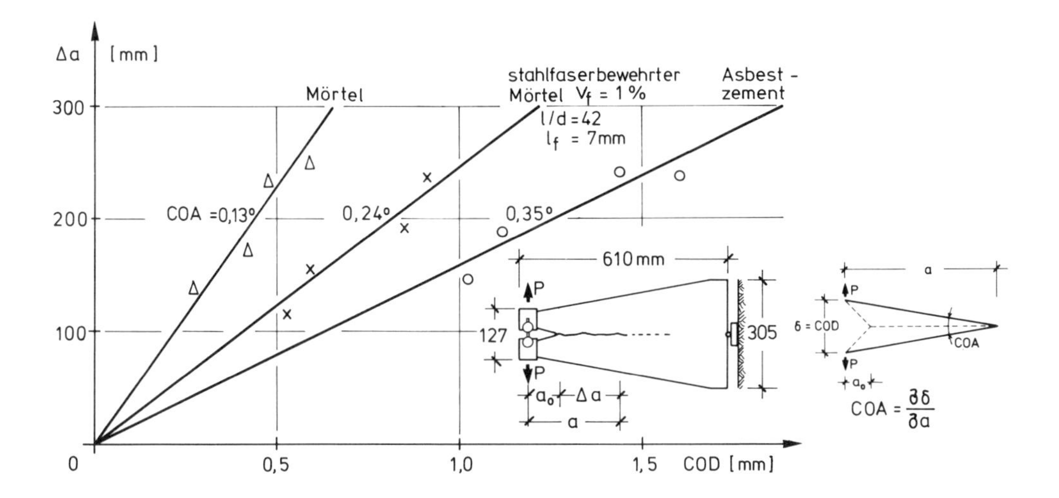

Bild 16
An „double cantilever beams" an zementgebundenen Werkstoffen gemessene Rißöffnungswinkel bei Rißfortschritt [14]

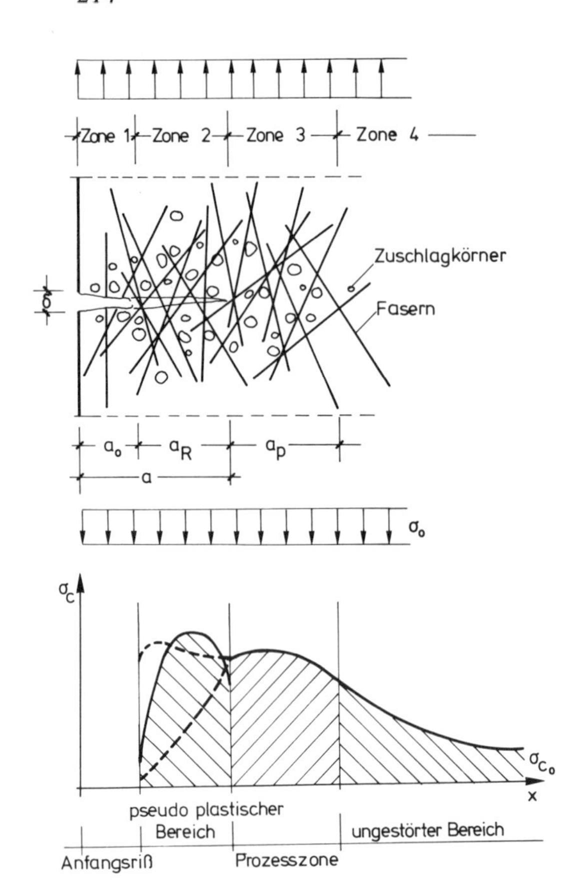

Bild 18
Die Spannungsverteilung vor der Rißwurzel bei Faserbeton

wenn die Restfaserlänge am Riß < $L_{cr}/2$ ist (L_{cr} vgl. Bild 19) und die Energieanteile aus dem Umbiegen der Fasern bei sich öffnendem Riß (Bild 20).

Zone 3:

In dieser Prozeßzone öffnen sich vorhandene Mikrorisse stabil. Die σ-ε-Beziehungen sind nichtlinear und unbekannt. Wie Bild 13 zeigt, bewirkt der Mörtel selbst auch einen Widerstand gegen die Rißausbreitung. Er ist jedoch, wie in Abschnitt 5 gezeigt wird, im Vergleich zu dem Anteil der Fasern in der Zone 2 verschwindend gering, so daß er vernachlässigt wird.

Zone 4:

In diesem ungestörten Bereich gelten die normalen σ-ε-Beziehungen.

In Abhängigkeit von der Rißöffnung bzw. den Dehnungen entlang der Rißachse werden die in Bild 18 gezeigten Spannungen hervorgerufen. Bei ansteigender Beanspruchung vergrößert sich die Rißöffnung. Bei einer bestimmten Rißöffnung sind die Fasern aus

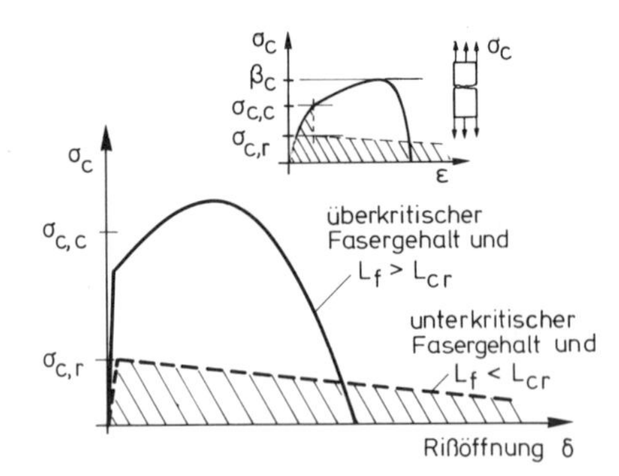

$\sigma_{c,c}$ – Rißspannung beim Reißen der Matrix (*c* für crack)
$\sigma_{c,r}$ – Restspannung nach dem Reißen der Matrix
L_{cr} – Mindestfaserlänge, damit Faserversagen und nicht Verbundversagen gewährleistet ist
L_f – Faserlänge

Bild 19
Typische Spannungs-Verschiebungs-Beziehung am Riß (σ_c: Composit-Spannung auf den Brutto-Querschnitt bezogen)

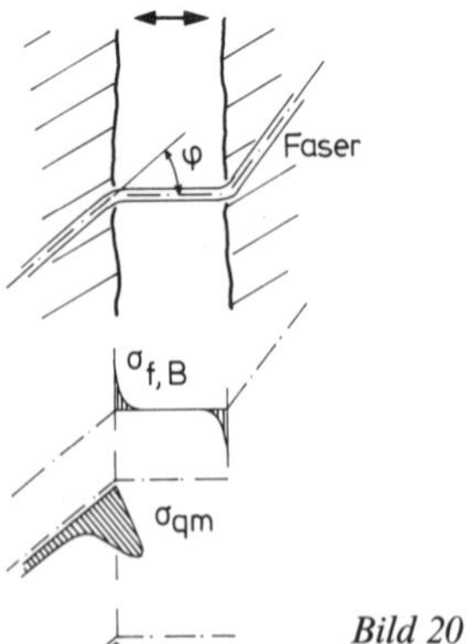

Bild 20
Verlauf der zusätzlichen Faserspannungen bei sich öffnendem Riß

einem Rißufer vollständig herausgezogen oder erreichen ihre Bruchfestigkeit, so daß sich Zone 1 vergrößert, während sich die voll ausgebildeten Zonen 2 und 3 und der Beginn von Zone 4 entlang der Rißachse verschieben.

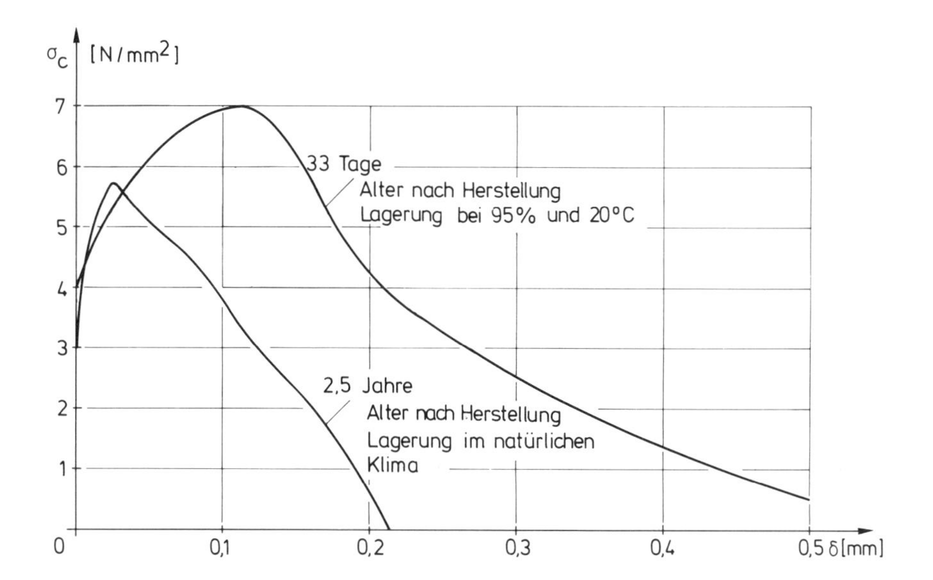

Bild 21
Gemessene Spannungs-Verschiebungs-Beziehungen von GFB mit unterschiedlichem Alter (Zugversuche an schmalen Proben)

Für die Berechnung der Rißverlängerung muß der Energieanteil der Faserbewehrung erfaßt werden. Dafür wird eine Spannungs-Verschiebungs-Beziehung $\sigma - \delta$ am Rißufer benötigt (Bild 21), die aber für Glasfaserbeton zeit- und umgebungsabhängig ist. Die Gründe dafür sind:

Die Festigkeit des Gleit- und Haftverbunds der Fasern nimmt mit der Zeit zu, da die inneren Fasern der Faserbündel erst nach und nach durch in die Bündel wachsende Hydrationsprodukte „befestigt“ werden. Die Faserfestigkeit selbst nimmt in der Zement-Mörtel-Matrix mit zunehmendem Alter ab. Die Faserbündel sind bei Glasfaserspritzbeton in der Ebene in allen Richtungen gleichmäßig verteilt, so daß Fasern, die nicht rechtwinklig zum Riß liegen, bei der Rißöffnung zusätzliche Querbeanspruchungen erfahren. Die Größe dieser Spannungen hängt wesentlich von der Härte der Einspannung am Rißrand ab. Mit fortschreitendem Alter wird die Einspannung des Faserbündels härter und es gibt zunehmend Faserbrüche an den Rißwänden (Bild 20).

Deshalb ist es unumgänglich, die σ-δ-Beziehung für GFB experimentell zu ermitteln (Bild 21).

Die Energie e an der Stelle x des Rißufers mit der Verschiebung $\delta(x)$ beträgt damit

$$e_{(x)} = \int_0^{\delta_x} \sigma_{(\delta x)} \, \mathrm{d}\delta \tag{4.10}$$

und die Gesamtenergie entlang der Rißfront a

$$A = \int_0^a \int_0^{\delta_x} \sigma_{(\delta x)} \, \mathrm{d}\delta \, \mathrm{d}a \tag{4.11}$$

Bei einer Rißverlängerung um da und einer momentanen Rißöffnung $\delta_{(a)}$ beträgt die Energieänderung, die dem Rißwiderstand R der Zone 2 entspricht.

$$\mathscr{R} = \frac{\partial \mathscr{A}}{\partial a} = \int_0^{\delta_a} \sigma_{(\delta)} \, \mathrm{d}\delta \tag{4.12}$$

5 Berechnung des Rißwiderstandes und des Rißfortschrittes

Unter Verwendung der genannten COA-Werte und mit Bild 21 nach Gl. (4.12) berechnete Verläufe des *Rißwiderstandes* in Abhängigkeit von der Rißerweiterung a zeigen die Bilder 22 bis 24. Man erkennt den Einfluß des Probenalters und des Rißöffnungswinkels COA sowie den großen Unterschied zwischen GFB und reinem Mörtel.

Für die Berechnung des *Rißfortschrittes* sind im allgemeinen Fall, d.h. bei beliebig geformten Bauteilen, Lösungsansätze notwendig, die nur mit Hilfe der FE-Methode gefunden werden können. Dieses Vorgehen wird in [11] für elastisch-plastische Werkstoffe beschrieben. Eine Lösung, die von einem bekannten Verlauf der R-Kurve für den untersuchten Fall ausgeht, wird in [15] gezeigt.

Um die Berechnung des Rißfortschritts am Beispiel vorzuführen, genügt hier der Sonderfall der unendlichen Scheibe mit Innenriß im einaxialen Zugspannungsfeld entsprechend Bild 7. Für sie gilt näherungsweise [16]

$$\mathscr{G}_{el} = \frac{2\pi\sigma^2 a}{E} \tag{5.1}$$

mit der momentanen Rißlänge $a = a_0 + \Delta a$.

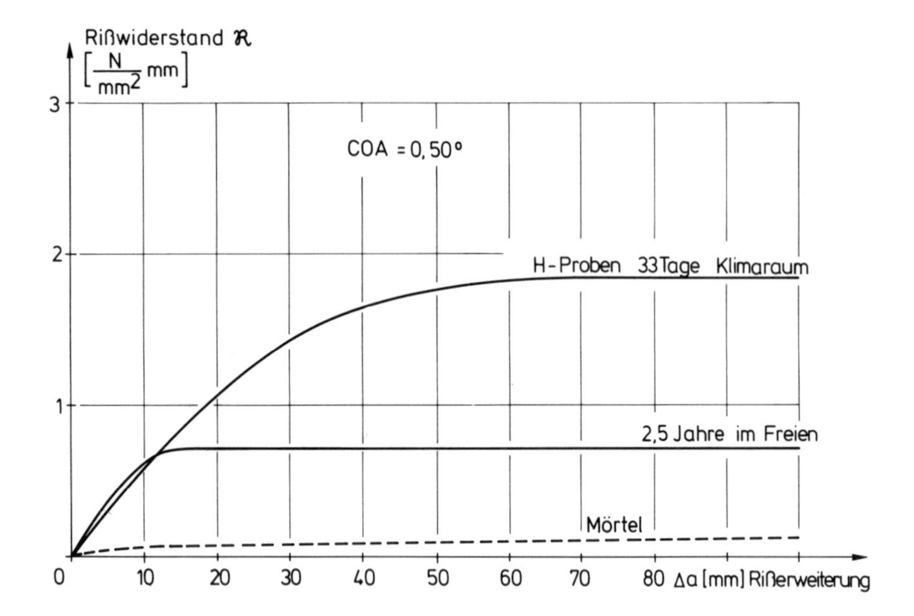

Bild 22
$\mathcal{R}$-Kurven von GFB mit Cemfil *I*
($V_f = 4{,}5$ Vol.-% und $l_f = 50$ mm)
bei einem Rißöffnungswinkel COA = 0,50°
und für unterschiedliches Alter

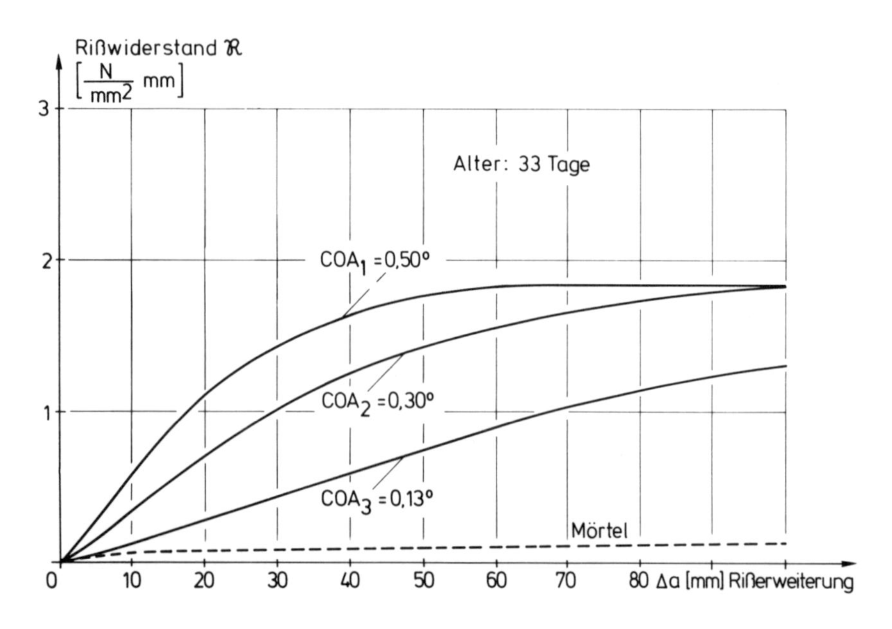

Bild 23
$\mathcal{R}$-Kurve von GFB für ein Betonalter
von 33 Tagen im Klimaraum
und bei unterschiedlichen Rißöffnungswinkeln

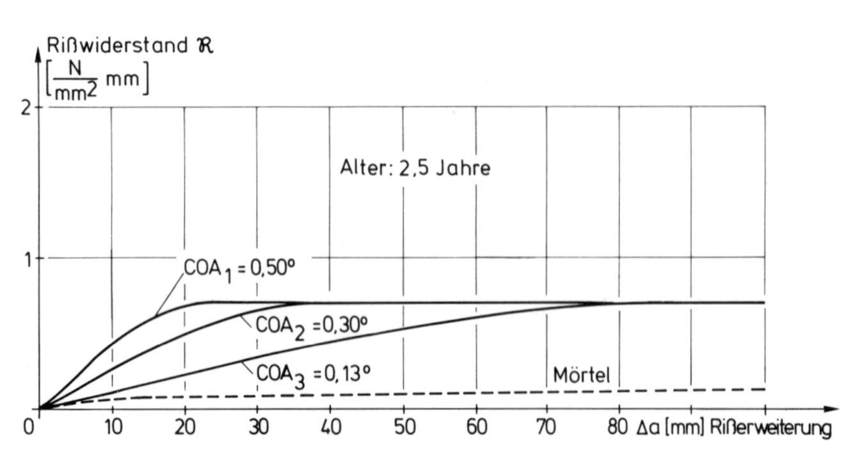

Bild 24
$\mathcal{R}$-Kurven von GFB für ein Betonalter
von 2,5 Jahren im natürlichen Klima
und bei unterschiedlichen Rißöffnungswinkeln

Für eine gegebene Spannung σ und die Anfangsrißlänge a_0 kann so $\mathcal{G}$ in Abhängigkeit von der Rißerweiterung Δa aufgetragen werden (Bild 25). Eine Rißerweiterung findet statt, sobald die Funktion $f = 0$ entsprechend Bild 25 ist. Es folgt dann aus Gl. (4.10)

$$\mathcal{G} = \mathcal{R} \tag{5.2}$$

und nach Bild 25 kann am Schnittpunkt der $\mathcal{G}$-Geraden und der $\mathcal{R}$-Kurve die zu σ_1 gehörige Rißerweiterung Δa_1 abgelesen werden.

Die Rißerweiterung wird instabil, und die Scheibe versagt bei kraftgesteuerter Belastung unter $\max \sigma$, wenn die Gl. (4.9) erfüllt ist

$$\frac{\partial \mathcal{G}}{\partial a} = \frac{\partial \mathcal{R}}{\partial a} \tag{5.3}$$

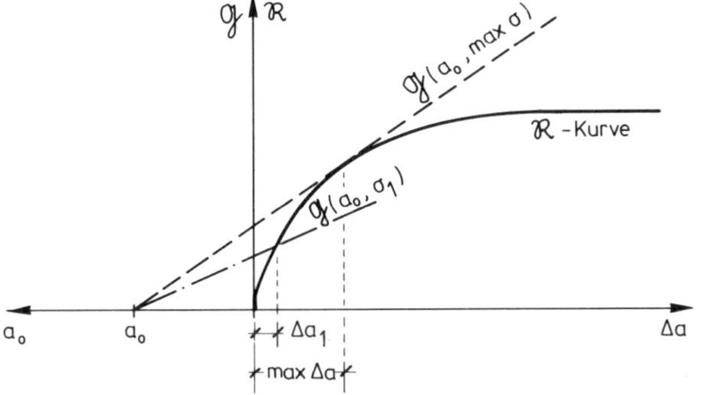

Bild 25
Ermittlung des Rißwachstums Δa unter steigender Last ausgehend von der Anfangsrißlänge a_0

Nach Bild 25 kann die zu $\max \sigma$ gehörige Rißerweiterung $\max \Delta a$ am Berührungspunkt der $\mathcal{G}$-Geraden und der $\mathcal{R}$-Kurve abgelesen werden.

Das Bild 26 enthält das *gesammelte Ergebnis* derartiger Berechnungen für Glasfaserbeton und Mörtel. Ausgehend von der Anfangsrißlänge a_0 kommt man über die linke Seite des Diagramms zu der maximal ertragbaren Spannung und dann rechts zu der zugehörigen Rißerweiterung Δa bzw. der Gesamtrißbreite $a = a_0 + \Delta a$ unmittelbar vor dem Versagen. Deutlich erkennt man, um wieviel *günstiger GFB gegenüber reinem Mörtel* ist, selbst bei dem als untere Schranke angesehene COA = 0,13°, aber auch daß GFB mit zunehmendem Alter seine Qualitäten einbüßt.

6 Zusammenfassung, Bewertung und Ausblick

In den Abschnitten 4 und 5 wurde gezeigt, daß die Bruchmechanik das geeignete Werkzeug zur Berechnung des Rißfortschritts in Beton sein kann. Es gelang damit der Nachweis der aus Versuchen bekannten Verbesserung der Eigenschaften des Mörtels durch die Beimischung von Fasern, also vor allem des erhöhten Rißwiderstandes.

Dieser erhöhte Rißwiderstand macht Glasfaserbeton für die baupraktische Anwendung dort interessant, wo man der Gefahr entgegenwirken will, daß von örtli-

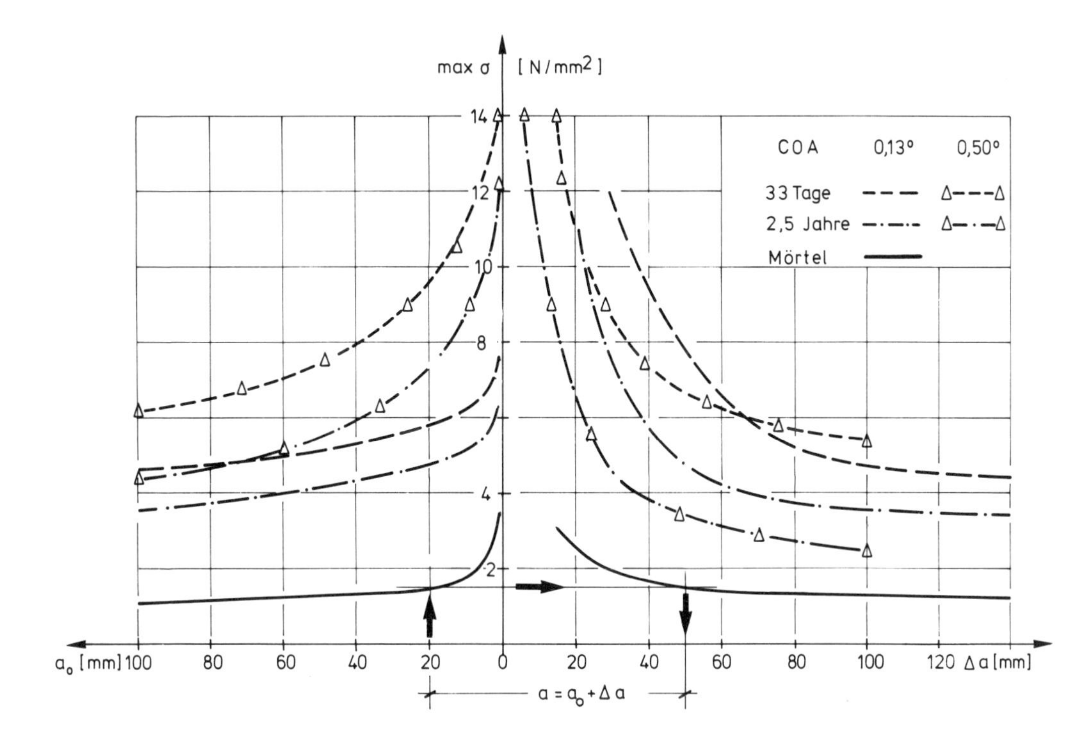

Bild 26
Einfluß der Anfangsrißlänge a_0 auf die maximal ertragbare Spannung des einaxialen Spannungsfeldes und die zugehörige stabile Rißerweiterung für Mörtel (COA = 0,13°) und GFB (0,13° ≦ COA ≦ 0,5°) mit unterschiedlichem Alter

Tabelle 1
Rißabstände und Rißbreiten von stabstahlbewehrtem GFB (V_f = 4,5 Vol.-% und l_f = 50 mm) sowie die berechneten Rißbreiten nach CEB

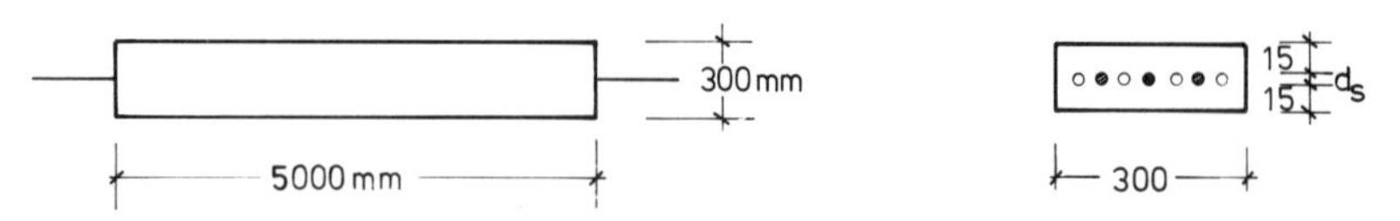

				Rißabstand			Rißbreite						Anzahl der Beobachtungen
				Gerechnet nach CEB	Gemessen		Gerechnet nach CEB		Gemessen				
Bewehrung	μ_s	σ_c [N/mm²]	σ_s [N/mm²]	$\bar{s}_r$ [mm]	$\bar{s}_r$ [mm]	s_{sr} [mm]	$\bar{w}$ [mm]	w_{95} [mm]	$\bar{w}$ [mm]	s_w [mm]	w_{95} [mm]	max w [mm]	
1∅16	1,43	4,1	286	130	7,7	7,2	0,1	0,17	0,014	0,006	0,024	0,05	257
4∅8	1,68	3,1	180	80	8,6	5,6	0,08	0,136	0,012	0,005	0,020	0,04	232
2∅12	1,75	5,0	286	128	8,0	4,8	0,11	0,187	0,014	0,008	0,027	0,04	373

$\bar{s}_r$ mittlerer Rißabstand; $\bar{w}$ mittlere Rißbreite; s_{sr} Standardabweichung der Rißabstände; s_w Standardabweichung der Rißbreite

chen Spannungsspitzen ein Versagen ausgeht, und wo man feinst verteilte Risse anstrebt. So kann mit GFB beispielsweise das Entstehen von Rissen im Verankerungs- oder Umlenkbereich von Bewehrungsstäben erheblich verzögert oder die Rißbildung selbst bei hohen Zwangsbeanspruchungen unschädlich gemacht werden.

Dies wird durch eine der in der Einleitung genannten Versuchsreihen über die „hybride" Verwendung von GFB eindrucksvoll bestätigt [3]. Die Tabelle 1 zeigt die Zusammenfassung von Rißbreitenmessungen an zugbeanspruchten stabstahlbewehrten GFB-Scheiben (Querschnitt 300/40 mm², Länge 5000 mm). Die Versuchsergebnisse sind verglichen mit den Werten, die sich nach CEB-Model-Code rechnerisch für Stahlbeton ohne Fasern ergeben. Man erkennt, daß die mittleren Rißbreiten durch die Beimischung von 4,5 Vol.-% Fasern zum Mörtel auf etwa $\frac{1}{10}$ zurückgehen. Es wird nicht nur ein sehr feines, sondern auch sehr gleichmäßiges Rißbild erreicht. Unter Gebrauchslast ergeben sich mittlere Rißbreiten unabhängig vom Stabdurchmesser um 0,014 mm bei einem Variationskoeffizienten um 60%, so daß als 95%-Fraktile aus einer Gesamtheit von ca. 900 Einzelrißmessungen ein W_{95} = 0,027 mm angegeben werden kann. Die günstige Wirkung des GFB wird aber erst voll deutlich, wenn noch mitgeteilt wird, daß die Versuchskörper schon im unbelasteten Zustand dadurch stark gerissen waren, daß das Schwinden des Mörtels durch die Bewehrungsstäbe behindert wird und die Rißbreiten bei gleichzeitig durchgeführten Versuchen an stabstahlbewehrten Mörtel-Scheiben ohne Fasern größer als 1,0 mm waren.

Allerdings ist dieser Schwindeinfluß, weil er sehr stark streut, der Grund, warum in Tabelle 1 rechnerische CEB-Werte zum Vergleich herangezogen wurden. Wegen dieses Schwindens gelang es auch bisher nicht, die Tabelle 1 bruchmechanisch nachzurechnen. Wir hoffen aber bis zur Festschrift zum 70. oder 80. Geburtstag von G. Rehm so weit zu sein.

7 Literatur

[1] Schlaich, J. und Menz, W.: The application of glass fibre reinforced concrete for shell structures. IASS-Symposium, Oulu, Finnland 1980.
[2] Schlaich, J.: Die Glasfaserbetonschale für die Bundesgartenschau 1977 in Stuttgart. Betontag 1977, Hamburg.
[3] Menz, W.: Verbundsysteme mit Glasfaserbeton. SFB 64-Mitteilungen 67/1984.
[4] Schlaich, J. und Menz, W.: Faserbetone. Baukalender 1979, Werner-Verlag.
[5] Schlaich, J. und Menz, W.: Untersuchungen an Glasfaserbeton in Zusammenhang mi einem Versuchsbau 1981/82. Forschungsbericht für die Forschungsgemeinschaft Bauen und Wohnen, Stuttgart.

[6] Oakley, D.R. und Proctor, A.B.: The Use of accelerated ageing procedures to predict the long term strength of GRC Composites. Cement and Concrete Research, Vol. 11, No. 3, May 1981.
[7] Bijen, J.: Durability of Some Glass Fiber Reinforced Cement Composites. ACI Journal, July–August 1983.
[8] Majumdar, A. und Singh, B.: Building Research Establishment. Current Paper CP 94/75, October 1975.
[9] Wittmann, F.H. und Zaitsev, Y.B.: Simulation of crack propagation and failure of concrete. Matériaux et Construction, Vol. 14, No. 83.
[10] Schmidt, M.: Stahlfaserspritzbeton. Eigenschaften, Herstellung und Prüfung. Beton 9/83.
[11] Doltsinis: Zur Berechnung des Rißfortschrittes in inelastischen Tragwerken. Zamp, Zeitschrift für angewandte Mathematik und Physik, Vol. 30, 1979.
[12] Griffith: The Theory of Rupture. 1st Int. Congress Appl. Mech. Delft (1924).
[13] Irwin: Fracture. Handbuch der Physik. Vol. 6, Springer Berlin 1958.
[14] Visalvanich, K. und Naaman, A. E.: Fracture Model for Fiber Reinforced Concrete. ACI Journal, March–April 1983.
[15] Bloom, J. H.: Elastisch-plastische Versagensanalyse. MPA-Seminar, Stuttgart, 1982.
[16] Hahn, H. G.: Bruchmechanik. Teubner-Verlag, Stuttgart.

Verhalten des Betons im verformungsgesteuerten axialen Zugversuch

Hans-Wolf Reinhardt

1 Einleitung

Lange Zeit wurde die Zugfestigkeit des Betons vernachlässigt. Sie galt – und gilt in vielen Fällen noch stets – als eine unzuverlässige Eigenschaft des Betons. Schwind- und Temperaturspannungen sorgen dafür, daß die für mechanische Belastungen nutzbare Zugfestigkeit manchmal nur ein Bruchteil der wirklichen Festigkeit ist. Andrerseits ist es möglich, durch sorgfältige Überlegungen die zufälligen Spannungen zu analysieren und die wirklichen Reserven der Zugfestigkeit zu bestimmen.

Unbewußt macht man bei jeder Betonkonstruktion von der Zugfestigkeit des Betons Gebrauch. Bei nicht auf Schub bewehrten Balken und Platten ist die Schubtragfähigkeit von der Zugfestigkeit abhängig; der Durchstanzwiderstand wird hauptsächlich von der Zugfestigkeit bestimmt; das Verbundverhalten von gerippten Stäben wird maßgeblich davon beeinflußt; die nachträgliche Befestigung von Elementen mit Hilfe von Ankern und Dübeln ist ohne Zugfestigkeit nicht denkbar, um nur einige Beispiele zu nennen.

Galt bisher die Aufmerksamkeit der Zugfestigkeit als einer Spannungsgröße, so nimmt heute das Interesse zu im Hinblick auf die Verformung und die Bruchenergie. Denn Schwind- und Temperaturprobleme sind häufig eher ein Verformungsproblem als ein Spannungsproblem, und die Frage der Bruchenergie stellt sich bei der Schematisierung des Verhaltens des Betons bei der Berechnung mit Hilfe von finiten Elementen. Die Kenntnis des Elastizitätsmoduls, der Zugfestigkeit und der Bruchenergie ist nötig und ausreichend, um das Materialverhalten zu schematisieren [1], [2].

Die Bestimmung der vollständigen Spannungs-Dehnungs-Linie des Betons ist ein erster Schritt in dieser Richtung. Sie liefert die genannten drei Größen für den speziellen Fall der einmaligen Belastung bis zum Bruch. Häufig wiederholte Belastung im Zugbereich oder im Zug-Druck-Bereich sind der allgemeinere Belastungsfall, und die Frage stellt sich dann, wie die genannten Eigenschaften dadurch beeinflußt werden. Im folgenden wird die Versuchsmethode für den verformungsgesteuerten Zugversuch beschrieben, und es werden einige Versuchsresultate mitgeteilt und analysiert.

2 Versuchsprogramm

2.1 Prüfmethode

Im Stevin-Laboratorium der TH Delft wurde vor einigen Jahren eine Prüfanlage entwickelt für die Bestimmung der Ermüdungsfestigkeit des Betons unter Zug- und Zug-Druckbelastung [3]. Diese Anlage wurde für die laufenden Untersuchungen angepaßt.

Die Belastung wird mit einem 100-kN-Hydraulikzylinder aufgebracht, der mit einem elektro-magnetischen Servoventil gesteuert wird. Das gewünschte Steuersignal, z. B. konstante Verformungsgeschwindigkeit, sinusförmige Belastung, gegebenes Lastspektrum, wird in einem Mikroprozessor generiert und während des Versuchs mit den Meßergebnissen eines Weg- oder Kraftaufnehmers verglichen. Das Servoventil sorgt dafür, daß das erzielte mit dem gewünschten Signal übereinstimmt. Da Beton auf Zug sich relativ spröde verhält, muß die Steuerung entsprechend empfindlich reagieren. Es zeigte sich im Lauf der Versuche, daß ein Ventil mit einer Durchflußmenge von 3,8 l/min und einer Reaktionszeit von 6,5 ms ausreicht für Verformungsgeschwindigkeiten zwischen 0,08 und 0,17 μm/s. Der Probekörper wird mit zwei Kugelgelenken und einer Kraftmeßdose in Serie geschaltet. Die Kugelgelenke bestehen aus Kegelrollenlagern, die vorgespannt werden können, so daß kein Spiel beim Übergang von Zug- auf Druckbelastung auftritt. Diese gelenkige Lagerung sorgt dafür, daß die Belastung zentrisch eingetragen wird. Mit Hilfe eines speziellen Eichgeräts konnte festgestellt werden, daß die Exzentrizität tatsächlich nicht mehr als 1 mm beträgt. Sobald jedoch ein Riß im Probekörper entsteht, wird die Belastung exzentrisch. Dies ist bei Ermüdungsversuchen kaum ein Nachteil, da dann ohnehin bereits die Bruchlastspielzahl fast erreicht ist. Für die Bestimmung der vollständigen Spannungs-Dehnungs-Linie ist dies jedoch nicht akzeptabel. Deshalb wurde eine Parallelführung entwickelt, die auch nach Anriß eines Probekörpers die Rotation verhindert. Die Ergebnisse von Dehnungsmessungen haben die Wirksamkeit dieser Einrichtung bestätigt.

Bisher wurden drei Verformungs-Zeit-Funktionen für die Belastung verwendet. Die erste ist die konstante Verformungsgeschwindigkeit von 0,08 bis 0,16 µm/s bei einer Meßlänge von 35 mm. Die zweite ist die wiederholte Belastung zwischen einer festen Unterspannung, die entweder eine Zug- oder eine Druckspannung sein kann, und einer Oberspannung, die auf der Hüllkurve (abfallender Ast des Spannungs-Dehnungs-Diagramms) liegt. Die Steuerung geschieht dabei so, daß die Verformungsrichtung automatisch umgekehrt wird, sobald die Kraft um 450 N abfällt (Obergrenze) oder die Untergrenze erreicht wird. Die dritte Verformungsart besteht aus einer Anzahl von Lastspielen zwischen einer festen Untergrenze und einer festen Obergrenze, die ca. 75% der Zugfestigkeit beträgt. Die Verformung nimmt dabei ständig zu, so daß schließlich die Hüllkurve erreicht wird. Daraufhin wird entsprechend der zweiten Verformungsart verfahren.

2.2 Probekörper

In einer ersten Versuchsreihe wurden zylindrische Probekörper verwendet, die zu guten Resultaten führten [4]. Der Nachteil dabei ist allerdings, daß die Dehnung und eine eventuelle Rißöffnung nur an der Oberfläche der Probe gemessen werden können. Dies war der Grund, um in der jetzigen Versuchsreihe Probekörper mit rechteckigem Querschnitt zu verwenden. Die Abmessungen sind $60 \times 50\ mm^2$ im Querschnitt und 250 mm Länge. Zur Festlegung der Bruchzone werden die Probekörper an zwei Seiten eingesägt, so daß ein Nettoquerschnitt von 50×50 mm verbleibt. Bei einem Größtkorn des Betons von 8 mm erscheint dies ausreichend.

Die Probekörper werden stehend in der Form von 300 mm breiten Scheiben hergestellt, um den Betonquerschnitt so homogen wie möglich zu machen. Nach 14tägiger Feuchtlagerung werden die Proben auf Maß gesägt und auch die seitlichen Sägeschnitte angebracht. Danach trocknen die Proben in der Prüfhalle bei ca. 20 °C und 60% rel. Luftfeuchtigkeit 40 bis 45 Tage lang aus.

Zur Prüfung werden die Körper in der Versuchsanlage ausgerichtet und mit einem Epoxydharz an die Belastungsplatten geklebt.

2.3 Versuchsmaterial

Als Versuchsmaterial wurde ein Beton mit folgender Zusammensetzung verwendet (Tabelle 1).

Die mechanischen Eigenschaften des Betons nach 28 Tagen Normlagerung ergaben sich wie folgt (Tabelle 2). Die jeweilige Standardabweichung ist in Klammern angegeben.

2.4 Meßgeräte

Zur Messung der Verformungen wurden sowohl Dehnmeßstreifen ($5 \times 20\ mm^2$) wie mechanisch-elektrische Meßbügel verwendet. Letztere haben einen größeren Meßbereich als Dehnmeßstreifen, sind jedoch bei sehr kleinen Verformungen ungenauer. Die Meßlänge betrug 35 mm für die Bügel in der Bruchzone und 100 mm für die Bügel in der Probekörperachse. Je 5 bis 11 dieser Meßgeräte waren an der Vorder- und Rückseite angebracht. Auf diese Weise war es möglich, den Dehnungsverlauf über die Breite der Probe zu ermitteln. Seitlich wurden zwei induktive Wegaufnehmer befestigt, die zur Verformungssteuerung dienten.

Tabelle 1
Zusammensetzung des Betons

Zement:	375 kg/m^3	Portlandzement B (etwa PZ 35)
Wasser:	187,5 kg/m^3	(Wasserzementwert 0,50)
Zuschlag:	1268 kg/m^2	Sand 0/4, quarzitisches Material
	541 kg/m^3	Feinkies 4/8, quarzitisches Material

Tabelle 2
Mechanische Eigenschaften des Betons, Mittelwert und Standardabweichung

Würfeldruckfestigkeit (150 mm)	f_c = 47,1 N/mm^2	(s = 2,0 N/mm^2)
Spaltzugfestigkeit	f_{spl} = 3,2 N/mm^2	(s = 0,23 N/mm^2)
Elastizitätsmodul	E = 39000 N/mm^2	(s = 3250 N/mm^2)

3 Versuchsergebnisse

3.1 Einmalige Belastung

Die direkten Ergebnisse eines verformungsgesteuerten Zugversuches sind die Belastungen als Funktion der Verformung. Bild 1 gibt die Spannungsverformungskurve exemplarisch wieder für einen Probekörper. Die Spannung σ ist dabei die Kraft pro Nettoquerschnitt. Die Verformung δ ist in μm angegeben und nicht in % oder ‰. Der Grund dafür ist, daß im Bereich des abfallenden Astes der σ-δ-Linie keine gleichmäßige Dehnung mehr auftritt, sondern daß sich eine Rißzone ausbildet, die sich weiter verformt, während der Rest der Probe elastisch zurückfedert. Bis zum Erreichen der Zugfestigkeit könnte man auch die Dehnung auftragen als δ/l_0 (l_0 = Meßlänge = 35 mm), während danach die örtliche Verformung als geeignete Meßgröße erscheint. In Bild 1 ist zu erkennen, daß die σ-δ-Linie erst linear verläuft, kurz vor Erreichen der Zugfestigkeit f_t abbiegt und nach Erreichen von f_t steil abfällt. Die Spannung nimmt dann stetig weiter ab, bis die Probe bei einer Gesamtverformung von etwa 175 μm vollständig getrennt ist.

Bild 2 gibt einen Eindruck von dem Verformungsverlauf über die Breite der Probe. Jeder Meßpunkt ist dabei der Mittelwert aus der Messung an der Vorder- und Rückseite der Probe. Bis auf die Messungen 4 und 5 ist ein gleichmäßiger Verformungsverlauf zu sehen, d. h. eine zentrische Verformung der Probe. Messung 4 und 5 weisen auf eine kleine Exzentrizität hin, die nach Erreichen der Zugfestigkeit auftritt. Offensichtlich beginnt der erste Anriß an der oberen Seite und eilt der Verformung an der Unterseite vor. Bei der Messung 6 ist der Verlauf bereits wieder gleichmäßig.

3.2 Wiederholte Belastung

Bild 3 ist das Ergebnis eines Versuches mit wiederholter Belastung, wobei die Unterspannung eine Zugspannung von 0,2 N/mm^2 ist. Es ist zu sehen, daß bei jedem Lastspiel eine charakteristische Schleife entsteht. Die Steigung des Ent- sowie des Belastungsastes nimmt

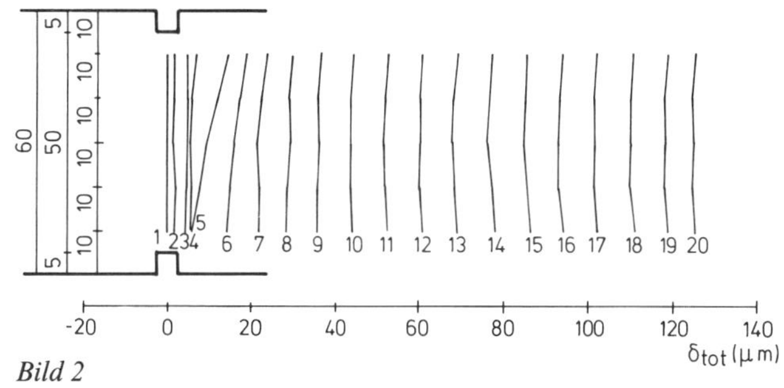

Bild 2
Verformungsverlauf über die Breite der Probe (Zahlen stimmen mit Verformungsschritten in Bild 1 überein)

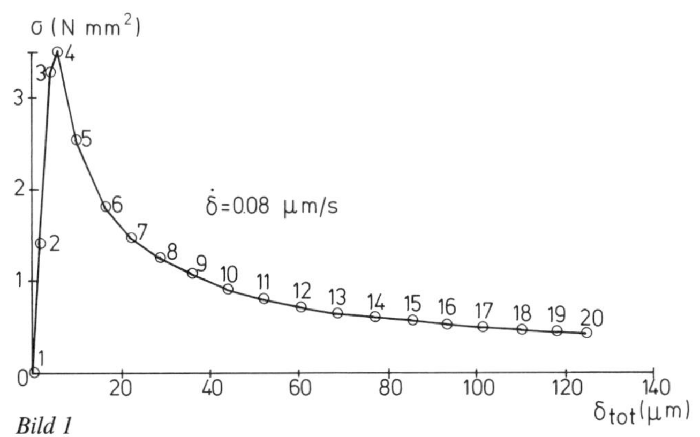

Bild 1
Spannungs-Verformungs-Linie mit konstanter Verformungszunahme

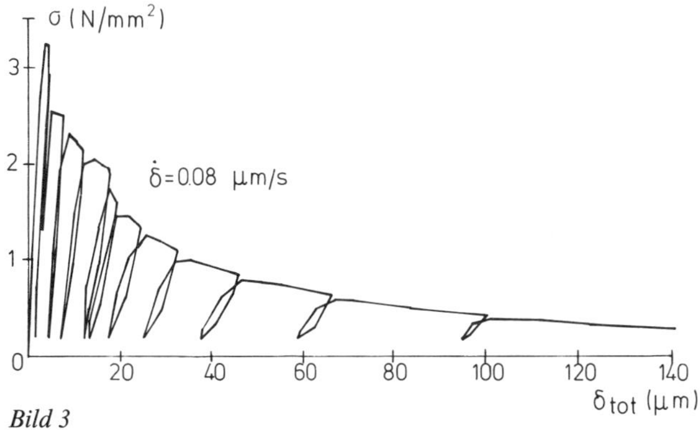

Bild 3
Spannungs-Verformungs-Linie bei wiederholter Belastung im Zugbereich

ständig ab, während die bleibenden Verformungen zunehmen. Folgt man der Definition von DOUGILL [5] so deutet der Flächeninhalt der Schleifen auf Materialdämpfung und Reibungsverluste hin, die bei der Verformung von Rißflächen entstehen können, während die Zunahme der bleibenden Verformung und die daran verbundene Energieaufnahme mit einer Schädigung des Materials zusammenhängen.
Im folgenden Bild 4 sind entsprechende Resultate wiedergegeben mit dem Unterschied, daß die Unterspannung eine Druckspannung von 3,3 N/mm^2 ist. Nach Erreichen der Zugfestigkeit verläuft die Entlastung beinahe linear mit derselben Steifigkeit wie bei der ersten Zugbelastung. Der Verlauf der zweiten Belastung weicht bereits etwas von der Erstbelastung ab. Mit zunehmender Anzahl von Lastwiederholungen stellt sich ein charakteristischer Verlauf ein: ausgehend von der Hüllkurve zunächst eine steile Spannungsabnahme, die in eine langgezogene Rückverformung übergeht bei einer geringen Druckspannung; danach eine Zunahme des Verformungswiderstands, bis die ursprüngliche Steifigkeit erreicht ist; anschließend Druckentlastung und erneute Zugbelastung mit geringem Widerstand bis zum Erreichen der Hüllkurve. Dieser Verlauf läßt vermuten, daß Rißentlastung, Rißschließung und erneute Rißerweiterung die Merkmale sind, die zu einem solchen Verlauf führen. Dabei braucht „Riß" nicht unbedingt einen sichtbaren Riß zu bedeuten, sondern kann auch die Ausbreitung einer Rißzone oder die zunehmende Rißfläche in einer Rißzone sein.

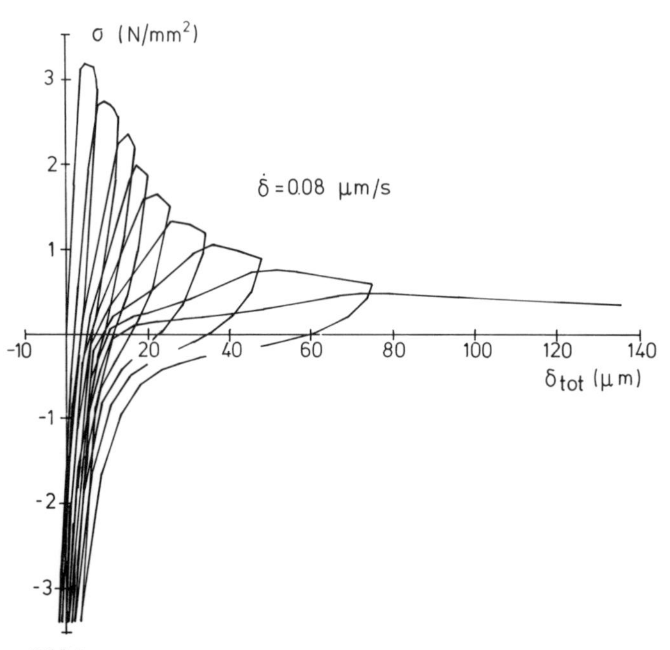

Bild 4
Spannungs-Verformungs-Linie bei wiederholter Belastung im Zug-Druck-Bereich

3.3 Hüllkurve der Spannungs-Verformungs-Linie

Das primäre Resultat der experimentellen Untersuchungen ist die Spannungs-Verformungs-Linie im verformungsgesteuerten Zugversuch, vor allem im Hinblick auf die Frage, ob diese von der Belastungsgeschichte abhängig ist. Um dies zu prüfen, sind im Bild 5 die Hüllkurven der Spannungs-Verformungs-Linien in ein Diagramm gezeichnet. Es zeigt sich dabei, daß die gegenseitigen Unterschiede so klein sind, daß kein signifikanter Einfluß der Belastungsgeschichte erkennbar ist. Dies ist ein Ergebnis, das die Verwendung der Spannungs-Verformungs-Linie in weiteren Betrachtungen, vor allem im Hinblick auf die Modellierung des Ermüdungsverhaltens des Betons, vereinfacht. Dies Ergebnis steht im Einklang mit einachsigen Versuchen an Beton auf Druckbelastung, wo ebenfalls gefunden wurde, daß die Belastungsgeschichte die Hüllkurve nicht beeinflußt [6]. Zur Deutlichkeit muß jedoch bemerkt werden, daß diese Schlußfolgerung innerhalb der Versuchsbedingungen gilt, d. h. für einen Beton mit konstanter Temperatur und Feuchtigkeit und etwa gleicher Verformungsgeschwindigkeit. Sobald sich die Verformungsgeschwindigkeit über zwei Zehnerpotenzen ändert, dürfte sich auch die Spannungs-Verformungs-Linie merkbar ändern.

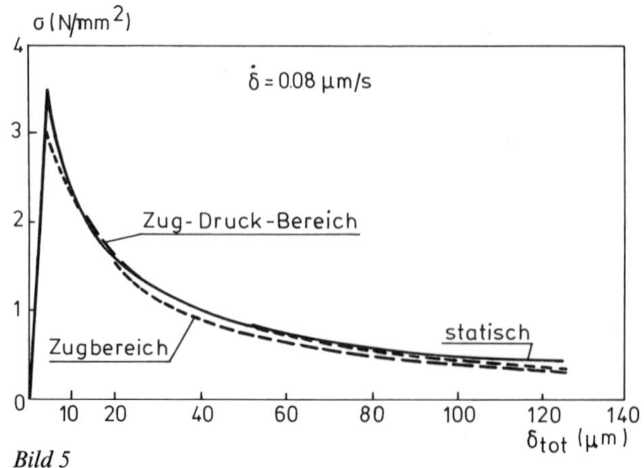

Bild 5
Hüllkurve bei unterschiedlicher Belastungsgeschichte

4 Diskussion der Ergebnisse

Wie im Beginn schon erwähnt, ist die Zugfestigkeit eine Eigenschaft des Betons, die bei der Berechnung der Tragfähigkeit von Betonkonstruktionen meist vernachlässigt wird. Trotzdem gibt es einige Fälle, die stark von der Zugfestigkeit beeinflußt werden und wo letztlich der Beton die Tragfähigkeit bestimmt. In diesen Fällen zeigt sich, daß eine ausschließliche Behandlung mit Hilfe der Festigkeitslehre nicht zum Ziel führt. So ist es damit zum Beispiel nicht erklärlich, weshalb ein kleiner Balken eine größere Schubtragfähigkeit besitzt als ein (geometrisch) maßstäblich vergrößerter Balken. Eine Analyse mit Hilfe der linear elastischen Bruchmechanik hat gezeigt, daß die relative Tragfähigkeitsabnahme mit wachsender Größe auf die Zunahme des Spannungsintensitätsfaktors zurückzuführen ist [7].

Nach den Untersuchungen von HILLERBORG und Mitarbeitern [8] ist die Anwendung der linearen Bruchmechanik auf Beton nicht möglich, es sei denn bei großen Abmessungen. Der Grund liegt in der Tatsache, daß sich Beton nicht linear elastisch verhält, so wie es die lineare Bruchmechanik voraussieht. Im Bereich eines Risses bildet sich immer eine Zone aus, die über die Elastizitätsgrenze gedehnt ist und damit eine geringere Spannung besitzt, als aus dem Gesetz von HOOKE folgt. Ähnliche Erscheinungen sind schon lange bei duktilen Werkstoffen bekannt, wo bei Spannungsspitzen Fließen auftritt und die Spannungsspitzen damit abgeflacht werden. Für duktile Werkstoffe wurde die lineare Bruchmechanik erweitert. Es wird dabei angenommen, daß sich eine Fließzone an der Rißspitze ausbildet, deren Größe von der Fließgrenze des Materials und dem Elastizitätsmodul abhängt [9]. Innerhalb dieser Fließzone sind die Spannungen gleich der Fließgrenze, unabhängig von der Verformung, die innerhalb dieser Zone nicht gleich ist.

Im Unterschied zu Werkstoffen mit einer ausgeprägten Fließgrenze fällt bei Beton die Spannung ab, nachdem einmal der höchste Punkt der Spannungs-Verformungslinie erreicht ist. Die Anwendung der für duktile Werkstoffe erarbeiteten elastisch-plastischen Bruchmechanik sind damit nicht ohne weiteres auf Beton übertragbar. Wohl gelten qualitativ einige Gesetzmäßigkeiten: die Fließzone ist größer bei höheren Spannungen und relativ niedrigerer Streckgrenze. Da die Fließzone eine gewisse Abmessung besitzt, ist der Einfluß dieser Zone größer bei kleinen Abmessungen des Bauteils, d. h. bei kleinen Abmessungen kann sich die Fließzone bereits über den ganzen Querschnitt ausgedehnt haben, während bei einem großen Querschnitt dies erst einige Prozent des Querschnitts ausmacht. Im zweiten Fall würde eine Berechnung mit Hilfe der linearen Bruchmechanik zu einer guten Näherung der Tragfähigkeit führen, während sie diese im ersten Fall stark überschätzen würde.

Ein zweiter Gesichtspunkt ist die stabile Rißausbreitung. In einem angerissenen Zugstab nimmt die Spannung ständig zu bis zu einem kritischen Punkt. Die lineare Bruchmechanik ist in der Lage, den Punkt des instabilen Rißwachstums anzugeben, der bei gegebener Rißlänge und Geometrie erreicht ist, sobald der Spannungsintensitätsfaktor gleich dem kritischen Wert (einer Materialkenngröße) ist. Nach dieser Theorie breitet sich der Riß vorher nicht aus. Bei Beton entspricht dies nicht den Beobachtungen; vielmehr breitet sich dort ein Riss erst stabil (d. h. im Gleichgewicht) aus, bevor es zum instabilen Rißwachstum kommt. Diese Erscheinung hängt ebenfalls mit dem Entstehen einer Fließzone zusammen.

Die Idee der elastisch-plastischen Bruchmechanik auf Beton anzuwenden bedeutet, das Spannungs-Verformungsverhalten des Betons unter axialer Zugbelastung auf eine angemessene Art auf die Rißzone zu übertragen. HILLERBORG und Mitarbeiter [8] schematisieren hierzu die Spannungs-Verformungs-Linie durch eine gerade Linie oder durch zwei geknickte gerade Linien, Bild 6.

Die Fläche unter diesen Linien ist die Bruchenergie G_F. Auch BAZANT [2] nähert die wirkliche Spannungs-Verformungs-Linie durch ein Dreieck an. Die Form des Dreiecks ist festgelegt durch den Elastizitätsmodul, die Zugfestigkeit und die Bruchenergie. Die Steigung des fallenden Astes ist damit auch gegeben. Diese Näherung ist sicherlich grob, wenn man Bild 5 vergleicht; sie wurde mangels genauer experimenteller Daten so gewählt und auch aus der Überlegung, daß die Bruchenergie eine wichtigere Größe ist als die Verformung δ_0, wobei keine Spannung mehr übertragen werden kann. Für statische Belastungsfälle hat sich eine derartige Schematisierung des Betons bewährt, und es wur-

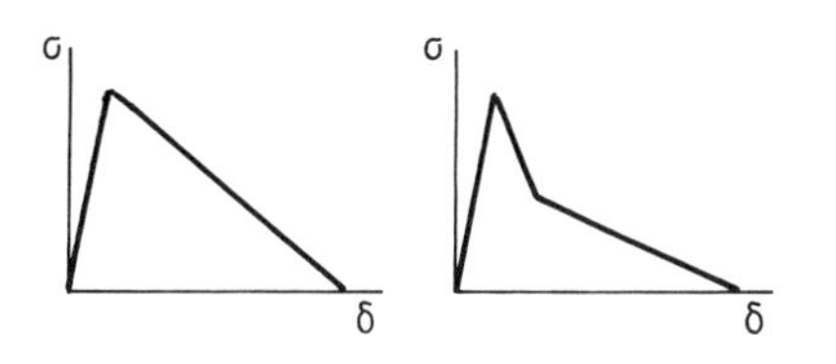

Bild 6
Schematisierung der Spannungs-Verformungs-Linie durch HILLERBORG u. a. [8]

den in Finite-Elemente-Berechnungen gute Ergebnisse erzielt [10].
Zurückkommend auf die Resultate des Zugversuches wurde ein mittlerer G_F-Wert von 122 N/m gefunden mit einem Variationskoeffizienten von 13%. Diese Größenordnung stimmt gut mit Ergebnissen aus der Literatur überein. Für die Berechnung wurde die gesamte Energie unter der Spannungs-Verformungs-Linie berücksichtigt, also angenommen, daß diese Energie vollständig bei der Bildung eines Risses aufgezehrt wird. Mikrorisse, die zusätzlich innerhalb der Meßlänge auftreten, sind also stillschweigend zum Hauptriß gerechnet.
Der fallende Ast der Spannungs-Verformungs-Linie kann gut mit einer Funktion folgender Form angenähert werden:

$$\sigma/f_t = 1 - (\delta/\delta_0)^k$$

wobei δ_0 die Verformung ist, wobei keine Spannung mehr übertragen wird. Aus den Versuchen folgt $\delta_0 =$ 175 µm und $k = 0{,}31$. Der lineare Belastungsast ist dabei als eine Gerade mit der Steigung E angenommen.
Für die Behandlung *wiederholter Belastung* ist die obengenannte Schematisierung nicht ausreichend, gibt sie doch keinen Anhaltspunkt für den Verlauf der Entlastung und erneuter Wiederbelastungen. Eine Ergänzung, die auf den Zugspannungsbereich beschränkt ist und die pro Lastwechsel dissipierte Energie vernachlässigt, ist in Bild 7 dargestellt. Zwei Größen sind dabei von der Rißöffnung abhängig: der Rißöffnungswiderstand C (N/mm³) und der Spannungsabfall $\Delta\sigma$. Wie aus dem Bild hervorgeht, nimmt der Rißöffnungswiderstand ständig ab von ca. 1200 N/mm³ am Beginn auf ca. 30 N/mm³ bei einer Rißöffnung von 100 µm. Dieser Abfall kann am besten mit der Funktion

$$\frac{C}{C_0} = 0{,}05\left[1 - \frac{\delta - \delta_e}{\delta_0 - \delta_e}\right] + 0{,}95\left[1 - \frac{\delta - \delta_e}{\delta_0 - \delta_e}\right]^{35}$$

angenähert werden. Dabei ist δ die Gesamtverformung, δ_e die bei der höchsten Spannung erreichte Verformung und δ_0 diejenige Verformung, bei der die Zugspannung Null geworden ist.
Der Spannungsabfall (Anteil, um den die vorige Spannung nicht mehr erreicht wird) ist am Anfang größer und wird danach kleiner. Eine gute Näherung ergibt

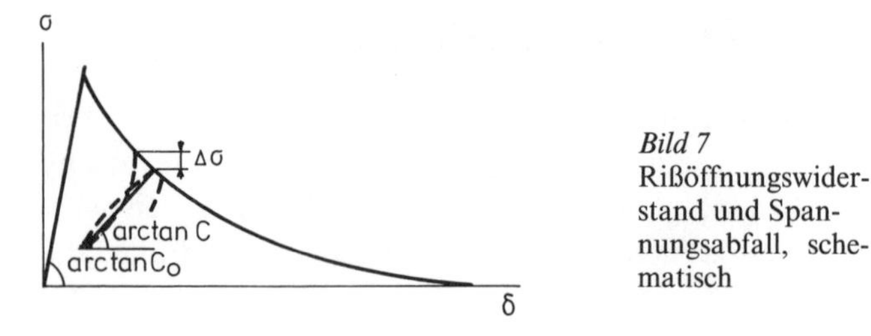

Bild 7
Rißöffnungswiderstand und Spannungsabfall, schematisch

sich, wenn der relative Abfall $\Delta\sigma/\sigma$ konstant mit 8% angenommen wird.
Um den Verlauf bei Zug-Druck-Belastung beschreiben zu können, bedarf es noch einer weiteren sorgfältigen Analyse der Versuchsresultate.

5 Ausblick

Das Ziel dieser Betrachtungen war, das Verhalten des Betons unter Zug ähnlich darzustellen, wie es unter Druck schon geraume Zeit bekannt ist. Eine Besonderheit dabei ist, daß das Zugversagen des Betons eine lokal begrenzte Erscheinung ist, was es erforderlich macht, zwischen Dehnung und Rißöffnung zu unterscheiden. Bei Druck treten zwar auch lokale Mikrorisse auf, die jedoch gleichmäßiger verteilt sind, so daß bis zum Bruch von einer Stauchung gesprochen werden kann. Um das Verhalten unter Zugrissen besser zu verstehen, wurde kurz auf die Bruchmechanik eingegangen. Die vorliegenden Ergebnisse werden in der kommenden Zeit weiter ausgewertet. Vor allem sollen sie gebraucht werden, um den Rißfortschritt unter Schwingbelastung mit Hilfe eines bruchmechanischen Modells zu beschreiben.

6 Bedankung

Ohne die tatkräftige Unterstützung der Herren Dr. Ir. H. A. W. Cornelissen, Ing. G. Timmers, W. van Veen und Ir. J. Moelands hätte das Forschungsprogramm nicht ausgeführt werden können. Ihnen gilt hier mein besonderer Dank.

7 Literatur

[1] Petersson, P. E.: Crack growth and development of fracture zones in plain concrete and similar materials. Report TVBM-1006, Lund, Sweden, 1981.

[2] Bazant, Z. P., Oh, B. H.: Crack band theory for fracture of concrete. RILEM Materials and Structures 16 (1983), no. 93, pp. 155–177.
[3] Cornelissen, H. A. W., Timmers, G.: Fatigue of plain concrete in uniaxial tension and in alternating tension-compression. Experiments and results. Report no. 5-81-7, Stevin Laboratory, Delft University of Technology, Delft, 1981.
[4] Reinhardt, H. W., Cornelissen, H. A. W.: Post-peak cyclic behaviour of concrete in uniaxial tensile and alternating tensile and compressive loading. Cement and Concrete Research 14 (1984), no. 2, pp. 263–270.
[5] Spooner, D. C., Dougill, J. W.: A quantitative assessment of damage sustained in concrete during compressive loading. Mag. Concrete Research 27 (1975), no. 92, pp. 151–160.
[6] Sinha, B. P., Gerstle, K. H., Tulin, L. G.: Stress-strain relations for concrete under cyclic loading. J. ACI 61 (1964), no. 2, pp. 195–211.
[7] Reinhardt, H. W.: Maßstabseinfluß bei Schubversuchen im Licht der Bruchmechanik. Beton- und Stahlbetonbau 76 (1981), no. 1, pp. 19–21.
[8] Hillerborg, A., Modéer, M., Petersson, P. E.: Analysis of crack formation and crack growth in concrete by means of fracture mechanics. Cement and Concrete Res. 6 (1976), pp. 773–782.
[9] Schwalbe, K. H.: Bruchmechanik metallischer Werkstoffe. Hanser, München, 1980.
[10] Hillerborg, A.: Examples of practical results achieved by means of the fictitious crack model. Preprints William Prager Symposium on Mechanics of Geomaterials: Rocks, concrete, soils. Ed. by Z. Bazant, Northwestern University, Evanston, Illinois. September 1983, pp. 611–614.

Verfestigung und Versprödung von Beton durch tiefe Temperaturen

Ferdinand S. Rostásy

1 Einleitung

Erd- und Petroleumgase werden zunehmend zu wichtigen alternativen Energieträgern. Die Lagerung von natürlichen und auch von technischen Gasen erfolgt meist bei Atmosphärendruck und im verflüssigten, also tiefkalten Zustand. Damit erfahren die Werkstoffe eines Lagertanks, abhängig von dessen Bauweise, sowohl im Betrieb als auch in gewissen Störfällen extrem tiefe Temperaturen (Beispiel: verflüssigtes Erdgas (LNG) besitzt bei Normaldruck eine Siedetemperatur von rd. $-165\,°C$).

Die Lagerung von LNG birgt gewisse Risiken in sich. Das zuverlässige Beherrschen der Risiken erfordert vertiefte Kenntnisse über das Verhalten der Konstruktionswerkstoffe bei tiefen Temperaturen, da erst dann das Bauteilverhalten vorhergesagt werden kann. Dabei interessiert aus Sicherheitsgründen vorrangig die Frage, ob die Werkstoffe bei Tieftemperatur neben der zu erwartenden Verfestigung auch eine Versprödung erfahren.

2 Beanspruchungen und deren Nachbildung in der Materialforschung

Bauwerke hoher Sicherheitsrelevanz, LNG-Tanks gehören hierzu, sind durch außergewöhnliche Zustände, vor allem in Störfällen, gekennzeichnet. Aus diesem Grund muß sich die Werkstofforschung an den Beanspruchungen von Betrieb und Störfällen orientieren. Die Banspruchungen umfassen mechanische (statische und dynamische Zustände) und thermische Einwirkungen vor dem Hintergrund des tatsächlichen Werkstoffzustands.

Die Einwirkungen auf den Werkstoff Beton hängen nachhaltig von der Bauweise eines LNG-Tanks ab. Die Entwicklungslinien von LNG-Tanks weisen zu Doppelmanteltanks, deren mehrfache Sicherheitsredundanz dem dichtbesiedelten Mitteleuropa gerecht wird [1], [2] und [3].

Bild 1 zeigt Bodenbereiche von zylindrischen Doppelmanteltanks. Der Außentank wird durch eine Kugelschale und Sohlplatte geschlossen. Der Innenbehälter, das eigentliche LNG-Behältnis aus Nickelstahl oder Spannbeton, ist oben offen, aber allseitig durch Wärmedämmung umschlossen: Prinzip Thermosflasche. Natürlich gibt es zahlreiche Varianten, Vergleiche müssen unterbleiben.

Welche Beanspruchungen der Beton im Regelbetrieb und in außergewöhnlichen Lastfällen erfahren kann, hängt nicht nur von der Bauweise, sondern auch von den Auslegungskriterien ab. Weil sich letztere jedoch von Fall zu Fall beträchtlich unterscheiden, sind die folgenden Bemerkungen eher qualitativer Natur.

Der Innenbehälter ist im Betrieb und in außergewöhnlichen Zuständen im stationären Temperaturzustand und so kalt wie das LNG selbst. Der Außenbehälter ist im Betrieb ebenfalls im stationären Temperaturzustand und weist, vergröbert dargestellt, die Umgebungstemperatur auf. Er wird kryogene Temperaturen

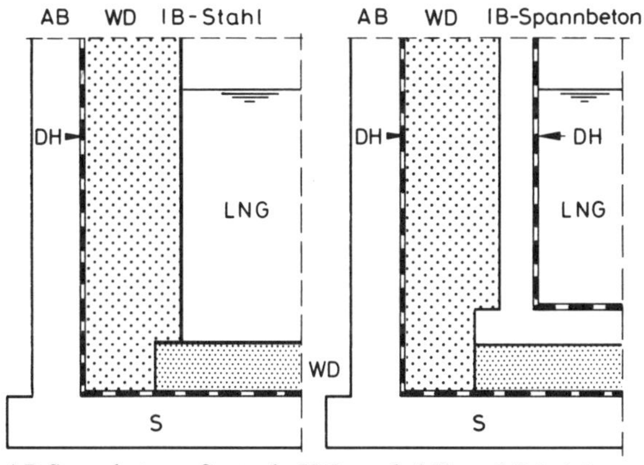

AB Spannbetonaußentank; IB Innenbehälter; S Schalplatte; WD Wärmedämmung; DH Dichthaut

Bild 1
Ausschnitte der Bodenbereiche von LNG-Doppelmanteltanks (schematisch)

nur in den Störfällen „Lokaler Kälteschock“ und „Globaler Kälteschock“ erfahren. Ein lokaler Kälteschock kann durch Leckage des Innentanks entstehen und auf die Sohlplatte und/oder Außenwand einwirken. Ein globaler Kälteschock entsteht im Gefolge des hypothetischen Totalversagens des Innenbehälters. Der Kälteschock folgt einer Temperatursprungfunktion und ist deshalb mit instationären Temperaturen verbunden.
Aus den geschilderten Zuständen leiten sich die Prüfbedingungen ab: Das mechanische und thermische Verhalten von Beton ist sowohl bei Normal- als auch bei variabler Tieftemperatur zu erforschen. Dabei sind sowohl stationäre Temperaturzustände (isothermische) als auch zyklisch und schockbedingt veränderliche zu untersuchen. Auch das Verhalten von vorbeanspruchtem oder unbeanspruchtem Beton nach einer Tieftemperaturbeanspruchung ist von Bedeutung, um die frostbedingte Entfestigung zu erforschen [4].
Das Studium des Hochtemperaturverhaltens von Beton zeigte, daß hohe Aufheizgeschwindigkeiten entfestigend wirken. Da aber hohe Temperaturen per se entfestigen, niedrige hingegen zur Verfestigung führen, muß die Wirkung hoher, schockartiger Abkühlgeschwindigkeiten von Grund auf überlegt werden. Hohe Abkühlgeschwindigkeiten erzeugen in Bauteilen und Probekörpern steile Temperaturgradienten und hohe Eigenspannungen, deren Intensität von geometrischen, mechanischen und thermischen Bedingungen abhängt. Mikro- und Makrorisse sind die Antwort. Damit wirkt sich eine hohe Abkühlgeschwindigkeit wohl weniger auf die mechanischen Eigenschaften des Betons aus, sondern verändert den strukturellen Zustand eines Bauteilbereichs durch Lockerung und Risse.
Andererseits weiß man, daß die Abkühlgeschwindigkeit den Phasenübergang und den Wassertransport in den Zementsteinporen beim Frieren und damit den inneren Druckaufbau beeinflußt. Wichtig ist jedoch die Erkenntnis, daß man den Einfluß hoher Abkühlgeschwindigkeit auf die mechanischen Eigenschaften des Betons durch Eintauchen von Probekörpern in Flüssigstickstoff wenig erhellt: der Probekörper bestimmt seine Antwort selbst. Glücklicherweise erfährt dieses Problem eine Entschärfung: Kälteschockversuche zeigten [5], daß die Abkühlgeschwindigkeit nur innerhalb der ersten Zentimeter Beton unterhalb der Schockfläche hoch ist und mit der Tiefe sehr rasch abnimmt.
Ob bei Materialuntersuchungen die Gleichzeitigkeit von tiefen Temperaturen und hohen Beanspruchungsgeschwindigkeiten infolge dynamischer Lastfälle berücksichtigt werden muß, ist bislang strittig.
Nach Diskussion der Einwirkungen ist es erforderlich, dem Werkstoffzustand zum Zeitpunkt einer Einwirkung Aufmerksamkeit zu widmen. Es wird später gezeigt werden, daß Verfestigung und Versprödung von Beton infolge Tieftemperatur nachhaltig von dessen Feuchte und Festigkeit abhängen. Die Betonbauteile von LNG-Tanks sind dickwandig und außerdem einseitig durch stählerne Dichthäute belegt. WIEDEMANN [4] zeigte, daß jene Betonbereiche von Außenbehältern, die kryogene Temperaturen erfahren können, praktisch nicht austrocknen werden. Die Anmachwassermenge $W_0 = wZ$ bleibt im Linearbereich konserviert. Diese Aussage muß für einen Spannbetoninnentank, der i.w. dauernd tiefkalt ist, modifiziert werden: Bis zur Inbetriebnahme ist eine Austrocknung des Betons möglich, danach nicht mehr. Die im Bauwerk realen Feuchte- und Erhärtungsbedingungen müssen durch eine sehr lange Probenvorlagerung im versiegelten Zustand oder bei einer hygrisch äquivalenten relativen Luftfeuchte nachgebildet werden.

3 Festigkeit und Verformung bei Druckbeanspruchung

Die Verfestigung der Konstruktionswerkstoffe durch tiefe Temperaturen ist eine bekannte Tatsache und läßt sich für gedrückten Beton in isothermischen Druckversuchen nachweisen. Bild 2 zeigt typische Ergebnisse [4]. Die Probekörper wurden nach einwöchiger Feuchtelagerung, bei unterschiedlichen relativen Luftfeuchtigkeiten und bei 20 °C bis zur Prüfung im Alter von rd. 200 *d* gelagert. Damit wird sich in ihnen näherungsweise die zugehörige Gleichgewichtsfeuchte eingestellt haben, die durch die mittlere Feuchte u_m beschrieben wird.
Zunächst zur Beobachtung: Die Druckfestigkeit steigt mit abnehmender Temperatur und mit zunehmender Feuchte an. Der Beitrag der Tieftemperatur (TT) allein zur Verfestigung, läßt sich aus dem Verhalten der getrockneten Proben ableiten. Man erkennnt, daß die Verfestigung gegenüber Raumtemperatur (RT) gering bleibt, sofern das verdampfbare und damit gefrierfähige Wasser ausgetrieben wurde. Wesentlich ist also das Zusammenwirken von Feuchte und Temperatur.

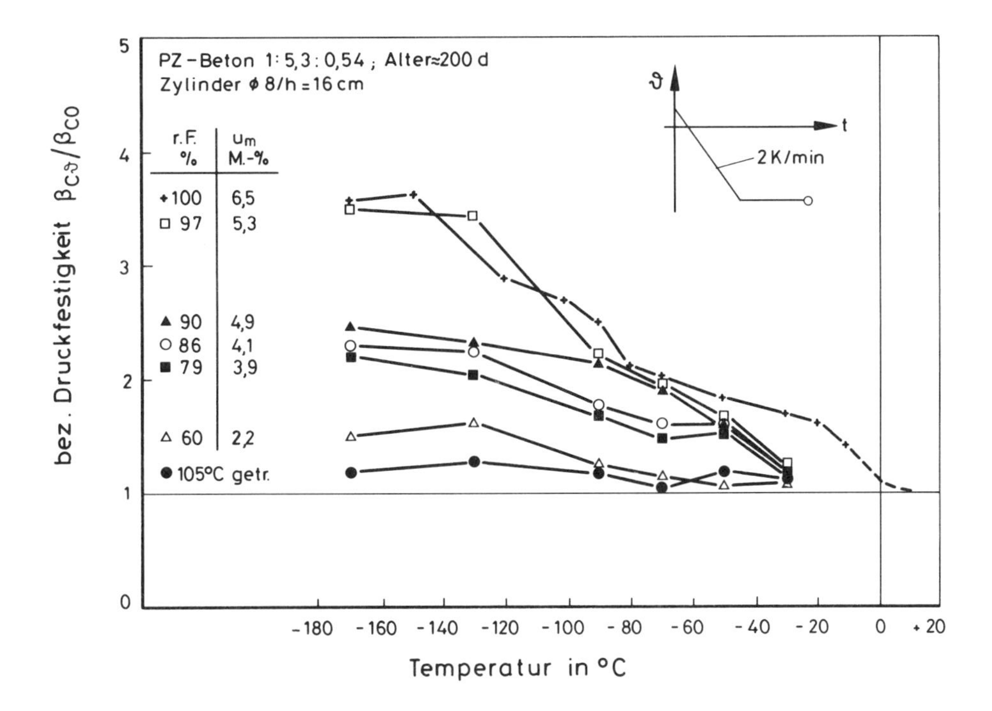

Bild 2
Bezogene Zylinderdruckfestigkeit von Beton in Abhängigkeit vom Lagerungsklima und von der Temperatur

Bild 3 zeigt die Verfestigung abhängig von u_m und von der Tieftemperatur. Man erkennt, daß sich die Verfestigung über den gesamten Bereich der mittleren Feuchten und Tieftemperaturen erstreckt. Auch im Bereich von $-90\,°C$ bis $-170\,°C$ findet für jeden Feuchtegehalt eine entsprechende Festigkeitszunahme statt.

Die klassische Methode der Verfestigung von Beton bei Normaltemperatur besteht in der Verminderung des Wasserzementwertes. Dies gilt, wie Bild 4 zeigt, für tiefe Temperaturen gleichermaßen; allerdings mit der Maßgabe, daß der Festigkeitsverlust infolge Zunahme des W_0/Z-Werts durch die abnehmende Temperatur zunehmend kompensiert wird. Bild 4 soll nur phänomenologisch die Rolle des Poreneises erhellen. Eine praktische Folgerung ist hieraus nicht ableitbar.

Die Zementart und Festigkeitsklasse üben, ebenso wie die Zuschlagart – sehr dichter und fester, natürlicher Zuschlag vorausgesetzt –, auf die bezogenen Werte keine merklichen Einflüsse aus.

Für die Bauteilbemessung ist nicht nur die Betondruckfestigkeit, sondern auch die Spannungsdehnungslinie des gedrückten Betons von Bedeutung. Bild 5 zeigt σ-ε-Linien von im Klima 20/100 vorgelagerten Proben in Abhängigkeit von der Temperatur. Mit Abnahme der Temperatur bis rd. $-80\,°C$ wächst nicht nur die Zylinderdruckfestigkeit, sondern auch die ihr zugehörige Dehnung ε_C an. Bis zu dieser Tempe-

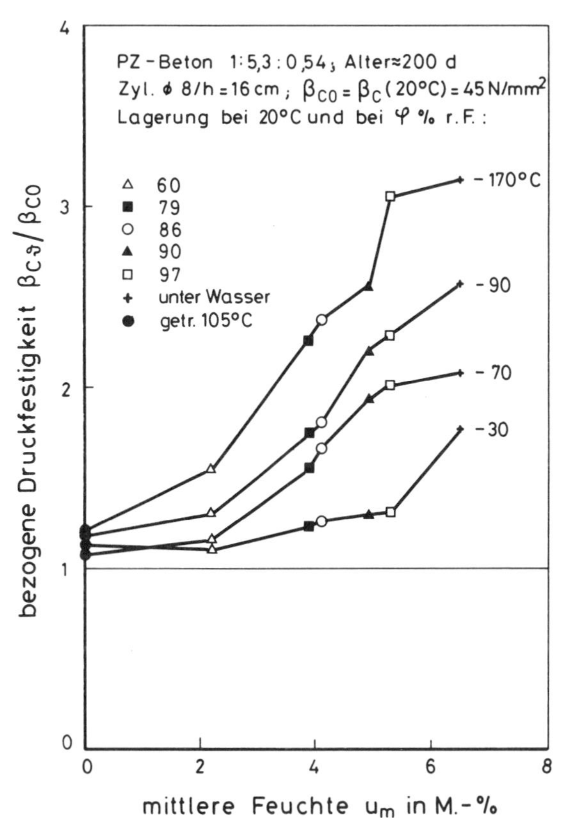

Bild 3
Druckverfestigung von Beton abhängig von der Gleichgewichtsfeuchte bei verschiedenen Temperaturen

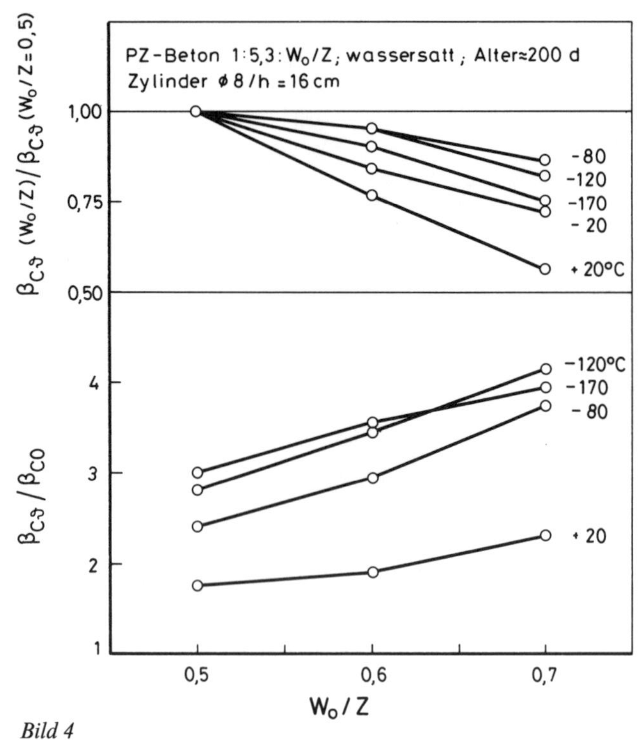

Bild 4
Druckverfestigung von Beton abhängig vom Wasserzementwert und von der Temperatur

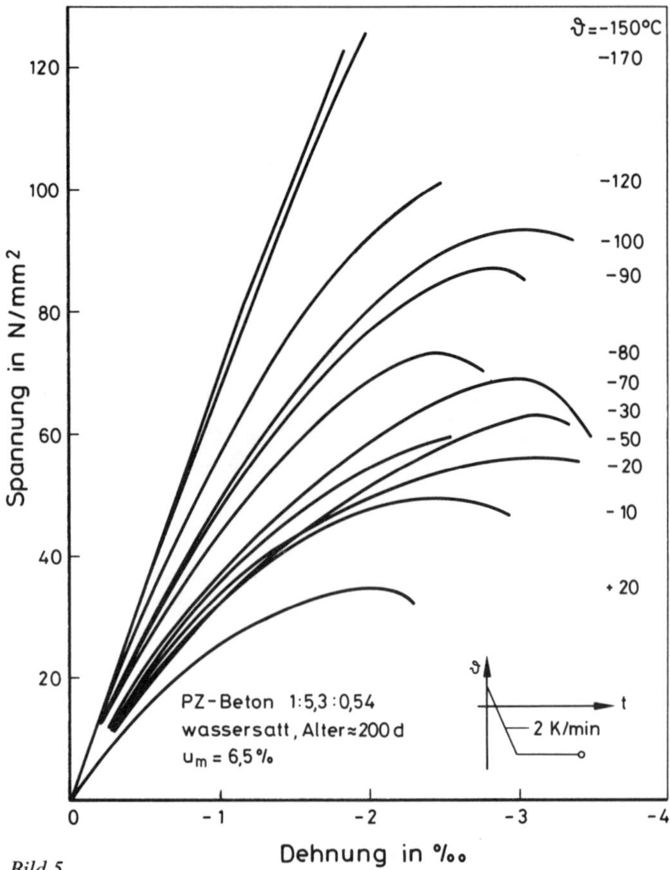

Bild 5
Spannung-Dehnungslinien von zentrisch gedrücktem Beton abhängig von der Temperatur

ratur konnten in den dehngesteuerten Versuchen sowohl ein Spannungsmaximum als auch ein abfallender Ast des σ-ε-Diagramms beobachtet werden. Ab $-80\,°C$ verändert sich die σ-ε-Linie zunehmend mit der Tendenz zu linearelastisch-sprödem Verhalten.
Schließlich zeigt Bild 6 einige mechanische Kennwerte, die man aus Bild 5 ableiten kann. Wesentlich sind zwei Erkenntnisse: Die Dehnung ε_C unter der Zylinderdruckfestigkeit sinkt unter dem Einfluß tiefer Temperaturen nicht unter den bei Raumtemperatur gemessenen Wert ab. Die Formänderungsenergie A (Integral über $\sigma(\varepsilon)$ bis β_C) beträgt stets ein Mehrfaches des Raumtemperaturwerts. Sie ist jedoch, insbesondere ab $\vartheta < -90\,°C$, hauptsächlich elastischer Natur, der plastische Anteil verschwindet. Die Tendenz zum spröden Verhalten bei tiefer Temperatur ist vor allem bei sehr feuchtem Beton ausgeprägt.
In der Pilotstudie wurde der Frage nachgegangen, ob eine schnellablaufende Belastung gepaart mit tiefer Temperatur eine zusätzliche Betonversprödung erzeugt. Probekörper wurden bei $+20$ und $-165\,°C$ mit folgenden Dehngeschwindigkeiten geprüft: 0,03‰/s und 200‰/s. Die hohe Dehngeschwindigkeit erzeugte sowohl bei RT als auch bei TT eine Verfestigung von

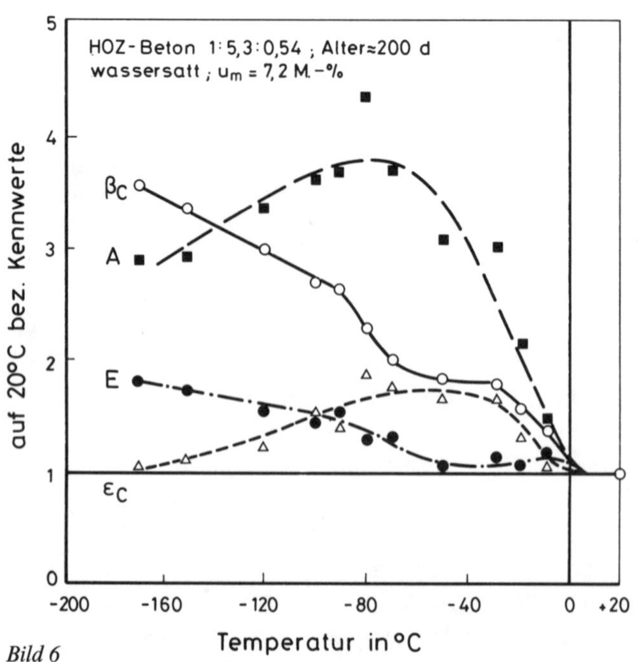

Bild 6
Bezogene mechanische Kennwerte von zentrisch gedrücktem Beton abhängig von der Temperatur

rd. 24 bzw. 16% der Zylinderdruckfestigkeit. Gleichzeitig nahmen dabei die Dehnungen ε_{CO} und $\varepsilon_{C\vartheta}$ zu: Eine zusätzliche dehngeschwindigkeitsabhängige Versprödung tritt also nicht ein. Die Spaltzugfestigkeit blieb i. w. unbeeinflußt von $\dot{\varepsilon}$.

4 Festigkeit und Verformung bei Zugbeanspruchung

Ebenso wie bei Normaltemperatur spielt die Betonzugfestigkeit in Massivbauteilen bei Tieftemperatur eine wichtige komplementäre Rolle. Darüber hinaus gewinnt sie an Bedeutung bei der Beherrschung von Zwang- und Eigenspannungen sowie bei der Rißbreitenbeschränkung bei Kälteschock [5].

Die Entwicklung der Spaltzugfestigkeit in Abhängigkeit von Temperatur und Betonfeuchte entspricht Bild 2. Allerdings ist der Grad der Verfestigung geringer. Er beträgt für wassersatten Beton mit $W_0/Z = 0,5$ bei $-170\,°C$ rd. 1,5 bis 2,5. Bild 7 zeigt die Verfestigung bei Zugbeanspruchung von wassersattem Beton, abhängig vom W_0/Z-Wert.

Umfassende Zugversuche an Normal- und Leichtbeton wurden von Bamforth et al. [6] durchgeführt (Bild 8). Biegezugfestigkeit und Zugbruchdehnung nehmen ungefähr linear mit der Tieftemperatur zu. Dies gilt auch für die axiale Zugfestigkeit, die auch bei $-165\,°C$ ungefähr die Hälfte der Biegezugfestigkeit beträgt. Die Reduktion der axialen Zugfestigkeit gegenüber der Biegezugfestigkeit ist von der Betonstruktur abhängig und rißbruchmechanisch erklärbar (s. [7] und [8]). Sie wird von der Tieftemperatur nicht beeinflußt.

5 Erklärung der Verfestigung

Bereits aus den Versuchsergebnissen war ableitbar, daß die Verfestigung von Beton infolge tiefer Temperaturen im engen Zusammenhang mit dem Porenwasser des Zementsteins steht. Eine werkstoffphysikalisch befriedigende Erklärung der Verfestigung muß am anormalen Gefrierverhalten von Wasser ansetzen, das einer physikalischen Wechselwirkung mit der Oberfläche der Porenwände unterworfen ist.

Setzer et al. [9], [10] zeigten, daß das Porenwasser, abhängig vom Porenradius, eine Gefrierpunktsdepression gegenüber 0 °C von ungebundenem, reinem Wasser erfährt: Je kleiner der Porenradius, um so niedriger ist die Gefriertemperatur. Damit koexistieren im Porensystem Wasser, Wasserdampf und Eis über einen weiten Bereich der Tieftemperatur. Dies bedeutet aber auch, daß innerhalb dieses Bereichs nacheinander Pha-

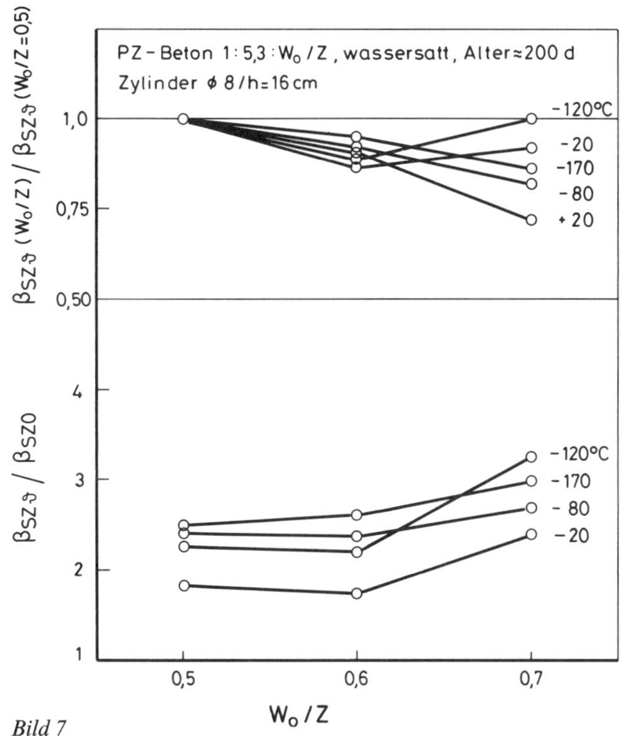

Bild 7
Zugverfestigung (Spaltzugfestigkeit) von Beton abhängig vom Wasserzementwert und von der Temperatur

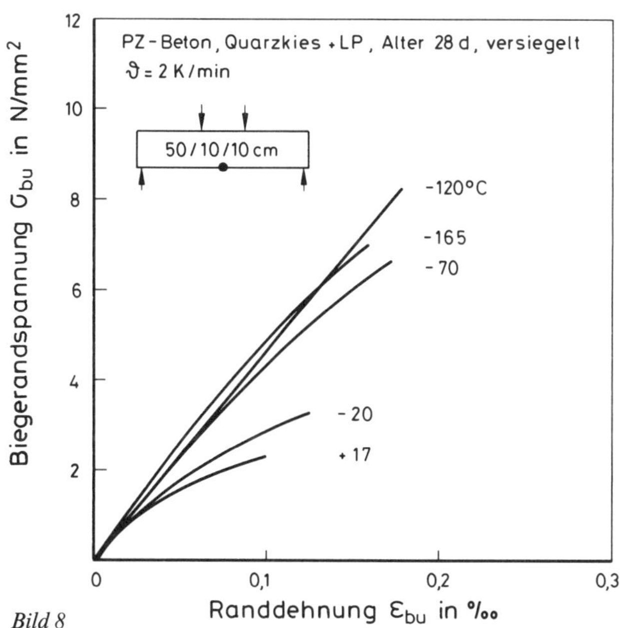

Bild 8
Spannung-Dehnungslinien von Beton bei Biegung abhängig von der Temperatur [6]

senübergänge Wasser–Eis ablaufen. Ausgeprägte Enthalpieänderungen wurden bei sehr feuchtem Zementstein zwischen -10 und $-25\,^\circ C$ sowie bei $-43\,^\circ C$ festgestellt.

Es ist bekannt, daß Beton mit $W_0/Z > 0{,}5$ und bei hoher Wassersättigung einen Frostschaden erleidet, der sich nach Durchlaufen des Frostzyklus $+20/-170/+20\,^\circ C$ u.a. als irreversible Expansion ausweist. Diese bleibende Dehnung ist eine Gefügeaufweitung, die das Volumen von Mikrorissen darstellt (Bild 9).

Nach SETZER [9] kann der Frostschaden durch Sprengwirkung infolge der Volumenzunahme des Eises nicht erschöpfend erklärt werden. Viel mehr verantwortlich sind Druckdifferenzen zwischen miteinander verbundenen Poren unterschiedlicher Größe, in denen die Phasenübergänge dementsprechend temperaturverschoben ablaufen. Da das Wasser in den groben Poren zuerst gefriert, findet ein Wassertransport von den feinen zu den groben Poren statt, der insbesondere bei großer Abkühlgeschwindigkeit zu Drücken und Mikrorissen führt.

Bild 9 zeigt die thermischen Dehnungen während eines Frosttauwechsels. Man erkennt, daß zwischen $-10\,^\circ C$ und $-50\,^\circ C$ eine deutliche Ausdehnung einsetzt. Der Verlauf der Gefrierpunktlinie [11] zeigt an, in welchen Porengrößen der Phasenübergang abhängig von der Temperatur stattfindet.

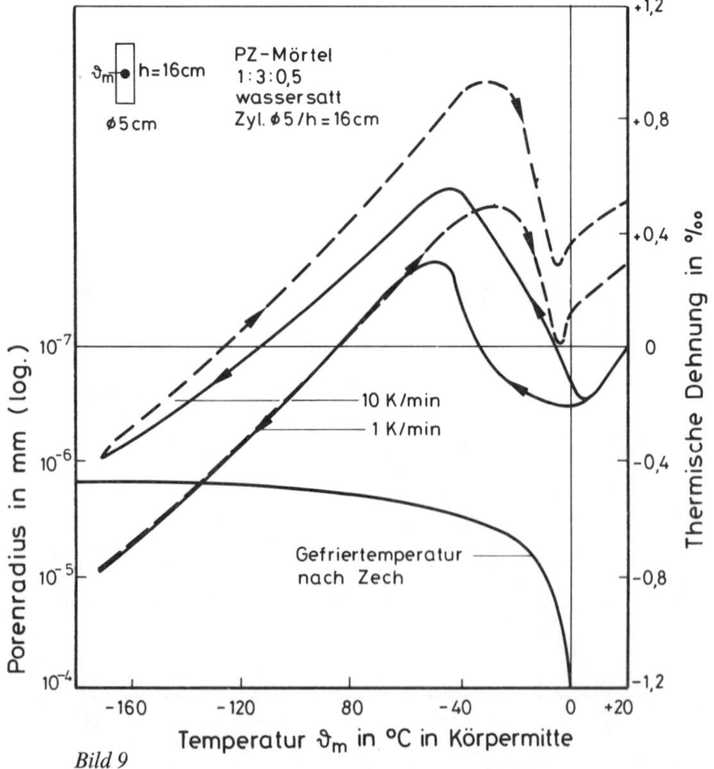

Bild 9
Thermische Dehnung von Zementmörtel abhängig von der Temperatur und der Abkühl- bzw. Erwärmungsgeschwindigkeit

Wenn bei der Abkühlung feuchten Betons innere Drücke und Mikrorisse entstehen, so ist vor diesem Hintergrund die Verfestigung schwer zu verstehen. Folgende Teilbeträge sind denkbar: Die Mikrorißbildung infolge Spannungen geht von Poren aus, die größere Durchmesser als 10^{-5} bis 10^{-4} mm aufweisen. Sind diese von Eis erfüllt, das zug- und drucktragfähig ist, so wird die Mikrorißbildung gebremst. Diese Verfestigung ist für die Kontaktzone Matrix-Zuschlag, die die Schwachstelle des Betongefüges darstellt, vorteilhaft. Aus Bild 10 kann man ableiten, daß die Mikrorißbildung beim tiefkalten Beton später als bei Normaltemperatur einsetzt. Die gegenüber dem Zuschlagkorn stärkere Kontraktion der Matrix verhindert das Entstehen von Verbundrissen in der Kontaktzone.

6 Beschreibungsmöglichkeiten der Verfestigung

6.1 Zur Notwendigkeit

Der Konstrukteur benötigt für Vorprojekt, Fallstudien und Ausführungsnachweise Materialangaben. Dies oft zu einem Zeitpunkt, zu dem Versuchsergebnisse noch nicht vorliegen. Er wird dann gezwungen sein, die Materialdaten des Betons auf Basis der Literatur und/oder Erfahrung abzuschätzen. Hierzu soll Hilfestellung geleistet werden. Mehr als grobe Ansätze sind nicht zu erwarten. Die Ansätze sind auch als Modelle zur Erzeugung von Arbeitslinien zu verstehen.

6.2 Maßgebende Betonfeuchte

Für Spannbetonbauteile von LNG werden hochfeste Betone der Festigkeitsklasse B45 ($W_0/Z \approx 0{,}45$) gewählt. Erfordert die Art des Nachweises den Einbezug der Tieftemperaturfestigkeit, so ist neben der maßgebenden Zylinderdruckfestigkeit β_{COm} (20 °C) im Nachweisalter t auch die Betonfeuchte u_m zu bestimmen. Geht man davon aus, daß der Beton an einem Stahlliner nicht austrocknet, so gilt

$$W(t) \approx W_0 = wZ.$$

Bei einem W_0/Z-Wert von rd. 0,45 wird, auch bei verhinderter Austrocknung, eine vollständige Hydrata-

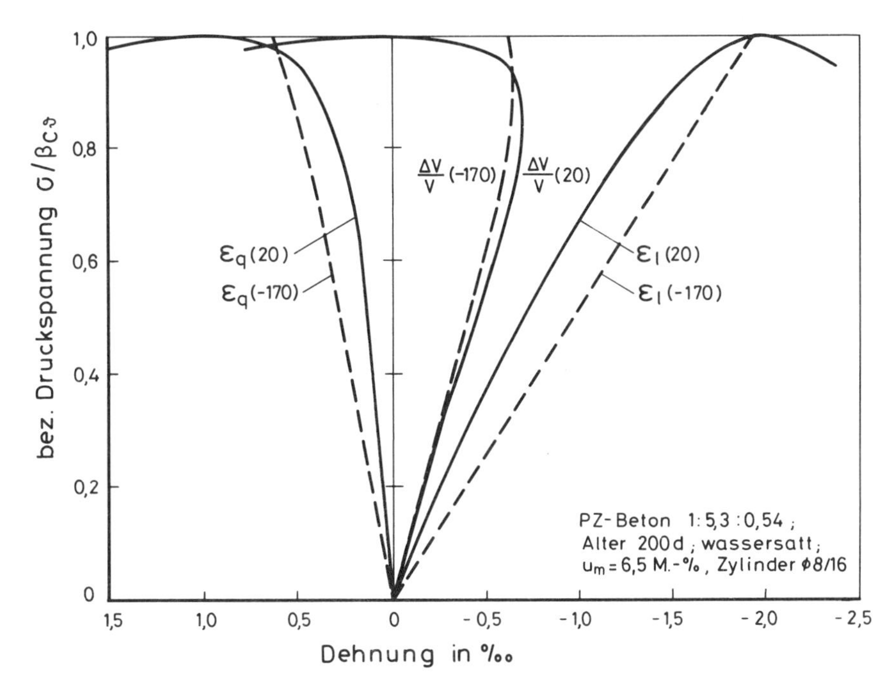

Bild 10
Bezogene Spannungen und Dehnungen für mittig gedrückten Beton bei RT und TT

tion nicht erreicht [12]. Diesen Umstand kann man bei der Bestimmung des nichtverdampfbaren Wassers durch einen Hydratationsgrad $m \approx 0{,}8$ berücksichtigen. Die mittlere Betonfeuchte wird im Versuch nach folgender Beziehung errechnet:

$$u_m = \frac{G(t) - G_{tr}}{G_{tr}} \, 100 \text{ M.-\%} \tag{1}$$

Hierin stellt G_{tr} das Trockengewicht der Probe dar. Definiert man nun mit K die Zuschlagmasse und das Hydratwasser zu

$$\text{chem}\, W \approx m\, 0{,}24\, Z, \tag{2}$$

so kann man die Betonfeuchte für übliche Betonzusammensetzungen abschätzen (Alter $t > 1$ a):

$$u_m \approx \frac{Z(w - m\,0{,}24)}{K + Z(1 + m\,0{,}24)} \, 100 \approx \frac{w - 0{,}19}{6} \, 100 \text{ M.-\%} \tag{3}$$

Die Bestimmung von u_m für austrocknende Betonbereiche kann nur über eine Diffusionsrechnung erfolgen [4], [13].

6.3 Verfestigungsansatz für Beton auf Druck

Die Druckverfestigung von Beton ist von mehreren Forschern angenähert worden. Aufgrund der Versuche läßt sich mittlere Zylinderdruckfestigkeit wie folgt darstellen:

$$\beta_{C\vartheta m} = \beta_{COm} + \Delta\beta_{CO} \approx \beta_{COm} + 12\, u_m \left\{1 - \left[\frac{\vartheta + 170}{170}\right]^2\right\} \tag{4}$$

Die Druckdehnung unter der Zylinderdruckfestigkeit kann man, entsprechend Bild 6, ebenfalls in der Form

$$\varepsilon_{C\vartheta} = \varepsilon_{CO} + \Delta\varepsilon_{C\vartheta} \tag{5}$$

approximieren. Bild 11 zeigt die Verläufe der Zuwächse von $\beta_{C\vartheta}$ und $\varepsilon_{C\vartheta}$. Für die Dehnung ε_{CO} bei RT kann der Wert -2‰ gewählt werden. Die Dehnung hängt von der Betonfeuchte nur wenig ab.
Die σ-ε-Linien für mittigen Druck kann man im aufsteigenden Ast als parabolisch verlaufend beschreiben:

$$\frac{\sigma}{\beta_{C\vartheta m}} = 1 - \left(1 - \frac{\varepsilon}{\varepsilon_{C\vartheta}}\right)^n \tag{6}$$

Die Hochzahl n, die für das Parabel-Rechteck-Diagramm von DIN 1045 den Wert 2 besitzt, ermöglicht die Transformation zum linear-elastischen Verhalten, das mit abnehmender Temperatur beobachtet wird:

$$n = 1 - \frac{\vartheta + 170}{170}; \quad 1 \leqq n \leqq 2 \tag{7}$$

Bild 12 zeigt das normierte σ-ε-Diagramm. Zwischen $n = 1$ und 2 liegen sämtliche σ-ε-Linien für mittigen

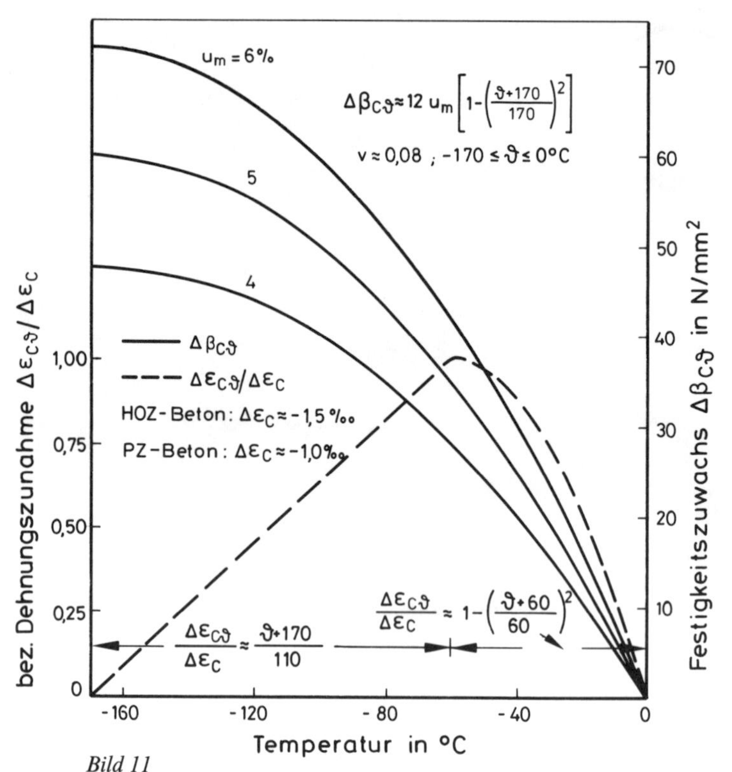

Bild 11
Zunahme der mittleren Zylinderfestigkeit und der Dehnung ε_C von Beton abhängig von der Temperatur (Rechenwerte)

Druck, abhängig von der Temperatur. Auf einen plastischen Anteil jenseits von $\varepsilon/\varepsilon_{C\vartheta} = 1$ sollte verzichtet werden.

6.4 Verfestigungsansatz für Beton auf Zug

Ein Verfestigungsansatz für gezogenen Beton kann für Rißuntersuchungen und für Zwangberechnungen notwendig werden. Die Versuche erlauben die Anknüpfung der mittleren zentrischen Zugfestigkeit an die mittlere Zylinderdruckfestigkeit in gewohnter Weise. Mit dem Ansatz des CEB-FIP-Model Code ergibt sich:

$$\beta_{Z\vartheta m} \approx 0{,}3\,\beta_{C\vartheta m}^{2/3} \tag{8}$$

Die σ-ε-Linien für mittigen Zug sind Gerade. Der Zugelastizitätsmodul und die Bruchdehnung nehmen mit abnehmender Temperatur zu. Nach [6] beträgt die Bruchdehnung ε_{Z0} bei $-165\,°C$ rd. 0,1‰ und bei RT rd. 0,075‰. Damit wird über die axiale Bruchdehnung wie folgt verfügt ($-170 \leqq \vartheta \leqq 0\,°C$):

$$\varepsilon_{Z\vartheta} = \varepsilon_{Z0} + \Delta\varepsilon_{Z\vartheta} \approx \varepsilon_{Z0} + \\ + \frac{\varepsilon_{Z0}}{3}\left(1 - \frac{\vartheta + 170}{170}\right) \tag{9}$$

Sowohl zwischen Biegezugfestigkeit und zentrischer Zugfestigkeit als auch zwischen den zugehörigen Bruchdehnungen herrscht auch bei TT das Verhältnis 2:1. Die Festigkeiten streuen beträchtlich: $v \approx 0{,}18$. Bild 13 zeigt Beispiele für σ-ε-Linien.

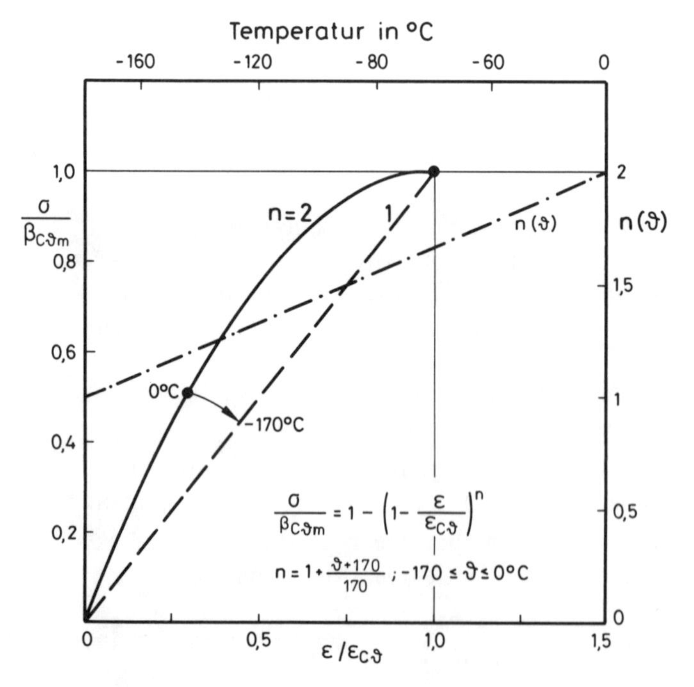

Bild 12
Normierte σ-ε-Linie von Beton für mittigen Druck bei Tieftemperaturbeanspruchung (Rechenwerte)

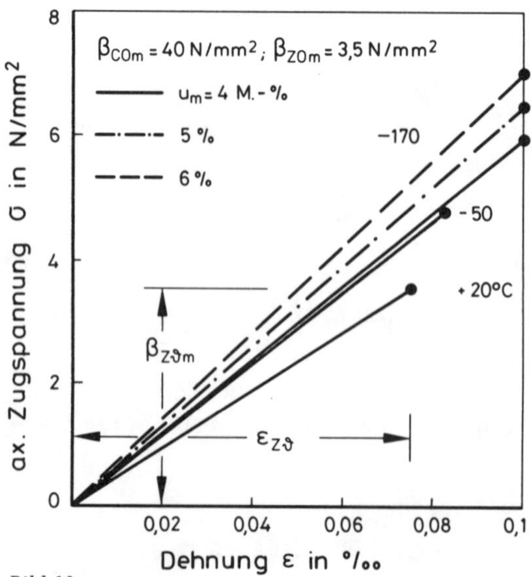

Bild 13
Arbeitslinien von Beton für zentrischen Zug abhängig von der Feuchte und Temperatur

6.5 Verfestigungsansatz für Betonrippenstahl BSt III

Kriterien zur Eignung, Prüfung und Überwachung von Betonrippenstahl für den Tieftemperatureinsatz befinden sich erst in der Entwicklung. Trotz noch herrschender Unsicherheit kann ein Verfestigungsansatz gewagt werden, der das Verhalten im elastoplastischen Teilbereich beschreibt.
Die σ-ε-Linie kann bilinear angesetzt werden, wobei eine temperaturlineare Zunahme des E-Moduls um 10% bis $-170\,°C$ angenommen werden darf. Die Streckgrenze nimmt mit abnehmender Temperatur zu, wobei die Verfestigung von der Stahlart abhängt. Versuche zeigen, daß die Verfestigung mit der Zähigkeit zusammenhängt. Grob gesprochen: Je ausgeprägter die Verfestigung, um so geringer ist die Zähigkeit bei $-170\,°C$.
Die Zunahme der Streckgrenze βS eines Betonrippenstahls der Festigkeitsklasse BSt 420500 kann wie folgt beschrieben werden:

$$\beta_{S\vartheta} = \beta_{S0} + \Delta\beta_{S\vartheta} \tag{10}$$

Hierin ist β_{S0} die Streckgrenze bei RT. Über den Zuwachs wird temperaturlinear verfügt ($-170 \leqq \vartheta \leqq 0\,°C$).

$$\Delta\beta_{S\vartheta} \approx \Delta\beta_S \left(1 - \frac{\vartheta + 170}{170}\right) \tag{11}$$

Der absolute Zuwachs $\Delta\beta_S$ sollte mit nicht mehr als 300 bis 500 Nmm^{-2} berücksichtigt werden, weil er dann auch nichttieftemperaturgeeignete Stähle umfassen würde.
Mit den Gln. (10), (11) und mit

$$E_{S\vartheta} \approx E_{S0} + \frac{E_{S0}}{10} \left(1 - \frac{\vartheta + 170}{170}\right) \tag{12}$$

ist die Streckdehnung $\varepsilon_{S\vartheta}$ fixiert. Die totale Stahldehnung ist nach DIN 1045 auf 5‰ bei RT begrenzt, die plastische Dehnung ε_{pl} beträgt hierin rd. 3‰. Über die plastische Dehnung muß der Konstrukteur abhängig vom Nachweisziel u.a. verfügen: $\text{tot}\,\varepsilon_s = 5‰$ oder $\varepsilon_{pl} = 5 - \varepsilon_S$. Ein tieftemperaturgeeigneter Betonstahl muß erheblich mehr leisten können als eine totale Dehnung von nur 5‰.

7 Bauteilzähigkeit bei Tieftemperatur

Mit der Verfestigung des Betons infolge Tieftemperatur geht zweifelsohne eine Versprödung einher, da die plastische Formänderungsarbeit abgebaut wird und bei $-170\,°C$ verschwunden ist. Aber bereits bei RT geht der Vergleich der plastischen Arbeiten von Betonstahl und Beton deutlich zu Ungunsten von Beton aus, wenn man die volumenbezogenen, also spezifischen plastischen Arbeiten aus den σ-ε-Linien der Stoffe einander gegenüberstellt.
Mit einem derartigen Vergleich wird man aber Verbundbaustoffen wie Stahlbeton und Spannbeton nicht gerecht. Nicht die spezifischen Arbeiten allein sind von Bedeutung, sondern die von Bewehrung und Betondruckzone gemeinsam geleistete plastische Arbeit ist maßgebend. Wie die plastische Arbeit eines gebogenen Stahlbetonstabs im Zustand II (Rißquerschnitt) von der Tieftemperatur beeinflußt wird, zeigt die nachfolgende Betrachtung. Diese kann nur qualitativer Natur sein, da sie von mehreren Vorgaben (Geometrie, Material und Grenzdehnungen) abhängt.
Die Zähigkeit eines Biegestabes wird durch das Verhältnis zwischen elastischer und plastischer Formänderungsarbeit im Grenzzustand der Tragfähigkeit gekennzeichnet. Dieses Verhältnis läßt sich für einen elastoplastischen Stab (Zweipunktquerschnitt, $M_u = M_S$) aus dessen Moment-Krümmungslinie ableiten (Bild 14):

$$\frac{W_{pl}}{W_{el}} = 2\left(\frac{\kappa_u}{\kappa_s} - 1\right) \tag{13}$$

Hierin bedeuten: κ_s, die Krümmung bei Fließbeginn und κ_u, die Grenzkrümmung, die zur vorgegebenen

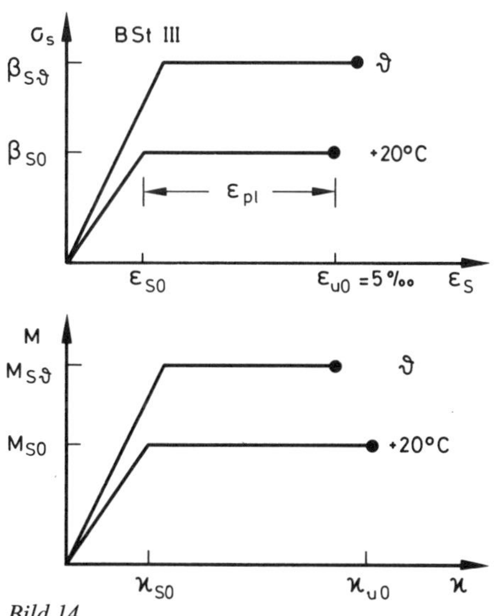

Bild 14
Schematische elastoplastische Spannung-Dehnungs- und Moment-Krümmungslinien abhängig von der Temperatur

Grenzdehnung ε_u gehört. Ein zähes Bauteilverhalten erfordert, daß die sog. Querschnittszähigkeit

$$\eta_Q = \frac{\kappa_u}{\kappa_s} > 1 \tag{14}$$

ist [14]. Für einen elastoplastischen Stahl kann man dessen Materialzähigkeit entsprechend darstellen (Bild 14):

$$\eta_{Ms} = \frac{\varepsilon_u}{\varepsilon_s} > 1 \tag{15}$$

Für Betonstahl BSt III ist $\eta_{Ms} = 2{,}5$ für die Stahlgrenzdehnung von 5‰ von DIN 1045.
Mit den Zähigkeitswerten η_{Ms} und η_Q läßt sich nun der Einfluß tiefer Temperaturen aufzeigen, wobei ein einseitig bewehrter Stahlbetonrechteckstab mit den Verfestigungsansätzen von Abschnitt 6 beispielhaft untersucht wird (Berechnung mit Programm PBMQ von U. Quast). Dabei wird der plastische Dehnanteil des Betonstahls III konstant mit 3‰ angesetzt.
Die Querschnittszähigkeit nimmt mit der Tieftemperatur ab (Bild 15). Der Vergleich mit der Materialzähigkeit zeigt, daß die Abnahme von η_Q für Bewehrungsgrade $\leqq 1{,}5\%$ i. w. durch den Stahl und weniger durch den Beton bewirkt wird. Den versprödenden Einfluß, der von der Betondruckzone ausgeht, erkennt man im Abstand zwischen η_{Ms} und η_Q; er ist für übliche Bewehrungsgrade $\leqq 1{,}5\%$ gering. Die Betonfeuchte übt bis $\mu \leqq 1{,}5\%$ einen nur geringen Einfluß aus (Bild 16). Wesentlich ist die Erkenntnis, daß die Zähigkeit eines Stahlbetonbauteils trotz Verfestigung und Versprödung des Betons nur wenig durch die Tieftemperatur abgebaut wird. Diese qualitative Aussage wird in kommenden Versuchen und Studien überprüft werden.

8 Zusammenfassung

Die Stahlbeton- und Spannbetonbauteile von LNG-Tanks werden, abhängig von deren Bauweise, in Betrieb und Störfällen extrem tiefe Temperaturen erfahren. Zur Vorhersage von Tragfähigkeit und Verformungsverhalten werden Werkstoffwerte benötigt. Insbesondere interessiert die Frage, wie sich die beobachtete Verfestigung und Versprödung von Beton in der Bauteilzähigkeit äußert. Nach Diskussion des Verhaltens von Beton bei Tieftemperatur werden Vorschläge zur Entwicklung von Arbeitslinien und zur Beurteilung der Querschnittzähigkeit unterbreitet.

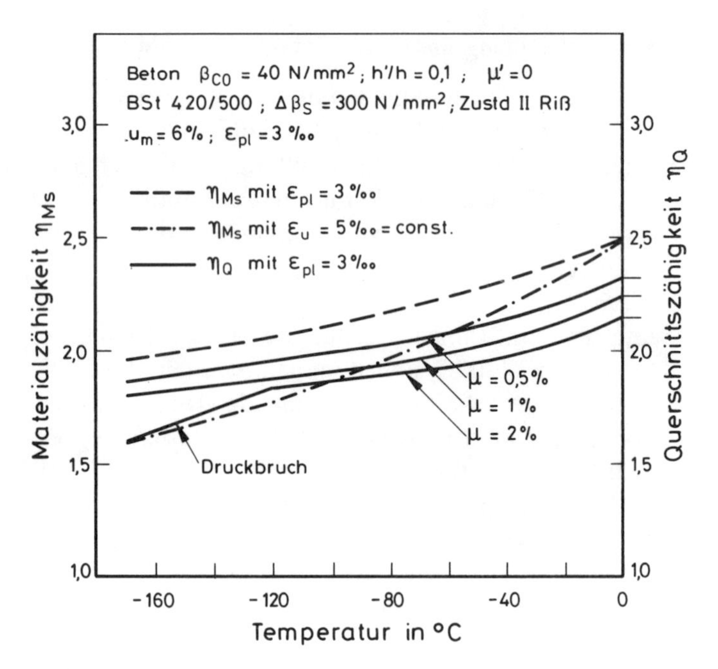

Bild 15
Material- und Querschnittszähigkeit abhängig von der Temperatur (Rechteckquerschnitt unter Biegung, Zustand II, im Riß)

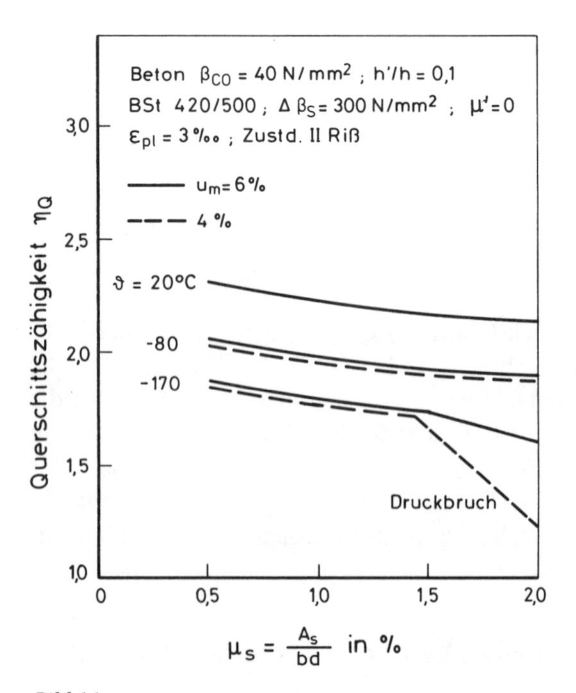

Bild 16
Querschnittszähigkeit abhängig von Temperatur, Bewehrungsgrad und Betonfeuchte

9 Literatur

[1] BRUGGELING, A.S.G.: Prestressed concrete for storage of liquefied gases. Viewpoint Publ. C&CA, London, 1981.

[2] BOMHARD, H.: Beton für Behältersysteme zur Speicherung flüssiger Gase – Wirklichkeiten, Möglichkeiten und Grenzen. Betonverlag 1981, Gesamtbericht, S. 407–422.

[3] Cryogenic Concrete. Proc. Sec. Int. Conference. Oktober 1983, Amsterdam.

[4] WIEDEMANN, G.: Zum Einfluß tiefer Temperaturen auf Festigkeit und Verformung von Beton. Diss. TU Braunschweig, 1982.

[5] SCHÄPER, M.: Tieftemperaturbeanspruchungen von Spannbetonsicherheitsbehältern bei Speichersystemen für verflüssigte Gase. Diss. Univ. Essen, 1984.

[6] BAMFORTH, P.B., MURRAY, W.T., und BROWN, R.D.: The application of concrete property data at cryogenic temperature to LNG tank design. Sec. Int. Conf. on Cryogenic Concrete, Amsterdam, Oct. 83.

[7] IVÁNYI, G.: Zugfestigkeit von Beton in örtlich veränderlichen Beanspruchungszuständen. Unv. Bericht des Instituts für Baustoffe, Massivbau und Brandschutz, TU Braunschweig, 1976.

[8] KÖNIG, G., und JAHN, M.: Über die verschiedenen Erscheinungsformen der Betonzugfestigkeit und ihre Bedeutung für das Tragverhalten von Massivbauten. Beton- und Stahlbetonbau 1983; H. 9, S. 243–247 und H. 10, S. 281–286.

[9] SETZER, M.J.: Einfluß des Wassergehalts auf die Eigenschaften des erhärteten Betons. DAfStb-Heft 280, 1977.

[10] STOCKHAUSEN, N., DORNER, H.W., ZECH, B., und SETZER, M.J.: Untersuchung von Gefriervorgängen im Zementstein mit Hilfe der DTA. Cem. and Concr. Res., Vol. 9, No. 6, S. 783/794, 1979.

[11] ZECH, B.: Zum Gefrierverhalten von Wasser im Beton. Diss. TU München, 1981.

[12] NEVILLE, A.M.: Properties of concrete. 3. Auflage, Pitman Publ. Ltd., 1981.

[13] HILSDORF, H.K.: Austrocknung und Schwinden von Beton. Festschrift H. RÜSCH, Verl. W. Ernst u. Sohn, Berlin, 1969.

[14] KEINTZEL, E.: Zähigkeitskriterien von Stahlbetonhochbauten in deutschen Erdbebengebieten. Diss. Univ. Karlsruhe (TH), 1981.

Das deutsche Gütesicherungssystem bei Betonstählen

Dieter Rußwurm

1 Ausgangslage

Das System der Gütesicherung muß im Zusammenhang mit dem Umfeld gesehen werden.
Der Verbrauch an Betonstählen in der BR Deutschland ist über die letzten Jahre hin gesehen nahezu konstant gewesen und bewegte sich in beträchtlichen Größenordnungen. Rechnet man alle Betonstähle (Betonstabstähle und Matten) zusammen, so hat sich seit 1970 ein Verbrauch von ca. 43 (!) Mio. Tonnen ergeben. Diese stattliche Menge setzt sich hauptsächlich aus den Sorten BSt 420/500 (Stäbe) und BSt 500/550 RK (Matten) zusammen. Insgesamt sind jedoch neben den 6 Sorten gemäß DIN 488, Ausgabe 1977, noch weitere Sorten durch Zulassungen abgedeckt (III RUS, III RTS, IV U, IV RUS, IV RTS, IV K, Matte 630, Sonderdyn-Matten, GEWI-Stahl).
Was die Zahl der Hersteller angeht, so ist der aktuelle Stand entsprechend der Mitteilung des Instituts für Bautechnik (1/84) folgender:

Stahlsorte	Gesamtzahl der Herstellwerke	BRD	EG	Nicht-EG-Länder
Stabstahl	63	12*)	29	19
Matte	57	42	11	4

*) noch produzierende Werke: 8

Daraus geht eindeutig hervor, daß sich zumindest bei Stabstahl der europäische Markt für Betonstähle sehr stark auf die BR Deutschland konzentriert.
Die beabsichtigte Erweiterung der EG wird zusätzliche Werke auf den Markt ziehen.
Der Vertrieb des Betonstahls erfolgt für den Kunden völlig anonym (Gattungshandel). Der Verbraucher kann in der Regel den Hersteller nicht bestimmen; lediglich bei Großbaustellen wird werksspezifisch bestellt. Meist findet auf den Händlerlagern eine Durchmischung statt.
Betonstahl gehört zu den Einfachprodukten im Bereich des Stahlhandels. Der Preis des Betonstahls ist starken Schwankungen unterworfen, zwischen Spitzenpreisen von ca. DM 1000,–/t und Tiefstpreisen von bis ca. DM 305,–/t bewegen sich die Notierungen. Der Markt ist zeitweise unberechenbar, er ähnelt dann mehr einem Warentermingeschäft.
Bei den starken Preisschwankungen und insgesamt niedrigem langfristigem Niveau ist es unvermeidbar, daß in den Werken die kaufmännische Seite gegenüber der Technik dominiert.
Betonstabstähle werden vorzugsweise in sog. Ministahlwerken hergestellt. Als Ausgangsmaterial dient Schrott, dessen chemische Zusammensetzung sich laufend verändert und durch Begleitelemente wie z. B. Cu verschlechtert. Das an sich erwünschte „recycling" eines wertvollen Materials hat aber seine Grenzen. Leider existieren bislang keine gesicherten Erkenntnisse darüber, in welchem Umfang die unbeabsichtigten Legierungselemente im Betonstahl vorhanden sein können. Der Weg zum schweißgeeigneten Betonstahl hat hier einer sehr bedenklichen Entwicklung der letzten Jahre einen Riegel vorgeschoben.
Betonstahlmatten werden aus Walzdraht hergestellt, der weltweit vertrieben wird. Während man für die einwandfreie Fertigung Drahtfestigkeiten zwischen 380 bis 480 (N/mm^2) gut verarbeiten kann, ist das Angebot zwischen 330 bis 560 (N/mm^2) angesiedelt.
Damit existieren sowohl bei Stabstahl als auch bei Matten starke qualitative Abhängigkeiten zu Ausgangswerkstoffen.
Wegen des Zwanges zur Rationalisierung haben manche Herstellwerke modernste Anlagen zur Herstellung von Betonstählen in Betrieb. Werden diese Anlagen auch ingenieurmäßig geführt, so sind ideale Voraussetzungen für die Herstellung von Betonstahl gegeben.
Die zur Anwendung kommenden Technologien bei

Stabstahl befinden sich im Umbruch: Während sehr lange der TORSTAHL marktbeherrschend war (bis 1968) und anschließend der warmgewalzte Stahl (bis 1981), wird seit neuestem mehr und mehr der wärmebehandelte Stahl (TEMPCORE) hergestellt. Parallel hierzu wird – als Folge der geforderten Schweißeignung in Zukunft noch der mikrolegierte Stahl zur Anwendung kommen.
Der klassische warmgewalzte Betonstahl wird unter dem Aspekt der Neugestaltung der DIN 488 völlig verschwinden.
Bei den Matten hat sich die herkömmliche Arbeitsweise über das Kaltwalzen/-ziehen nicht weiterentwikkelt. Eine Betonstahlmatte aus warmgewalzten gerippten Drähten ist erst in Erprobung.
Die Anforderungen an die Eigenschaften des Betonstahls als Bewehrung im Stahlbetonbau werden als bekannt vorausgesetzt. Viele zusätzlich auftretende undefinierte Beanspruchungen (unkontrolliertes Schweißen, mech. Verletzungen, enges Biegen, Rückbiegen etc.) müssen aber bei der Festlegung der Anforderungen berücksichtigt werden.
Ein Hauptproblem stellt die Realisierung der Überwachung in dem großen geographischen Raum dar, der von völlig unterschiedlichen Qualitätsphilosophien geprägt wird: So erfreulich prinzipiell die Realisierung eines europäischen Marktes ist, so schwierig stellt sich die Situation aus der Sicht der Güteüberwachung dar. Was den Betonstahl angeht, ist es gänzlich unwahr, wie gelegentlich in internationalem Rahmen zu hören ist, daß die BR Deutschland die Gütebestimmungen als Handelshemmnisse benutzt. Das liberale System auf dem Betonstahlsektor hat zu der voran geschilderten Aufteilung der Zulieferer geführt. Andere europäische Länder sind hier wesentlich restriktiver, wie aus eigenen Erfahrungen bestätigt werden kann.
Inwieweit hier Harmonisierungsbestrebungen, z. B. über die Baubedarfsrichtlinie der EG-Behörden eine Verbesserung bringen, ist fraglich. Die vorliegenden Texte stellen wenig Sachwissen, dafür aber einen ausgefeilten Stand der Bürokratie unter Beweis.

2 Grundlagen des Systems

2.1 Vorschriften

In der BR Deutschland ist die Überwachung von Baustoffen, für die der Nachweis einer ständigen ordnungsgemäßen Herstellung erforderlich ist, in § 24 der Musterbauordnung (MBO) geregelt.
Darin wird ausgesagt, daß die Überwachung von Überwachungsgemeinschaften oder Prüfstellen vorgenommen wird. Diese Stellen bedürfen der Anerkennung durch die oberste Bauaufsichtsbehörde.
Der Nachweis der Überwachung muß durch ein Überwachungszeichen erfolgen.
Die allgemeinen Bestimmungen der Musterbauordnung bzw. der korrespondierenden Bauordnungen der Bundesländer (z. B. 1 BayBO Art. 25) werden durch DIN 18200 präzisiert. Danach besteht Überwachung aus Eigen- und Fremdüberwachung; letztere setzt sich aus

- Erstprüfung,
- Regelprüfung und gegebenenfalls
- Sonderprüfung

zusammen. Die Erstprüfung dient der Feststellung, ob die Herstellwerke personell und einrichtungsmäßig die Voraussetzung für eine ordnungsgemäße Herstellung besitzen, und ist aus Werksinspektion und Werkstoffprüfung zusammengesetzt.
In periodischen Abständen wird durch die fremdüberwachende Stelle die Regelprüfung vorgenommen; dabei ist die Handhabung der Eigenüberwachung zu prüfen, die Ergebnisse zu bewerten und selbst am Erzeugnis Prüfungen vorzunehmen.
Über die Ergebnisse der Prüfungen sind Berichte zu verfertigen. Sonderprüfungen sind gleichsam Ahndungsmaßnahmen bei Nichteinhaltung der Vorschriften.
Die Eigenüberwachung durch das Werkslabor dient der kontinuierlichen Überwachung der Einhaltung der Anforderungen am Fertigungslos.
Es gibt auch andere Systeme der Überwachung, z. B. den bei Schweißarbeiten im Stahlbau – und zukünftig auch bei Schweißarbeiten an Betonstählen – üblichen Eignungsnachweis. Während bei dem aus MBO und DIN 18200 hergeleiteten System eine laufende behördliche Aufsicht mit der Möglichkeit des Eingreifens gegeben ist, stellt sich das System mit Eignungsnachweis als „geschlossenes System" dar. Hier sind ausschließlich Beziehungen zwischen Betrieb und den den Eignungsnachweis führenden Stellen vorgesehen.
Die in DIN 55355 vorgestellten Grundelemente für Qualitätssicherungssysteme wurden für Betonstähle nicht übernommen.

2.2 Technische Gegebenheiten

Der Qualitätsstand eines Produktes ist das Ergebnis eines in Wechselwirkung befindlichen Regelkreises.

Dieser besteht aus folgenden Komponenten:
- Anforderungen an das Produkt
- Prüfpläne
- Produktionsbedingungen

Die Anforderungen an das Produkt resultieren aus den Sicherheitsanforderungen der jeweiligen Bauweise. Für Stahlbeton sind diese in Form von deterministischen Größen (Sicherheitsbeiwerten) in DIN 1045 festgelegt. Eine neue Entwicklung hin zur probabilistischen oder semi-probabilistischen Sicherheitstheorie ist in der Grusibau [1] enthalten. Diese Regeln werden, da sie auch im internationalen Bereich um sich greifen, sicher in Zukunft größere Bedeutung gewinnen.

Aus dem Gesamtkomplex „Sicherheit" ergeben sich für den Werkstoff Anforderungen an seine Eigenschaften, z. B. die Streckgrenze, die in geeigneter Form definiert werden müssen.

Im Rahmen der DIN 488 ist man dazu übergegangen, als Qualitätsdefinition das 5%-Quantil der Grundgesamtheit festzulegen. Exakt lautet diese Definition der Qualität

$$W(p_x \leq 0{,}05) \geq 0{,}90$$

Auf diese Weise hat man eine punktuelle Definition der Qualität erstellt. Das Bild 1 läßt erkennen, welche Folgen daraus resultieren. Durch die Festlegung nur eines Parameters (Quantil) können recht unterschiedliche Angebote als bedingungsgemäß betrachtet werden: Sie unterscheiden sich insbesondere durch die kleinsten Einzelwerte.

Auf der anderen Seite des Regelkreises stehen die Produktionsbedingungen. Diese bedingen die Qualität des „Angebotes". Es spielen sehr viele Parameter eine Rolle, deren wesentlichste sind:

technologische Ausrüstung des Werkes
personelle Besetzung (Zahl und Qualifikation)
werksinterne Produktionssteuerung
Preis des Vormaterials
Erlössituation
Eigenkapitalausstattung des Werkes
Risikobereitschaft der Werksleitung
Gesamtökonomische Situation
Umwandlungskosten als Funktion der Auslastung.

Alle diese Größen führen zu einem „Angebot" des Produktes, das nicht stationär ist, sondern dessen Qualität zeitabhängig schwankt.

Zwischen den genannten Größen (Angebot, Anforderung) steht als „Filter" der Prüfplan. Der Zusammenhang ist in Bild 2 schematisch verdeutlicht.

Man erkennt daraus:
- Der Filter $W_A\,(p_x)$ muß wirksam werden, wenn ein „Angebot" erscheint, das den Anforderungen nicht entspricht.
- Die intensivste Prüfung muß im Bereich der Mindestanforderungen vorgenommen werden, da hier die Entscheidung am schwierigsten ist.

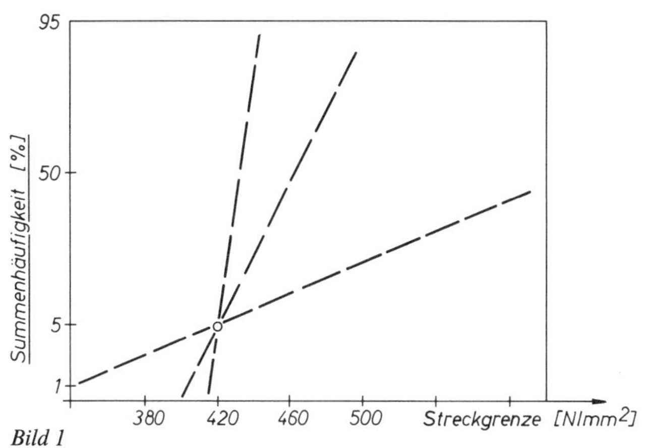

Bild 1
Beispiele für bedingungsgerechte Einhaltung des 5%-Quantils bei unterschiedlich zu erwartenden Kleinstwerten

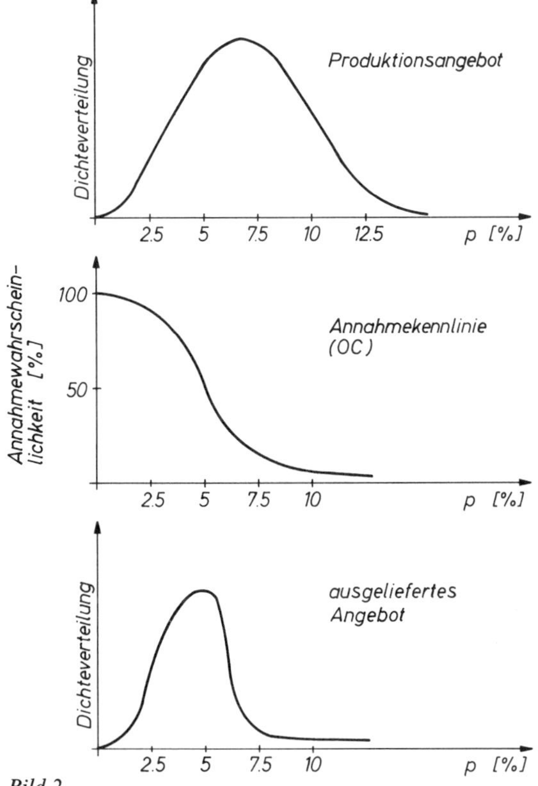

Bild 2
Schematische Darstellung der Filterwirkung eines Prüfplanes auf das Produktionsangebot

- Der Filter müßte strenggenommen nicht generell festgelegt, sondern jeweils an das Angebot angepaßt sein, mit der Maßgabe, die Mindestqualität realisieren zu lassen.

Dieses System ist mit gehörigem mathematischem Aufwand geschlossen darstellbar; es lassen sich dazu bei stark vereinfachten Annahmen Beispiele rechnen. KÜHLMEYER [2] hat insbesondere die Wechselwirkung zwischen Angebot und Prüfplan mehrfach durchleuchtet.

Theoretisch sähe das optimale System wie folgt aus: Die vorerst unbekannte Qualität des Angebots wird grob abgeschätzt. Dann müßte in Abstimmung auf die Anforderungen der geeignete Prüfplan ermittelt und eine Entscheidung über das Angebot getroffen werden. Dieser Vorgang muß, da das Angebot nicht qualitativ gleich ist, kontinuierlich wiederholt werden; mit anderen Worten: es gibt keinen gleichbleibenden Prüfplan. Die Praxis sieht sich unter ökonomischen Zwängen und verfährt anders: Der Hersteller wünscht nicht, daß zuviel seiner Produktion verworfen wird. Das ist ein direkter Verlust. Diese sogenannte Rückweiserate der Produktion sollte bei gleichem Prüfplan $< 1\%$ bleiben: Weiterhin befürchtet man Reklamationen mit teueren Folgeschäden.

Aus diesem Grunde wird sich der Hersteller auf ein Angebot einstellen, das für ihn zu große Risiken vermeidet. Das ist natürlich die „wohlmeinende" Interpretation eines gewinnmaximierenden Unternehmens.

2.3 Die Rolle der Statistik im Rahmen der Güteüberwachung von Betonstählen

2.3.1 Grundgedanken

Wie bereits angedeutet, hat sich bei Betonstählen im Rahmen der DIN 488, Teil 6, die Statistik festgesetzt. Man kann sogar sagen, daß diese Norm eine der ersten Werkstoffnormen war, die mit statistischen Hilfsmitteln ausgestattet wurde.

Setzt man die Statistik ein, so muß man sich im klaren sein, was damit bezweckt werden soll. Der Vorteil der Statistik wird gemeinhin darin gesehen, daß man aufgrund einer Prüfung von wenigen Stichproben eine Aussage über die Grundgesamtheit machen kann.

Als Grundgesamtheit kann das Prüflos oder aber eine andere Fertigungseinheit (z. B. die Produktion eines Werkes innerhalb eines Monats) angesehen werden.

Der Vorteil aus der Anwendung der Statistik ergibt sich aber nur, wenn man neben exakten Prüfergebnissen der Stichproben auch Kenntnis von der (theoretischen) Wahrscheinlichkeitsverteilung des geprüften Merkmals hat.

Im allgemeinen wird hier bei variablen Größen die bekannte Normal-(GAUSS-)Verteilung und bei attributiven Größen die Binomialverteilung zugrunde gelegt. Diese erlauben einfache Rechenvorgänge.

Ob aber eine der genannten Verteilungen vorliegt, ist sehr schwer zu testen. Die bekannten Tests auf Normalität sind aufwendig.

Die Prüfgrößen

SHAPIRO–WILK [3]

$$W = \frac{(\Sigma a_k \cdot (x_n + 1 - k - x_n))^2}{\Sigma (x - x_i)^2} \tag{1}$$

KOLMOGOROFF [4]

$$\tilde{D} = \frac{\max |F_B - F_E|}{n} \tag{2}$$

sind stark von der Probenzahl abhängig und lassen bei großen Probenzahlen selbst auf hohem Signifikanzniveau keine eindeutige Aussage zu.

Das bewährteste Mittel Normalität zu testen besteht darin, die Summenhäufigkeitslinie in einem Wahrscheinlichkeitspapier zu zeichnen.

Liegt ein Histogramm aus vielen Werten vor, kann die Kurve direkt gezeichnet werden, liegt nur eine Stichprobe beschränkten Umfangs vor, so geben z. B. HENNING, WARTMANN [5] die Auftragspositionen, die vereinfacht auch nach

$$p = \frac{i}{n + 1} \tag{3}$$

errechnet werden können.

Somit lauten einige Voraussetzungen für die Anwendung der Statistik:

- Die (theoretische) Wahrscheinlichkeitsverteilung muß bekannt sein.
- Die Stichproben müssen repräsentativ sein.
- Das betreffende Merkmal muß homogen sein, d. h. eine Vermischung darf nicht stattgefunden haben.

Das Kriterium für die Entscheidung, ob ein Los bedingungsgemäß ist, liegt im Vergleich zwischen dem Schätzwert des Quantils und dem charakteristischen oder dem Nennwert des Werkstoffes; bei der Streckgrenze von Betonstahl III RU z. B. der Wert 420 N/mm^2.

Als Schätzfunktionen bieten sich eine ganze Reihe an: Ganz allgemein muß bei allen Schätzungen unterschieden werden, ob die Standardabweichung σ bekannt ist oder durch deren Schätzwert s (aus der Stichprobe) abgeschätzt wurde.
Ganz allgemein lauten dann die Formeln:

$$x_{p,s} = \bar{x} - k_s \cdot s \quad \text{bzw.} \tag{4}$$

$$x_{p,\sigma} = \bar{x} - k_\sigma \cdot \sigma \tag{5}$$

wobei $\bar{x}$ der Stichprobenmittelwert und s die Stichprobenstreuung ist.
Einige Schätzmethoden lauten:

A. Schätzung mit – frei zu wählender – Aussagewahrscheinlichkeit W
(Bei Betonstählen ist W zu 0,9 festgelegt)

$$x_{P,s} = \bar{x} - T^{-1}_{f(0,9|-U_p \cdot \sqrt{n})} \cdot \frac{1}{\sqrt{n}} \tag{A1}$$

$$x_{p,\sigma} = \bar{x} - \left(U_p + \frac{U_w}{\sqrt{n}}\right) \cdot \sigma \tag{A2}$$

(bei 5%-Quantil: $U_p = 1{,}645$; $W = 0{,}9$; $u_w = 1{,}282$)
Die Schätzung (A1) erfolgt durch die Verwendung der nicht zentralen t-Verteilung [6].

B. Schätzung mit Prediktorverteilung

Ein durch Rackwitz [7] für Werkstoffe übernommenes Verfahren schätzt x_p als Quantil der Prediktorverteilung, so daß der Erwartungswert von p ($E(p)$) der Unterschreitungswahrscheinlichkeit P gleich der Unterschreitungswahrscheinlichkeit des zu schätzenden Quantils ist.
Die Formel für die Benutzung von s bzw. σ lautet:

$$x_{p,s} = \bar{x} - T_f^{-1}(p) \cdot \sqrt{\frac{n+1}{n}} \cdot s \tag{B1}$$

$$x_{p,\sigma} = \bar{x} - u_p \cdot \frac{n+1}{n} \cdot \sigma \tag{B2}$$

T_f^{-1} ist die inverse Funktion der t-Verteilung

C. Schätzung mit Sicherheitsrelevanz

Bei dieser von Struck [8] vorgestellten Schätzung wird der Bemessungswert, der ja auch als Quantil anzusehen ist, an die Sicherheitsanforderungen für bauliche Anforderungen angebunden. Der Wert U_p wird angesetzt zu

$$U_p = U'_p = \alpha_R \cdot \alpha_x \cdot \beta_{\text{erf}} - \frac{\Delta\beta}{\alpha_R \cdot \alpha_x}$$

Dabei sind:

α_x: Streuungseinfluß des betrachteten Merkmals relativ zu den anderen den Werkstoffwiderstand beeinflussenden Größen (hier: $\alpha_x = 1{,}0$ nach „Grusibau")

α_R: Streuungseinfluß des Bauteilwiderstandes im Verhältnis zur Einwirkung (hier: $\alpha_R = 0{,}8$, nach „Grusibau")

β: Sicherheitsindex (aus „Grusibau")

$\Delta\beta$: zulässige Abweichung des Sicherheitsindex β (hier: $\Delta\beta = 0{,}5$)

Diese Größen können sodann für die Gleichungen nach A. und B. verwendet werden, so daß sich folgendes ergibt:

$$x_{p,s} = \bar{x} - T_f^{-1}(0{,}9\, U'_p) \cdot \frac{1}{\sqrt{n}} \cdot s \tag{C1}$$

$$x_{p,\sigma} = \bar{x} - \left(U'_p + \frac{U_w}{\sqrt{n}}\right) \cdot \sigma \tag{C2}$$

Ferner mit $p' = \varnothing\,(-\alpha_R \cdot \alpha_x \cdot \beta_{\text{erf}})$:

$$x_{p,s} = \bar{x} - T_f^{-1}(p') \cdot \sqrt{\frac{n+1}{n}} \cdot s \tag{C3}$$

$$x_{p,\sigma} = \bar{x} - U'_p \cdot \sqrt{\frac{n+1}{n}} \cdot \sigma \tag{C4}$$

2.3.2 Anwendung auf praktische Fälle:

Wahrscheinlichkeitsverteilungen

In den Bildern 3 bis 5 sind für einige *einfachste* Verteilungen die zugehörigen Summenhäufigkeitslinien in Gausssches Wahrscheinlichkeitspapier eingetragen:

Bild 3:

Es wurde eine Rechteck-(Gleich-)Verteilung angenommen. Im Gaussschen Netz könnte man auf eine zweigipfelige Normalverteilung schließen.
Die Normalitätstests bringen eine negative Aussage bei kleinen Probenzahlen.
Eine Dreieckverteilung stellt sich in weiten Bereichen als Normalverteilung dar.

Bild 4:

Die „Hausdachverteilungen" ergeben eine „wunderschöne" Übereinstimmung mit einer Normalverteilung; Tests stets positiv.

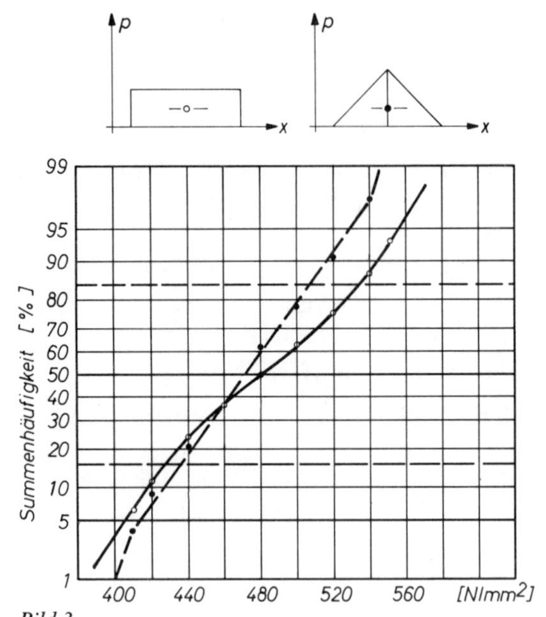

Bild 3
Darstellung „einfachster“ Verteilungen im GAUSSschen Summenhäufigkeitsnetz

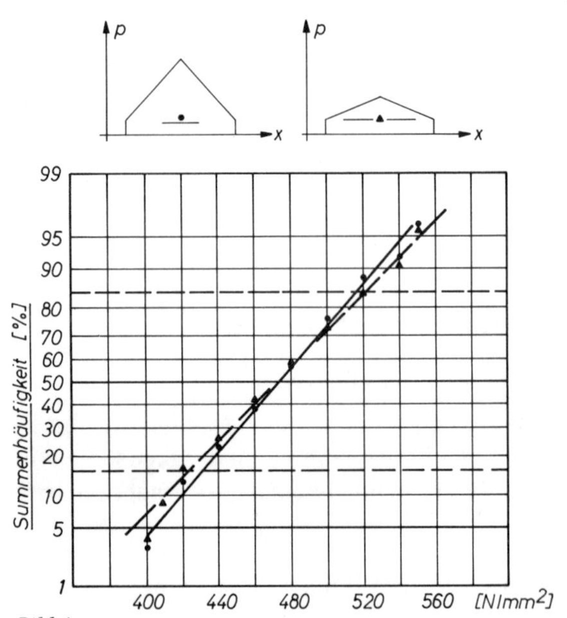

Bild 4
Darstellung „einfachster“ Verteilungen im GAUSSschen Summenhäufigkeitsnetz

Bild 5:

Halbkreis- und Parabelverteilung sind auch in weiten Bereichen nicht von einer Normalverteilung zu unterscheiden.
In allen Beispielen wurden folgende Grenzen eingehalten:

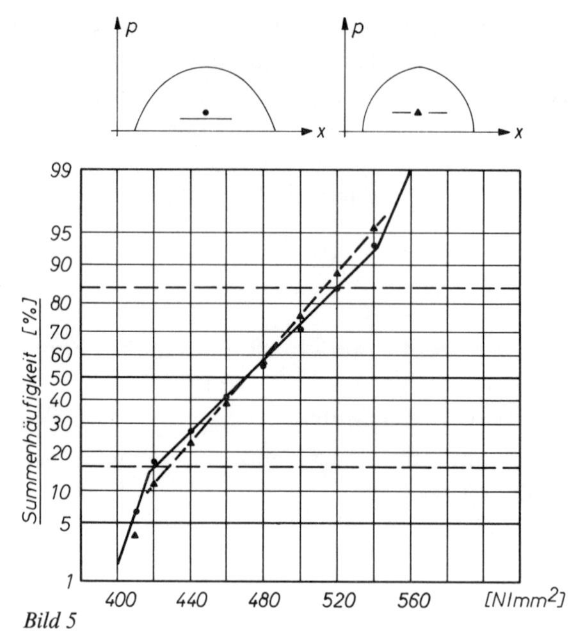

Bild 5
Darstellung „einfachster“ Verteilungen im GAUSSschen Summenhäufigkeitsnetz

$$x_1 = 400 \text{ N/mm}^2$$
$$x_n = 560 \text{ N/mm}^2$$

Das 5%-Quantil beträgt (gerundet):

Gleichverteilung:	408 N/mm^2
Dreieckverteilung:	414 N/mm^2
„Hausdachverteilung“: a)	405 N/mm^2
b)	402 N/mm^2
Kreisverteilung:	405 N/mm^2
Parabelverteilung:	412 N/mm^2

Schätzungen

I. Schätzungen des 5%-Quantils der „einfachen“ Verteilungen

Aus den Histogrammen ist es leicht möglich, die Standardabweichungen σ für die genannten Verteilungen zu ermitteln. Das 5%-Quantil errechnet sich dann über U_σ bei Stichproben umfangen von 5, 15 und 100 (beispielshafte Probenzahlen) zu den in Tabelle 1 angegebenen Werten.
Die mit (A2) mit $W = 0{,}9$ errechneten Werte liegen auf der sicheren Seite; die Formel (B2) liefert z. T. zu hoch angesetzte Schätzwerte.
Die Formeln (C2) und (C4) hängen stark vom Sicherheitsindex ab. Die Tabelle 1 enthält auch die exakten Werte des 5%-Quantils sowie den für alle als Mittelung über die NV errechneten Wert.

Tabelle 1
Schätzwerte für Quantile (5%) bei einfacher Verteilung

Schätzformel	Probenzahl n Stichprobe	Art der Verteilung Rechteck	Dreieck	Hausdach 1	Hausdach 2	Halbkreis	Parabel
A2	5	377	407	388	381	384	394
	15	389	416	399	392	395	404
	100	398	422	407	401	403	411
B2	5	397	421	406	399	402	410
	15	401	424	410	404	406	414
	100	404	426	412	406	408	416
C2 $\beta = 2{,}7$	5	324	370	341	330	334	349
	15	336	378	352	341	345	359
	100	345	385	360	350	354	367
C2 $\beta = 4{,}3$	5	266	329	289	273	279	300
	15	277	337	299	284	290	309
	100	286	343	307	293	298	317
C4 $\beta = 2{,}7$	5	370	403	382	374	377	388
	15	377	407	388	380	383	393
	100	379	409	390	383	386	395
C4 $\beta = 4{,}3$	5	306	357	325	312	317	334
	15	316	364	333	322	326	342
	100	320	367	338	326	330	346
5%-Quantile		408	414	405	402	405	412
Kleinster Einzelwert		← 400 →					
5%-Quantile (gemitteltes σ für NV)		← 412 →					

II. Angabe der k_s- und k_σ-Werte für unterschiedliche Schätzverfahren und Probenzahlen

In der Tabelle 2 sind für die verschiedenen Berechnungsverfahren die k_s- und k_σ-Werte für unterschiedliche Probenzahlen zusammengestellt.
Man erkennt die starke Abhängigkeit von den Probenzahlen aber auch vom Schätzverfahren.

III. Anwendung der Schätzverfahren auf reale Werte der Überwachungspraxis von Betonstählen

Für die Überwachungsergebnisse eines Werkes werden fortlaufend die Ergebnisse von 15 Entnahmen (n_i = 20) ausgewertet. Die Ergebnisse enthält Tabelle 3.
Die sicherheitsrelevanten Berechnungen liegen alle auf der „sicheren" Seite; ein sehr starker Einfluß entsteht durch die Wahl des Sicherheitsindex β.
Die Methoden (A1) und (B1) (Abschätzung des 5%-Quantils mit genormtem Wert $W = 0{,}9$ für Prediktorverteilung) erbringen die beste Annäherung an den Wert des 5%-Quantils, der sich aus der Gesamtheit der Proben bei zeichnerischer Darstellung ergibt (x_5 = 431,5 N/mm²). Für die Gesamtheit brachte (A1) und

Tabelle 2
k_s- und k_σ-Werte für die verschiedenen Schätzverfahren

Probenzahl	Nichtzentrale t-Verteilung $w = 0{,}90$; $p = 0{,}05$		Prediktor-Verteilung		Sicherheitsrelevante Betrachtung $\Delta\beta = 0{,}5$; $\beta = 4{,}3$; $w = 0{,}90$		Sicherheitsrelevante Betrachtung $\Delta\beta = 0{,}5$; $\beta = 2{,}7$; $w = 0{,}90$		Sicherheitsrelevante Betrachtung (Prediktorverteilung) $\beta = 4{,}3$		$\beta = 2{,}7$	
	k_s A1	k_σ A2	k_s B1	k_σ B2	k_s C1	k_σ C2	k_s C1	k_σ C2	k_s C3	k_σ C4	k_s C3	k_σ C4
2	13,1	2,55	7,73	2,01	43,5	4,96	29,9	3,69	36,5	3,97	8,03	2,49
3	5,31	2,38	4,13	1,90	–	4,80	–	3,52	–	–	–	–
5	3,40	2,22	2,33	1,80	7,51	4,63	5,21	3,36	9,43	3,76	3,80	2,37
10	2,52	2,05	1,92	1,72	5,88	4,46	4,08	3,19	5,01	3,61	2,95	2,26
15	2,33	1,97	1,81	1,70	5,41	4,39	3,75	3,11	4,27	3,55	2,70	2,23
20	2,21	1,93	1,77	1,68	5,17	4,34	3,58	3,07	3,97	3,52	2,60	2,21
30	2,08	1,88	1,73	1,67	4,92	4,29	3,40	3,02	3,72	3,49	2,50	2,19
50	1,96	1,83	1,70	1,66	4,69	4,24	3,24	2,97	3,52	3,47	2,42	2,18
100	1,86	1,77	1,67	1,65	4,48	4,19	3,09	2,91	3,44	3,45	2,37	2,17
500	1,74	1,70	1,65	1,645	4,24	4,12	2,91	2,84	3,42	3,44	2,33	2,16

(A2) ein positives Ergebnis, obwohl bei der Einzelbetrachtung der Entnahmen bei (A1) insgesamt 4 und bei (A2) 2 nicht bedingungsgemäße Bewertungen abgegeben wurden. Das Bild 6 macht deutlich, wie stark die unterschiedlichen Schätzungen differieren. Der wirklich vorhandene Wert des 5%-Quantils bleibt bei all diesen Schätzverfahren unbekannt. Der kleinste Einzelwert entspricht einem 0,18%-Quantil bei Annahme einer Normalverteilung.
Neben den erwähnten Problemen mit der Anwendung der Statistik soll noch ein weiteres, besonders delikates erwähnt werden:
So ist z. B. jeder Anwender überzeugt, daß das 5%-(Ausschuß)-Quantil sich auf *alle* Eigenschaften des Betonstahls bezieht. Mit anderen Worten: max. 5% der gelieferten Menge des Betonstahls entsprechen hinsichtlich der geforderten Eigenschaften nicht den Anforderungen.
Die Norm legt aber das Quantil nur für die jeweilige Eigenschaft fest. Es ist leicht einsehbar, daß je mehr Eigenschaften gefordert werden und für alle das 5%-Quantil gilt, der Gesamtausschußanteil beträchtlich wird. Ein Beispiel: Haben alle 8 geforderten Eigenschaften bei Betonstahlmatten exakt 5% Ausschuß, so beträgt die Wahrscheinlichkeit, in einer beliebigen

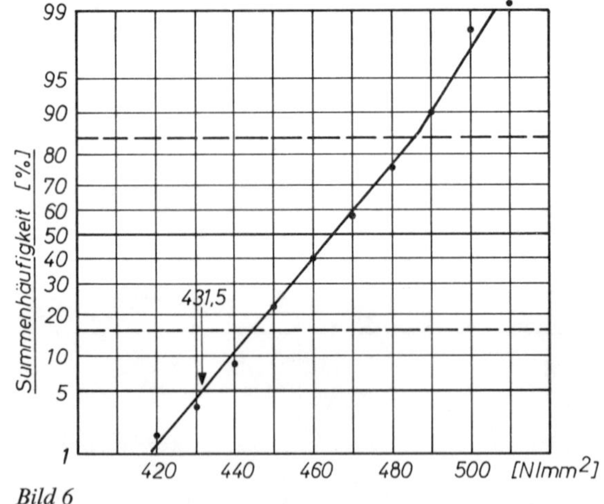

Bild 6
Summenhäufigkeitsverteilung der Auswertungen von Tabelle 3 ($n = 300$)

Probe mindestens einen Schlechtwert (welcher Art auch immer) zu finden, ca. 33%. Eine wahre Freude für reklamationslüsterne Kunden.
Nichtsdestotrotz hat die Statistik viele Vorteile gebracht; man sollte sie jedoch stets als Ingenieur einsetzen und von ihr nicht Wunderdinge erwarten. In vielen Fällen müssen aber noch die Zusammenhänge klarer erkannt werden als bislang.

Tabelle 3
Anwendung der Schätzverfahren auf reale Werte

Mittelwert [N/mm²]	Standardabweichg. (s) [N/mm²]	Kleinster Einzelwert (x_1) [N/mm²]	5%-Quantile errechnet nach:					
			A_1	B_1	C_1 $\beta = 4{,}3$	C_1 $\beta = 2{,}7$	C_3 $\beta = 4{,}3$	C_3 $\beta = 2{,}7$
472	14	433	441	447	399	422	416	436
456	23	420	405	415	337	374	365	396
468	21	423	422	431	359	393	385	413
453	21	409	407	416	344	378	369	398
462	20	410	418	427	357	390	383	410
464	19	425	422	430	366	396	388	415
460	14	420	429	435	387	410	404	424
465	20	427	421	429	362	393	385	413
462	16	433	427	434	379	405	398	420
470	25	428	415	426	341	381	371	405
475	18	436	435	443	382	410	403	428
462	12	441	435	441	400	419	414	431
476	17	444	438	446	388	415	408	432
478	15	451	445	451	400	424	418	439
461	14	429	430	436	388	411	405	425
465	19,2	409	431	433	382	408	399	420
$n = 20 = \text{const.}$			$n = 300 = \Sigma n$					

3 Realisierung der Gütesicherung auf dem Gebiet des Betonstahls

3.1 Vorbemerkung

Die Entwicklung der Güteüberwachung bei Betonstählen ist in der Bundesrepublik Deutschland auf eine relativ kurze Zeitspanne beschränkt. Vor dem Jahr 1972 waren die Betonstähle aufgrund von Zulassungen im Handel. Die Güteüberwachung wurde durch Prüfrichtlinien erfaßt.
Seit 1972, dem Erscheinungstermin der Betonstahlnorm DIN 488, ist im Teil 6 dieser Norm die Güteüberwachung für die Stähle dieser Norm geregelt. Obwohl der Teil 6 von DIN 488 nur eine Vornorm geblieben ist, hat er Schrittmacherdienste für mehrere andere Normen sowie für die Zulassungsregeln anderer Betonstähle gehabt.
Wesentliche Beiträge zu dieser Vornorm stammen von Rehm [9].

Nachfolgend wird das System 1972 und die derzeit diskutierte Fassung von Teil 6 der DIN 488 dargestellt.

3.2 DIN 488, Teil 6, Ausgabe 1972

Schwerpunkt der Erstprüfung ist eine einmalige Untersuchung am Produkt, die bei Nicht-Bestehen beliebig häufig wiederholt werden darf. Als Bewertungskriterien bei der Erstprüfung ist nur die Einhaltung aller Nennwerte gefordert sowie der Nachweis des 5%-Quantils mit 50%iger statistischer Sicherheit.
Die Norm liefert noch keine eindeutige Definition der (Grenz-)Qualität. Das 5%-Quantil ist bei einigen Merkmalen benannt, teilweise ist es unklar formuliert, teilweise fehlt es ganz.
In einer, dem derzeitigen Kenntnisstand nicht mehr entsprechenden Weise wird die Rolle der Annahmekennlinie überbetont: Sie wird als die *eigentliche,* die ausschließlich die Qualität der Produktion bewirkende Größe dargestellt. Allerdings wußte man bereits von

der Wichtigkeit einer sog. gesteuerten Produktion, die man als Voraussetzung für die Prüfpläne zugrunde legte. Einen Nachweis für das Vorhandensein einer gesteuerten Produktion führte man aber nicht ein.
Diese Gegebenheiten führten zu der Regelung der Eigenüberwachung nach zwei Systemen, einem mit und einem ohne Vorinformation.

System ohne Vorinformation

Es wurden je Prüfeinheit (Schmelze, Tagesproduktion bei Matten) Mindestprobenzahlen festgelegt und das Bewertungskriterium wie folgt vorgegeben:

Probenzahl: n	Annahmewahrscheinlichkeit
$\bar{x}_n \geq x_{\text{nenn}} + v$	A1
$x_1 \geq x_{\text{nenn}}$	A2

Der sog. Vorhaltewert „v" setzt sich seinerseits aus einem Annahmefaktor k und einer Standardabweichung zusammen. Bei der Streckgrenze des Stahles beispielsweise beträgt $v = 15$ N/mm². Geht man von einer mittleren Schmelzenstreuung von 10 N/mm² aus, so beträgt $k = 1{,}5$. Der Wert „v" ist eigenschaftsspezifisch festgelegt. Damit hat man es im Prinzip mit einem $(\bar{x},\sigma)$-Plan zu tun, der kombiniert ist mit einem Mindestwertplan. Die Annahmewahrscheinlichkeit A ist das Produkt A1 · A2.
Hätte man keinerlei Annahme für σ getroffen, dann müßte man, bezogen auf ein freigewähltes Signifikanzniveau und der Streuung der Stichprobe sehr hohe k_s-Werte (z. B. $k = 5{,}31$ bei $s = 1 - \alpha = 0{,}90$; $n = 3$) in Kauf nehmen. Die Folge: eine Verhinderung der Produktion von Betonstahl oder eine volkswirtschaftlich nicht zu verantwortende Überqualität.
Das gewählte Kriterium hat sich, wie nachgewiesen werden konnte, aber in der Praxis gut bewährt [10]. Die Bilder 7 und 8 zeigen Summenhäufigkeitsverteilungen der wichtigsten Merkmale.

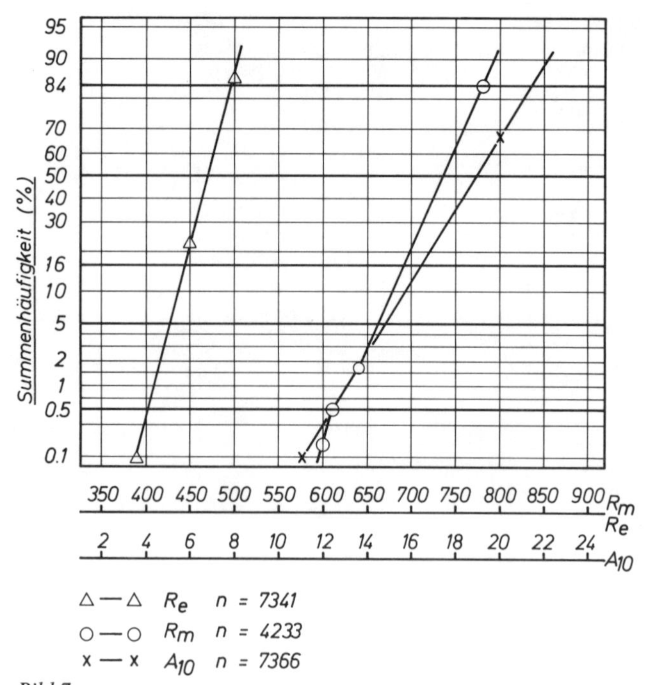

Bild 7
Qualitätsniveau von Betonrippenstahl BSt 420/500 RU (mehrere Werke)

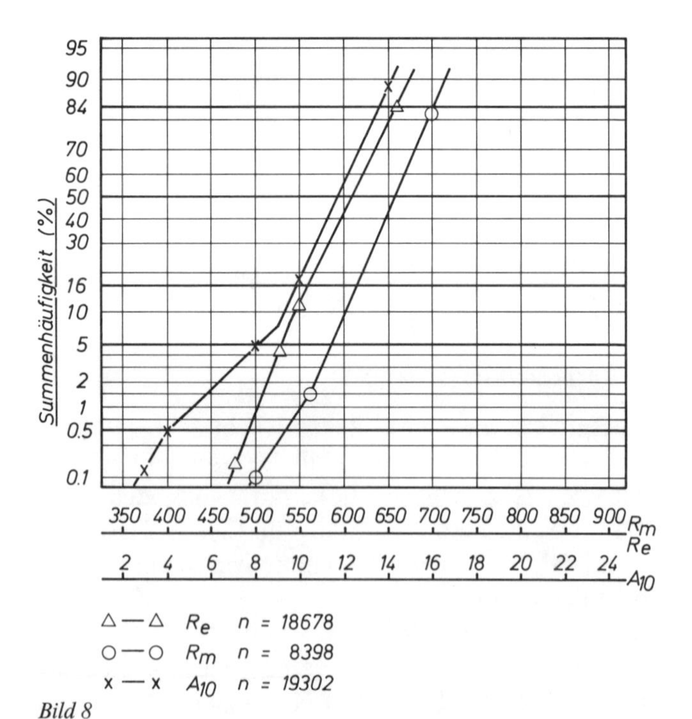

Bild 8
Qualitätsniveau von Betonstahlmatten BSt 500/550 RK

System mit Vorinformation

Das System mit Vorinformation ist ein zukunftsträchtiges Konzept für eine moderne, an den Realitäten orientierte Werkstoffprüfung. Das Prinzip beruht darauf, zuerst ein Produktionsniveau eines Herstellers sicherzustellen und dann die laufende Prüfung auf dieses Niveau bezogen und nicht an der Anforderung zu orientieren. Eine Sicherung gegen das Abdriften des „Angebotes" bietet eine untere Schwelle in Gestalt eines 0,1%-Quantils.

Das Verfahren wurde in der Regel aber nicht angewandt, weil es als kleinen, aber wichtigen Mangel, eine Selbststeuerung zu laufend besserer Qualität enthielt. Nichtsdestotrotz kann man diese Tatsache seiner Hereinnahme in eine Norm nicht genug hervorheben, weil hier erstmalig die klassischen „statischen" Betrachtungsweisen der Statistik verlassen wurden.

Fremdüberwachung

Das sehr liberale System der Güteüberwachung weist der Fremdüberwachung eine große Bedeutung zu.
Die fremdüberwachenden Stellen haben dabei mehrere Aufgaben zu übernehmen. Zum einen stellen sie aufgrund der Möglichkeit unabhängig das Produkt zu prüfen für den Hersteller ein neutrales Korrektiv dar. Insbesondere ist dabei die Prüfung außerhalb des Herstellerwerkes nützlich, weil alle möglichen systematischen Fehler ausgeschaltet sind.
Von größter Wichtigkeit in diesem Zusammenhang ist die korrekte Probennahme durch die Beauftragten der Prüfstelle.
Durch ihre Objektivität kann die Prüfstelle ferner dafür sorgen, daß wettbewerbsverzerrende Unterschiede im Qualitätsniveau und im Umfang der Prüfungen vermieden werden.
Diese Aufgabe ist von größter Bedeutung im internationalen Bereich, in dem z. T. gänzlich unterschiedliche Regelungen der Güteüberwachung üblich sind.
Die Prüfstellen befinden sich in einer Mittlerrolle zwischen Hersteller und Verbraucher. Das für den Verbraucher anonyme Produkt erfährt durch das Gütesiegel der Prüfstelle eine Aufwertung.
Auf der anderen Seite wird der Hersteller vor unberechtigten Reklamationsforderungen (wie sie insbesondere zu Zeiten des Preisverfalls an der Tagesordnung sind) durch die Einschaltung der Prüfstelle bewahrt. Bei berechtigten Reklamationen ist durch die Einschaltung der Prüfstelle am ehesten Klarheit über Umfang und Folgen (auch bautechnisch) zu gewinnen.
Nicht zuletzt dient die Tätigkeit der Fremdüberwachung der öffentlichen Sicherheit: Die Tatsache, daß die Quote des Totalversagens im Baubereich erfreulich niedrig liegt, hat auch hier seine Ursache.
Es sollte als selbstverständlich gelten, daß in Anbetracht dieser Aufgaben, die Auswahl der Prüfstellen nach strengsten Maßstäben hinsichtlich der Voraussetzungen erfolgt: Neben den prüftechnischen Kenntnissen sollten umfangreiches technologisches Wissen sowie Erfahrungen im Bereich des Stahlbetonbaus, d. h. über die Verwendung des Produktes, vorliegen.
Eine weltweite Resonanz hatte die Kennzeichnungspflicht der Betonstähle gemäß DIN 488. Damit gilt ein Betonstahl nur dann als normgemäß, wenn er ein Werkkennzeichen des Instituts für Bautechnik trägt und der Fremdüberwachung unterliegt. Diese Forderung gilt für den Zuständigkeitsbereich der deutschen Bauaufsicht: Im Drittlandshandel wird die DIN 488 nur als Bestellnorm genutzt.
Kunde und Lieferer müssen sich aber über die fehlenden Überwachungsschritte im klaren sein.

3.3 Entwurf DIN 488, Teil 6, Fassung 1983

Der Neuentwurf für DIN 488, Teil 6, stellt die logische Weiterentwicklung des bisherigen Standes dar. Die bewährten Elemente wurden belassen, wo nötig modifiziert, sowie einige neue Gedanken hinzugefügt. Wesentlich ist die eindeutige Definition der Qualität als Quantil bei einer vorgegebenen statistischen Sicherheit von $s = 1 - \alpha = 0{,}90$.
Man ist sich klar geworden, daß die statistische Betrachtung der Qualität auf die Eigenschaft abgestimmt werden muß.
Der Verbraucher kann z. B. davon ausgehen, daß nicht 5% der Betonstähle beim Biegen brechen, sondern kann eine quasi totale Biegefähigkeit als gegeben ansehen. Die unvermeidlichen singulären Fehler, die zum vereinzelten Versagen führten (bei großtechnischer Produktion nicht auszuschließen), müssen demzufolge prüftechnisch anders bewertet werden.
Die zentrale Stellung der Annahmekennlinie quasi als „Qualitätsdefinition" ist logischerweise verschwunden.
Bei der Erstprüfung wurde eine wesentliche Verbesserung eingeführt, insofern als sie in drei Einzelteile aufgesplittet wurde:
- die Werksinspektion (verbessert)
- die Werkstoffprüfung (wie bisher)
- den langfristigen Qualitätsnachweis (neu).

Die Werksinspektion mußte wegen der sich laufend weiterentwickelnden Technologien in ihrer Bedeutung gehoben werden. Der Werkstoffprüfung kommt wie bisher die Aufgabe zu, in entsprechend umfangreicher Weise die Qualität des Produktes zu ermitteln und zu beurteilen.
Damit war aber – insbesondere dann, wenn nach mehrmaliger Wiederholung ein positives Ergebnis zu-

stande kam – keine Aussage über den Qualitätsstand der zu erwartenden laufenden Produktion möglich.
Dieses Übel wird durch die Einführung eines langfristigen Qualitätsnachweises behoben, insofern, als der Hersteller innerhalb von ca. einem Jahr nachweisen muß, daß die Anforderungen permanent eingehalten werden. Der Hersteller erhält somit nach Abschluß von Werksinspektion und Materialsprüfung nur ein vorläufiges Werkkennzeichen. Die endgültige Entscheidung fällt nach der dritten Stufe.
Für die Eigenüberwachung gilt im Prinzip die bisherige Regelung mit einem

- festgelegten System mit Vorhaltemaß und einem
- offenen System mit Vorinformation.

Die Regelungen wurden aber etwas modifiziert:
Das Vorhaltemaß muß über den Nachweis der Binnenstandardabweichung im Prüflos an die realen Gegebenheiten angepaßt werden.
Das System mit Vorinformation ist nicht durch einen *definitiven* Prüfplan in der Norm enthalten, da man sich über die Vielzahl an Möglichkeiten klar war.
Die Festlegung derartiger Prüfpläne muß nach ausreichendem Nachweis der sog. Vorinformationen einvernehmlich zwischen Herstellwerk und Prüfstelle erfolgen.
Die Fremdüberwachung hat ihre Aufgaben wie bislang üblich zugewiesen bekommen. Zusätzlich wurde ein Katalog von sog. Ahndungsmaßnahmen eingeführt. Diese Maßnahmen dienen dazu, bei Nicht-Einhaltung des Qualitätsniveaus adäquat und sachbezogen richtig reagieren zu können.

4 Zusammenfassung

Die Güteüberwachung von Betonstählen ist wegen der komplexen Bedingungen eine schwierige Aufgabe.
Die wesentlichen Probleme sind durch den internationalen Handel, die unterschiedlichen und sich stets ändernden Technologien sowie die derzeit schwierige Situation auf dem Sektor der Stahlindustrie gegeben. Ferner sind Fragen genereller Art im Zusammenhang mit der Anwendung der Statistik und speziell der Prüfpläne von Bedeutung. Das bisher bestehende System führte zu einem befriedigenden Qualitätsniveau und arbeitet somit im Interesse der Öffentlichkeit sehr erfolgreich.
Wegen der besseren Einsichten in die statistischen Zusammenhänge der Änderung der Stahlsorten und Technologien weist der Entwurf für die neue DIN 488 (Teil 6) einige wesentliche Verbesserungen auf, die das System wirksamer und anpassungsfähiger gestalten.

5 Literatur

[1] Grundlagen zur Festlegung von Sicherheitsanforderungen für bauliche Anlagen, DIN. Beuth Verlag GmbH, 1981.
[2] Kühlmeyer, M.: Construction of an Economic and Secure Inspection and Production Strategy, Frontiers in Statistical Quality Control. Physica Verlag, Würzburg, 1981.
[3] Shapiro, S. S. and Wilk, M. B.: An Analysis of Variance Test for Normality. Biometrica (1965) 52, 3 and 4, p. 591.
[4] Kolmogoroff, A.: Confidence limits for unknown distribution function. Ann. Math. Statist. 23 (1952), p. 525.
[5] Henning, H. J. und Wartmann, R.: Stichproben kleinen Umfangs im Wahrscheinlichkeitsnetz.
[6] Mesnikow, G. J. and Liebermann, G. J.: Tables of the noncentral *t*-distribution. Stanford University Press. Stanford California.
[7] Rackwitz, R.: Zur Statistik von Eignungs- und Zulassungsversuchen für Bauteile. Bauingenieur 56 (1981) S. 103–107.
[8] Struck, W.: Bemessungswerte und charakteristische Wert von Bauteilwiderständen unter Einbeziehung von Versuchsergebnissen. Amts- und Mitteilungsblatt der BAM 11 (1981), Nr. 4, S. 316–325.
[9] Rehm, G. und Rehm, H.: Zur Frage der Prüfregelung bei der Qualitätssicherung von Bewehrungsstählen. Betonsteinzeitung, Heft 3, 1969.
[10] Rehm, G. und Russwurm, D.: Anmerkung zur Güte von Betonstählen. Betonwerk- und Fertigteil-Technik Heft 1/1977.

Erwähnt:

Musterbauordnung; Bayerische Bauordnung; DIN 18200; DIN 55355.

Verbesserung der Werkstoffeigenschaften von Betonstählen durch Optimierung der Herstellbedingungen

Horst Weitzmann
und Wilhelm Dening

1 Einleitung

Die Badischen Stahlwerke in Kehl am Rhein produzieren mit 2 Elektroöfen, 2 Stranggießanlagen und 2 Warmwalzwerken mehr als 700000 Tonnen Stahl im Jahr.
Das 1968 fertiggestellte Werk war das erste sogenannte Ministahlwerk in Deutschland und Modell für eine Vielzahl ähnlich konzipierter Produktionsstätten in aller Welt. Die beiden Warmwalzwerke sind für die Produktion von Betonstahl und Walzdraht ausgelegt. Der weitaus überwiegende Teil des Walzdrahtes wird zu Betonstahlmatten und Gitterträgern weiterverarbeitet.
Die Beschränkung auf eine Produktgruppe bietet die Möglichkeit, die Produktionsanlagen für die Herstellung dieser Produkte gezielt auszustatten und auf diese Weise bei optimaler Wirtschaftlichkeit eine gleichbleibend hohe Produktqualität zu garantieren.
Da die Produktionsmethoden ebenso wie die Qualitätsanforderungen einem stetigen dynamischen Wandel unterworfen sind, spielen Forschung und Entwicklung eine wichtige Rolle.
Als Ergebnis einer engen Zusammenarbeit mit Hochschulen und anderen Forschungseinrichtungen konnte so im Jahr 1973 die erste bauaufsichtliche Zulassung für den Betonstahl IV (KB IV) erreicht werden, der bereits drei Jahre später die erste Zulassung für einen voll schweißbaren Betonstahl IV S (KB IV S) folgte.
Seit einigen Jahren wird bei den Badischen Stahlwerken schweißbarer Betonstahl nach dem Tempcore-Verfahren hergestellt. Im Rahmen des vom BMFT geförderten Stahlforschungsprogrammes wurden in zwei Forschungsvorhaben die Möglichkeiten zur Optimierung einiger Materialeigenschaften, insbesondere der Dauerschwingfestigkeit und der Verbundeigenschaften, durch Variation der Herstellbedingungen und der Rippengeometrie untersucht. Über einige wesentliche Ergebnisse dieser Forschungsvorhaben wird nachfolgend berichtet.

2 Auswirkung des Kühlvorganges auf die Festigkeitseigenschaften wassergekühlter Betonstähle

Wassergekühlte Betonstähle durchlaufen nach dem letzten Walzgerüst eine Kühlstrecke, in der die Oberflächen von etwa 1000 °C auf unter 200 °C abgeschreckt werden. Das dabei entstehende Härtegefüge (Martensit) führt zu einer erheblichen Festigkeitssteigerung, so daß bei Einhaltung der Analysengrenzwerte für voll schweißbare Betonstähle die Festigkeitsanforderungen für BSt III S und IV S mit genügendem Spielraum eingehalten werden können.
Die im Querschnittsinneren der abgeschreckten Betonstähle verbleibende Walzhitze führt nach dem Abkühlvorgang zu einem „Selbstanlaßeffekt" der abgeschreckten Randzonen, wodurch dem Martensitgefüge eine hervorragende Dehnfähigkeit verliehen wird.
Die Zusammenhänge zwischen chemischer Zusammensetzung und Kühlintensität einerseits sowie der statischen Festigkeits- und Dehnungseigenschaften andererseits waren weitgehend bekannt (siehe z. B. [1]). Nicht grundsätzlich erforscht war dagegen die Auswirkung der genannten Einflußparameter auf die Schwingfestigkeit von Betonstählen. Ziel eines umfangreichen Forschungsprogrammes war es deshalb, die Herstellbedingungen für wassergekühlte Betonstähle unter Einbeziehung des Qualitätsmerkmales Schwingfestigkeit zu optimieren.
Am Beispiel einer der untersuchten III S-Schmelzen ist im nachfolgenden Bild 1 der Einfluß des Kühlvorganges auf die wesentlichen Festigkeitseigenschaften gezeigt. Der ungekühlte Nullversuch lag wegen der für die Schweißbarkeit geforderten Analysengrenzwerte naturgemäß weit unter der geforderten Mindeststreckgrenze von $R_e = 420$ N/mm². Durch den Kühlvorgang konnte die Streckgrenze jedoch bis auf über $R_e = 480$ N/mm² angehoben werden. Die Bruchdehnung

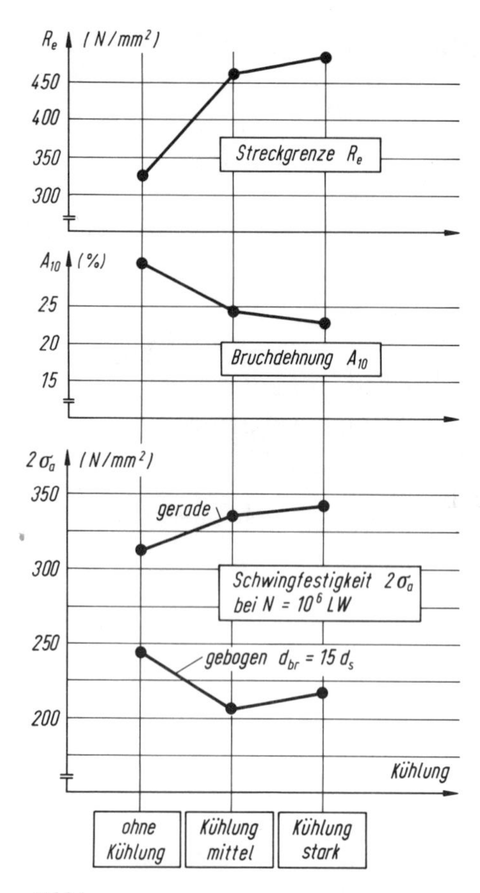

Bild 1
Einfluß von Kühlung und Kühlintensität auf die Festigkeits- und Verformungseigenschaften von Betonstählen. Stabdurchmesser $d_s = 16$ mm

fiel mit zunehmender Kühlintensität zwar ab, blieb mit $A_{10} = 22\%$ aber auch bei hoher Kühlintensität noch um mehr als das zweifache über dem geforderten Mindestwert.

Die Dauerschwingfestigkeit gerader Stäbe stieg mit der Kühlintensität ebenso wie die Streckgrenze an. Dieses Ergebnis stimmt mit den aus der Literatur bekannten Zusammenhängen für unbehandelte Stäbe überein. Überraschend waren dagegen zunächst die Ergebnisse im gebogenen Zustand: gegenüber dem ungekühlten Vergleichszustand nahm die Schwingfestigkeit von durch Kühlung verfestigten Betonstählen deutlich ab. Diese Abnahme deutet auf eine erhöhte Empfindlichkeit der unter Eigenspannungen stehenden martensitischen Randzonen gegenüber dem Bauschinger-Effekt (abgeminderte Festigkeit nach vorangegangener plastischer Verformung in entgegengesetzter Richtung) hin, der im Bereich von Biegungen für die Randzonen des Querschnitts praktisch immer zum Tragen kommt. Allerdings war die Auswirkung der Kühlintensität auf den Abfall der Schwingfestigkeit im gebogenen Zustand gering. Die Schwingfestigkeit nahm mit zunehmender Kühlintensität bei allen untersuchten Schmelzen eher wieder zu, was auf eine günstige Entwicklung des Eigenspannungszustandes zurückgeführt werden kann. Eine ausführliche Berichterstattung zu diesem Forschungsvorhaben enthält [2].

Für die Produktion und praktische Anwendung von Betonstählen bedeuten die gewonnenen Erkenntnisse, daß die im Hinblick auf die Schweißbarkeit festgelegten Analysengrenzwerte nicht ausgenützt werden müssen, sondern daß die geforderten Festigkeitseigenschaften durch eine möglichst hohe Kühlintensität sichergestellt werden können. Neben einer erhöhten „Schweißsicherheit" wird dadurch auch eine verbesserte Schwingfestigkeit erreicht. Außerdem kann aus diesem Ergebnis der Schluß gezogen werden, daß für einen höherfesten Betonstahl eher bessere Schwingfestigkeiten zu erwarten sind als für einen niederfesten, wenn es sich um eine voll schweißbare wassergekühlte Qualität handelt.

3 Einfluß der Rippenanordnung auf der Staboberfläche auf die Schwingfestigkeit sowie auf die Verbund- und Sprengwirkung

Die Anordnung der Rippen auf der Oberfläche von Betonstählen beeinflußt im wesentlichen die Verankerungswirkung im Beton und die Schwingfestigkeit. Die Festlegung der Rippenanordnung (Abstand und Höhe) bei der Einführung der Betonrippenstähle erfolgte im Hinblick auf die Verankerungswirkung im Beton (Rippenabstand und -höhe) sowie die grundsätzliche Eignung für dynamische Beanspruchung (sichelförmige Schrägrippen). Das Rippenmuster für den Betonstahl IV (alternierende Rippenneigung) wurde unter Beibehaltung der Werte für die bezogene Rippenfläche in erster Linie aus Kennzeichnungsgründen gewählt. Bei dynamischer Beanspruchung ist diese Rippung jedoch ungünstiger als die des Betonstahles III, so daß die nutzbaren Schwingbreiten für den Betonstahl IV gegenüber den für Betonstahl III festgelegten Werten nicht erhöht werden konnten. Ziel eines zweiten Forschungsprogrammes war es deshalb, ein Optimum der Rippenanordnung unter Berücksichtigung

der Verankerungswirkung im Beton und der Auswirkungen auf die Schwingfestigkeit der Betonstähle zu finden.
In den Bildern 2 und 3 sind der Einfluß von Rippenabstand c und Rippenneigung zur Stabachse β auf die Schwingfestigkeit und die Verbundeigenschaften gezeigt. Für alle untersuchten Varianten wurde die bezogene Rippenfläche f_R konstant gehalten, das heißt, mit abnehmenden Rippenabstand wurden auch die Rippenhöhen in gleichem Maße geringer. Ebenso wurde der Rippenquerschnitt bei allen Varianten gleich gehalten, so daß sich am Rippenfuß zumindest rechnerisch immer gleiche Kerbspannungsfaktoren ergaben.
Das Bild 2 zeigt, daß sowohl mit abnehmendem Rippenabstand c als auch mit abnehmender Rippenneigung β die Schwingfestigkeit gerader und gebogener Stäbe zunimmt.

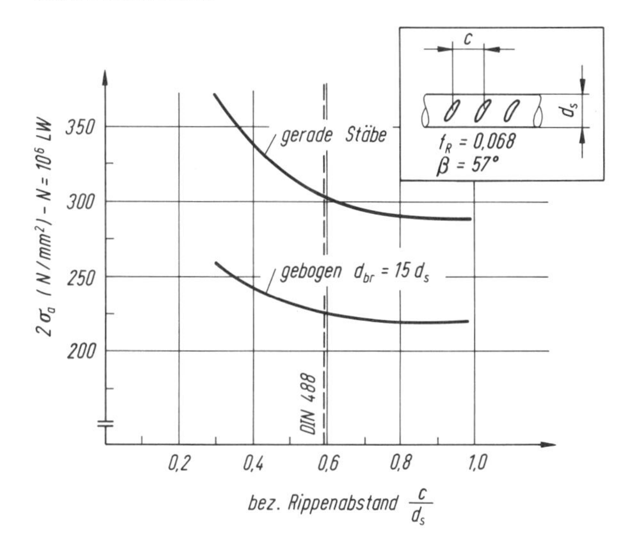

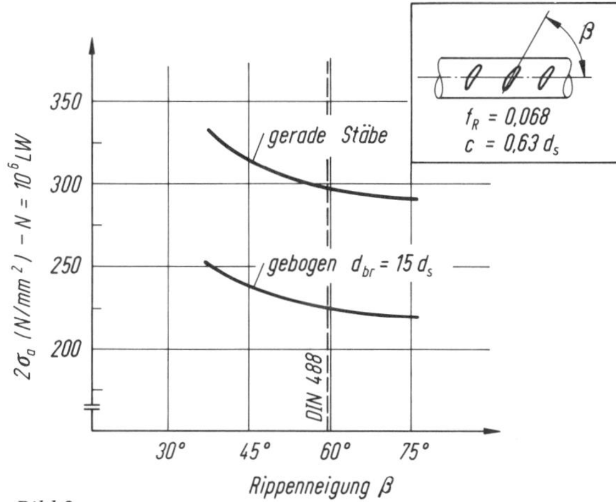

Bild 2
Einfluß von Rippenabstand und Rippenneigung auf die Schwingfestigkeit von Betonstählen (Ergebnisse erzielt an 25-mm-Stäben)

Das Bild 3 zeigt beispielhaft die Auswirkung unterschiedlicher Rippenanordnungen auf die Verbundeigenschaften. Insgesamt ist der Einfluß von Rippenabstand und Rippenneigung auf Verbund und Sprengwirkung, verglichen mit den für diese Anwendungskenngrößen üblichen Streuungen gering. Das heißt, dem Qualitätsmerkmal Schwingfestigkeit kommt bei der Beurteilung der Rippenanordnung eine größere Bedeutung zu, sofern die bezogene Rippenfläche gleich bleibt.
Zusammenfassend kann aus den Versuchen der Schluß gezogen werden, daß bei deutlich geringeren Rippenabständen und Rippenhöhen sowie geringfügig kleineren Rippenneigungen als in DIN 488 festgelegt sowohl bessere Ergebnisse für die Schwingfestigkeit als auch für das Verbundverhalten erzielt werden können. Dies konnte sowohl in umfangreichen weiterführenden Dauerschwingversuchen sowie in Biege- und Schubversuchen unter Verwendung von Betonstählen mit optimierter Rippenausbildung gezeigt werden.
Als Sekundäreffekt ergibt sich für kleinere Rippenabstände und Rippenhöhen bei gleicher bezogener Rippenfläche ein größerer Kernquerschnitt bzw. ein geringerer Querschnittsanteil in den Rippen. Da zur Kraftaufnahme nur der Kernquerschnitt zur Verfügung steht, führt dieser Effekt zu einer besseren Materialausnutzung und damit zu einer wirtschaftlicheren Betonstahlherstellung.
Wenn die gefundene optimale Rippenanordnung auf der Staboberfläche wegen der in der DIN 488 festgelegten Rippung für Betonstähle derzeit auch nicht unmittelbar in die praktische Produktion umgesetzt werden kann, so ermöglichen die gewonnen Erkenntnisse unter Ausnutzung des in der DIN 488 gegebenen Spielraumes doch die Produktion eines Betonstahles mit bestmöglicher Schwingfestigkeit.

4 Zusammenfassung

Die Erkenntnisse aus umfangreichen Untersuchungen zum Einfluß der chemischen Zusammensetzung und der Kühlbedingungen, insbesondere auf die Schwingfestigkeit wassergekühlter Betonstähle und zum Einfluß der Rippenanordnung auf die Verankerungswirkung und Schwingfestigkeit von Betonstählen, ermöglichen eine weitere gezielte Verbesserung der Werkstoffeigenschaften von Betonstählen durch Opti-

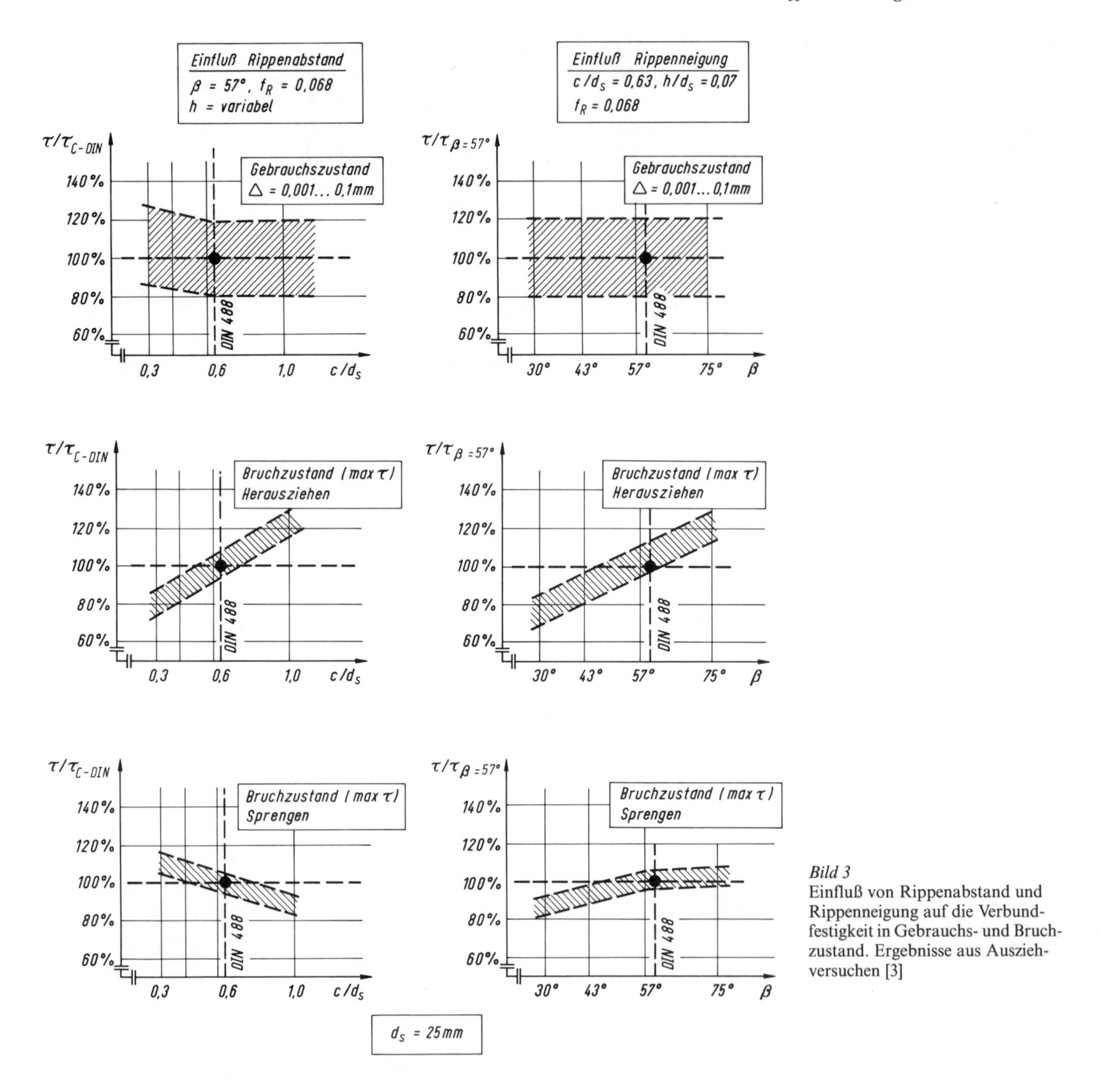

Bild 3
Einfluß von Rippenabstand und Rippenneigung auf die Verbundfestigkeit in Gebrauchs- und Bruchzustand. Ergebnisse aus Ausziehversuchen [3]

mierung der Kühlbedingungen und der Rippenanordnung. Neben dieser Verbesserung der Anwendungsbedingungen lassen sich auch diese Weise Betonstähle verbessern und auch wirtschaftlicher herstellen.
Das aufgezeigte Beispiel macht deutlich, daß durch konsequente anwendungsbezogene Forschung und Entwicklung eine stetige Weiterentwicklung und Verbesserung der Produktqualität erreicht wird.

5 Literatur

[1] Economopoulos, Resen, Lessel und Steffes: Application of the Tempcore Process to the Fabrication of High Yield Strength Concrete-reinforcing Bars. C.R.M. No 45, December 1975.

[2] Martin, Schiessl und Schwarzkopf: Optimierung der Gebrauchseigenschaften von wassergekühlten Betonstählen. IBS Nr. 874/83.

[3] Martin, Schiessl und Schwarzkopf: Optimierung der Rippenausbildung hochfester Betonstähle. IBS Nr. 784/83.

Festigkeit und Dauerhaftigkeit als Kriterien für die Entwicklung neuer Zementarten

Gerd Wischers

1 Was ist Zement?

Zement ist aufgrund seiner sehr hohen Festigkeit der wichtigste und der charakteristischste Vertreter der sogenannten hydraulischen Bindemittel. Dabei handelt es sich um ein sehr feinkörniges Pulver, das nichtmetallischer, anorganischer Natur ist, das mit Wasser angemacht wird, das dann selbständig erhärtet und das nach der Erhärtung auch unter Wasser dauerhaft fest und raumbeständig ist. Träger der Erhärtung sind in erster Linie Calciumsilicate, jedoch vermögen auch Calciumaluminate und andere Verbindungen hydraulisch zu erhärten. Andere im Bauwesen häufig verwendete und ebenfalls mit Wasser angemachte, pulvrige Bindemittel unterscheiden sich dadurch von Zement, daß sie entweder nicht selbständig erhärten, wie z. B. Kalk, der die Kohlensäure der Luft zur Erhärtung benötigt, oder daß sie nach dem Erhärten nicht wasserfest sind, wie z. B. Gips.

2 Aufgaben des Zements im Beton

Zement wird ganz überwiegend zur Herstellung von Normalbeton verwendet. Vorrangige Aufgabe des mit Wasser angemachten Zements ist es dabei, die einzelnen Zuschlagkörner vollflächig und dauerhaft zu verkleben und die auch in einem verdichteten Zuschlaggemisch stets noch vorhandenen Lücken und Hohlräume weitgehend zu füllen. Letzteres kann auf zweifache Weise geschehen, nämlich erstens dadurch, daß der Zement die Kornzusammensetzung des Zuschlaggemisches im Feinstbereich in günstiger Weise ergänzt, und zweitens dadurch, daß die durch die chemisch-mineralogische Reaktion des Zements mit dem Anmachwasser entstehenden Hydratationsprodukte ein größeres Volumen als der ursprüngliche Zement einnehmen und so die ursprünglich mit Anmachwasser gefüllten Poren im Beton teilweise ausfüllen.

Zement muß im erhärteten Beton somit insbesondere drei Anforderungen erfüllen, nämlich

- hochfestes Verkleben der Zuschlagkörner,
- Ausfüllen von zuvor mit Wasser gefüllten Poren durch Hydratationsprodukte und
- Ergänzen der Kornzusammensetzung des Zuschlags im Feinstbereich zur Minderung des Wasseranspruchs und Verbesserung der Verarbeitbarkeit.

Ferner bewirkt der Zement den Korrosionsschutz des Bewehrungsstahls im Beton, weil schon zu Beginn der Reaktionen des Zements mit dem Anmachwasser ein danach fortwährendes, hochalkalisches Milieu entsteht, in dem sich auf der Stahloberfläche eine Schutzschicht (Passivschicht) bildet.

Alle Betoneigenschaften müssen dauerhaft vorhanden sein, d. h. über Jahrzehnte ihre Funktion ohne wesentliche Einbußen erfüllen, und zwar auch bei den Umweltbedingungen, wie sie üblicherweise hierzulande vorliegen. Dazu zählen bei Außenbauteilen Trocknen und Durchfeuchten, Temperaturwechsel mit Frostbeanspruchung sowie chemische Einwirkungen durch Gase, z. B. Kohlenstoffdioxid, und durch angreifende Lösungen, wie z. B. den sogenannten „sauren“ Regen.

Insbesondere bei der Forderung nach Dauerhaftigkeit ist zu berücksichtigen, daß Beton seinem Wesen nach ein poröser Baustoff ist. Das Ausmaß der Porosität und die Porengrößenverteilung hängen von der Zusammensetzung des Betons, von den Eigenschaften des Zements und von der Nachbehandlung und Lagerung des Betons ab. Eine insgesamt hohe Porosität oder eine ungünstige Porengrößenverteilung mindern praktisch alle Betoneigenschaften. Allerdings kann die negative Auswirkung auf die einzelnen Betoneigenschaften sehr unterschiedlich sein, z. B. nur mäßig auf die Festigkeit, jedoch sehr groß auf den Korrosionsschutz der Bewehrung oder den Frostwiderstand. Die Festigkeit ist daher kein allein ausreichendes Kriterium für die Qualität eines Betons.

3 Hydratationsmechanismen des Zements [1]

Nach dem Anmachen lösen sich die einzelnen Zementkörper nicht in dem Anmachwasser auf – wie etwa Zucker im Kaffee – sondern an den Reaktionsflächen zwischen Zementkörpern und Wasser bilden sich Hydratationsprodukte, die in den wassergefüllten Raum zwischen den Körnern hineinwachsen, siehe Bild 1. Diese Hydratationsprodukte sind außerordentlich feine, nur mit dem Elektronenmikroskop zu identifizierende, fasrige Kristalle. Sie bilden ein von submikroskopischen Poren durchsetztes Gel, das bei Portlandzement nach vollständiger Hydratation einen etwa 2,06fach so großen Raum einnimmt wie das ursprüngliche, unhydratisierte Klinkerkorn.
Der Zement vermag somit auch nach vollständiger Hydratation nur einen ganz bestimmten Teil der zunächst mit Anmachwasser gefüllten Räume mit Hydratationsprodukten auszufüllen. Wie groß der zu füllende Raum insgesamt ist, hängt vom Wasserzementwert ab. In Tabelle 1 sind die Verhältnisse für w/z-Werte von 0,40, 0,60 und 0,80 angegeben. Durch Anheben des Wasserzementwertes von 0,40 auf 0,80 steigt der im frischen Zementleim mit Anmachwasser gefüllte Porenraum von 55,4 auf 71,3 Vol.-%, d.h. auf das 1,3fache. Der Anteil der auch nach vollständiger Hydratation verbleibenden Kapillarporen steigt jedoch von 8,0 Vol.-% bei $w/z = 0{,}40$ auf 40,8 Vol.-% bei $w/z = 0{,}80$, d.h. um mehr als das 5fache.
Übersteigt die Kapillarporosität größenordnungsmäßig 25 Vol.-%, so muß damit gerechnet werden, daß die

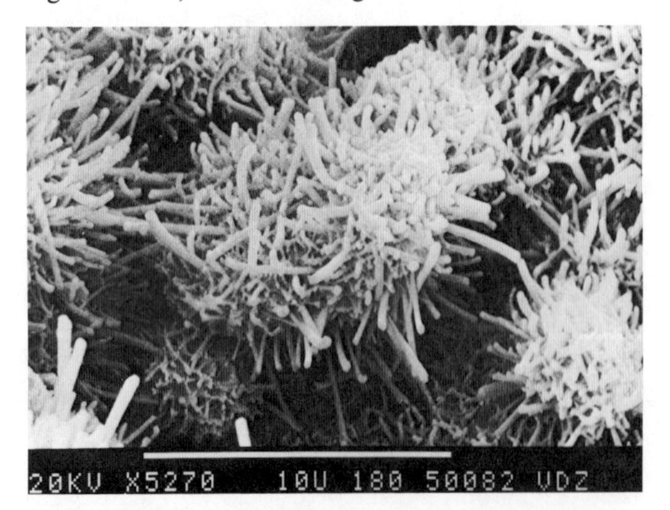

Bild 1
Rasterelektronenmikroskop (REM)-Aufnahme von hydratisierenden Portlandzementkörnern. Hydratationsdauer 12 Tage; Wasserzementwert 0,80

Tabelle 1
Volumenanteile im frischen Zementleim und im vollständig hydratisierten Zementstein (Verhältnisse bei Portlandzement)

Frisch angemachter Zementleim			Vollständig hydratisierter Zementstein	
Wasserzementwert	Zement Vol.-%	Wasser Vol.-%	Hydratationsprodukte Vol.-%	Verbleibende Kapillarporen Vol.-%
0,40	44,6	55,4	92,0	8,0
0,60	35,0	65,0	72,0	28,0
0,80	28,7	71,3	59,2	40,8

Poren untereinander verbunden sind und daß dadurch der Zementstein sehr durchlässig ist. Die Durchlässigkeit hängt jedoch nicht nur vom Gesamtporenraum, sondern auch von der Porengrößenverteilung ab. So tritt eine große Durchlässigkeit erst bei einem größeren Gesamtporenvolumen auf, wenn die Poren überwiegend sehr fein sind.
Die zuvor genannten Verhältnisse gelten für eine vollständige (100%ige) Hydratation. Nun ist die Hydratation ein chemisch-mineralogischer Vorgang, der nicht in Stunden abläuft, sondern Tage, Wochen, Monate und bei langsam reagierenden Zementbestandteilen oder bei groben Zementkörnern sogar Jahre und Jahrzehnte dauert, vorausgesetzt, daß ständig ein ausreichendes Feuchtigkeitsangebot vorliegt. Es gibt somit zwischen dem Anmachen und der vollständigen Hydratation alle möglichen Zwischenstadien, wie das in Bild 2 schematisch dargestellt ist. Bei einem Hydratationsgrad von 50% ist der volumenmäßige Anteil der Kapillarporen natürlich viel größer als bei 100%, d.h. der Zementstein ist weniger dicht.
Bei gleichen Zementausgangsstoffen ist es möglich, den Hydratationsfortschritt durch feineres Mahlen erheblich zu steigern. Eine höhere Mahlfeinheit verursacht naturgemäß deutlich höhere Herstellkosten, jedoch sind damit mehrere zementtechnologische Vorteile verbunden, wie das aus der schematischen Darstellung in Bild 3 anschaulich hervorgeht. Wenn jedes Zementkorn noch einmal geteilt wird, dann wächst die reaktionsfähige Oberfläche um rund ein Drittel und der mittlere Abstand der einzelnen Körner wird bei gleichem Wasserzementwert deutlich kleiner. Die größere spezifische Oberfläche bewirkt einen größeren

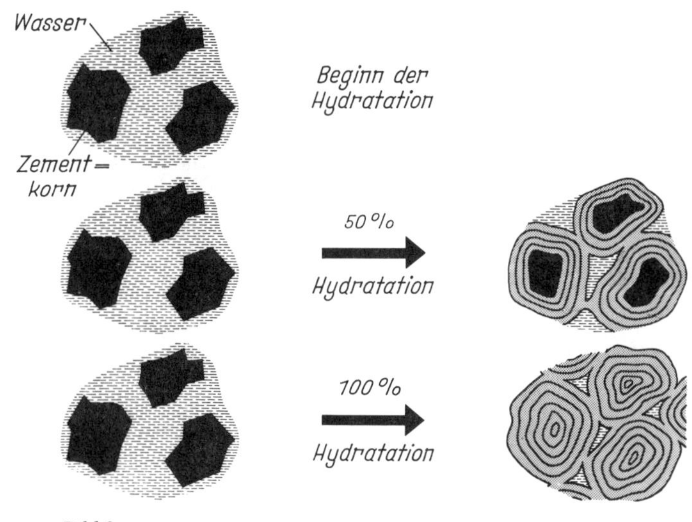

Bild 2
Schematische Darstellung des Einflusses des Hydratationsgrades auf das Gefüge und die Dichtigkeit von Zementstein

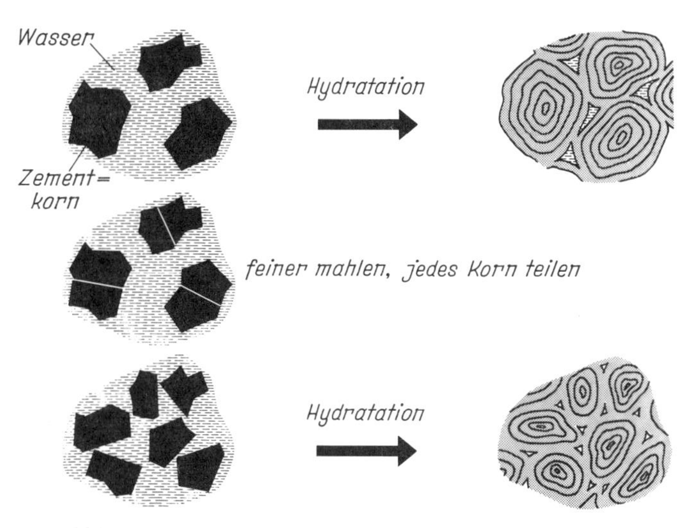

Bild 3
Schematische Darstellung des Einflusses der Mahlfeinheit auf das Gefüge und die Porengrößenverteilung von Zementstein

chemisch-mineralogischen Reaktionsumsatz je Zeiteinheit. Der geringere Abstand zwischen den einzelnen Zementkörnern bewirkt, daß der Abstand bei gleichem Kristallwachstum deutlich schneller überbrückt wird. Aus diesen beiden Gründen wird das Zementsteingefüge schneller dicht und fest. Der Gesamtporenraum eines Zementsteins aus sehr fein gemahlenem Zement ist zwar nach vollständiger Hydratation etwa gleich groß wie der aus grober gemahlenem Zement gleicher Ausgangsstoffe, jedoch verteilt sich dieser Gesamtporenraum auf eine wesentlich größere Zahl feinerer Poren, was eine geringere Durchlässigkeit und eine höhere Festigkeit bewirkt. Feineres Mahlen ist daher eine der zementtechnologischen Maßnahmen, um trotz Verwendung von Ausgangsstoffen, die langsamer oder auch deutlich weniger hydratisieren als Portlandzementklinker, Zemente herzustellen und zu entwickeln, die den konventionellen Zementen hinsichtlich Festigkeit und Dauerhaftigkeit gleichwertig sind.

4 Zementausgangsstoffe

Die Träger der Erhärtung in hydraulischen Bindemitteln sind in erster Linie Calciumsilicate. Die Zementausgangsstoffe müssen daher in einem ausreichenden Maße reaktionsfähigen Kalk CaO und/oder reaktionsfähige Kieselsäure SiO_2 enthalten. Außerdem kann auch reaktionsfähige Tonerde Al_2O_3 mit reaktionsfähigem Kalk festigkeitsbildende und porenfüllende Hydratationsprodukte bilden. Wenn die reaktionsfähigen Bestandteile im Zement sehr kalkreich sind, also z. B. ein Kalk-Kieselsäure-Verhältnis von 2,5 oder noch höher haben, dann läuft die Reaktion mit dem Anmachwasser relativ stürmisch ab. Vereinfachend kann man die Zementausgangsstoffe nach dem Kalk-Kieselsäure-Verhältnis und nach deren mengenmäßigem Anteil in vier Hauptbestandteile unterteilen.

4.1 Hydraulische Bestandteile

Hydraulische Bestandteile bestehen zu mehr als zwei Dritteln aus Calciumsilicaten, und ihr Kalk-Kieselsäure-Verhältnis liegt oberhalb von 2,0. Der klassische Vertreter der hydraulischen Bestandteile ist der Portlandzementklinker, der sogar ein Kalk-Kieselsäure-Verhältnis von größenordnungsmäßig 3 aufweist und dadurch unter Abspalten von Calciumhydroxid sehr schnell hydratisiert. Portlandzementklinker wird durch Brennen bis zur Sinterung einer entsprechend zusammengesetzten Rohmehlmischung bei Temperaturen um 1450 °C gebrannt. Bild 4 ist eine Rasterelektronenmikroskop(REM)-Aufnahme von gemahlenem Portlandzementklinker, bei dem allerdings aus aufnahmetechnischen Gründen die beim Mahlen entstehenden und für die Eigenschaften sehr wichtigen Teilchen unter 10 μm weitgehend entfernt worden sind.

4.2 Latent hydraulische Bestandteile

Latent hydraulische Bestandteile bestehen ebenfalls zu mehr als zwei Dritteln aus reaktionsfähigem CaO und

Bild 4
REM-Aufnahme von gemahlenem Portlandzementklinker, bei dem Anteile unter 10 µm zuvor weitgehend entfernt wurden

Bild 5
REM-Aufnahme von vorzerkleinertem Hüttensand, der aus einer feuerflüssigen Hochofenschlacke durch Abschrecken mit Wasser (Granulieren) entstanden ist und dadurch im glasigen Zustand vorliegt

reaktionsfähigem SiO_2. Sie sind jedoch kieselsäurereicher, d. h. ihr Kalk-Kieselsäure-Verhältnis braucht nur oberhalb 1,0 zu liegen. Dabei darf der Gehalt an MgO dem CaO-Gehalt hinzugerechnet werden. Bei Stoffen solcher Zusammensetzung sind die darin enthaltenen CaO- und SiO_2-Anteile nur dann hydraulisch reaktionsfähig, wenn sie in glasigem Zustand vorliegen, d. h. zunächst geschmolzen und dann durch schnelles Kühlen in glasigem Zustand erstarrt sind. Latent hydraulische Zementbestandteile müssen daher zu mehr als zwei Dritteln glasig sein. Der klassische Vertreter dieser Bestandteile ist schnell gekühlte, glasige Hochofenschlacke geeigneter Zusammensetzung, die man als Hüttensand bezeichnet. Es können jedoch auch glasige Aschen aus Kohlekraftwerken solche Zusammensetzung und Eigenschaften aufweisen, obwohl sie meist ein Kalk-Kieselsäure-Verhältnis von deutlich unter 1,0 haben und obwohl die dafür erforderliche Schmelzfeuerung aus Umweltschutzgründen immer mehr zugunsten von Feuerungen mit niedrigerer Temperatur zurückgedrängt wird. Die Bezeichnung „latent" hydraulisch besagt, daß diese Stoffe zwar selbständig hydraulisch erhärten können, daß hierzu jedoch eine Anregung angebracht ist, damit solche Reaktionen in technisch nutzbaren Zeiträumen ablaufen. Bild 5 ist eine REM-Aufnahme von vorzerkleinertem Hüttensand.

4.3 Puzzolanische Bestandteile

Kennzeichen puzzolanischer Stoffe ist ein bestimmter Anteil an hydraulisch reaktionsfähiger, amorpher (d. h. nichtkristalliner), meist glasiger Kieselsäure. Quarz besteht zwar auch aus reiner Kieselsäure; er ist jedoch kristallin und kann daher nicht an hydraulischen Reaktionen teilnehmen. Puzzolanische Bestandteile enthalten im allgemeinen keine oder nur sehr geringe Anteile an reaktionsfähigem Kalk; sie sind deshalb keine selbständig erhärtenden Bestandteile. Sie können nur dann festigkeitsbildende Calciumsilicathydrate bilden, wenn reaktionsfähiger Kalk in einem Bindemittelgemisch unmittelbar oder mittelbar zur Verfügung gestellt wird, z. B. durch das bei der Hydratation des Klinkers abgespaltene Calciumhydroxid. Weil die puzzolanischen Stoffe kein reaktionsfähiges CaO haben, das den Motor für die hydraulischen Reaktionen darstellt, reagieren alle Puzzolane sehr langsam, und zwar insbesondere im Vergleich zu Klinker, aber auch im Vergleich zu Hüttensand. Die klassischen Vertreter der Puzzolane, die nach dem nahe dem Vesuv gelegenen italienischen Städtchen Pozzuoli benannt werden, sind die natürlichen Puzzolane vulkanischen Ursprungs, wie z. B. der Traß. Künstliche Puzzolane sind unter anderem gebrannter Ton (gemahlenes Ziegelmehl) und glasige Flugasche [2]. Der Anteil reaktionsfähiger Kieselsäure in den vorgenannten Stoffen kann sehr verschieden sein. Er reicht von wenigen Prozent bis zu etwa 50%, d. h. Puzzolane enthalten im allgemei-

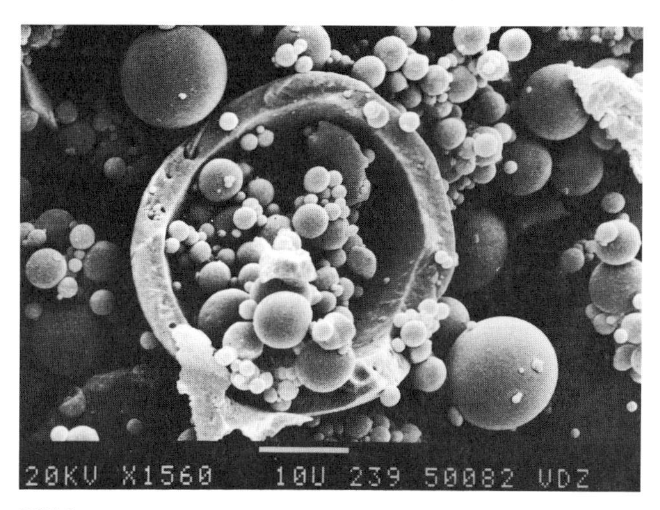

Bild 6
REM-Aufnahme einer Flugasche aus einer Schmelzfeuerung, die aufgrund ihres glasigen Zustands einen relativ hohen Anteil an reaktionsfähiger Kieselsäure aufweist

Bild 7
REM-Aufnahme von gemahlenem, festem Kalkstein; Calcit-Kristalle ($CaCO_3$) mit der für Kalkstein typischen rhomboedrischen Form und gebrochenes feinkörniges Material

nen einen beträchtlichen Anteil an nicht reaktionsfähigen Bestandteilen. Eine Ausnahme hiervon macht lediglich der bei der Ferrosiliciumherstellung anfallende SiO_2-Staub, der aus zu über 90% amorpher, reaktionsfähiger Kieselsäure besteht, die jedoch aufgrund ihrer sehr großen Feinheit einen sehr großen Wasseranspruch aufweist. Bild 6 ist die REM-Aufnahme einer qualitativ hochwertigen Flugasche aus einer Schmelzfeuerung im Anlieferungszustand.

4.4 Füller

Füller sind weitgehend inerte Bestandteile, d.h. sie liefern keine oder nur wenig festigkeitsbildende oder porenfüllende Calciumsilicate. Füller sind in jedem Zement in zumindest sehr kleinen Mengen enthalten. Auch verhalten sich diejenigen Bestandteile, die aufgrund ihrer geringen Reaktionsgeschwindigkeit oder ihres großen Korndurchmessers nicht vollständig reagieren, praktisch wie Füller. Bei geeigneter stofflicher Beschaffenheit und geeigneter Korngröße können Füller die Verarbeitbarkeit (Wasseranspruch, Fließfähigkeit, Wasserhaltevermögen) deutlich verbessern. Wenn sie eine ausreichend hohe Eigenfestigkeit aufweisen und ihr Verbund mit den Hydratationsprodukten gut ist, wirken sie auch im Zementsteingefüge stabilisierend und raumausfüllend. Als Füller wird bevorzugt Kalkstein verwendet, weil das Tricalciumaluminat des Zementklinkers mit dem Calciumcarbonat an der Oberfläche des Kalksteins zu reagieren vermag und dadurch einen guten Verbund bewirkt. Bild 7 ist eine REM-Aufnahme von gemahlenem festen Kalkstein mit einem $CaCO_3$-Gehalt von über 97 Gew.-%.

5 Zementtechnologische Maßnahmen

Mit Hilfe sogenannter zementtechnologischer Maßnahmen werden im Zementwerk aus den zuvor genannten Ausgangsstoffen Zemente hergestellt, die innerhalb einer Festigkeitsklasse vergleichbare 28-Tage-Festigkeiten aufweisen, die jedoch entsprechend ihrer Zusammensetzung und Kornverteilung je nach Zementart mehr oder weniger voneinander abweichende, charakteristische Eigenschaften aufweisen.
Grundsätzlich kann man die nachfolgend skizzierten drei verschiedenen zementtechnologischen Maßnahmen unterscheiden; sie werden in der Praxis meist in Kombination angewendet. Außerdem gehört zu den zementtechnologischen Maßnahmen auch das Homogenisieren und Vergleichmäßigen des Zements. Dabei versteht man unter Homogenisieren, daß alle Teile einer Lieferung möglichst gleiche Eigenschaften aufweisen, und unter Vergleichmäßigen, daß aufeinanderfolgende Lieferungen möglichst gleichbleibende Eigenschaften aufweisen.

5.1 Zusammensetzung des Zements

Die verschiedenen Zementarten entstehen durch unterschiedliche Zusammensetzung aus den Ausgangsstoffen. Alle Zementarten enthalten dabei Klinker, der Portlandzement zu mehr als 95 Gew.-%. Der mögliche Anteil der anderen Ausgangsstoffe richtet sich nach deren Leistungsfähigkeit im Hinblick auf die Anforderungen an den Zement. Dabei können zwar gewisse Eigenheiten der Ausgangsstoffe durch die anderen zementtechnologischen Maßnahmen teilweise kompensiert werden, jedoch sind dabei den in technisch-wissenschaftlichen Beiträgen gern genannten theoretischen Möglichkeiten technisch-wirtschaftliche Grenzen gesetzt. Unter Beachtung dieser Grenzen kann der Anteil des Füllers im Zement 20 Gew.-%, der der Puzzolane 35 Gew.-% und der des Hüttensands 80 Gew.-% wohl nur in Ausnahmefällen übersteigen. Im allgemeinen wird der Anteil dieser Ausgangsstoffe im Zement sogar um jeweils ein Viertel unter den genannten Werten liegen. Schwankende Eigenschaften in den Ausgangsstoffen lassen sich durch Verändern des Anteils um bis zu ±15 Gew.-% um den Mittelwert teilweise kompensieren.

5.2 Kornverteilung des Zements

Neben der Zusammensetzung des Zements ist dessen Kornverteilung eine in mancher Hinsicht noch stärker wirkende Einflußgröße auf die Eigenschaften des Zements, und zwar sowohl auf die rheologischen Eigenschaften des frisch angemachten Zementleims als auch auf die Erhärtung und die Eigenschaften des erhärteten Zementsteins. Die spezifische Oberfläche nach BLAINE ist ein Integralwert, der zwar für die Überwachung der Produktion recht brauchbar ist, der jedoch hinsichtlich der Eigenschaften nur eine begrenzte Aussagekraft hat. Aufschlußreicher ist die heute mit einem Laser-Granulometer bestimmte Kornverteilung, die dann als Korngrößensummenlinie in einem RRS-Netz dargestellt wird, siehe Bild 8.

Bei gleicher spezifischer Oberfläche kann der Anstieg der Geraden im RRS-Netz recht unterschiedlich sein. Der Anstieg der Geraden hat dabei einen gegenläufigen Einfluß auf die Festigkeit und die Verarbeitbarkeit: je steiler der Anstieg ist, desto höher ist die bei gleicher spezifischer Oberfläche erreichte Festigkeit; je flacher der Anstieg ist, desto günstiger ist die Verarbeitbarkeit. Eine günstige Verarbeitbarkeit erfordert andererseits einen Mindestanteil sehr feiner Körner, z. B. einen Durchgang von mindestens 5 Gew.-% bei 1 µm. Ist der Durchgang bei 1 µm deutlich größer, dann geht der Einfluß des Anstiegs der Geraden auf die Verarbeitbarkeit zurück. Frühhochfeste Zemente, wie z. B. PZ 55, weisen einen hohen Durchgang bei 1 µm und zudem einen relativ steilen Anstieg der Geraden auf, was bei guter Verarbeitbarkeit zu einer hohen Anfangs- und Endfestigkeit beiträgt.

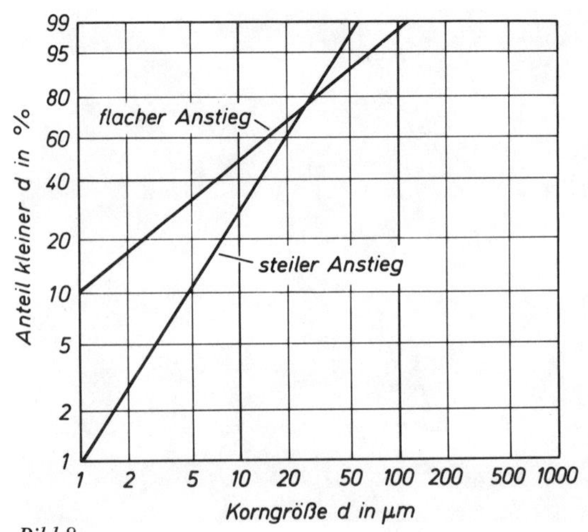

Bild 8
Darstellung der Korngrößenverteilung im RRS-Netz. Zwei Zemente mit gleicher spezifischer Oberfläche, jedoch unterschiedlichem Anstieg der Korngrößensummengeraden (schematisch)

Neben der unterschiedlichen Mahlbarkeit der verschiedenen Zementausgangsstoffe hat auch die Mahlanlage einen Einfluß auf die Kornverteilung. Der Anstieg der Geraden läßt sich jedoch in gewissen Grenzen verändern, z. B. durch Änderung der Umlaufzahl bei Sichterumlaufmahlanlagen. Werden alle Ausgangsstoffe der Mühle aufgegeben und dann gemeinsam vermahlen, so stellt sich eine geschlossene Kornverteilung ein. Durch eine getrennte Vermahlung der einzelnen Ausgangsstoffe lassen sich die Zementeigenschaften noch weitergehend optimieren. Hiervon wird im Ausland häufiger Gebrauch gemacht. Allerdings sind dann anschließend besondere Maßnahmen zum Homogenisieren und Vergleichmäßigen erforderlich, die beim gemeinsamen Vermahlen eo ipso gegeben sind.

5.3 Chemische Optimierung

Verschiedene Stoffe vermögen die Geschwindigkeit der hydraulischen Reaktionen zu beschleunigen oder zu verlangsamen. Das früher zur Beschleunigung häufig verwendete Calciumchlorid scheidet heute wegen seiner korrosionsfördernden Wirkung auf den Bewehrungsstahl aus. Generell wird die Hydratation durch hochbasische Stoffe (Alkalien) angeregt. Das kann bei Klinker über das erwünschte Maß hinausgehen, ist bei

Hüttensand und trägen Puzzolanen jedoch erwünscht. Durch moderne Öfen kann heute der Klinker – falls erwünscht – von diesen im Rohmaterial enthaltenen Stoffen entlastet werden, die dann für die Verwendung bei Zementen mit träge reagierenden Ausgangsstoffen zur Verfügung stehen.
Die chemische Optimierung hängt von Menge, Art und Eigenschaften der Ausgangsstoffe und von deren Kornverteilung ab. Die chemische Optimierung im Zementwerk betrifft nicht nur die Erhärtung, sondern auch das rheologische Verhalten des Zementleims, insbesondere sein Ansteifen und Erstarren. Hierzu dient die Optimierung des als Erstarrungsregler zugesetzten Calciumsulfats hinsichtlich Menge und Modifikation.
Die chemische Optimierung ist so komplexer Natur, daß sie zuverlässig nur in einem Zementwerk mit entsprechenden Prüfeinrichtungen und Fachpersonal ausgeführt werden kann. Die Zugabe von Stoffen im Betonwerk oder auf der Baustelle, die in den Reaktionsmechanismus des Zements eingreifen, stellt daher immer ein nicht vollständig kontrollierbares Risiko dar.

6 Prüfverfahren für neue Zementarten

Für die konventionellen Zemente – das sind vor allem Portlandzement und hüttensandhaltige Zemente – reichen die in der Zementnorm angegebenen Prüfverfahren und Anforderungen zur Beurteilung und Bewertung aus. Während Wasseranspruch, Erstarren, Raumbeständigkeit und Festigkeit des Zements gemessen werden, darf man unterstellen, daß bei Einhalten der zugehörigen Anforderungen andere, nicht überprüfte Eigenschaften aufgrund der Hydratationsmechanismen von hydraulischen und latent hydraulischen Zementausgangsstoffen stets gegeben sind. Das gilt insbesondere für das Ausfüllen von zuvor mit Wasser gefüllten Poren durch die Hydratationsprodukte und die dadurch bewirkte Dichtigkeit. Letztere ist eine der Voraussetzungen für die Dauerhaftigkeit, d. h. eine jahrzehntelange Widerstandsfähigkeit gegen Witterungseinflüsse und Frost und ein ebenso langer Korrosionsschutz der Bewehrung.
Schon bei puzzolanischen Bestandteilen – sowohl bei natürlichen Puzzolanen als auch bei Flugasche, SiO_2-Staub und anderen – kann man nicht ohne weiteres davon ausgehen, daß damit hergestellte Zemente, die die Festigkeitsanforderungen erfüllen, dann auch ohne weiteren Nachweis alle Anforderungen an die Dauerhaftigkeit erfüllen. Das liegt daran, daß Puzzolane im Gegensatz zu Klinker und Hüttensand stets nur zu einem Teil aus reaktionsfähigen Stoffen bestehen (der Rest ist meist inert), daß sie nicht selbständig erhärten, sondern dazu einen Kalkspender benötigen, daß die Reaktion ihrer Natur nach langsam abläuft und daher gegen mangelnde Nachbehandlung empfindlich ist, daß das Volumen der Hydratationsprodukte per Saldo vermutlich kleiner ist als das von Klinker, weil bei der Hydratation des Puzzolans das bei der Hydratation des Klinkers abgespaltene Calciumhydroxid verbraucht wird, und daß der Verbrauch des Calciumhydroxids die Carbonatisierung begünstigt.
Ähnliche Vorbehalte bestehen auch gegen Zemente mit einem über 5 Gew.-% hinausgehenden Fülleranteil, weil Füller, wenn überhaupt, dann nur ganz wenig porenfüllende Hydratationsprodukte liefern. Beim Vergleich Kalksteinfüller mit Puzzolan ist allerdings zu beachten, daß der Fülleranteil meist nur größenordnungsmäßig 15 Gew.-% beträgt, während der Puzzolananteil häufig das Doppelte ausmacht. Im Vergleich zu den konventionellen Zementen sind füllerhaltige Zemente und auch Zemente aus natürlichem Puzzolan meist wesentlich feiner gemahlen, wodurch hinsichtlich deren Dichtigkeit ein günstigeres Zementsteingefüge entsteht. Allerdings können gerade feinporige Gefüge aufgrund ihrer hohen Saugfähigkeit sensibel gegen Frostbeanspruchung reagieren, so daß eine diesbezügliche Überprüfung angezeigt ist. Im Forschungsinstitut der Zementindustrie werden daher heute bei der Entwicklung neuer Zementarten die nachfolgenden Untersuchungen im Hinblick auf die Festigkeit und die Dauerhaftigkeit durchgeführt:

- Verarbeitbarkeit, insbesondere Wasseranspruch, Fließfähigkeit, Ansteifen und Erstarren,
- Festigkeit; Anfangsfestigkeit, 28-Tage-Festigkeit, Festigkeit in höherem Alter,
- Gefüge; Untersuchungen mittels Rasterelektronenmikroskop und Röntgenbeugung,
- Carbonatisierung, Porosität und Durchlässigkeit von Sauerstoff in Abhängigkeit von der Nachbehandlung,
- Frostwiderstand von Zementmörtel und Beton.

Ein Teil der genannten Prüfverfahren oder Untersuchungsmethoden sind in der Werkstoffwissenschaft oder Materialprüfung gebräuchlich oder sogar genormt. Noch nicht allgemein verbreitet sind die in den letzten Jahren im Zementforschungsinstitut entwickel-

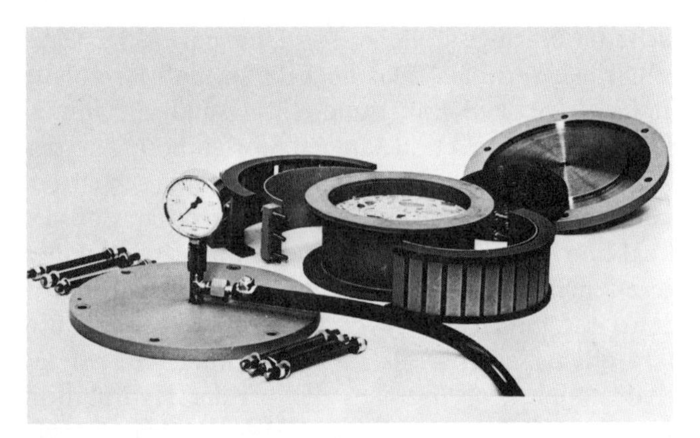

Bild 9
Meßzelle zur Bestimmung der Gasdurchlässigkeit von Betonscheiben mit 15 cm Durchmesser und 5 cm Dicke

Bild 10
Gesamte Versuchseinrichtung zur Bestimmung der Gasdurchlässigkeit bei verschiedenen Drücken. Messen des Gasdurchflusses je Zeiteinheit mit Blasenzählgerät

ten oder weiterentwickelten Prüfverfahren für die Duchlässigkeit von Sauerstoff und für den Frostwiderstand.

6.1 Durchlässigkeit von Sauerstoff

Hierzu wurde im Rahmen eines europäischen Ringversuches eine spezielle Meßzelle entwickelt [3], die in Bild 9 abgebildet ist. Die Betonprobekörper, die in die Meßzelle eingesetzt und mit einer angepreßten Kunststoffmanschette an den Zylinderflächen abgedichtet werden, haben einen Durchmesser von 15 cm und eine Dicke von 5 cm. Sie können nach der jeweiligen Nachbehandlung nach beiden Seiten austrocknen und stellen quasi zwei gegeneinander betonierte Betonüberdeckungsschichten von jeweils 2,5 cm dar.
Die eingespannten und abgedichteten Probekörper werden auf der einen Seite einem Sauerstoffüberdruck ausgesetzt, der um jeweils weitere 0,5 bar gesteigert wird, nachdem sich eine konstante Durchflußgeschwindigkeit eingestellt hat. Die Durchflußmenge je Zeiteinheit wird bei den einzelnen Druckstufen (maximal 2,5 bar) in cm^3/s mit einem Blasenzählgerät gemessen. Die gesamte Versuchseinrichtung zeigt Bild 10.
Aus solchen Messungen werden dann Durchlässigkeitskoeffizienten errechnet, die in der Größenordnung von 10^{-15} bis 10^{-19} m^2 liegen und die einen besseren Vergleich hinsichtlich des Einflusses der zu untersuchenden Zementart und der Nachbehandlung ermöglichen. Bei den Untersuchungen wird eine stets gleiche Betonzusammensetzung aus Rheinkiessand mit einem Zementgehalt von 300 kg/m^3 und einem Wasserzementwert von 0,60 untersucht, die den Festlegungen der Dauerhaftigkeits-Richtlinie des Deutschen Ausschusses für Stahlbeton entspricht. Die Untersuchungen zeigen, daß es auch mit Puzzolanen und Füllern möglich ist, neue Zementarten zu entwickeln, die bei geeigneten zementtechnologischen Maßnahmen und ausreichender Nachbehandlung Sauerstoff-Durchlässigkeitskoeffizienten aufweisen, wie sie für konventionelle Zementarten üblich sind.

6.2 Frostwiderstand von Zementmörtel und Beton

In den letzten Jahrzehnten wurden im Zementforschungsinstitut umfangreiche Untersuchungen über den Frostwiderstand und den Frost-Tausalz-Widerstand von Beton für Straßen, Brücken und Wasserbauwerke durchgeführt [4]. Aufbauend auf diesen Erfahrungen wurden zwei Prüfverfahren für den Einfluß neuer Zementarten auf den Frostwiderstand entwikkelt [5]. Das in der Durchführung wesentlich weniger aufwendige Prüfverfahren an Zementnormmörtel mit automatischer Frost-Tauwechsel-Prüfanlage und Prüfung der Resonanzfrequenz befindet sich noch in der abschließenden Probelaufphase. Angeeicht werden soll dieses Verfahren an die Ergebnisse von 10-cm-Prüfkörpern aus Beton mit frostsicherem Rheinkiessand und einer Zusammensetzung gemäß der Dauerhaftigkeits-Richtlinie, die im Alter von 28 Tage (7 Tage feucht, dann an Luft) 100mal in Messingkästen (siehe Bild 11) unter Wasser bei –15 °C eingefroren und

Bild 11
Messingbehälter, in dem zwei 10-cm-Betonwürfel unter Wasser eingefroren und aufgetaut werden

nach 16 Stunden Frost dann 8 Stunden lang bei +20 °C aufgetaut werden. Festgestellt wird der Gewichtsverlust nach jeweils weiteren 10 Frost-Tauwechseln.
Betone mit einem Zementgehalt von 300 kg/m³ und einem Wasserzementwert von 0,60 wiesen bei konventionellen Zementen nach 100 Frost-Tauwechseln einen Gewichtsverlust von in der Regel bis zu 5 Gew.-% auf, in einzelnen Fällen zwischen 5 und 10 Gew.-%. Es ist möglich, auch mit Zementen aus Puzzolanen und Füllern ein gleich gutes Ergebnis zu erreichen, jedoch bedarf dies einer sorgfältigen Auswahl dieser Zumahlstoffe und darauf abgestimmter zementtechnologischer Maßnahmen. Ziel der noch nicht abgeschlossenen Forschung ist es derzeit, Kriterien und möglichst einfache Prüfverfahren für die Auswahl und die laufende Überwachung geeigneter Puzzolane und Füller zu entwickeln.

Bild 12
10-cm-Betonwürfel mit unterschiedlicher Schädigung durch 100 Frost-Tauwechsel. Angaben der Schädigung durch Gewichtsverlust in Prozent

7 Zusammenfasung

Aufgabe des Zements im Beton ist das hochfeste Verkleben der einzelnen Zuschlagkörner, das Ergänzen der Kornzusammensetzung des Zuschlags im Feinstbereich zur Minderung des Wasseranspruchs und zur Verbesserung der Verarbeitbarkeit sowie das Ausfüllen von zuvor mit Anmachwasser gefüllten Lücken zwischen den festen Bestandteilen durch Hydratationsprodukte. Die Zementausgangsstoffe lassen sich in hydraulische, latent hydrauliche und puzzolanische Bestandteile sowie Füller unterteilen. Der aus technischen Gründen mögliche Gewichtsanteil an den verschiedenen Ausgangsstoffen im Zement geht von 100% bei Klinker über 80% bei Hüttensand hinunter auf 30 bis 35% bei Puzzolan und auf 15 bis 20% bei geeigneten Füllern. Während bei konventionellen Zementen aus Klinkern und Hüttensand die Dauerhaftigkeit aufgrund der Hydratationsmechanismen dieser Ausgangsstoffe gewährleistet ist, wenn sie die Festigkeitsanforderungen erfüllen, bedarf dies weitergehender Untersuchungen bei der Entwicklung neuer Zementarten aus Puzzolanen und Füllern. Dazu zählen neben Verarbeitbarkeit und Festigkeit eingehendere Gefügeuntersuchungen, die Porosität und Sauerstoffdurchlässigkeit, die Carbonatisierungsgeschwindigkeit und der Frostwiderstand.

8 Literatur

[1] Wischers, G. und Richartz, W.: Einfluß der Bestandteile und der Granulometrie auf das Gefüge des Zementsteins. In: beton 32 (1982), S. 337–341 und 379–386.
[2] Richartz, W.: Zusammensetzung und Eigenschaften von Flugaschen. In: Zement-Kalk-Gips 37 (1984), S. 62–71.
[3] Gräf, H. und Grube, H.: The influence of curing on the gas permeability of concrete with different compositions. In: RILEM Seminar on the durability of concrete structures under normal outdoor exposure. Hannover (March 1984).
[4] Bonzel, J. und Siebel, E.: Neuere Untersuchungen über den Frost-Tausalz-Widerstand von Beton. In: beton 27 (1977), S. 153–158, 205–211 und 237–244.
[5] Siebel, E.: Laboratory investigations of the frost resistance of concrete and their correlation with field performance. in: RILEM seminar on the durability of concrete structures under normal outdoor exposure. Hannover (March 1984).

Befestigungstechnik

Moderne Befestigungstechnik – Stand der Entwicklung –

Otto Wagner

1 Einleitung

Das Bauen wird heute weitgehend vom Massivbau beherrscht. Die Vorteile des Stahlbetons liegen u. a. in der beliebigen Formgebung des Werkstoffs Beton und – wenn er sorgfältig hergestellt wird – in seiner großen Dauerhaftigkeit und Wartungsfreiheit. Einer seiner Nachteile zeigt sich nach Fertigstellung des Rohbaus, wenn ohne ausreichende Vorplanung bei Industrie- und Kraftwerksbauten aber auch im normalen Hochbau Ergänzungen oder Einbauten, wie z. B. Maschinen, Rohrleitungen, Konsolen für Krananlagen, vorgehängte Fassaden, untergehängte Decken u. ä. kraftschlüssig mit Massivbauteilen verbunden werden sollen. Gleiches gilt auch für Mauerwerksbauten. Insbesondere im Kernkraftwerksbau vergehen zwischen Planung bzw. Erstellung des Rohbaus und der späteren Montage einer Fülle von Einbauten oft Jahre, so daß bei der Rohbauerstellung die genauen Befestigungsstellen teilweise noch nicht absehbar sind.
Für solche Verankerungen im Ingenieurbau hat sich in den letzten Jahrzehnten eine neue Befestigungstechnik entwickelt, bei der Dübel der verschiedensten Art in vorgebohrte Löcher gesetzt und mit verhältnismäßig einfachen Mitteln bei geringen Verankerungstiefen große Lasten eingetragen werden können.
Da allgemein anerkannte Regeln der Technik für derartige Dübelsysteme noch nicht existieren, muß nach den gesetzlichen Bestimmungen (Bauordnungen) der Nachweis der Brauchbarkeit durch allgemeine bauaufsichtliche Zulassungen geführt werden. Hierfür wurde im Jahre 1973 der Sachverständigenausschuß „Ankerschienen und Dübel“ gebildet, der das für die Erteilung der Zulassungen zuständige Institut für Bautechnik in Berlin berät. Im April 1975 wurden die ersten Zulassungen für Metalldübel erteilt.

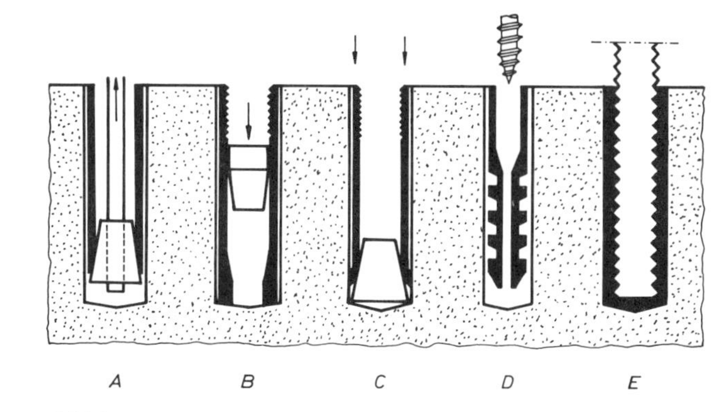

Bild 1
Standarddübeltypen

2 Typen von Dübeln

Die von zahlreichen Firmen bisher entwickelten Dübelarten lassen sich i. W. in 5 Gruppen einteilen. Bild 1 zeigt drei Typen von Metalldübeln, die durch einen Konus im Bohrlochgrund gespreizt und dadurch verankert werden. Auch beim Typ D wird eine Kunststoffhülse durch Eindrehen einer Schraube gespreizt. Verbund- oder Klebeanker vom Typ E bestehen aus einem Gewindestahl, der durch einen Kunstharzmörtel in das Bohrloch eingeklebt wird.

2.1 Dübeltyp A

Typ A ist ein Dübel mit kraftkontrollierter zwangsweiser Spreizung, dessen Teile meist aus verzinktem oder auch nichtrostendem Stahl bestehen. Er besteht aus einer Dübelhülse, einem Gewindekonus (oder zwei), einer Unterlegscheibe und einer Sechskantschraube oder einem Gewindebolzen mit Mutter (Bild 2). Die Dübelhülse ist durch Längsschlitze in Zungen unterteilt. Beim Anziehen der Schraube oder Mutter wird die Dübelhülse durch Einziehen des Konus gespreizt. Durch Aufbringen eines Drehmoments mit einem geeichten Drehmomentenschlüssel kann ein bestimmter Spreizdruck erzeugt werden. Die Drehmomente sind in den Zulassungen vorgeschrieben und genau einzuhalten.

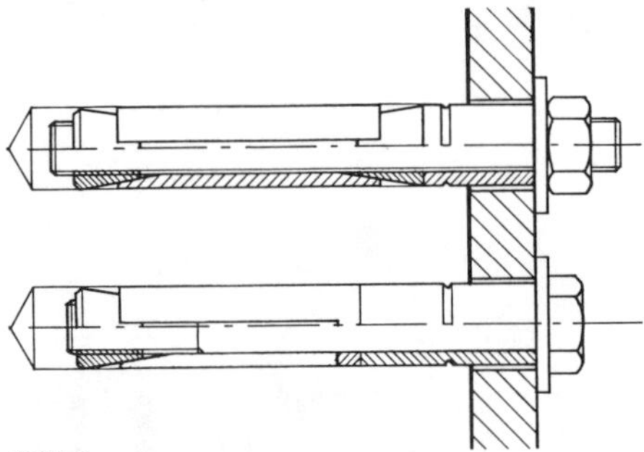

Bild 2
Dübel Typ A mit kraftkontrollierter zwangsweiser Spreizung

2.2 Dübeltyp B

Typ B ist ein Dübel aus verzinktem oder nichtrostendem Stahl mit wegkontrollierter zwangsweiser Spreizung. Er besteht aus einer Dübelhülse mit geschlitztem Spreizteil und Innengewinde und einem Spreizkonus (Bild 3). Der Spreizkonus besteht aus einem zylindrischen Körper mit konischem Ansatz, der die gleiche Neigung wie die Hülse im Spreizteil aufweist. Die Spreizung erfolgt durch Einschlagen des Spreizkonus in die Dübelhülse. Der Einschlagweg ist in der Zulassung vorgeschrieben und muß mit einer entsprechenden Lehre kontrolliert werden. Zusätzlich ist der feste Sitz der Dübel stichprobenweise mit einem Probebelastungsgerät zu überprüfen.

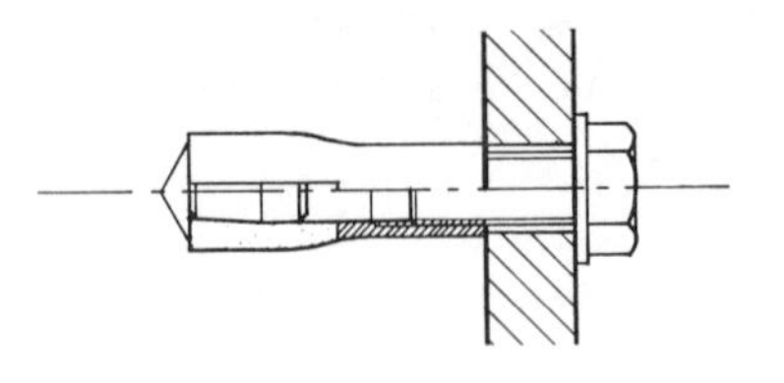

Bild 3
Dübel Typ B mit wegkontrollierter zwangsweiser Spreizung

2.3 Dübeltyp C

Typ C ist ein Selbstbohrdübel mit wegkontrollierter zwangsweiser Spreizung (Bild 4). Die Dübelhülse hat im Spreizbereich eine geschlitzte Bohrkrone mit Zähnen, die zunächst zur Herstellung des Bohrlochs dient. Zur Verspreizung des Dübels wird die Hülse mit einem Bohrhammer über einen Konus getrieben, der sich auf dem Bohrlochgrund abstützt. Der Sitz des Konus in der Hülse muß mit einem Kontrollstab überprüft werden. Die aufgebrachte Spreizkraft ist wie bei Typ B stichprobenweise nachzuprüfen.

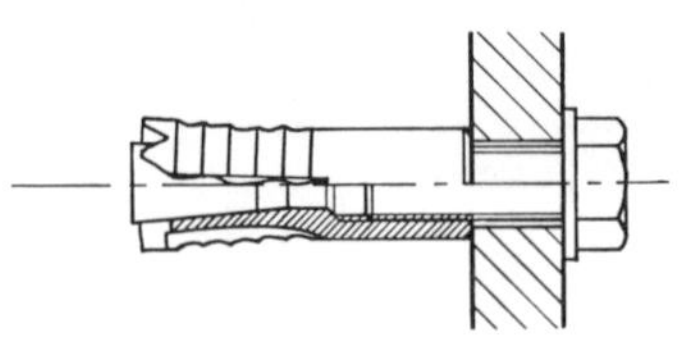

Bild 4
Dübel Typ C, Selbstbohrdübel mit wegkontrollierter zwangsweiser Spreizung

2.4 Dübeltyp D

Typ D besteht aus einer Dübelhülse aus Polyamid und einer zugehörigen Spezialschraube. Der gezahnte Spreizteil der Hülse ist geschlitzt und hat i. d. R. federnde Sperrzungen. Der Dübel wird durch Eindrehen der Schraube in die Hülse gespreizt. Die Schraubenspitze muß das Ende der Hülse durchdringen.
Kunststoffdübel sind derzeit im wesentlichen nur für die Befestigung von Fassadenbekleidungen in Beton und verschiedenen Mauerwerksarten zugelassen.

2.5 Dübeltyp E

Typ E besteht aus einem Gewindestahl mit Mutter, Unterlegscheibe und Mörtelpatrone. Seine Wirkungsweise beruht auf der Ausnutzung des Verbundes zwischen Gewindestahl, Reaktionsharzmörtel und Beton (Bild 5). Die Mörtelpatrone besteht aus einer oder mehreren Glasampullen, gefüllt mit Quarzzuschlagstoff, Reaktionsharz und Härterstäbchen.
Zum Setzen des Dübels wird die Patrone in das Bohrloch eingeführt und der Gewindestahl mit einer Hammerbohrmaschine eingetrieben. Das Bohrloch muß dabei bis zur Betonoberfläche vermörtelt werden. Bis zur Belastung des Verbundankers sind je nach Temperatur des Verankerungsgrundes bestimmte Wartezeiten einzuhalten.

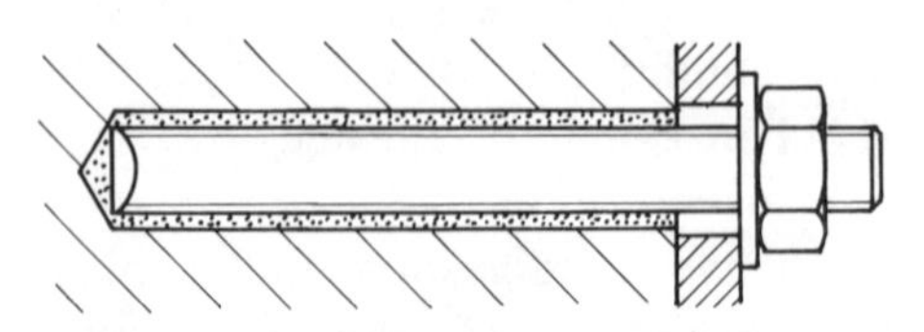

Bild 5
Dübel Typ E, Verbund- oder Klebeanker

2.6 Andere Dübelarten

Neben den v. g. Standardtypen wurden eine Reihe von Spezialdübeln entwickelt, von denen Bild 6 einige Bei-

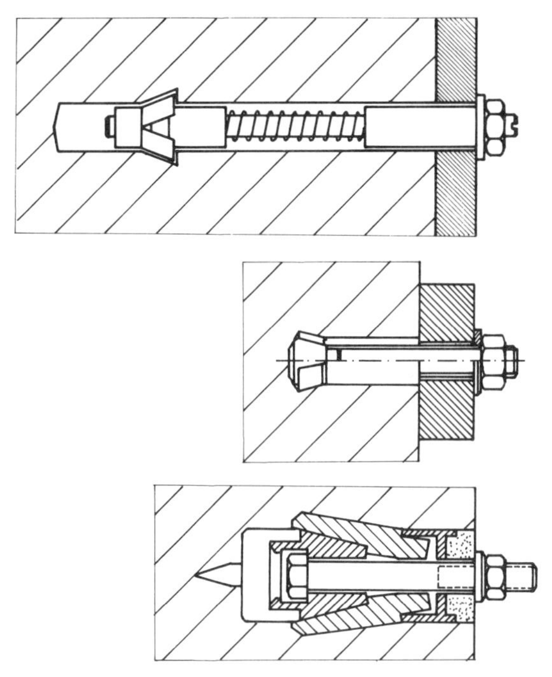

Bild 6
Spezialdübel mit Hinterschneidung

spiele zeigt. Die gezeigten Typen zeichnen sich dadurch aus, daß mit besonderen Bohrgeräten im Bohrloch Hinterschneidungen hergestellt werden, wodurch die Dübel ohne Spreizdruck verankert und damit kleinere Abstände erreicht werden können.

3 Tragfähigkeit, Kurz- und Langzeitverhalten

Die Verankerung der Spreizdübel wird durch radial zum Bohrloch wirkende Spreizkräfte erzeugt. Der Verankerungsgrund wird im Spreizbereich örtlich sehr hoch beansprucht, wobei insbesondere tangentiale Ringzugspannungen und radiale Druckspannungen entstehen.
Die stählernen Dübelhülsen werden durch den Spreizdruck im Spreizbereich in den Beton eingepreßt, wodurch eine gewisse Hinterschneidung entsteht, die durch Kriechen des Betons weiter vergrößert wird.
Kunststoffdübel verursachen durch die verhältnismäßig weiche, verformbare Hülse wesentlich geringere Spreizkräfte als Stahldübel gleichen Durchmessers. Sie beanspruchen den Verankerungsgrund erheblich weniger und sind daher auch für Mauerwerk geeignet. Ihre axialen Traglasten sind im Vergleich zu Stahldübeln gering.

Die sog. Verbundanker des Typs E sind im unbelasteten Zustand praktisch spreizdruckfrei, so daß relativ kleine Abstände der Dübel untereinander und vom Rand möglich sind.
Die in den Besonderen Bestimmungen der Zulassungsbescheide festgelegten zulässigen Beanspruchungen beruhen auf umfangreichen Versuchsergebnissen und gelten nur für die dort genannten Anwendungsbedingungen hinsichtlich Festigkeit des Verankerungsgrundes, Bauteilabmessungen, Verankerungstiefe, Abstandsregelungen u. a.
In einer Dissertation [4] ist versucht worden, ein theoretisches Verfahren zur Berechnung der axialen Traglast von Stahlspreizdübeln zu entwickeln, das es ermöglichen soll, mit wenigen Versuchen zu einer Aussage zu kommen.
Alle zugelassenen Dübel dürfen auf zentrischen Zug, Abscheren und Schrägzug unter beliebigem Winkel beansprucht werden, wobei die zulässigen Werte für die drei Beanspruchungsrichtungen aus Vereinfachungsgründen jeweils gleich groß sind. Für Kunststoffdübel ist jedoch eine ständig wirkende Zugbelastung (z. B. infolge Eigenlast) nur als Schrägzug zulässig, der mit der Dübelachse mindestens einen Winkel von 10° bilden muß.
Die zugelassenen Dübel dürfen nur für Verankerungen unter vorwiegend ruhender Belastung verwendet werden. Verankerungen in der aus Lastspannungen erzeugten Zugzone des Betons sind nicht zulässig. Dies gilt nicht für Dübel Typ A, B und C kleinerer Durchmesser zur Verankerung hängender Decken nach DIN 4121 und DIN 18168 und statisch vergleichbare Befestigungen.
Von Dübelverbindungen wird in der Regel eine Dauerhaftigkeit ohne zeitliche Begrenzung erwartet. Die großen Spreizdrücke von Spreizdübeln können zu Kriechen und Relaxation des Verankerungsgrundes bzw. des Dübelwerkstoffs (bei Typ D) und damit zu einem Abfall der Traglast und mit der Zeit eintretenden Verschiebungen bzw. zum Lockern der Verbindung führen.
In zahlreichen Versuchsreihen wurde das Schlupfverhalten unter konstanter Zugbeanspruchung verfolgt. Es zeigte sich, daß der meist eintretende Anfangsschlupf nach wenigen Tagen abklingt. Die Zulassungsbescheide geben entsprechende Werte für das Verschiebungsverhalten unter Kurzzeit- und Dauerbelastung in Höhe der zulässigen Lasten an.
Über Jahre laufende Versuche unter verschiedenen at-

mosphärischen Auslagerungsbedingungen ergaben keine signifikanten Veränderungen der Traglasten. Die zulassungsgemäß ausgeführten Dübelverbindungen können daher nach den bisherigen Erfahrungen als auf lange Zeit sichere Konstruktionen angesehen werden.
Bleibt anzumerken, daß bei auf Zug beanspruchten Dübelverankerungen die Zugfestigkeit des Betons in Anspruch genommen und damit vom Grundprinzip des Betonbaus abgewichen wird. Aus diesem Grund und zur Abdeckung zusätzlicher Unsicherheiten dieser speziellen Verbindungsart werden die zulässigen Beanspruchungen gegenüber den 5%-Fraktilenwerten der Zulassungsversuche mit einem höheren Sicherheitsfaktor, i. d. R. $v = 3$, abgesichert.

4 Versagensarten

Beim Setzen der Dübel bzw. unter Beanspruchung bis zum Versagen können folgende Versagensarten auftreten (Bild 7):

a) Spalten bzw. Aufsprengen bei zu geringen Achs- oder Randabständen oder sehr feingliedrigen Bauteilen. Damit dies unterbleibt, schreiben die Zulassungen bestimmte Achs-, Eck- und Randabstände und Bauteilabmessungen vor.
b) Herausziehen, i. d. R. bei Kunststoffdübeln vom Typ D, die durch Reibschluß verankert werden, z. T. auch bei Stahldübeln und Verbundankern vom Typ E.
c) Ausbildung eines Bruchkegels durch trichterförmiges, sprödes Ausbrechen des den Dübel umgebenden Betons, im wesentlichen bei Stahlspreizdübeln mit zu geringer Verankerungstiefe.
d) Stahlversagen, d. h. Bruch des Bolzens, insbesondere infolge Ermüdungsbeanspruchung, Biegung oder Querzug.

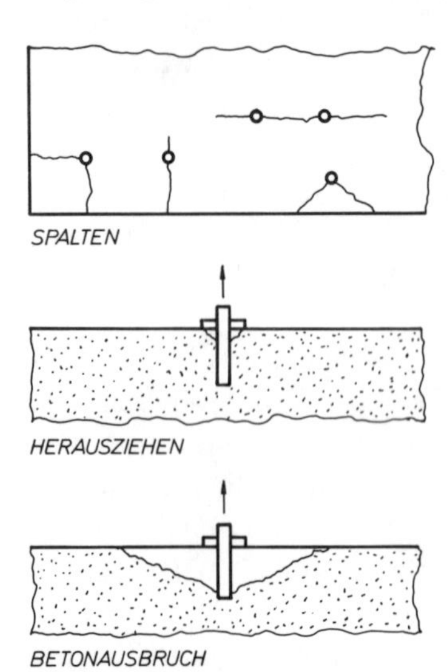

Bild 7
Versagensarten von Dübelverbindungen

5 Einfluß der Bewehrung

Die in der Druckzone eines Stahlbetonbauteils allenfalls vorhandene Bewehrung kann die Ausbildung eines flachen Bruchkegels unter Zugbeanspruchung nicht verhindern.
Auch eine zusätzlich eingelegte Oberflächenbewehrung verbessert das Tragverhalten nur in geringem Maße, sie kann allenfalls einen schlagartigen, spröden Bruch verhindern und nach Erreichen der Höchstlast eine gewisse Resttragfähigkeit, die i. d. R. unterhalb der zulässigen Last liegen wird, sichern.
In Sonderfällen kann durch eine sog. Rückhängebewehrung in Form von Bügeln der Ausbruchkegel ins Innere des Bauteils rückverankert werden. Damit läßt sich zwar nur eine geringe Steigerung der Bruchlast, jedoch eine deutliche Erhöhung der Resttragfähigkeit erreichen. Der Nachteil solcher Zusatzbewehrungen besteht darin, daß sie im voraus geplant und bei der Bauausführung exakt an den späteren Verankerungsstellen eingebaut werden müssen. Für solche Fälle dürften jedoch andere Verankerungsarten, wie z. B. Ankerschienen oder Ankerplatten, sinnvoller sein.
Die in den Zulassungsbescheiden festgelegten Werte für Achs- und Randabstände liegen auf der sicheren Seite und verhindern somit ein Aufspalten des unbewehrten Betons. Die Auswertung zahlreicher Versuche zeigt, daß durch eine orthogonale Oberflächenbewehrung bei gleicher Tragfähigkeit engere Achs- und Randabstände möglich sind. Da in der Betondruckzone üblicherweise keine oder nur eine geringe Bewehrung vorhanden ist, müßte eine zusätzliche Bewehrung angeordnet werden. Auch hierfür wären im Einzelfall besondere Nachweise zu erbringen.

6 Einschränkungen im Bereich der Zugzone

In Abschnitt 3 wurde bereits darauf hingewiesen, daß die Verankerung von Dübeln in der aus Lastspannungen erzeugten Zugzone i. a. nicht zulässig ist. In der

Zugzone von Stahlbetonbauteilen muß bekanntlich mit der Bildung von Rissen bis zu einer Rißbreite von 0,4 mm gerechnet werden. Diese oft erst nach dem Setzen der Dübel aufgehenden Risse rücken die Bohrlochwandungen der im Riß sitzenden Dübel auseinander, was zu einem Abfall der Spreizkraft und damit der Tragfähigkeit des Dübels führt. Die Verschiebungswege unter Last nehmen zu.

Besonders ausgeprägt ist dieser Effekt bei Kunststoffdübeln, die primär durch Reibschluß und bei Verbundankern, die durch Verbund zwischen Kunstharzmörtel und Beton im Bohrloch verankert sind.

Für den bereits erwähnten Sonderfall der hängenden Decken nach DIN 4121 und DIN 18168 sowie für statisch vergleichbare Befestigungen gestatten die Zulassungen der meisten zwangsweise spreizenden Metalldübel auch die Verankerung in der Zugzone. Geregelt ist allerdings nur der Lastbereich bis max. 1,5 kN. Daneben sind eine Reihe von kleineren Spezialdübeln zugelassen, die nur für die Befestigung von Unterkonstruktionen leichter Deckenbekleidungen und Unterdecken verwendbar sind.

Derzeit ist eine Entwicklung im Gange, die darauf abzielt, in Zukunft nur noch Dübel zuzulassen, die sowohl für eine Verankerung in der Druckzone, als auch in der Zugzone geeignet sind. Dabei soll eine bestimmte Lastverschiebungscharakteristik in der gerissenen Zugzone gefordert werden. Die zulässigen Lasten sollen mindestens 50% derjenigen für die Druckzone betragen.

7 Anordnung von Dübelgruppen

Die in den Zulassungsbescheiden enthaltenen Mindestabstände von Einzeldübeln untereinander sind aufgrund von Versuchen so festgelegt, daß eine gegenseitige Beeinflussung des Tragverhaltens ausgeschlossen ist. Der Wunsch, möglichst große Lasten konzentriert in Stahlbetonbauteile zu verankern, führte zur Entwicklung von Dübelgruppen oder Mehrfachbefestigungen. Innerhalb der Gruppe werden die Mindestabstände unterschritten, wodurch sich die Spaltkräfte der Einzeldübel überlagern.

Bei Zweiergruppen kommt es ab einem Achsabstand von etwa der dreifachen Setztiefe, bei Vierergruppen ab einem Achsabstand von etwa der vierfachen Setztiefe zur Ausbildung eines gemeinsamen Bruchkegels und damit zu einer entsprechenden Reduzierung der Tragkraft.

Eine weitere Einflußgröße auf die Tragkraft von Dübelgruppen ist die Streuung der Bruchlasten der Einzeldübel. Ebenso kann ein unterschiedliches Verschiebungsverhalten der Einzeldübel zu einer Kräfteumlagerung führen und sich damit traglastmindernd auswirken.

Für die meisten kraftkontrolliert spreizenden Metalldübel ist die Anwendung von Dübelgruppen (meist Dübelpaare) aufgrund der durchgeführten Versuche zugelassen. Die Mindestabstände können den Zulassungsbescheiden entnommen werden.

8 Planung und Ausführung

Dübelverbindungen sind ingenieurmäßig zu planen und unter Beachtung der Zulassungsbestimmungen zu bemessen. Es sind prüfbare Berechnungen und Konstruktionszeichnungen anzufertigen, nach denen sich die Ausführung der Dübelarbeiten zu richten hat.

Besondere Bedeutung kommt der Montage und der Kontrolle der Dübelarbeiten auf der Baustelle zu, die nur von sachkundigem Personal durchgeführt werden dürfen. Wegen der Bedeutung, die der Einhaltung des vorgeschriebenen Bohrlochdurchmessers zukommt, müssen Spezialbohrer verwendet werden, die den Anforderungen des Merkblattes des Instituts für Bautechnik entsprechen. Alle Dübel dürfen nur als serienmäßig gelieferte Befestigungseinheit verwendet werden. Der Austausch von Einzelteilen ist unzulässig. Während der Herstellung der Dübelverbindungen sind Aufzeichnungen über die ordnungsgemäße Montage zu führen. Bleibt noch anzumerken, daß für tragende Verbindungen nur Dübel verwendet werden dürfen, deren Herstellung einer Überwachung unterliegt, was auf dem Lieferschein mit dem einheitlichen Überwachungszeichen zu bestätigen ist.

9 Untersuchung von Schadensfällen

Die meisten bisher aufgetretenen Schadensfälle sind auf Montagemängel zurückzuführen.

Ein besonders krasser Fall ist vor einigen Jahren aus Schweden bekanntgeworden. Beim Bau zweier Kernkraftwerke wurden etwa 30% der rund 20000 Dübel nicht auf die erforderliche Verankerungstiefe gesetzt. Wo immer beim Bohren Bewehrungsstähle getroffen wurden, kürzte man die Dübel von 150 auf etwa 40

bis 50 mm. Die Folgen für die Tragwirkung der Dübel liegen auf der Hand. Bemerkenswert ist, daß dieses krasse Fehlverhalten erst nachträglich durch zusätzliche Kontrollen entdeckt wurde.
Besonders gravierende Mängel wurden in Deutschland an Dübelverbindungen festgestellt, die vor Erteilung der ersten bauaufsichtlichen Zulassungen ausgeführt worden waren.
Ende 1979 stürzte in einem Münchner U-Bahnhof ein Bereich von etwa 150 m² der untergehängten Putzdecke mit einem Flächengewicht von 55 kg/m² ab. Mehrere Personen wurden verletzt. Die Decke war im Jahre 1970 mit auf zentrischen Zug beanspruchten Kunststoffdübeln in der Zugzone der Stahlbetondecke verankert worden.
Nach den Feststellungen des Gutachters (Prof. REHM) war für den Einsturz das Zusammentreffen einer Reihe von Montagemängeln maßgebend. Die Bohrlochdurchmesser waren um bis zu 10% zu groß, die Schraubendurchmesser fast 20% zu klein. Teilweise waren Schrauben mit dem Hammer eingeschlagen worden. Die Einschraubtiefe der Schrauben betrug statt 40 mm nur zwischen 23 und 38, i. M. 32 mm. Der Einsturz wurde schließlich in einem Bereich ausgelöst, in dem ein Teil der Dübel in Rissen bis zu max. 0,4 mm Breite saßen.
Eine Reihe weiterer Einstürze untergehängter Decken, die an nicht zugelassenen auf zentrischen Zug beanspruchten Kunststoffdübeln aufgehängt waren, gab Veranlassung, derartige Konstruktionen stichprobenweise zu untersuchen. Besonderer Wert wurde dabei darauf gelegt, alle aus der Montage herrührenden Gegebenheiten zu erfassen und einen Einblick in die übliche Ausführungsqualität zu erlangen.

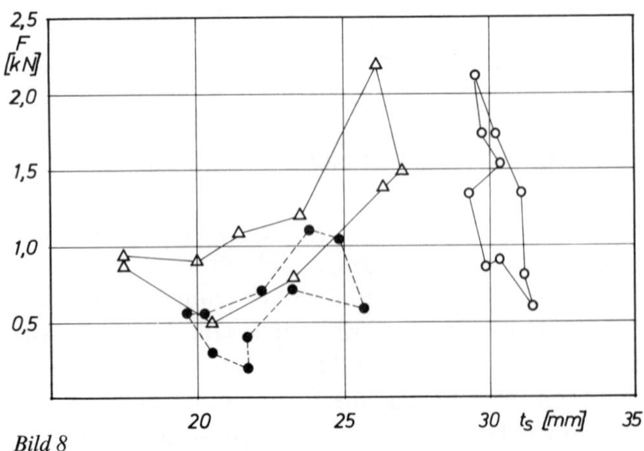

Bild 8
Ausziehversuche an Kunststoffdübeln in Massivdecken

Als Beispiel ist für drei gleiche Dübeltypen vom Durchmesser 6 mm das Ergebnis der Ausziehversuche aus 3 Massivdecken (Normalbeton) in Bild 8 dargestellt. Je Decke wurden zehn Dübel an verschiedenen Stellen gezogen. Der Sollwert der Verankerungstiefe betrug jeweils 35 mm. Besonders beachtenswert ist weniger die insbesondere wegen der zu kleinen Einschraubtiefe zu geringe Traglast als die erschreckend hohe Streuung der Einzelwerte.
In allen Fällen, in denen von unqualifizierten nicht ausreichend überwachten Firmen vergleichbare Fehler produziert wurden, muß zwangsweise mit ähnlichen Schäden gerechnet werden.

10 Zusammenfassung

Die Entwicklung der Dübeltechnik hat im letzten Jahrzehnt einen hohen Stand erreicht. Ordnungsgemäß erstellte Verbindungen gewährleisten ein hohes Maß an Sicherheit und Dauerhaftigkeit. Voraussetzung hierfür sind eine sorgfältige ingenieurmäßige Vorplanung und peinliche Beachtung der in den Zulassungsbescheiden enthaltenen Besonderen Bestimmungen.
Gefahren entstehen dann, wenn unqualifiziertes Personal am Werke ist und ausreichende Koordination zwischen Planenden und Ausführenden fehlt. Eine sorgfältige Überwachung und Kontrolle ist unverzichtbar.
Als Nachteil muß die Vielzahl der verschiedenen Typen und Systeme mit unterschiedlichen Abmessungen und Anwendungsbedingungen angesprochen werden. Für gleichartige Aufgaben gibt es in der Befestigungstechnik sehr unterschiedliche Lösungen. Eine Reduzierung der Sortenvielfalt verbunden mit einer Klassifizierung nach Lastgruppen mit zugeordneten Abstandsregelungen wäre sehr zu begrüßen.

11 Literatur

[1] Kolloquium „Spreizdübel im Betonbau“ am 23.2. 1973, TU Darmstadt.
[2] SCHRAGE, J. und SCHORN, H.: Dübel für tragende Konstruktionen, Beton- und Fertigteiltechnik 8/1973.
[3] SELL, R.: Tragfähigkeit von mit Reaktionsharzmörtelpatronen versetzten Betonankern und deren Berechnung, Bautechnik 10/1973.
[4] WAGNER-GREY: Experimentelle und theoretische Untersuchungen zum Tragverhalten von Spreizdübeln in Beton, Dissertation TU München, 1976.
[5] PLANCK, A.: Bautechnische Einflüsse auf die Tragfähigkeit von Kunststoffdübeln, Mitteilungen aus der BAM, BMT 6, 1977.

[6] Internationales Kolloquium, „Verankern mit Dübeln", am 15.9. 1977, TU Darmstadt.
[7] WAGNER-GREY: Nachträglich eingesetzte Dübel für tragende Verbindungen, Bauingenieur 53/1978.
[8] Seminar Befestigungstechnik im Ingenieurbau, am 5.2. 1979, Haus der Technik, Essen.
[9] Kolloquium „Befestigungstechnik im Bauwesen", im Mai 1979, TU Stuttgart.
[10] LANG, G. und VOLLMER, H.: Dübelsysteme für Schwerlastverbindungen, Bautechnik 6/1979.
[11] BAUER, C. O.: Befestigungstechnik – Wettbewerb von Systemen, Bauingenieur 55/1980.
[12] KRAUSS, K.: Fragen zu Dübelverbindungen, Tagungsbericht 6/1982 der Landesvereinigung der Prüfingenieure Baden/Württ.
[13] REHM, G.: Aufgaben und Ziele der Forschung sowie Umsetzung der Ergebnisse in der Praxis, Betonwerk und Fertigteiltechnik 9/1982.
[14] HENSCHKER, R.: Anforderungen des Planers an die Befestigungstechnik, Betonwerk und Fertigteiltechnik 9/1982.
[15] SEGHEZZI, D.: Wirtschaftliche und sichere Befestigungssysteme für die Baupraxis, Betonwerk und Fertigteiltechnik 9/1982.
[16] Verzeichnis der allgemein bauaufsichtlich zugelassenen Dübel und Ankerschienen. Mitteilungen des Instituts für Bautechnik Nr. 4, August 1983.

Befestigen mit Hinterschnittankern

Artur Fischer

1 Stand der Technik

Dübel als Befestigungsmittel für den nachträglichen Ausbau gewinnen im Zuge der Rationalisierung zunehmend an Bedeutung. Maßgebend hierfür sind die durch Massenproduktion niedrigen Stückpreise sowie die einfache und zeitsparende Handhabung bei der Anwendung.

Die Aufgabe dieser Befestigungsmittel besteht darin, die über die Anschlußkonstruktion angreifenden Kräfte sicher und dauerhaft in den Ankergrund einzuleiten. Die dafür erforderliche Verbindung mit dem Ankergrund kann durch Verspreizen oder Einmörteln der Dübel im Bohrloch erfolgen.

Metallspreizdübel werden in Abhängigkeit der Spreizprinzipien in 3 Gruppen eingeteilt (Bild 1). Dabei wird zwischen kraftkontrolliert spreizenden Dübeln (Gruppe A) und wegkontrolliert spreizenden Dübeln, bei denen der Konus in die Hülse (Gruppe B) und die Hülse auf den Konus (Gruppe C) eingetrieben wird, unterschieden.

Die vorgenannten Metallspreizdübel übertragen die eingeleiteten Kräfte vorwiegend durch Reibung auf den Ankergrund und sind somit auf die Erzeugung einer Spreizkraft beim Verankern angewiesen.

Bei der Erzeugung dieser Spreizkraft wird bei Dübeln der Gruppe A in Abhängigkeit des Verformungswiderstandes des Ankergrundes und bei Dübeln der Gruppe B und C in Abhängigkeit der Spreizgeometrie eine Verformungsmulde im Ankergrund hergestellt.

Diese bei der Verspreizung erzeugte Verformungsmulde kann als Durchmesservergrößerung angesehen werden, die bei der Übertragung der eingeleiteten Kräfte mitwirkt. Der Einfluß dieser Verformungsmulde auf das Tragverhalten der genannten Metallspreizdübel ist jedoch wegen des unterschiedlichen Verformungsverhaltens des Ankergrundes nicht exakt zu erfassen.

Wie sicher Dübel als Befestigungsmittel sind, hängt außer der ordnungsgemäßen Produktfunktion vom Kenntnisstand des Anwenders sowie der Ausführung der Montage ab. Eine unsachgemäße Anwendung oder fehlerhafte Montage kann somit bei der Überlagerung mit anderen ungünstigen Komponenten zum Schaden führen. Sicherheit als Voraussetzung ist deshalb von besonderer Bedeutung, wenn Dübel als Befestigungsmittel zur Verankerung tragender Konstruktionen verwendet werden.

In diesem bauaufsichtlich relevanten Bereich sind Neuentwicklungen vorhanden, die durch ihre Konstruktion dem Sicherheitsgedanken verstärkt Rechnung tragen. Die Wirkungsweise dieser Befestigungsmittel beruht in einer formschlüssigen Verbindung mit dem Ankergrund.

Für den Einbau wird ein Bohrloch hergestellt, welches im Ankergrund einen größeren Durchmesser als am Bohrlochmund besitzt (Bild 2). Die Herstellung dieser

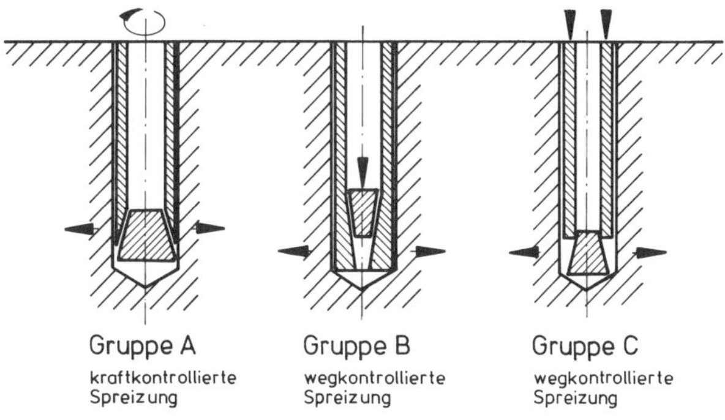

Bild 1
Spreizprinzipien von Dübeln [1]

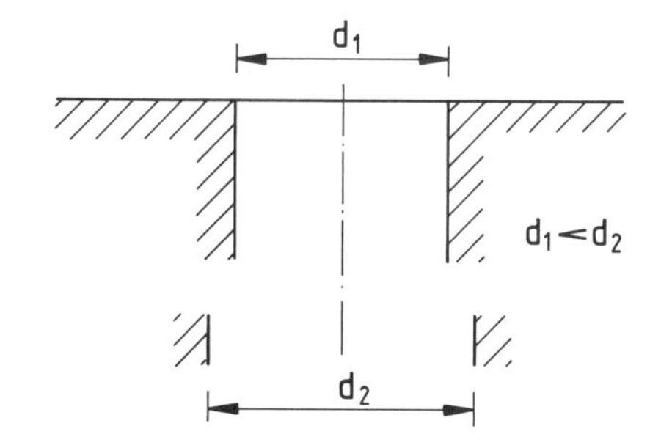

Bild 2
Bohrloch mit Hinterschneidung

Bohrlocherweiterung im Ankergrund, die zum Unterschied zu der bei den Dübeln der Gruppe A, B und C genannten Durchmesservergrößerung mit einem Bohrer erfolgt, nennt man Hinterschneidung, und die in diese Bohrlöcher eingesetzten Befestigungsmittel werden Hinterschnittanker genannt.

2 Systembeschreibung und -anwendung

2.1 Geschichtliche Entwicklung

Das Verankern in hinterschnittenen Bohrlöchern zählt zu den ältesten Befestigungstechniken die wir kennen. Schon vor ca. 1500 Jahren wurden in Steine Löcher gemeißelt, die sich im Innern schwalbenschwanzförmig erweitern.

Heute wird dieses Verfahren der Bohrlochherstellung nur noch vereinzelt, wie zum Beispiel beim Einmörteln von Geländerpfosten oder bei der Verankerung von schweren Fassadenplatten angewandt.

Im Zuge der Rationalisierung in der Befestigungstechnik geht man deshalb dazu über, die Herstellung der hinterschnittenen Bohrlöcher maschinell durchzuführen. Begonnen wurde dabei mit dieser Bohrlochherstellung in Baustoffen mit relativ geringer Druckfestigkeit. Aus diesem Grunde wurden auch die ersten Hinterschnittanker für die Verankerung in Gas- und Schaumbeton entwickelt.

Moderne Bohrgeräte und spezielle Bohrverfahren haben es später ermöglicht, hinterschnittene Bohrlöcher auch in Baustoffen mit hoher Druckfestigkeit herzustellen. Somit war es möglich, den Anwendungsbereich zu erweitern und Hinterschnittanker für die Verankerung in Beton zu entwickeln.

Für die Verankerung tragender Konstruktionen wurden die ersten Zulassungsbescheide für Befestigungsmittel mit hinterschnittenem Bohrloch im Jahre 1979 vom Institut für Bautechnik (IfBt) erteilt. Zwischenzeitlich liegen für 4 verschiedene Ankersysteme Zulassungsbescheide vor (Tabelle 1).

2.2 Konstruktionsmerkmale

Die derzeitig vorhandenen Hinterschnittanker unterscheiden sich sowohl bezüglich der Form des Hinterschnittes sowie der Hinterschnittanordnung. Zusätzlich wird noch zwischen direktem und indirektem Formschluß mit dem Ankergrund unterschieden.

Tabelle 1
Zulassungsbescheide für Befestigungsmittel mit hinterschnittenem Bohrloch [2]

Benennung	Antragsteller	Zulassungs-Nr.	Ersterteilung
Injections-Anker FIM u. FIH	Fischer	Z-21.3-61	13. Juli 1979
SYBA-Anker	Syba	Z-21.2-42	10. Dez. 1979
Gasbeton-Dübel HGS*	Hilti	Z-21.2-235	19. Jan. 1982
Einspann-anker Ultra-Plus	Liebig	Z-21.1-186	22. Jun. 1982
Zykon-Anker FZA	Fischer	Z-21.1-218	15. Aug. 1983

* Baugleich mit SYBA-Anker

Bei dem auf Bild 3 dargestellten Ankersystem erstreckt sich die Bohrlochhinterschneidung in etwa auf die gesamte Bohrlochtiefe. Die formschlüssige Verbindung mit dem Ankergrund erfolgt indirekt über einen zusätzlichen Füllstoff, der in das Bohrloch eingebracht wird. Im Gegensatz dazu wird bei dem Ankersystem gemäß Bild 4 die Bohrlochhinterschneidung nur in einem Teilbereich der Bohrlochtiefe vorgenommen, in die der Anker direkt eingebracht wird.

Bezüglich der Hinterschnittanordnung wird zwischen der bei der Hinterschneidung erzielten konischen Er-

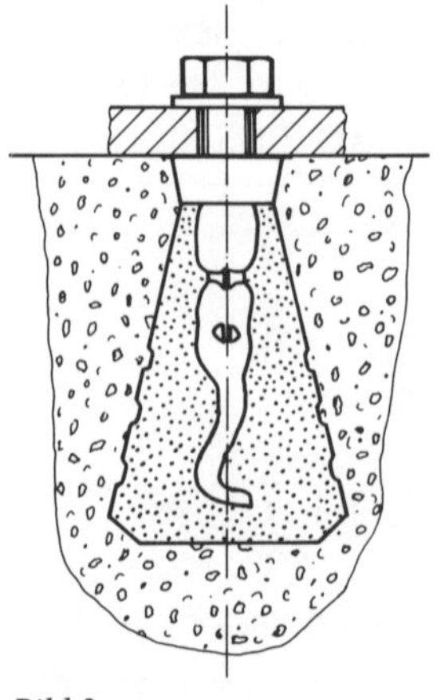

Bild 3
fischer-Injections-Anker FIM

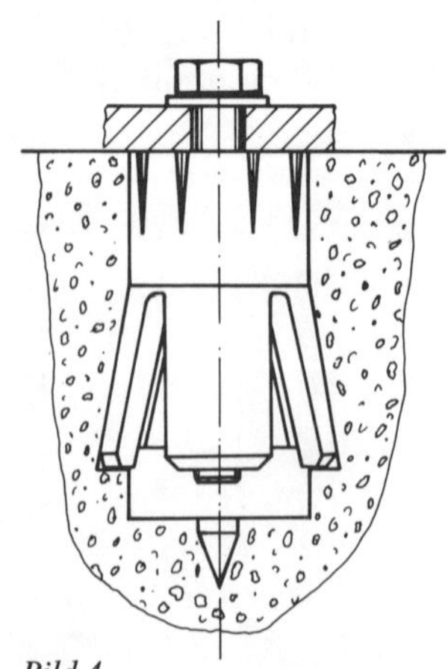

Bild 4
Syba-Anker bzw. Hilti-Gasbeton-Dübel HSG

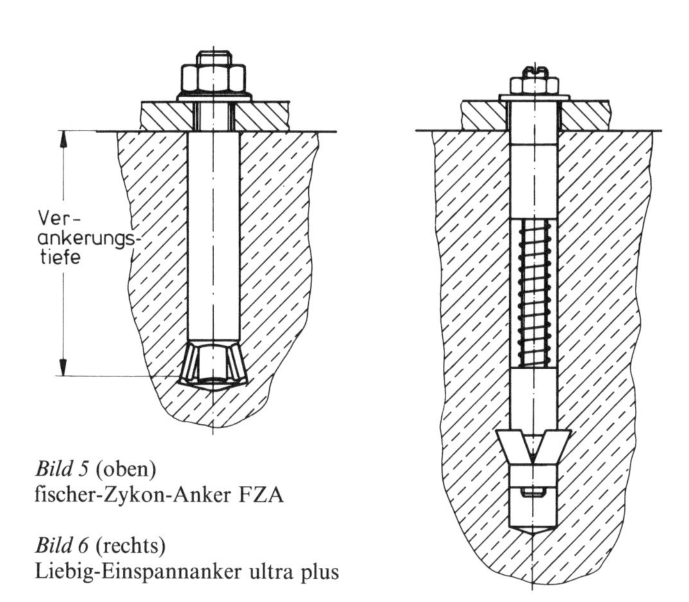

Bild 5 (oben)
fischer-Zykon-Anker FZA

Bild 6 (rechts)
Liebig-Einspannanker ultra plus

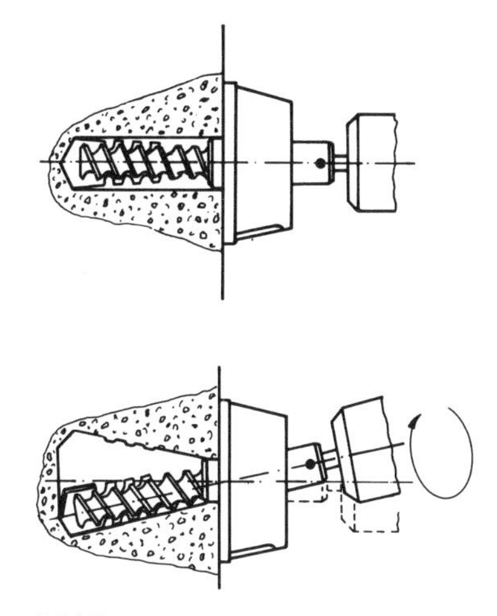

Bild 7
Bohrlochherstellung für fischer-Injections-Anker FIM

weiterung zur Bohrlochtiefe (Bild 5) und zur Oberfläche gerichtet (Bild 6) unterschieden.
Eine ausführliche Systembeschreibung und die Betrachtung der verschiedenen Anwendungskriterien wird im weiteren an den gemäß Bild 3 und Bild 5 dargestellten Ankern vorgenommen.

2.3 Anwendungsbereich

Mit Hinterschnittankern kann derzeitig in Vollbaustoffen verschiedener Druckfestigkeit verankert werden. Durch Zulassungsbescheide ist dabei die Anwendung in 2 verschiedenen Baustoffarten geregelt.
Bei dem auf Bild 3 dargestellten fischer-Injections-Anker (FIM) handelt es sich um ein Befestigungsmittel, welches aus einer Stahl-Innengewindehülse und einem Kunststoff-Dichtflansch besteht. Gemäß Zulassungsbescheid [3] kann mit diesem Element in sämtlichen genormten oder zugelassenen Gasbetonsteinen und Gasbetonplatten verankert werden.
Bei Mauerwerk aus Gasbetonsteinen ist das Verankern in der Fuge und bei Gasbetonplatten das Verankern in der aus Lastspannungen erzeugten Zugzone zulässig. Diese Anwendungsmöglichkeiten beruhen vorwiegend auf der Wirkungsweise dieses im hinterschnittenen Bohrloch eingesetzten Ankers.
Bild 5 zeigt den fischer-Zykon-Anker (FZA), einen aus einer Spreizhülse und einem Konusbolzen bestehenden Stahlanker. Bei diesem Konstruktionsprinzip wird im derzeitigen Zulassungsbescheid [4] die Anwendung in Normalbeton bei einer Verankerungstiefe von 40 mm für unterschiedliche Gewindegrößen geregelt.

Eine Verankerung in der aus Lastspannungen erzeugten Zugzone von flächigen Stahlbetonteilen ist ebenso wie die Anordnung mit verringerten Achs- und Randabständen in der Druckzone im Zulassungsbescheid geregelt. Maßgebend für diese Anwendungen waren das günstige Tragverhalten des Ankers im hinterschnittenen Bohrloch.

2.4 Bohrlochherstellung

Voraussetzung für die ordnungsgemäße Anwendung von Hinterschnittankern ist ein auf das Konstruktionsprinzip des Ankers abgestimmtes hinterschnittenes Bohrloch. Es ist deshalb sehr wichtig, diese Bohrlochherstellung durch einfache und anwendungsgerechte Mittel zu erreichen. In Abhängigkeit der Ankergrundart werden dabei verschiedene Formen der Hinterschneidung erzeugt.
Beim fischer-Injections-Anker erfolgt die Bohrlochherstellung einschließlich der Hinterschneidung mit demselben Bohrgerät (Bild 7). Mit einem speziell auf den Ankergrund Gasbeton abgestimmten Bohrer wird bis zum Anliegen der Bohrglocke am Ankergrund gebohrt. Anschließend wird durch eine kreisförmige Schwenkbewegung der Bohrmaschine um die Bohrglocke die Bohrlochhinterschneidung als kegelstumpfförmige Aussparung hergestellt. Die so erstellte Bohrlochhinterschneidung erstreckt sich in etwa auf die gesamte Bohrlochtiefe.
Beim fischer-Zykon-Anker wird ebenso wie beim fischer-Injections-Anker mit einem speziellen Bohrer bis zum Anliegen Bohrglocke am Ankergrund gebohrt

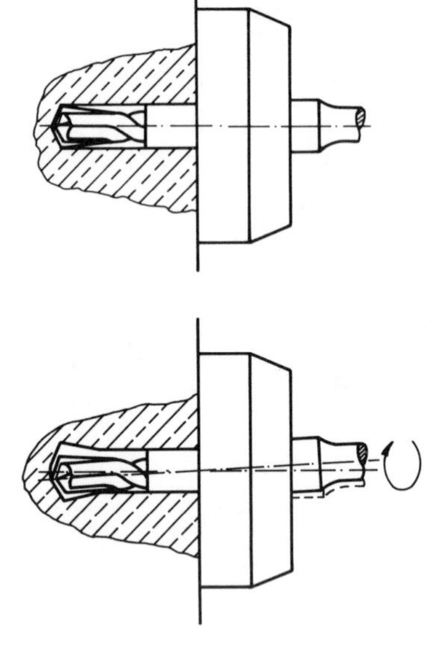

Bild 8
Bohrlochherstellung für fischer-Zykon-Anker FZA

(Bild 8) und durch eine kreisförmige Schwenkbewegung um die Bohrglocke die erforderliche Bohrlochhinterschneidung jedoch nur in der Tiefe des Bohrloches hergestellt.

2.5 Montageart

Bei der Montage von Hinterschnittankern wird eine formschlüssige Verbindung zwischen Ankergrund und Anker geschaffen, über die die am Anker angreifenden Kräfte in den Ankergrund eingeleitet werden. Diese formschlüssige Verbindung kann direkt durch den Hinterschnittanker oder indirekt durch einen, den Anker umgebenden zusätzlichen Füllstoff erfolgen.
Da von einer ordnungsgemäßen Montage die Funktionssicherheit der Anker abhängt, ist es wichtig, einfache und sichere Kontrollmöglichkeiten zu schaffen.
Beim fischer-Injections-Anker (Bild 9) wird durch den in das hinterschnittene Bohrloch eingesetzten Anker

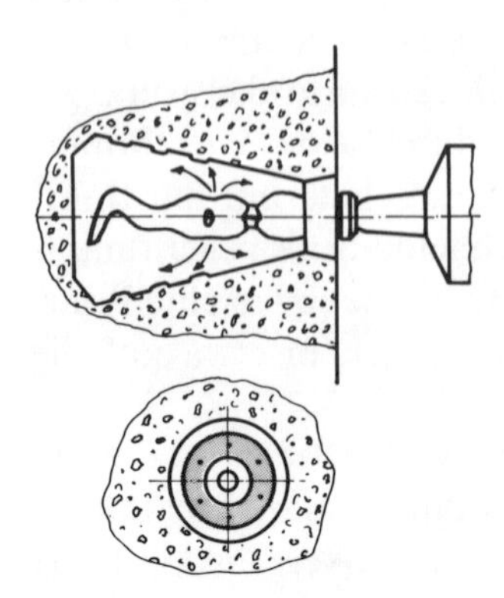

Bild 9
Montagevorgang und Funktionskontrolle beim fischer-Injections-Anker FIM

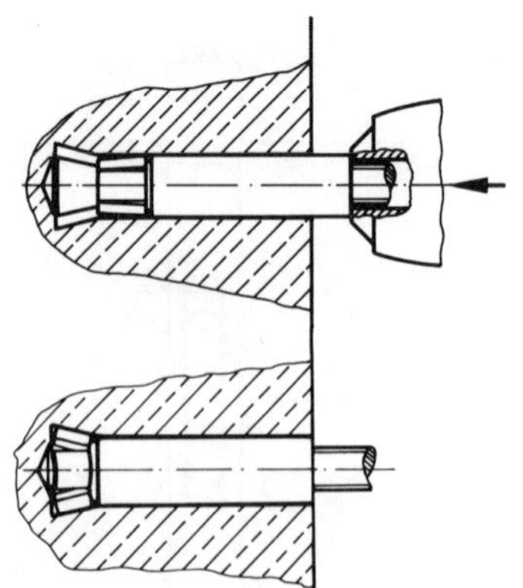

Bild 10
Montagevorgang und Tragfähigkeitskontrolle beim fischer-Zykon-Anker FZA

Injectionsmörtel als Füllstoff in das Bohrloch gepreßt. Eine vollkommene Verpressung des Bohrloches und somit eine formschlüssige Verbindung mit dem Ankergrund ist dann erreicht, wenn sich der Dichtflansch durch den Injectionsmörtel dunkel verfärbt.
Dieser Vorgang bei der Montage gilt als sichere Funktionskontrolle für den Anker, der nach Einhaltung einer bestimmten Abbindezeit belastet werden darf.
Die Montage beim fischer-Zykon-Anker (Bild 10) wird mit einem Einschlaggerät, mit welchem die Ankerhülse zur formschlüssigen Verbindung mit dem Ankergrund über den Konus in die Bohrlochhinterschneidung eingetrieben wird, vorgenommen. Der Anker ist ordnungsgemäß verankert und darf ohne zusätzliche Kontrolle belastet werden, wenn die Spreizhülse mindestens bündig mit dem Ankergrund ist.
Durch diese einfache Sichtkontrolle ist gewährleistet, daß das Bohrloch die erforderliche Bohrlochtiefe besitzt und die Schwenkbewegung beim Hinterschneiden ordnungsgemäß bis zum vorgegebenen Anschlag in der Bohrglocke ausgeführt wurde.

3 Tragverhalten

3.1 Vorbemerkung

Die dem Anwender zur Verfügung stehende Leistungsfähigkeit eines Ankers wird aufgrund der Tragfähigkeit und der dafür erforderlichen Anwendungskriterien ermittelt. Da der Zustand des Ankergrundes auf die genannten Kriterien von entscheidender Bedeutung ist, werden die Leistungsdaten für den ungerissenen und den gerissenen Ankergrund angegeben. Welches Tragverhalten ein Hinterschnittanker in Beton besitzt, wird am fischer-Zykon-Anker in Verbindung mit anderen Metallspreizdübeln aufgezeigt.

3.2 Anwendung in der aus Lastspannungen erzeugten Druckzone

Bei den Metallspreizdübeln der Gruppe A, B und C werden durch den Spreizvorgang bei der Montage Kräfte unbekannter Größen in den Ankergrund eingeleitet.
Steht für diese Spreizkraft keine ausreichend große Betonfläche pro Dübel zur Verfügung, so können beim Verspreizen Spaltrisse zum Rand oder zum benachbarten Dübel entstehen, die das Tragverhalten der Dübel entscheidend beeinflussen.
Beim fischer-Zykon-Anker werden aufgrund der Bohrlochgeometrie keine meßbaren Spreizkräfte beim Einbau in das hinterschnittene Bohrloch erzeugt [5]

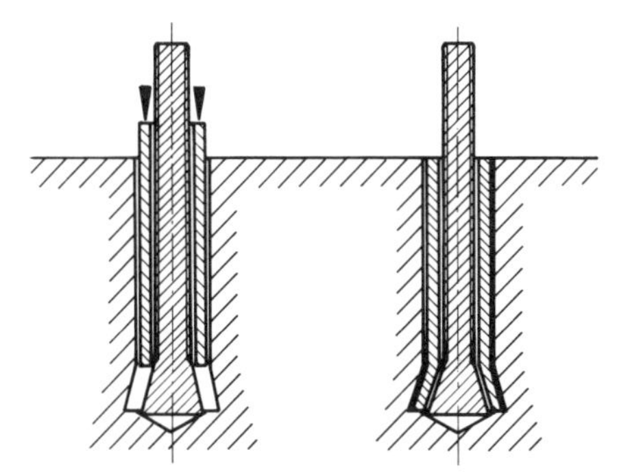

Bild 11
Spreizprinzip des fischer-Zykon-Ankers FZA

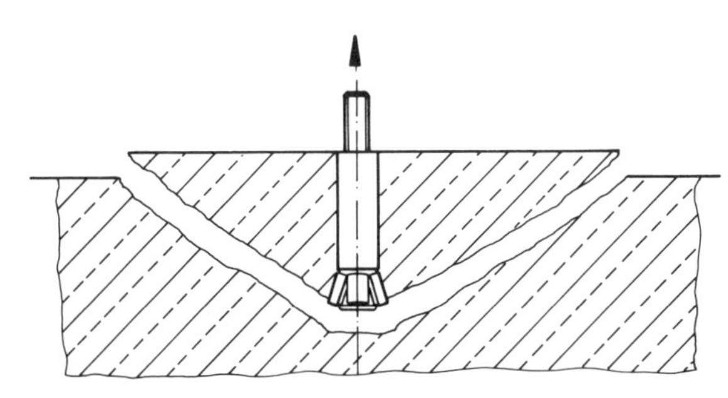

Bild 12
Kegelförmiger Betonausbruch als Versagensursache bei zentrischer Zugbeanspruchung

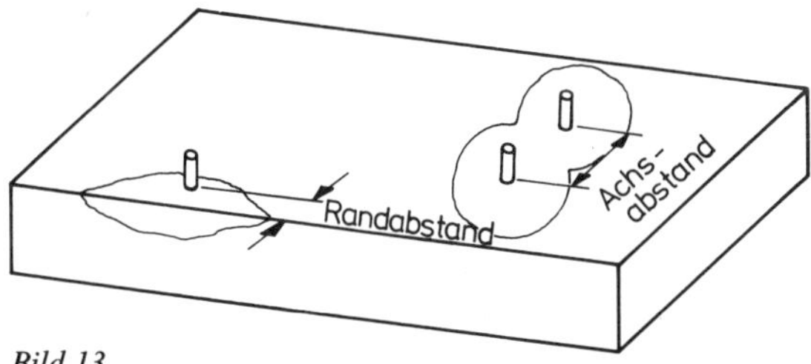

Bild 13
Beeinflussung des Betonausbruchs durch Bauteilrand bzw. Überschneidung

(Bild 11). Der Ankergrund wird somit erst beim Angriff einer äußeren Last über den Anker beansprucht. Steht einem Anker eine ausreichend große Betonfläche zur Verfügung, so entsteht bei der Belastung bis zur Bruchlast Betonausbruch als Versagensursache (Bild 12). Werden die für die maximale Tragfähigkeit der Anker erforderlichen Achs- und Randabstände unterschritten, so kann sich der auf Bild 12 dargestellte Ausbruchkegel eines Einzelankers nicht voll ausbilden (Bild 13), was eine Abminderung der maximalen Tragfähigkeit zur Folge hat.
Diese Zusammenhänge zwischen reduzierten Achs- und Randabständen bei reduzierter Last wurden für den fischer-Zykon-Anker bauaufsichtlich geregelt [4]. Dem Anwender ist es dadurch möglich, Abstände zum Bauteilrand bzw. Dübelgruppen mit einem inneren Achsabstand von 5 cm bei entsprechender Lastreduzierung auszuführen, was den Gegebenheiten in der Praxis sehr entgegenkommt.

3.3 Anwendung in der aus Lastspannungen erzeugten Zugzone

Bei der Anwendung von Dübeln in der aus Lastspannungen erzeugten Zugzone wird das Tragverhalten dieser Elemente vom Zustand des Ankergrundes im Verankerungsbereich beeinflußt.
Aus diesem Grunde sind sowohl Eignungsversuche und die Ermittlung des Tragverhaltens der Dübel in gerissenem Ankergrund durchzuführen.
Die Eignung wird bei den Dübeln der Gruppe A durch das Nachspreizverhalten und bei den Dübeln der Gruppe B und C durch die Spreizkraftreserven bestimmt. Da sich in der Praxis die Risse im Ankergrund im Laufe der Zeit mehrfach öffnen und schließen können, ist die Zuverlässigkeit der Eignungsparameter auf Dauer zu gewährleisten.
Das Tragverhalten der Dübel im Riß hängt im wesentlichen von der Aufrechterhaltung einer Reibkraft zwischen Dübel und Bohrlochwandung nach der Rißöffnung ab. Hierfür ist bei den Metallspreizdübeln der Gruppe A, B und C eine Spreizkraft erforderlich, die, wie bei der Eignung beschrieben, nach Öffnen der Risse noch vorhanden sein muß bzw. durch Nachspreizen neu erzeugt werden soll.
Die qualitative Beeinflussung des Tragsverhaltens durch Risse ist in Bild 14 dargestellt.
Dabei wurden Dübel der Gruppe A, B und C mit einem Außendurchmesser von ca. 16 mm in Rissen

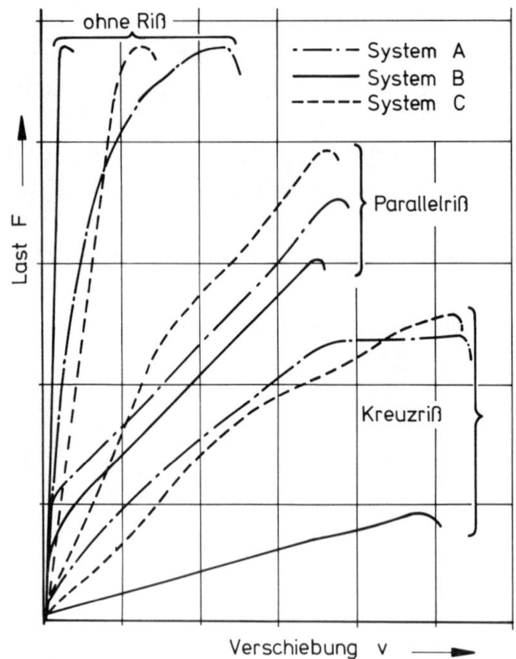

Bild 14
Qualitativer Verlauf der Last-Verschiebungskurven von Dübeln der Gruppe A, B und C in ungerissenem Beton, in Parallelrissen und in Kreuzrissen [6]

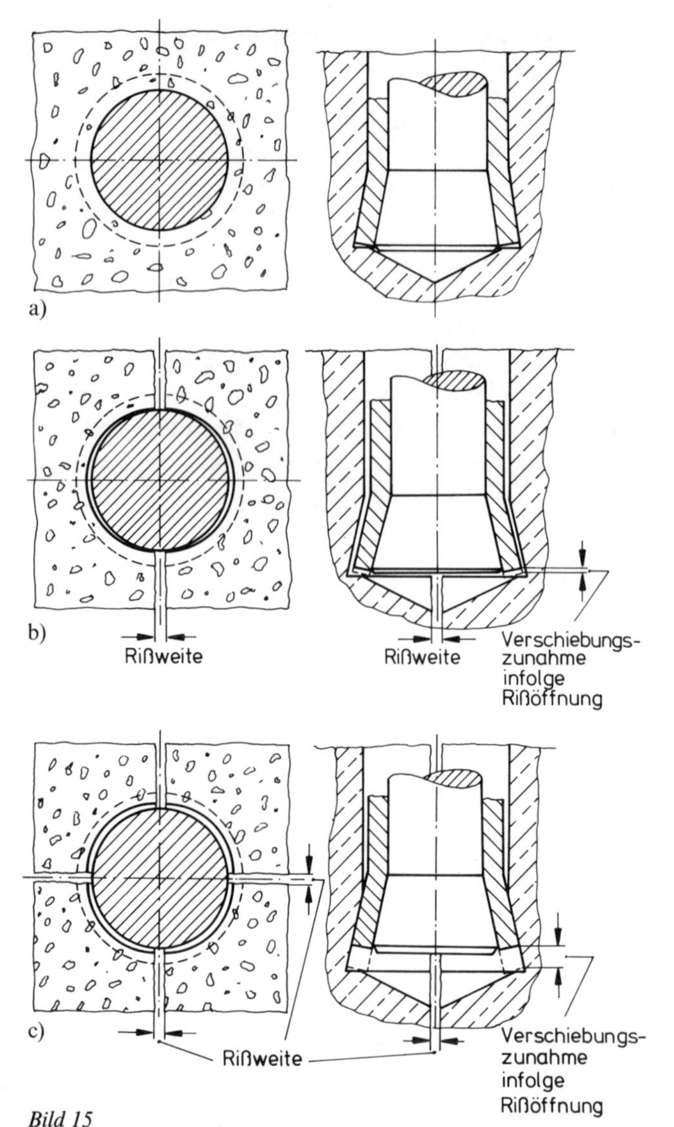

Bild 15
Spreizbereich des fischer-Zykon-Ankers FZA bei verschiedenen Ankergrundzuständen, a) ungerissener Ankergrund, b) Parallelriß durch Bohrlochmitte, c) Kreuzriß durch Bohrlochmitte

von ca. 0,4 mm Rißweite bei der Anwendung in B 25 untersucht. Die Ursache für den flachen Verlauf der Last-Verschiebungskurven gegenüber dem ungerissenen Ankergrund ist in der Aufhebung des räumlichen Spannungszustandes und dem damit verbundenen geringeren Verformungswiderstand des Ankergrundes zu sehen. Infolge Rißöffnung wird zudem die Kontaktfläche zwischen Bohrlochwandung und Dübelspreizteil vermindert.

Beim fischer-Zykon-Anker sind aufgrund der formschlüssigen Verbindung mit dem Ankergrund im Riß andere Tragmechanismen wie bei den Dübeln der Gruppe A, B und C vorhanden.

Es ist weder ein gesichertes Nachspreizen, noch eine vorhandene Spreizkraftreserve für das Tragverhalten maßgebend. Allein die geometrischen Verhältnisse beeinflussen das Tragverhalten der Anker im Riß.

In einem Parallelriß ändert sich die Lage des Ankers gegenüber dem ungerissenen Beton nur unwesentlich (Bild 15a und b). Theoretisch bleibt der Formschluß zwischen Anker und Ankergrund im Spreizbereich nahezu vollkommen erhalten, was bei der Ermittlung der maximalen Tragfähigkeit der Anker im geöffneten Riß bei normaler Rißweite einen Ausbruchkegel als Versagensursache zur Folge hat. Die Abminderung der erreichten Tragkraft im Riß gegenüber ungerissenem Beton beruht somit auf der Spaltung des Betonbruchkegels durch den Riß.

Im geöffneten Kreuzriß ändert sich die Lage des Ankers gegenüber dem ungerissenen Beton durch die geometrischen Verhältnisse bei der Rißöffnung (Bild 15c). Das Tragverhalten selbst wird bei normaler Rißweite jedoch nur etwa um das gleiche Maß des vorgenannten Paralleleinflusses gemindert. Für den fischer-Zykon-Anker FZA M 12 × 80 ist in Bild 16 das im Kurzzeitversuch ermittelte Tragverhalten dargestellt. Die Ergebnisse wurden in hochfestem Beton sowie bei einer Rißweite von ca. 0,3 mm ermittelt.

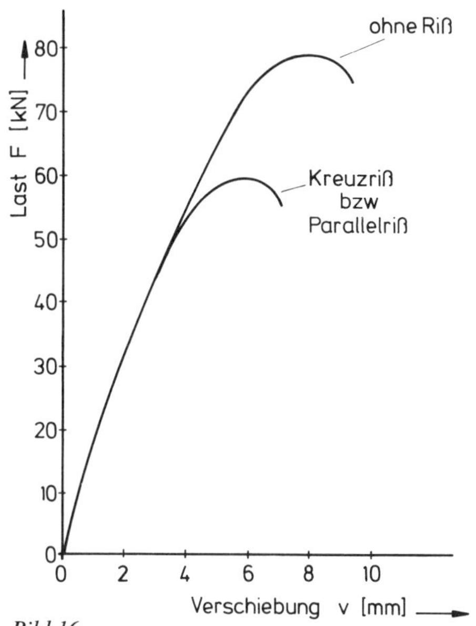

Bild 16
Last-Verschiebungskurven für fischer-Zykon-Anker FZA M 12 × 80 in hochfestem Beton sowie bei einer Rißweite von ca. 0,3 mm [7], [8]

Da die Abminderung zwischen ungerissenem und gerissenem Ankergrund relativ gering ist, können mit dem fischer-Zykon-Anker hohe zulässige Lasten in der Zugzone erreicht werden. Diese Lasten sind aus der Sicht des Ankers als Dauerlasten anzusehen, da beim wiederholten Öffnen und Schließen der Risse der Formschluß und somit die Funktionsfähigkeit der Anker erhalten bleibt.
Das Last-Verschiebungsverhalten gemäß Bild 16 zeigt beim fischer-Zykon-Anker bezüglich des Anstieges keine Unterschiede zwischen ungerissenem und gerissenem Ankergrund. Diese Tatsache ist bei der Betrachtung von Gruppenbefestigungen, bei denen einzelne Anker der Gruppe im gerissenen Ankergrund verankert sein können, von entscheidender Bedeutung für die Tragfähigkeit der Gesamtgruppe.

4 Zusammenfassung

Dübel als Befestigungsmittel für Bereiche mit hohen Sicherheitsanforderungen sind aus wissenschaftlicher Sicht noch ein sehr junger Anwendungsbereich. Die Klärung noch offener Fragen und die Bestätigung vorhandener Ansätze können dazu beitragen, in diesem Anwendungsbereich neue Einsatzmöglichkeiten für Dübel zu erschließen.
Hinterschnittanker bieten bei einfacher Handhabung und jederzeit erkennbarer Funktionsfähigkeit gute Voraussetzungen hohe Anforderungen zu erfüllen. Durch das spreizdruckfreie Verankern können sehr günstige Einbaubedingungen und durch die formschlüssige Verbindung mit dem Ankergrund hohe Lasten auch in kritischen Verankerungsbereichen sicher aufgenommen werden.

5 Literatur

[1] SCHRADE, J. und SCHORN, H.: Dübel für tragende Konstruktionen. In Betonwerk und Fertigteilwerk 39 (1973), Heft 8, Seite 585.

[2] Institut für Bautechnik: Verzeichnis der allgemein bauaufsichtlich zugelassenen Dübel und Ankerschienen, Stand 15. November 1983.

[3] Institut für Bautechnik: Zulassungsbescheid für fischer-Injections-Anker FIM und FIH, Zulassungs-Nr.: Z-21.3-61 vom 27. August 1982.

[4] Institut für Bautechnik: Zulassungsbescheid für fischer-Zykon-Anker FZA M 6 – M 10 Zulassungs-Nr.: Z-21.1-218 vom 15. August 1983.

[5] REHM, G.: Gutachtliche Stellungnahme zur Frage der Eignung von fischer-Zykon-Anker FZA 6 × 40, FZA 8 × 40 und FZA 10 × 40 für tragende Konstruktionen, München 1982.

[6] REHM, G. und LEHMANN, R.: Untersuchung mit Metallspreizdübeln in der gerissenen Zugzone von Stahlbetonbauteilen, Bericht der FMPA Baden-Württemberg, Otto-Graf-Institut, Stuttgart 1982.

[7] Prüfbericht Nr. II.4-14 369/1 der FMPA Baden-Württemberg, Otto-Graf-Institut. Ermittlung der Tragfähigkeit des fischer-Zykon-Anker FZA M 12 × 80, Januar 1984.

[8] Prüfbericht Nr. II.3-14 315/1 der FMPA Baden-Württemberg, Otto-Graf-Institut. Untersuchung des Tragsverhaltens von fischer-Zykon-Ankern FZA M 12 × 80 in Kreuzrissen, Januar 1984.

Befestigungsverfahren im Bauwesen, mit besonderer Berücksichtigung des Bolzensetzens

Hans Dieter Seghezzi

1 Entwicklung der Befestigungstechnik

Befestigungstechnik ist so alt oder fast so alt wie die Bautechnik. Schon VITRUV beschrieb in seinen 10 Büchern über Architektur im 1. Jahrhundert v. Chr. praktische Lösungen von Befestigungsaufgaben, wie das Abhängen von Zwischendecken. An den Ruinen des Kollosseums in Rom sind Dübellöcher zu sehen und bei den Ausgrabungen am Magdalensberg in Österreich wurde ein gehärteter Nagel gefunden, der aus der Zeit der Römer stammt. Diese Beispiele lassen sich fortsetzen.

Doch bis zur Mitte des 20. Jahrhunderts blieb die Befestigungstechnik beschränkt auf das Ausspitzen und Einmörteln und auf die Verwendung von Holzdübeln. Erst vor etwa 40 bis 50 Jahren setzte eine stürmische Entwicklung ein, die zur weiten Verbreitung der modernen Befestigungstechnik im Bauwesen geführt hat.

Unter Befestigen versteht man die Lösung folgender Aufgabe: An einem Teil A, dem Untergrund, muß ein Teil B, das zu befestigende Bauteil, befestigt werden. Somit ist Befestigungstechnik ein Teilgebiet der Verbindungstechnik, das dadurch charakterisiert ist, daß die zu verbindenden Teile A und B nicht gleichwertig sind. Im allgemeinen ist Teil A, der Untergrund, vorher vorhanden, hat häufig eine tragende Funktion und ist in der Regel schwer und daher schwierig zu transportieren. Dagegen wird das Teil B, das zu befestigende Bauteil, meistens nachträglich bereitgestellt und ist beweglicher und leichter.

Ihre starke Entwicklung verdankt die Befestigungstechnik den Veränderungen in der Bautechnik und dem Einfallsreichtum und Unternehmergeist der Industrie. In der Bautechnik waren dafür die Verwendung neuer Baustoffe, die Entwicklung der Baumethoden, die starke Zunahme des Ausbaus und die Veränderungen im Bauablauf maßgebend. Die Verwendung von Beton, Stahl, Stahlbeton, Leichtmetallen, Kunststoffen und Glas neben den althergebrachten Werkstoffen Stein und Holz führte zum Bedürfnis, Teile aus diesen Werkstoffen miteinander zu kombinieren, sie zu verbinden oder sie aneinander zu befestigen. Die verschiedenen Funktionen eines Bauwerks wie Übertragung statischer Kräfte, Wärmedämmung, Wärmespeicherung, Wetterschutz und Raumabschluß werden nicht mehr wie früher von einem einzigen Bauteil, z. B. einer Mauer übernommen, sondern verschiedenen Schichten oder Bauteilen zugeordnet. Diese müssen aneinander oder am tragenden Skelett befestigt werden. Durch die Entwicklung der Erschließung, der Heizung, Belüftung, Beleuchtung und der Entsorgung hat der technische Ausbau an Bedeutung gewonnen und der Befestigungstechnik zahlreiche Aufgaben gestellt. Schließlich hat sich auch der Bauablauf gegenüber früher geändert. Die Details des Ausbaues werden erst geplant, wenn der Rohbau bereits im Gang ist. Zudem sollen die Ausbauten so gestaltet werden, daß sie später ersetzt oder verändert werden können, denn der Innenausbau veraltet wesentlich schneller als das Gebäude selbst. Diese Forderungen schufen ein Bedürfnis für eine leichte, einfache und wirtschaftliche Montage und für flexible Anschlußpunkte sowie für Lösungen zum Ausgleich von Maßtoleranzen zwischen Untergrund und Anschlußteilen. Zur Erfüllung dieser Aufgaben wurden neue Lösungen der Befestigungstechnik geschaffen.

Beim Aufspüren der Entwicklung kann man häufig zwischen Ursache und Wirkung kaum unterscheiden. Befestigungstechnische Lösungen ermöglichten neue Wege in der Baumethodik, im Ausbau und im Bauablauf. Umgekehrt führten Entwicklungen in der Bautechnik zu neuen Anforderungen an die Befestigungstechnik. Die Entwicklung ist noch keineswegs abgeschlossen, aber es darf festgestellt werden, daß die Befestigungstechnik bereits heute einen wichtigen Platz in der Bautechnik einnimmt.

Bei der Vielfalt des Angebots an Befestigungsmethoden und -mitteln ist es für die Baupraxis sehr schwierig,

den Überblick zu behalten und die jeweils richtige Auswahl zu treffen. Deshalb ist der Ruf nach einer Systematisierung und Ordnung nicht zu überhören.
Im folgenden wird der Anwendungsbereich der Befestigungstechnik grob abgesteckt und eine Übersicht über die Befestigungsverfahren gegeben. Danach wird näher auf das in der Literatur seltener behandelte Verfahren des Bolzensetzens eingegangen.

2 Die Anwendungsbereiche der Befestigungstechnik

Ohne Übertreibung kann gesagt werden: wo gebaut wird, wird auch befestigt (Tabelle 1). Die Krafteinleitung geschieht an der vorderseitigen Oberfläche des Untergrundes (z. B. beim Schweißen), im Innern (z. B. beim Nageln, Dübeln oder Bolzensetzen) oder wird auf der vom Anschlußteil abgewandten Rückseite des Untergrunds (z. B. beim Nieten oder Klemmen) vorgenommen. Nach der Krafteinleitung richten sich auch Verfahren und Mittel, die zur Vorbereitung der Befestigung notwendig sind (z. B. die Reinigung der Oberfläche, das Bohren von Dübellöchern oder das Stanzen oder Bohren vor dem Vernieten). Verfahren und Mittel zur Vorbereitung haben häufig einen wesentlichen Einfluß auf die Wirtschaftlichkeit des Befestigens und auf die Qualität, Sicherheit und Zuverlässigkeit der Befestigungen.
Vielfalt zeichnet die Liste der zu befestigenden Bauteile aus, für welche die Befestigungstechnik Lösungen anbietet. Die am häufigsten vorkommenden Bauteile des Hochbaus sind in Tabelle 2 aufgeführt. Jedoch auch in anderen Bereichen des Bauwesens, wie dem Tiefbau, dem Straßenbau (Leitplanken, Lärmschutzwände, Beleuchtungen, Beschilderung), dem Wasserbau usw. ist die Befestigungstechnik zu Hause.

Tabelle 1
Häufige Untergründe

Stahl	
Beton	Normalbeton Leichtbeton
Mauerwerk	Vollziegel Hohlziegel Kalksandstein Betonsteine Gasbetonsteine, -platten
Leichte Trennwände	Gipskartonplatten Preßspanplatten Kunststoffplatten
Holz	
Leichtmetall	

Tabelle 2
Zu befestigende Teile im Hochbau

- Türrahmen
- Fensterrahmen
- Trennwände
- Rohre
- Leitungen
- Kabel
- Armaturen
- Einrichtungsgegenstände
- Schienen
- Heizkörper
- Isolierplatten, -matten
- Fassadenbekleidungen
- Abdichtungen
- Fußplatten, Kopfplatten
- Konsolen
- abgehängte Decken

3 Verfahren der Befestigungstechnik

9 Verfahren spielen im Bauwesen eine Rolle (Tabelle 3). Die Funktion einer Befestigung, nämlich ihre Fähigkeit, eine geometrische Anordnung von Untergrund und zu befestigendem Teil ständig aufrechtzuerhalten und die auftretenden Kräfte zu übertragen, beruht auf Formschluß (z. B. beim Nieten), Reibschluß (z. B. bei bestimmten Dübeltypen) oder einer chemischen Verbindung (z. B. beim Kleben), die Stoffschluß genannt sei. Häufig ist auch eine Kombination maßgebend. Z. B. wirkt bei vorgespannten Schrauben in Querrichtung die Reibung zwischen Untergrund und Anschlußteil, während in Achsrichtung durch die Gewindegänge und den Schraubenkopf Formschluß besteht. Bei Spreizdübeln in Beton wird die Tragfähigkeit bei achsialer Zugbeanspruchung zum überwiegenden Teil durch die Reibung zwischen Bohrlochwand und Dübeloberfläche herbeigeführt. Da jedoch beim Spreizen Teile des Dübels in den Beton eingedrückt werden, entsteht eine zusätzliche Halterung durch Formschluß.

Tabelle 3
Verfahren der Befestigungstechnik

Verfahren	Untergründe	Verfahrensart	Krafteinleitung	Funktionsweise
Schrauben	Stahl, Hohlwände, Holz, Leichtmetall	kombiniert/direkt	innen/Rückseite	Formschluß
Nieten	Stahl, Hohlwände, Leichtmetall	kombiniert	Rückseite	Formschluß
Einlegen	Beton	einlegen	innen	Formschluß
Nageln	Holz	direkt	innen	Reibschluß, Formschluß
Dübeln	Beton, Mauerwerk, Holz, Hohlwände	kombiniert	innen	Reibschluß, Formschluß, Stoffschluß
Bolzensetzen	Beton, Stahl	direkt/kombiniert	innen	Stoffschluß Reibschluß
Klemmen	Stahl, Leichtbeton	direkt	Rückseite	Formschluß, Reibschluß
Schweißen	Stahl, Leichtbeton	direkt	Vorderseite	Stoffschluß
Kleben	Stahl, Beton, Holz, Leichtmetall	direkt	Vorderseite	Stoffschluß

In allen Fällen kombinierter oder wechselnder Wirkung ist in Tabelle 3 nur die wichtigste Funktionsweise eingetragen.
Für die Anwendung eines Befestigungsverfahrens sind die 3 übrigen Merkmale aus Tabelle 3 ebenfalls wichtig. Zum einen sind die Verfahren auf bestimmte Untergrundmaterialien beschränkt. Zum anderen erzwingt die Verfahrensart (Direktverfahren, kombinierte Verfahren und Einlegeverfahren) eine entsprechende Anpassung des Bau- und Befestigungsablaufes. Bei Einlegeverfahren beispielsweise müssen die Befestigungselemente, wie Kopfbolzen oder Ankerschienen, bereits bei der Erstellung des Untergrundes angebracht werden, während das Anschlußteil erst viel später befestigt wird. Einlegeverfahren erfordern daher eine frühzeitige Vorausplanung und genaue Festlegung der Befestigungspunkte. Kombinierte Verfahren bestehen aus mehreren hintereinander ablaufenden Schritte, z. B. dem Bohren eines Loches, dem Einbringen und Verspreizen des Dübels und dem Anschrauben des zu befestigenden Bauteils. Dagegen wird bei den Direktverfahren, z. B. beim Nageln, das zu befestigende Bauteil ohne Vorbereitung des Untergrundes in einem Schritt befestigt.
Bei der Auswahl eines Verfahrens und der Wahl der Befestigungsmittel sind verschiedene Betrachtungen anzustellen, für welche Tabelle 4 als Checkliste dienen kann. Der Planer wird geneigt sein, sich vornehmlich mit den Anforderungen an die Befestigung zu beschäftigen und ihnen die entsprechenden Eigenschaften der Befestigungen gegenüberzustellen.
Aber auch der Wirtschaftlichkeit muß eine gewichtige Rolle zugemessen werden, obwohl der Anteil der Befestigungen an den Baukosten weniger als 1% ausmacht. Dabei müssen neben dem Preis des Befestigungsmaterials der Zeitaufwand für Planung, Vorbereitung und Ausführung und die Kosten für Abschreibung und Instandhaltung von Geräten und Werkzeugen berücksichtigt werden. Diese können von Verfahren zu Verfahren und von Produkt zu Produkt stark variieren.
Für die praktische Anwendung sind die Merkmale der Verfahren und die Verfügbarkeit von Befestigungselementen an der Baustelle zu bedenken. Verfahren mit hohem Lärmpegel beispielsweise sind bei Umbauar-

Tabelle 4
Kriterien für die Auswahl von Befestigungsverfahren und -mitteln

Untergrund / zu befestigendes Teil	
Beanspruchung / Anforderungen	Lasten (stat., dyn., Schock) Lastrichtung Abstände (Rand-, Achs-) Risse Temperatur Korrosion
Eigenschaften der Befestigung	
Merkmale des Befestigungs-verfahrens	Antriebsenergie Arbeitssicherheit Handhabung/Platzbedarf Umweltbelastung Arbeitskomfort
Wirtschaftlichkeit	Materialkosten Personalkosten Amortisation und Verzinsung von Geräten/Werkzeugen Instandhaltung von Geräten/Werkzeugen
Verfügbarkeit	Liefertermine Zugang zu Bezugsquellen Reparaturservice für Geräte
Beratung, Haftung, Schulung	

beiten in bewohnten Gebäuden nicht erwünscht. Die Unabhängigkeit von Stromquellen kann im Rohbau oder im Freigelände ein gewichtiger Vorteil für batteriebetriebene Geräte oder kartuschenbetriebene Bolzensetzgeräte sein.
Häufig übersehen bei der Auswahl werden Aspekte der Schulung, der Beratung, des Geräteservices und der Haftung, welche vom Lieferanten geboten werden. Für den optimalen Einsatz von Befestigungsverfahren ist die Beratung zur Erzielung einer richtigen Befestigung und das Training des ausführenden Personals ebenso wichtig wie der angebotene Reparaturdienst. Die vom Lieferanten übernehmbare Haftung kann insofern nicht außer acht gelassen werden, als Schäden infolge mangelhafter Befestigungen bis zu 10% und mehr der gesamten Bausumme ausmachen können, obwohl die Befestigungen selbst weniger als 1% der Bausumme kosteten. Mit diesem Hinweis sei nochmals die Bedeutung des richtig gewählten Befestigungsverfahrens und der richtig bemessenen und richtig ausgeführten Befestigung unterstrichen.

4 Bolzensetzen

Bolzensetzen ist ein weitverbreitetes Befestigungsverfahren für die Untergrundmaterialien Stahl und Beton, das wegen seiner Unabhängigkeit von Stromquellen sehr flexibel verwendbar und zudem wirtschaftlich, schnell und vergleichsweise umweltfreundlich ist. Ursprünglich mit hohen Gefahren für den Bedienenden und für die Umgebung belastet, wurde das Verfahren in den 50er und 60er Jahren durch die Einführung des Kolbenantriebes, die Entwicklung hochfester und gleichzeitig zäher Setzbolzen und die Einführung von Geräten mit Einhandbedienung praxisgerecht und sicher gemacht. Bei Betonuntergrund allerdings sind gewisse Nachteile geblieben, die den Verwendungsbereich einschränken.

4.1 Verfahrensprinzip und Produkte

Befestigungssysteme zum Bolzensetzen bestehen aus Geräten, Kartuschen als Energieträger und Nägeln oder Bolzen, deren Kopf in der Regel mit einem Gewinde zum Anschrauben des zu befestigenden Bauteils versehen ist. Bei den älteren Geräten nach dem Schußprinzip (Bild 1) wird die Kartuschenenergie direkt auf den Setzbolzen übertragen [1], der nach Beschleunigung im Lauf beim Auftreffen auf die Oberfläche des zu befestigenden Materials oder des Untergrundes eine Geschwindigkeit von 250 bis 500 m/sec und eine kinetische Energie von 300–1000 Joule aufweist. In dieser hohen Energie liegt eine inherente Gefahrenquelle für den Bedienenden und die Umgebung, der man durch Schutzkappen und weitere Sicherheitseinrichtungen am Gerät Rechnung zu tragen versucht.
Ganz im Gegensatz dazu wird bei den Kolbengeräten, von denen 3 verschiedene Ausführungsprinzipien vorkommen (Bild 1), die Energie der Kartusche auf einen Kolben übertragen, der seinerseits den Setzbolzen mit Geschwindigkeiten von weniger als 100 m/sec in den Untergrund eintreibt. Die kinetische Energie des Setzbolzens ist dabei mit 20–50 J vergleichsweise gering und entsprechend niedrig ist die Unfallgefahr. Der Kolben selbst, der rund mit 95% der gesamten kineti-

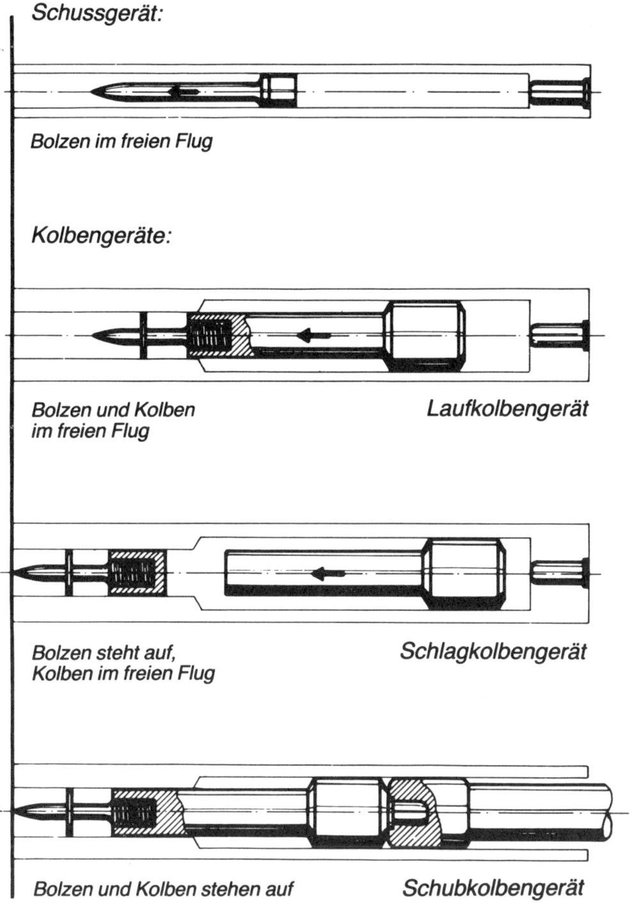

Bild 1
Typen von Bolzensetzgeräten, prinzipielle Darstellung, aus [1]

schen Energie beaufschlagt wird, kann das Gerät nicht verlassen und stellt somit keine Gefahr für die Umgebung dar.
Um eine gute Befestigung zu erreichen, müssen die Komponenten des Systems (Gerät, Kartusche und Setzbolzen) in ihren Eigenschaften aufeinander abgestimmt sein. Insbesondere muß die Energie der gewählten Kartusche dem Eintreibwiderstand entsprechen, welcher dem Setzbolzen beim Durchdringen des zu befestigenden Bauteils und beim Eindringen in den Untergrund entgegenwirkt. Dieser Forderung wird durch die Verwendung unterschiedlich starker Kartuschen und durch eine Leistungsregulierung am Gerät Rechnung getragen, welche durch Regulierung des Brennraumes verwirklichbar ist.
Verschiedene, eingebaute Sicherheitseinrichtungen, die der DIN 7260 [2] entsprechen müssen, verhindern ein unabsichtliches Auslösen und Abfeuern in den freien Raum. Zur Qualität der Befestigung, aber auch zur erhöhten Arbeitssicherheit, trug die Entwicklung von zähen Setzbolzen wesentlich bei, welche an Zuschlagstoffen im Beton oder beim versehentlichen Auftreffen auf Kanten von Stahlprofilen nicht brechen, sondern abgebogen werden.

4.2 Anwendungsbereiche

Setzbolzen sind für Befestigungen auf Baustählen mit Festigkeiten unter 650 N/mm^2, auf Stahlguß, auf Schwerbeton, Leichtbeton und auf Mauerwerk geeignet, sofern dieses mit einem Zementmörtel verputzt ist. Die zulässigen Lasten pro Befestigung liegen im Bereich von 1,7 bis 5 kN auf Stahl und von 200 bis 700 N bei Beton. Bolzensetzen ist auf Beton ein Verfahren für leichte Befestigungen, auf Stahl für leichte bis mittlere Befestigungen.
Im Direktverfahren, also in einem Arbeitsgang, können Bauteile am Untergrund angenagelt werden, wobei ausreichend zähe Bauteile nicht einmal vorgebohrt sein müssen. Diese einfache Handhabung, verbunden mit Schnelligkeit, Umweltfreundlichkeit und Wirtschaftlichkeit ist der Hauptvorteil des Verfahrens. Im kombinierten Verfahren kann zunächst ein Bolzen gesetzt werden, an dessen Kopf in einem zweiten Schritt das zu befestigende Teil angeschraubt oder eingehängt wird. Diese Befestigungen lassen sich nachträglich wieder lösen.
Bolzensetzen findet Anwendung bei der Befestigung von Blechen, insbesondere Profilblechen (Bild 2) [3],

Bild 2
Befestigung von Trapezblechen mit Setzbolzen

von Holzteilen wie Latten, Brettern, Platten etc., von Isoliermaterialien, von Putzträgern und von Profilen oder Bauteilen aus Stahl und Aluminium. Auch zum Verbinden von Teilen kann das Verfahren eingesetzt werden, beispielsweise bei Verbundkonstruktionen zur Verbindung des Stahlträgers mit der Betondecke. Hierbei werden die Schubkräfte über speziell entwickelte Verbundbügel und 2 Setzbolzen übertragen (Bild 3).
Die Anwendung des Verfahrens auf Stahl wird begrenzt durch den Eindringwiderstand, der mit zunehmender Festigkeit des Stahls und wachsender Bauteildicke (Flanschdicke) so groß wird, daß die Setzbolzen beim Eindringen abscheren (Bild 4). Lange Nägel, wie sie für das Befestigen von dickeren Holz- oder Isoliermaterialien verwendet werden, knicken bereits vor Erreichen der Scherfestigkeit aus. Dadurch wird die Anwendungsgrenze aus Bild 4 in Richtung kleinerer Festigkeitswerte des Grundmaterials und geringerer Flanschdicken herabgesetzt.
Befestigungen sind innerhalb der abgesteckten Anwendungsgrenzen einwandfrei, wenn die Stahlbauteile, auf denen befestigt wird, mindestens 6 mm dick sind. Sind sie dünner, können Befestigungen auch möglich sein, jedoch muß dies im Einzelfall geprüft werden.
Was die Begrenzung der Anwendungen auf dem Untergrundwerkstoff Beton anbetrifft, so spielen verschiedene Faktoren eine Rolle, die sich in 2 Gruppen zusammenfassen lassen: die große Streuung der Traglast und das schlechte optische Aussehen von Befestigungen bzw. Fehlbefestigungen (Ausfälle).
Wie in Abschnitt 4.4 noch gezeigt werden wird, ist der Variationskoeffizient der Traglast relativ groß, weshalb immer mit sehr kleinen Werten und sogar mit dem Wert 0 gerechnet werden muß. Diese große Streuung schließt auf Beton die Einzelbefestigung nahezu aus und beschränkt dort die Anwendung des Verfahrens auf Mehrfachbefestigungen mit linien- und flächenartiger Anordnung der Befestigungspunkte (z. B. Schienen, Latten, abgehängte Decken).

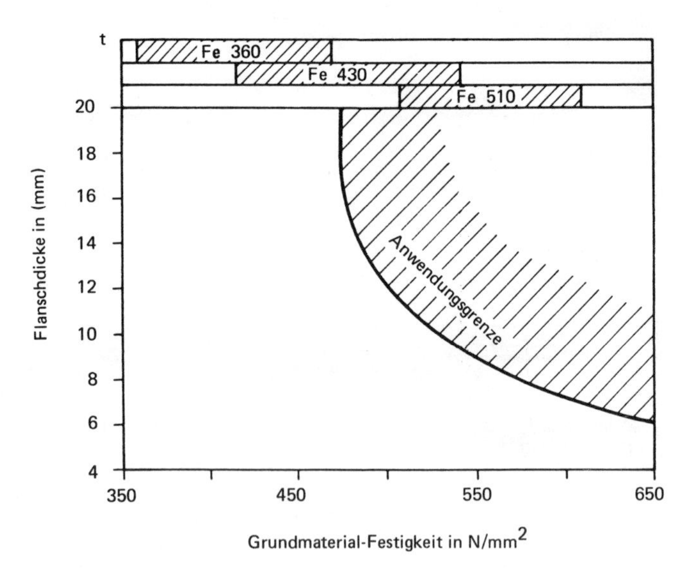

Bild 4
Anwendungsgrenze von Setzbolzen auf den Baustählen Fe 360, Fe 430 und Fe 510

Bild 3
Befestigen von Verbundbügeln und verlorener Schalung auf Stahlträgern, vor dem Vergießen der Betondecke

Beim Eintreiben des Setzbolzens wird Material in die Poren und Hohlräume des Betons verdrängt. Sind nicht genügend Poren vorhanden oder ist der Beton zu spröde, so kommt es zu Rissen und zu „Ausplatzungen" an der Oberfläche. Die Befestigung hat dann optisch ein schlechtes Aussehen (was allerdings bei Rohbauten und bei später nicht mehr sichtbaren Befestigungen nicht stören muß) oder das Befestigungselement ist im Untergrund nicht verankert (Ausfall). Hier sind dem Verfahren Grenzen gesetzt, die von der Verdichtung des Betons, seiner Festigkeit, der Härte seiner Zuschlagstoffe, seinem Alter und dem Wasser-/Zementfaktor bei der Betonherstellung abhängen.
Auf einen Vorteil des Bolzensetzens im Vergleich zum Dübeln in Bohrlöchern soll speziell hingewiesen werden. Werden Armierungseisen getroffen, so kann sich der Setzbolzen im Armierungseisen verankern. Zwar entsteht hierdurch eine Kerbe, aber wegen der eingetretenen Verfestigung des Armierungsstahls ist die Abminderung seines Lastverhaltens praktisch vernachlässigbar.

4.3 Tragverhalten in Stahl

Beim Eintreiben verdrängt der Schaft des gehärteten Setzbolzens das Untergrundmaterial Stahl in einem Kaltfließpreßvorgang nach der Seite, nach vorne und auch entgegen der Eintreibrichtung. Die starke Erwärmung an der Grenzfläche von Setzbolzen und Stahluntergrund, die ungefähr 900 °C erreicht, führt zum örtlichen Verlöten und Verschweißen. Zusätzlich kommt eine ringförmige Zugspannung im Stahluntergrund zustande, die zu hohen radialen Druckspannungen auf die Oberfläche des Setzbolzens führt. Bei achsialer Zugbelastung wird somit der Bolzen auch durch Reibung im Untergrund festgehalten. Zudem werden Setzbolzen für Stahl mit Kerben oder Randrierungen am Schaft versehen, welche einerseits die örtliche Verschweißung fördern, andererseits einen zusätzlichen Formschluß ergeben.

Das Tragverhalten wird somit in achsialer Richtung durch die kombinierte Wirkung von Reibschluß, Formschluß und Stoffschluß (Verschweißungen, Verlöten) zustandegebracht. Durch diese günstige Funktionsweise kommen bei Längs-, Schräg- und Querzugbeanspruchungen relativ hohe Versagenswerte für die Verankerung zustande, verbunden mit einer nur geringen Streuung (Bild 5). Aus einigen tausend Einzelwerten stammen die in Bild 6 eingetragenen Versagens-Mittelwerte $\bar{x}$ und die 5%-Fraktilenwerte. Aus diesen Werten errechnet sich für die zylindrische Grenzfläche zwischen Setzbolzen und Stahlträger eine Scherfestigkeit von 150 N/mm². Die in der Praxis vorkommenden Abstände zwischen benachbarten Setzbolzen sind groß

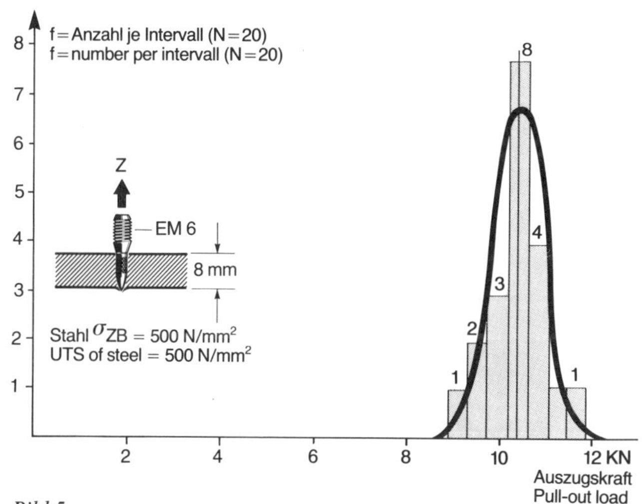

Bild 5
Häufigkeitsverteilung der Versagenswerte von Setzbolzen in Stahl, aus [4]

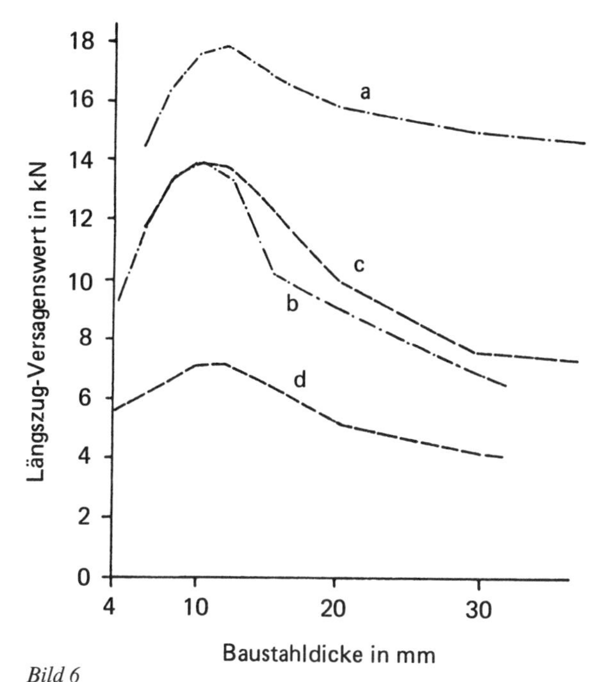

Bild 6
Versagenswerte von Setzbolzen bei Längszug in Baustahl unterschiedlicher Dicke
a) Mittelwerte für Setzbolzen mit Schaftdurchmesser von 4,5 mm
b) 5%-Fraktilenwerte für Setzbolzen mit Schaftdurchmesser von 4,5 mm
c) Mittelwerte für Setzbolzen mit Schaftdurchmesser von 3,7 mm
d) 5%-Fraktilenwerte für Setzbolzen mit Schaftdurchmesser von 3,7 mm

genug, so daß die Traglast durch die gegenseitige Beeinflussung der Setzbolzen nicht herabgesetzt wird. Auch bei Befestigungen in Randnähe bleibt die volle Traglast erhalten, sofern der Abstand vom Rand mindestens 10 mm beträgt, was üblicherweise ohnehin der Fall ist.

Bei statischer Querbeanspruchung versagt nicht die Verankerung im Stahluntergrund, sondern der Setzbolzen oder das zu befestigende Bauteil, z. B. wegen Überschreiten des erträglichen Lochleibungsdruckes. Je nach Lage des Kraftangriffspunktes ist bei Querbeanspruchung die Scherfestigkeit des Setzbolzens, deren 5%-Fraktilenwert im Bereich von 1,1 kN/mm² liegt, oder seine Biegefestigkeit (ca. 3,4 kN/mm²) maßgebend.

Für dynamische Belastungen sind, wie beim statischen Querzug, nicht die Verankerung im Stahluntergrund, sondern die Eigenschaften des Setzbolzens entscheidend. Bild 7 zeigt die Mittelwerte (WS = 50%), die 5%-Fraktilenwerte und die daraus errechneten zulässigen Lasten für die achsiale Zugbeanspruchung. Bei dynamischen Querbeanspruchungen liegen die ent-

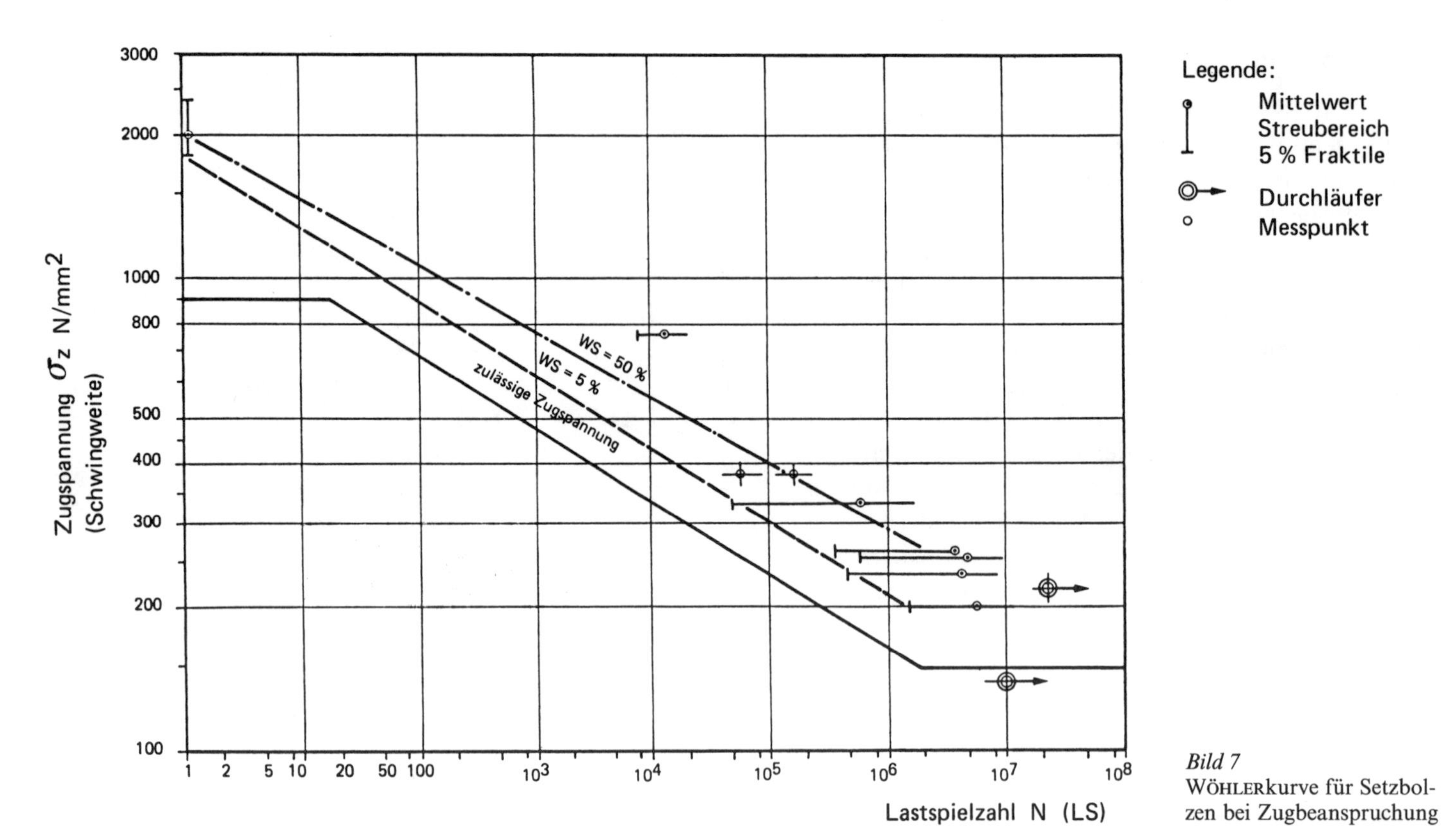

Bild 7
WÖHLERkurve für Setzbolzen bei Zugbeanspruchung

sprechenden Werte etwas höher, so daß man mit den Werten aus Bild 7 auf der sicheren Seite liegt.
Eine völlig andere Fragestellung bezieht sich auf die Beeinträchtigung des Tragverhaltens des Stahluntergrundes (Träger, Stützen etc.) durch das Anbringen einer Befestigung [4]. Entsprechende Forschungsarbeiten zeigen, daß grundsätzlich alle Befestigungsverfahren eine Abminderung des Tragverhaltens des Stahluntergrundes herbeiführen, jedoch diese beim Bolzensetzen am geringsten ist. Sie entspricht in etwa der Wirkung einer Stumpfschweißnaht. Dieses günstige Verhalten des Bolzensetzens ist damit zu erklären, daß in der Umgebung des eingetriebenen Setzbolzens eine Gefügeverfestigung des Stahluntergrundes stattgefunden hat, welche die Kerbwirkung des durch den Setzbolzen erzeugten Loches und die damit verbundene Querschnittsverminderung teilweise kompensiert. Dieses vorteilhafte Verhalten ist bei schwingungsbeanspruchten Stahlkonstruktionen von Bedeutung, wie z. B. bei Brücken, Kränen, Großgeräten, Fahrzeugen, Schiffen und Offshore-Konstruktionen.

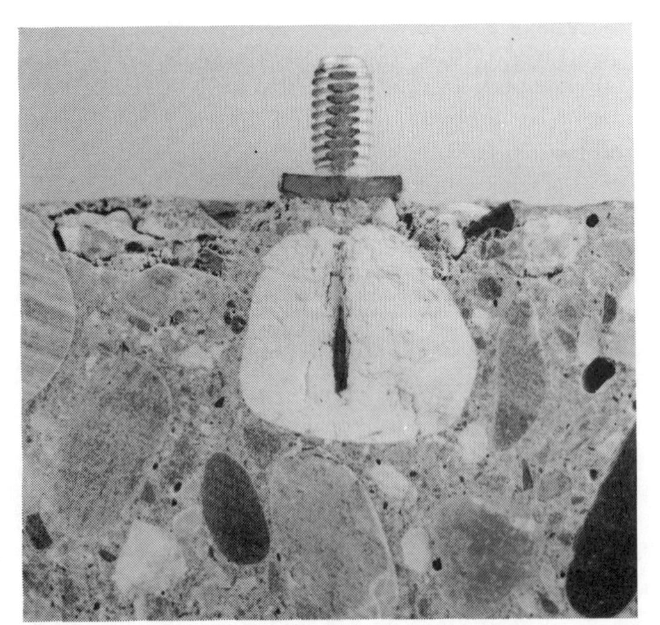

Bild 8
Durchdringen eines Setzbolzens durch weiche Zuschlagstoffe von Beton

4.4 Tragverhalten in Beton

Beim Eintreiben in Beton durchdringt der Setzbolzen die weicheren Betonbestandteile, wie Zementstein und weiche Zuschlagstoffe (Bild 8) und verdrängt deren Material nach der Seite in vorhandene Poren und Hohlräume. Dabei entstehen im Bereich der Bolzen-

spitze sehr hohe Temperaturen, bei denen ein Versintern des Betons mit der stark aufgerauhten Stahloberfläche des Setzbolzens stattfindet. Bei achsialer Zugspannung solcher Bolzen versagt nicht die Grenzfläche, sondern der Beton (Bild 9). Die Versagenswerte liegen zwischen 5 bis 10 kN.
Härtere Zuschlagstoffe und solche, welche unter ungünstigem Auftreffwinkel getroffen werden, können dagegen vom Setzbolzen nicht durchdrungen werden. Dieser wird vielmehr abgelenkt und gekrümmt. Dadurch entsteht zwar eine formschlüssige Verbindung. Sie ist jedoch ohne Nutzen, weil die sehr hohen Spannungen zu Rissen im Beton führen, welche dessen Tragfähigkeit stark herabsetzen. Bei einem gewissen Anteil ist die Resttragfähigkeit des Betons sogar auf den Wert 0 herabgesetzt (Ausfälle beim Setzen). Insgesamt ergibt sich aus diesem Grunde eine sehr starke Streuung der Versagenswerte für Längs-, Quer- und Schrägzug (Variationskoeffizient ca. 0,4, Bild 10).
Auf Beton sind daher Einzelbefestigungen nicht zu empfehlen, es sei denn jede Befestigung wird nachträglich kontrolliert. Mehrfachbefestigungen dagegen, bei denen eine Umlagerung der Kräfte zwischen den Befestigungspunkten stattfindet, sind sehr wohl möglich, da die Mittelwerte beim Bolzensetzen in Beton eine beachtliche Höhe erreichen. Je nach Setztiefe liegen sie bei 2 bis 10 kN. Wegen der großen Streuung der

Bild 9
Bruchkegel mit vorstehender Bolzenspitze beim Auszugsversuch in Beton

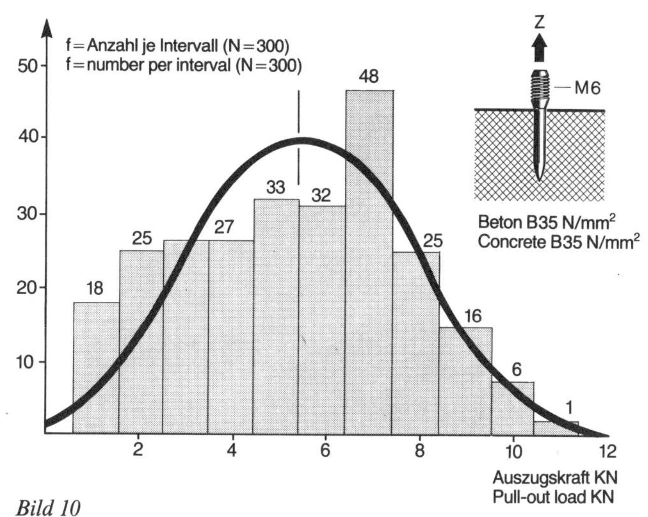

Bild 10
Häufigkeitsverteilung der Versagenswerte auf Beton, aus [4]

Einzelversagenswerte kommt es jedoch bei Mehrfachbefestigungen immer zu einer Überlagerung der beiden Verteilungskurven von Versagenswert und Belastung, weshalb übliche Sicherheitsbetrachtungen für eine Bemessung nicht mehr anwendbar sind [5]. Mit einem stochastischen Modell gelingt es dagegen, dem schritt-

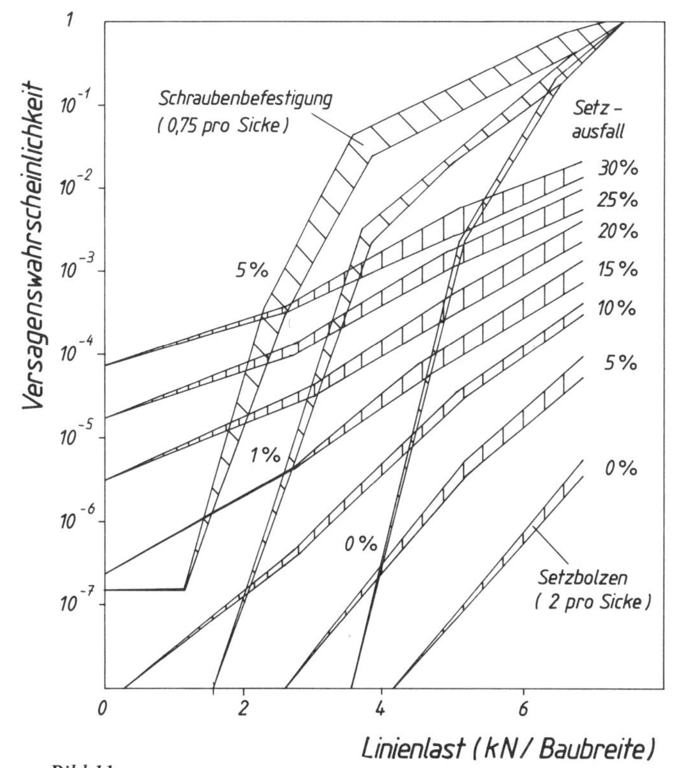

Bild 11
Wahrscheinlichkeit des totalen Versagens von Mehrfachbefestigungen mit Setzbolzen auf Beton und Schrauben in eingelegten Ankerschienen

Tabelle 5
Zulässige Lasten bei Mehrfachbefestigungen mit Setzbolzen auf Beton

Zulässige Tragkraft pro Befestigungselement (BE) in (N)

Befestigte Materialien	zugrunde gelegte Setzausfallrate %	Mindestanzahl BE / Linie oder Knoten ≥ 5	≥ 8
Abgehängte Befestigungen	0	600	700
Stahl- und Aluteile Dicke ≥ 1 bzw. 1,2 mm	≤ 25	450	650
Stahlbleche Dicke ≥ 0,71 mm Alubleche Dicke ≥ 0,90 mm	≤ 25	300	450
Stahlbleche Dicke ≥ 0,60 mm Alubleche Dicke ≥ 0,80 mm	≤ 25	200	300
Holzteile, Dicke ≥ 24,00 mm	≤ 25	400	600

weisen Versagen einzelner Befestigungspunkte mit zunehmender Last und dem jeweils erfolgenden Umlagern auf die restlichen, noch gesunden Befestigungspunkte Rechnung zu tragen [6, 7]. In einem solchen Modell kann auch die Tatsache berücksichtigt werden, daß ein nicht unbeträchtlicher Anteil der Befestigungen bereits vor dem Aufbringen der äußeren Last ausgefallen ist, d. h. den Versagenswert 0 hat.
Das Ergebnis einer solchen stochastischen Betrachtung und Berechnung ist in Bild 11 dargestellt, für den Fall der Befestigung von Profilblechen direkt auf Beton, ohne die praxisübliche Verwendung von eingelegten Stahlplatten. Zum Vergleich wurde dieselbe Betrachtung auch für Schrauben angestellt, welche wegen ihrer weit kleineren Streuung der Versagenswerte mit der üblichen Sicherheitsbetrachtung behandelt werden können. Durch einen solchen Vergleich gewinnt man Anschluß an Erfahrungswerte von bewährten Verfahren und kann die errechneten Versagenswahrscheinlichkeiten besser beurteilen. Es sei noch auf ein interessantes Nebenergebnis dieser Vergleichsbetrachtung hingewiesen. Obwohl man bei Schraubenverbindungen üblicherweise mit der Ausfallrate 0 rechnen kann, läßt sich nicht ausschließen, daß durch unsachgemäßes Verlegen ein Fehleranteil von 1% entstehen könnte. Beim Anstieg von Ausfall 0 auf Ausfallrate 1% erhöht sich jedoch im Falle der Schraubenverbindung die Versagenswahrscheinlichkeit bereits um den Faktor 100 bis 1000, während das Bolzensetzen, bei dem ohnehin mit einem hohen Anfangsausfall gerechnet werden muß, hinsichtlich Montagefehler praktisch unempfindlich ist.
Für die verschiedenen Anwendungsfälle führen die stochastischen Berechnungen zu unterschiedlichen Ergebnissen. Das Resultat ist jedoch in jedem Fall sehr einfach, nämlich eine zulässige Last, bezogen auf den einzelnen Befestigungspunkt (Tabelle 5).
Bei statischen und dynamischen Querzugbeanspruchungen kommt es entweder zum Versagen des befestigten Bauteils oder zu Verformungen des Betons und des Setzbolzens, dessen erträgliches Maß die zulässige Beanspruchung bestimmt. Bei statischen und dynamischen Querzugbeanspruchungen mit Versagen des Betons bzw. des Setzbolzens sind jedoch die Traglasten bei Querbeanspruchung höher als bei statischer Längszugbeanspruchung, so daß man bei Bemessungen mit den Längszugwerten immer auf der sicheren Seite liegt.
Im Gegensatz zum Bolzensetzen auf Stahl müssen beim Setzen auf Beton die Achs- und Randabstände beachtet werden. Der minimale Randabstand, bei dem man mit Sicherheit das Absprengen der freien Kante verhindert, beträgt 70 bis 100 mm. Bei Mehrfachbefestigungen mit Achsabständen unterhalb dieses Wertes muß auch eine Abminderung der Traglast pro Befestigungspunkt berücksichtigt werden.

5 Literatur

[1] Seghezzi, H. D.: Untersuchung der Bewegungsenergie als Ursache der Unfallgefahr beim Bolzensetzen. Sichere Arbeit, 17 (1964), S. 5–9.

[2] DIN-7260: Ausgabe 1974, Teil 1, Bolzensetzwerkzeuge – Begriffe, Konstruktion, Kennzeichnung.

[3] Seghezzi, H. D., Beck, F. und Thurner, E.: Profilblechbefestigung mit Setzbolzen – Grundlage und Anwendung. Der Stahlbau, 47 (1978), S. 225–233.

[4] Seghezzi, H. D.: Wirtschaftliche und sichere Befestigungssysteme für die Baupraxis. Betonwerk und Fertigteiltechnik, 1 (1983), S. 41–46 und 2 (1983), S. 117–123.

[5] Seghezzi, H. D.: Zur Frage der Zuverlässigkeit und Sicherheit bei Direkt- und Dübelbefestigungen. Schweizer Archiv, 11 (1971), S. 1–12.

[6] Wälti, P. und Thürlimann, B.: Probalistic Safety Model for the Design of Fastening System. Proceedings of Icossar, 81 (1981), S. 359–370.

[7] Patzak, M.: Zur Frage der Sicherheit von Setzbolzenbefestigungen in Betonteilen. Betonwerk und Fertigteiltechnik, 5 (1979), S. 308–314.

Wechselbeziehungen zwischen Befestigungstechnik und Stahlbetonbauweise

Rolf Eligehausen

1 Problemstellung

Die moderne Befestigungstechnik hat sich in den letzten 20 Jahren stürmisch entwickelt und ist in nahezu alle Bereiche des Bauwesens eingedrungen.
Die hauptsächlich verwendeten Befestigungsmittel sind in den Bildern 1 bis 3 dargestellt. Metallspreizdübel und Verbundanker (Bild 1) sowie Hinterschnittanker (Bild 2) werden in nachträglich erstellte Bohrlöcher eingesetzt und verankert. Die Verankerung erfolgt bei Dübeln durch Aufspreizen der Hülse, wobei Spreiz- und damit Haltekräfte geweckt werden. Man unterscheidet kraftkontrolliert spreizende Dübel, die durch Aufbringen eines Drehmomentes verankert werden und die bei Belastung nachspreizen können, sowie wegkontrolliert spreizende Dübel, die durch Einschlagen eines Konus in die Hülse bzw. Auftreiben der Hülse auf den Konus verankert werden und die nicht nachspreizen können. Bei Verbundankern erfolgt die Verankerung durch Vermörteln der Ankerstange mit

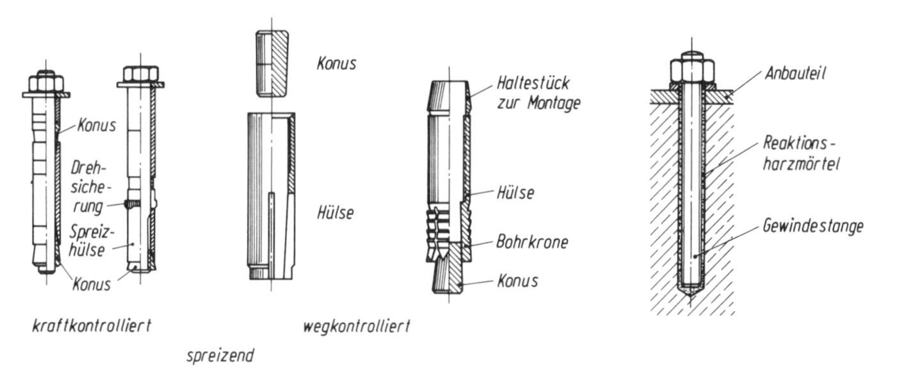

Bild 1
Nachträglich gesetzte Befestigungsmittel – Metallspreizdübel und Verbundanker

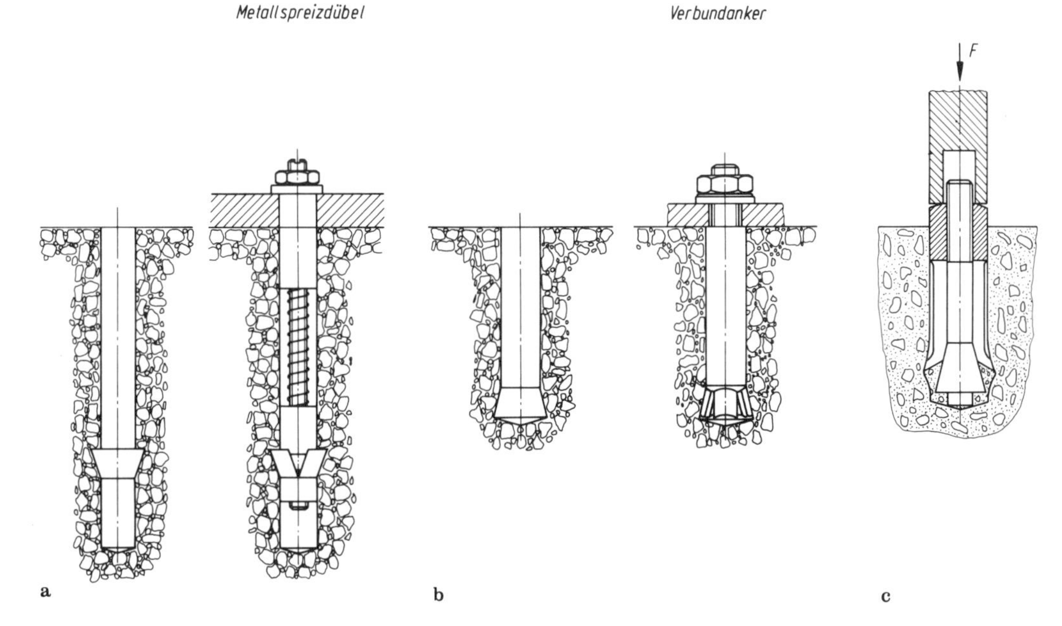

Bild 2
Nachträglich gesetzte Befestigungsmittel – Hinterschnittanker

Reaktionsharzmörtel. In den letzten Jahren wurden verschiedene Typen von Hinterschnittankern entwikkelt (Bild 2), die die Kräfte überwiegend durch mechanische Verzahnung infolge Hinterschneidung des Betons in den Ankergrund einleiten. Dabei kann die Hinterschneidung durch einen speziellen Bohrvorgang (Bild 2a, b) bzw. beim Setzen des Ankers (Bild 2c) erzeugt werden. Ankerschienen und Kopfbolzen (Bild 3) werden in die Schalung eingelegt und einbetoniert.

Die Anwendungsbedingungen für Befestigungselemente werden bisher in der Regel aus Versuchen abgeleitet, bei denen der Beton außer den Verankerungslasten keine sonstigen Lasten zu tragen hat und ungerissen ist. Dabei wird das Versagen der Befestigung durch Herausziehen (nur Dübel und Verbundanker), kegelförmigen Betonausbruch, Spalten des Betonkörpers oder Versagen des Stahls hervorgerufen (Bild 4).

Die häufig auftretende Versagensart „Betonausbruch" stellt die obere Grenze der Betontragfähigkeit dar. Der Ausbruchkegel (Bild 5) zeigt, daß Befestigungselemente örtlich die Zugfestigkeit des Betons ausnutzen. Die Betonzugfestigkeit wird jedoch in manchen Fällen auch von dem als Ankergrund dienenden Stahlbetonbauteil in Anspruch genommen. Dies gilt zum Beispiel für Stöße der Bewehrung durch Übergreifung (Bild 6) oder im Schubbereich von auf Schub unbewehrten Platten.

Ordnet man daher Befestigungselemente in Bereichen an, in denen hohe Zugspannungen im Beton aus der Tragwerkswirkung vorhanden sind, beeinflussen sich Befestigungselement und das als Ankergrund dienende Stahlbetonbauteil gegenseitig im Tragverhalten. Dabei kann sowohl die Tragfähigkeit der Verankerung als auch diejenige des Stahlbetonbauteils reduziert werden. Weiterhin können durch die Verankerungslasten oder durch sonstige Beanspruchungen Risse im Beton

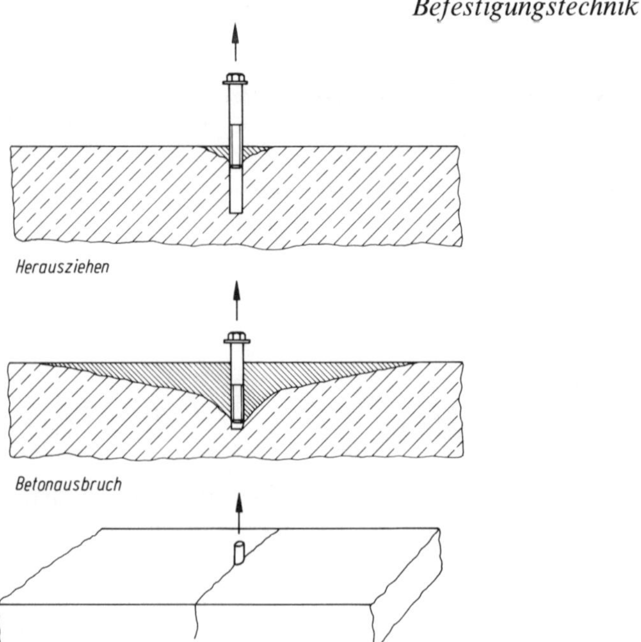

Bild 4
Versagensarten für Befestigungselemente

Bild 5
Typischer Ausbruchkegel eines Hinterschnittankers

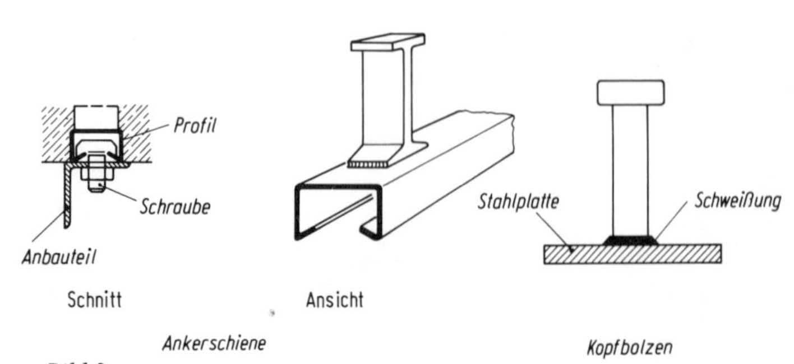

Bild 3
In die Schalung eingelegte Befestigungselemente – Ankerschienen und Kopfbolzenplatten

Bild 6
Bruch eines Übergreifungsstoßes ohne Querbewehrung (entnommen aus [12])

hervorgerufen werden, die das Tragverhalten der Befestigungselemente beeinträchtigen. Im folgenden werden diese Problemkreise an einigen typischen Beispielen erläutert und eventuell notwendige Konsequenzen für die Bemessung der Befestigungen bzw. die konstruktive Gestaltung der Stahlbetonbauteile aufgezeigt.

2 Verankerungen in der Druckzone

Bisher dürfen Verankerungen – von einigen Ausnahmen abgesehen – nur in der aus Lastspannungen erzeugten Druckzone von Stahlbetonbauteilen angeordnet werden. Dabei wird davon ausgegangen, daß nur in Haupttragrichtung des Stahlbetonbauteils eine Druckzone vorliegen muß. Diese einschränkende Vorschrift ist jedoch nicht ausreichend, um in allen Fällen eine ungünstige Beeinflussung des Tragverhaltens der Befestigung durch das Stahlbetonbauteil auszuschließen [1]. Die Begründung für diese Aussage und die notwendigen Konsequenzen werden am Beispiel der Verbundanker erläutert.

Einzellasten erzeugen bei Flächentragwerken Biegemomente in beiden Achsrichtungen, wobei bei zentrischer Zugbeanspruchung der Befestigung die Momente in Querrichtung mindestens ca. 80% der Momente in Haupttragrichtung betragen. Auch wenn in Haupttragrichtung eine Druckzone vorliegt, treten bei Wänden in Querrichtung Zugspannungen und damit möglicherweise Risse im Beton im Bereich der Befestigung auf (Bild 7). Bei Verankerungen in der Druckzone von überwiegend auf Biegung beanspruchten Bauteilen kann außerdem ein relativ großer Teil der Verankerungslänge in der Zugzone bzw. im gerissenen Beton liegen (Bild 8).

In den dargestellten Fällen ist mit einer signifikanten Abminderung der Tragfähigkeit der Verbundanker zu rechnen. Dies ist aus Bild 9 zu ersehen. Aufgetragen ist das Verhältnis der Bruchlast von auf zentrischen Zug beanspruchten Verbundankern im Riß zu der in Vergleichsversuchen gemessenen Bruchlast im ungerissenen Beton in Abhängigkeit von der Rißbreite auf der Bauteiloberseite. Die Versuche wurden an Biegeplatten und Dehnkörpern durchgeführt. Zunächst wurden

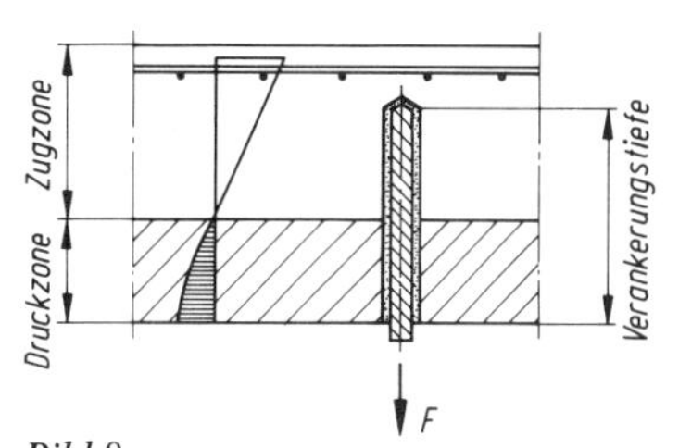

Bild 8
Verankerungen durch die Druckzone hindurch (nach [1])

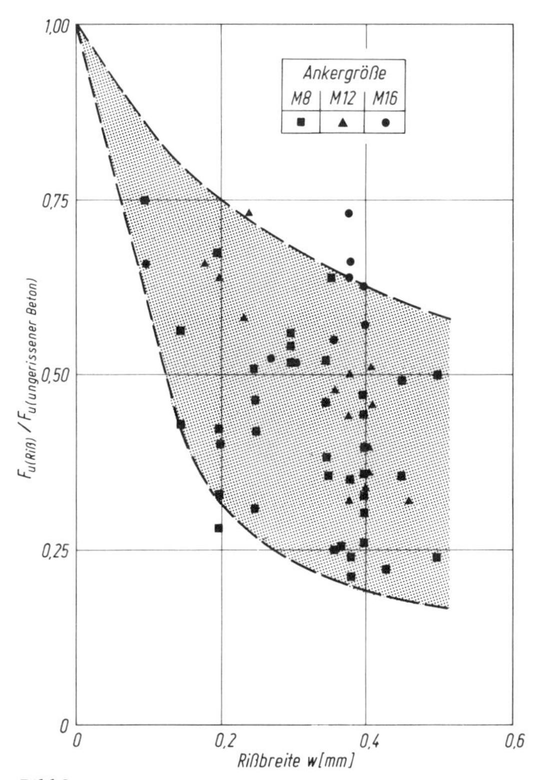

Bild 9
Verhältnis der Bruchlast von in Rissen angeordneten Verbundankern zur Bruchlast im ungerissenen Beton in Abhängigkeit von der Rißbreite (nach [2])

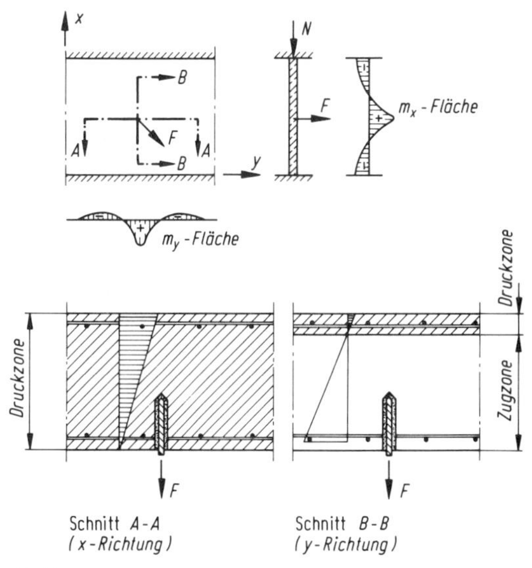

Bild 7
Verankerungen mit Verbundankern in einer Wand (nach [1])

Haarrisse erzeugt, die Anker in diese Risse gesetzt, die Risse durch Belasten der Probekörper aufgeweitet und anschließend die Anker bei geöffnetem Riß bis zum Bruch belastet. Bei einigen Ankern wurden 10^5 Lastwechsel etwa im Bereich der Gebrauchslast zwischengeschaltet. Die Versuchsergebnisse streuen sehr stark. Dies ist auf den zufälligen Verlauf des Risses über den Bohrlochumfang und über die Einbettungstiefe zurückzuführen. Die Tragfähigkeit der Anker im Riß nimmt mit zunehmender Rißbreite schnell ab. Sie beträgt nach den wenigen durchgeführten Versuchen bei der im Stahlbetonbau maximal als zulässig angesehenen Rißbreite von 0,4 mm nur das ca. 0,2- bis 0,6fache des im ungerissenen Beton zu erwartenden Wertes. Bei Lastwechseln sind noch geringere Tragfähigkeiten zu erwarten.

Um in den o. g. Anwendungsfällen (Bilder 7 und 8) trotz der Empfindlichkeit von Verbundankern gegenüber Rissen ausreichend sichere Verankerungen zu gewährleisten, wurden in [2] Anwendungsregeln ausgearbeitet, die in die Neuzulassungen der Verbundanker (u. a. [3]) übernommen wurden. Dabei wurde davon ausgegangen, daß Verankerungen in der „Druckzone" von Beton- und Stahlbetonbauteilen liegen. Der Nachweis, daß eine Druckzone vorliegt, ist in jedem Einzelfall unter Berücksichtigung der durch die Verankerungen eingeleiteten Lasten für die Haupttragrichtung des als Ankergrund dienenden Bauteils zu führen.

Um sichere Befestigungen zu gewährleisten, muß die Wahrscheinlichkeit für das Auftreten von Rissen im Bereich der Verankerung sehr gering sein. Dies kann durch Begrenzung der in Bauteilquerrichtung auftretenden Betonzugspannung erreicht werden. Risse können jedoch nicht mit absoluter Sicherheit ausgeschlossen werden, da die Zugfestigkeit örtlich niedrig sein kann oder Zwängungskräfte auftreten können, die hohe Zugspannungen hervorrufen. Daher muß durch entsprechende Maßnahmen sichergestellt werden, daß auch bei in Extremfällen eventuell auftretenden Rissen die Sicherheit der Verankerung noch ausreichend hoch ist.

Aufgrund der Überlegungen in [2] kann eine Zugspannung entsprechend 50% der 5%-Fraktile der zentrischen Zugfestigkeit nach [4] dann als zulässig angesehen werden, wenn gleichzeitig durch zusätzliche konstruktive Maßnahmen gewährleistet ist, daß die Sicherheit der Befestigung bei eventuell auftretenden Rissen mindestens $\gamma = 2{,}0$ (gegenüber $\gamma = 3{,}0$ im ungerissenen Beton) beträgt.

Die Einhaltung der zulässigen Zugspannung erforderte eine bestimmte Mindestbauteildicke. Diese wurde in [2] für ungünstige Anwendungsfälle errechnet und ist in Bild 10 in Abhängigkeit von der Last pro Verankerungspunkt aufgetragen. Zum Vergleich sind die erforderlichen Bauteildicken nach der bisherigen Zulassung (u. a. [5]) eingezeichnet, die nur im Hinblick auf bohrtechnische Gesichtspunkte festgelegt wurden. Die sich theoretisch ergebende minimale Bauteildicke entspricht bei Einzelbefestigungen etwa den Werten der bisherigen Zulassung (Bild 10a). Da die Höhe der Zugspannung von der Höhe der auftretenden Last abhängt, muß bei Ankergruppen mit mehreren etwa gleich hoch beanspruchten Ankern die Bauteildicke auf die Gesamtlast der Befestigung bezogen werden. Demgegenüber war bisher der Einzelanker maßgebend. Daher sind bei Ankergruppen größere Bauteildicken als bisher erforderlich (Bild 10b) bzw. bei vorgegebener Bauteildicke ist die Höhe der zu verankernden Last zu beschränken. Diese Einschränkung ist auch unter Beachtung der Belange der Praxis vertretbar.

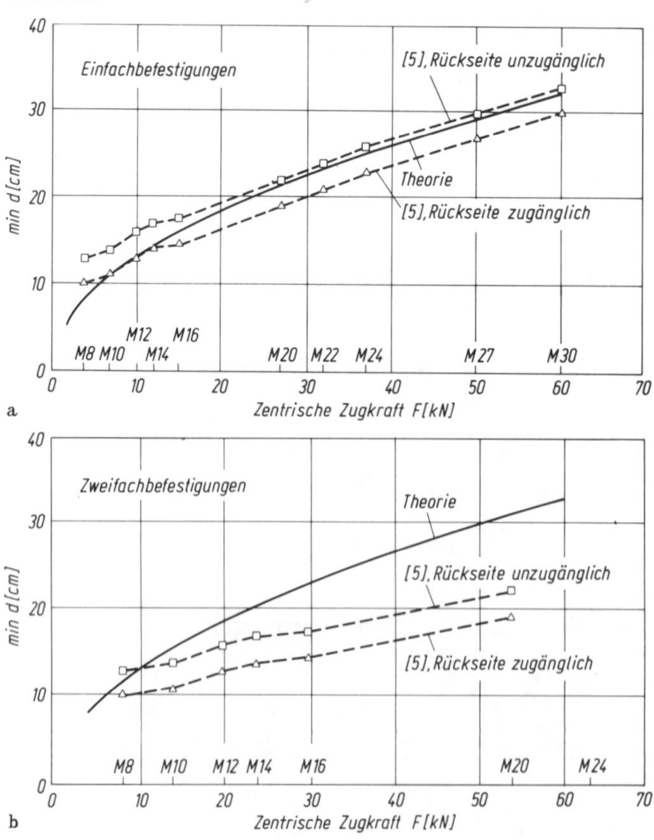

Bild 10
Minimale Bauteildicke in Abhängigkeit von der Last pro Verankerungspunkt (nach [2])

Die oben geforderte Sicherheit der Verankerung bei in Extremfällen auftretenden Rissen im Beton ($\gamma = 2{,}0$) kann z.B. durch Beschränkung der Rißbreiten auf $w \sim 0{,}1$ mm erreicht werden (vgl. Bild 9). Zur Einhaltung dieser Rißbreite ist je nach Stabdurchmesser ein Bewehrungsgehalt von $\mu = A_s/b \cdot h \sim 0{,}2\%$ bis 0,4% in beiden Achsrichtungen erforderlich [2]. Diese Bewehrungsgrade sind in vielen Fällen nicht vorhanden. Eine weitere Möglichkeit ist die Anordnung eines Teils der Verankerungslänge im ungerissenen Beton. Geht man davon aus, daß die Risse näherungsweise bis zur Querschnittsmitte reichen, die Anker bis nahe an die lastabgewandte Querschnittsseite geführt werden und in der Zugzone keine bzw. nur sehr geringe Verbundspannungen in der Fuge zwischen Kunstharzmörtel und Bohrlochwand wirken, beträgt die erforderliche Setztiefe nach [2] etwa das 1,5fache der z.B. in [5] geforderten Mindestwerte. Das Tiefersetzen der Anker um 50% gegenüber bisher ist ohne wesentliche Schwierigkeiten möglich.

Bei Dübeln, Kopfbolzen und Ankerschienen liegen insgesamt günstigere Verhältnisse als bei Verbundankern vor, weil bei diesen Befestigungselementen die Traglast geringer durch Risse beeinflußt wird als bei Verbundankern (vgl. Abschnitt 3).

Auch bei Anordnung von Verankerungen in der „Druckzone" bestehen also Wechselbeziehungen zwischen Befestigung und dem als Ankergrund dienenden Bauteil. Die mögliche ungünstige Beeinflussung der Traglast der Befestigung kann jedoch durch gezielte konstruktive Maßnahmen akzeptierbar gering gehalten werden. Allerdings schränken diese Maßnahmen die Anwendbarkeit der Befestigungselemente deutlich ein. Weiterhin ist dem Anwender die Lage der „Druckzone" im Bauteil meist nicht bekannt. Daher hat der zuständige Sachverständigenausschuß „Dübel und Ankerschienen" beschlossen, in Zukunft generell davon auszugehen, daß Befestigungen in der gerissenen Zugzone angeordnet sind.

3 Verankerungen in der Zugzone

3.1 Einfluß von Rissen im Ankergrund auf das Tragverhalten von Befestigungen

In der Zugzone von Stahlbetonbauteilen können Risse in einer Richtung bzw. z.B. in kreuzweise gespannten Platten in zwei Richtungen (Kreuzriß) auftreten. Dabei können die Befestigungselemente in Rissen bzw. neben ihnen angeordnet sein, wobei der erste Fall am ungünstigsten ist.

Zur Untersuchung des Einflusses von Rissen im Ankergrund auf das Tragverhalten von Befestigungselementen werden unterschiedliche Probekörper, nämlich Biegeplatten, Zugglieder (Bild 11) oder spezielle Körper zur Erzeugung von Kreuzrissen (Bild 12) benutzt. Üblicherweise werden zunächst Haarrisse erzeugt, die Verankerungselemente in oder neben diese gesetzt und die Risse durch Belastung der Probekörper auf die vorgesehene Breite aufgeweitet. Anschließend werden die Befestigungen bei geöffnetem Riß monoton bis zum Bruch belastet. Häufig werden auch eine schwellende Belastung der Verankerungen bei geöffnetem Riß oder mehrmaliges Öffnen und Schließen der Risse bei belasteter Verankerung zwischengeschaltet. Während eine schwellende Belastung der Befestigung als Zeitrafferversuch für eine Dauerlast angesehen werden kann, dient das Öffnen und Schließen der Risse zur

Bild 11
Prüfung von Dübeln in einem Dehnkörper (entnommen aus [8])

Bild 12
Prüfung von Dübeln in Kreuzrissen (entnommen aus [8])

Nachahmung des Einflusses einer veränderlichen Belastung des Bauteils.

Das Last-Verschiebungsverhalten und die Bruchlast werden durch den verwendeten Prüfkörper und die angewandte Prüfmethode zum Teil wesentlich beeinflußt. Die ungünstigsten Ergebnisse sind nach den Auswertungen [6], [7] bei Verwendung von Kreuzrißkörpern in Verbindung mit einem Öffnen und Schließen der Risse zu erwarten.

Befestigungselemente, die in der Zugzone angeordnet werden sollen, müssen für diesen Anwendungsfall geeignet sein. Dies bedeutet, daß sie auch in ungünstigen Anwendungsfällen – z. B. Lage in Kreuzrissen und veränderliche Belastung des Bauteils – ihre Funktion auf Dauer erfüllen müssen. Unter diesem Gesichtspunkt ist der Einsatz von Verbundankern und wegkontrolliert spreizenden Einschlagankern – zumindest bei den bisher üblichen Setztiefen – kritisch zu beurteilen. Demgegenüber sind sehr gut nachspreizende und den Beton hinterschneidende Dübel sowie Kopfbolzen und Ankerschienen im allgemeinen als geeignet anzusehen.

Ein Vorschlag für die Prüfung der Eignung von Dübeln, die für Anwendungen in der Zugzone vorgesehen sind, ist in [7] enthalten.

Risse im Ankergrund vermindern die Tragfähigkeit von Befestigungselementen (Bilder 13 und 14). Aufgetragen ist jeweils die Bruchlast im Riß bezogen auf den im ungerissenen Beton zu erwartenden Wert in Abhängigkeit von der Rißbreite im Verankerungsbereich. Bild 13 gilt für geeignet nachspreizende Metalldübel und Bild 14 für Hinterschnittanker und Kopfbolzen. Die Bruchlast im ungerissenen Beton wurde nach der in [9] angegebenen empirischen Gleichung berechnet, wobei die rechnerischen Bruchlasten sowohl für Dübel als auch für Hinterschnittanker und Kopfbolzen sehr gut mit den in Vergleichsversuchen gemessenen Werten übereinstimmen. Als Versuchskörper dienten Zugglieder mit einer näherungsweise konstanten Rißbreite über die Körperdicke. Die Befestigungselemente mit einer Verankerungstiefe von ca. 80 mm wurden nach dem Öffnen des Risses durch eine monoton ansteigende zentrische Zugkraft bis zum Ver-

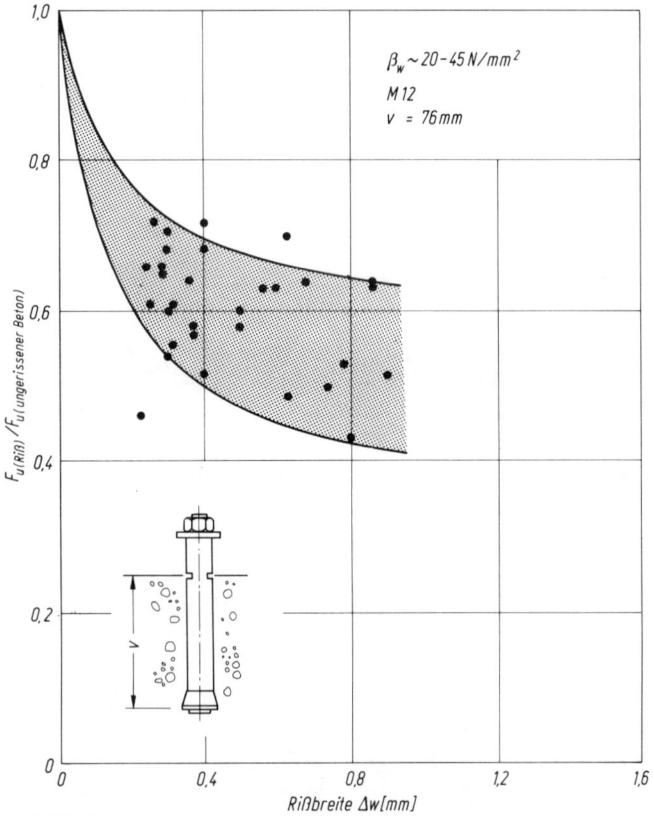

Bild 13
Verhältnis der Bruchlast von in Rissen angeordneten nachspreizenden Metalldübeln zur Bruchlast im ungerissenen Beton in Abhängigkeit von der Rißbreite

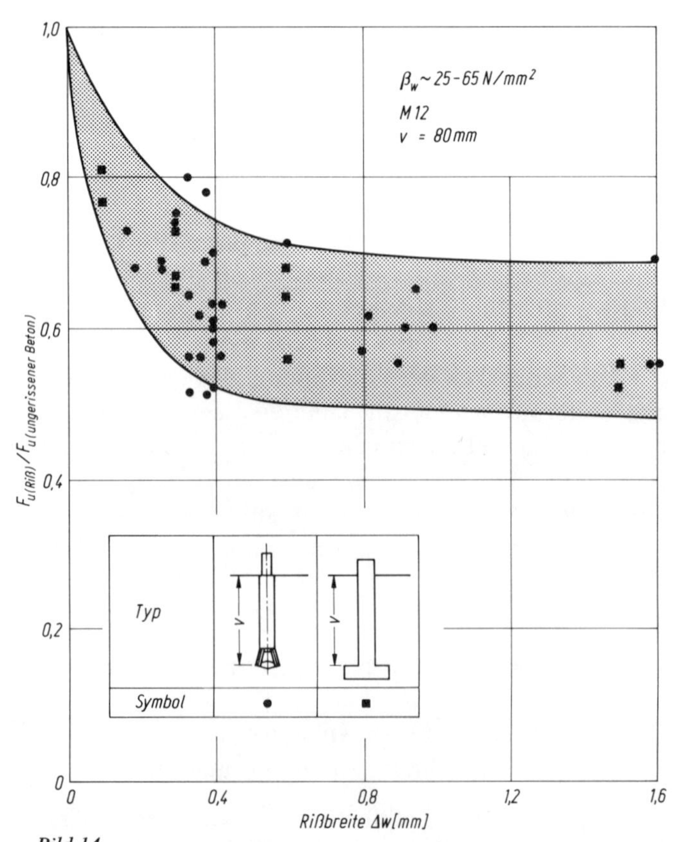

Bild 14
Verhältnis der Bruchlast von in Rissen angeordneten Hinterschnittankern und Kopfbolzen zur Bruchlast im ungerissenen Beton in Abhängigkeit von der Rißbreite

sagen belastet, das im allgemeinen durch kegelförmigen Betonausbruch hervorgerufen wurde.
Die Bruchlast von Befestigungselementen im Riß nimmt bis zu einer Rißbreite von etwa 0,4 mm schnell und bei breiteren Rissen langsam ab. Sie beträgt bei der im Stahlbetonbau maximal als zulässig angesehenen Rißbreite von 0,4 mm das ca. 0,5- bis 0,7fache des für ungerissenen Beton geltenden Wertes. Dabei ist kein wesentlicher Einfluß der unterschiedlichen Befestigungssysteme (nachspreizende Dübel, Hinterschnittanker oder Kopfbolzen) zu erkennen. Während die Bruchlast von Metallspreizdübeln bei größeren Rißbreiten weiter absinkt, bleibt sie bei Hinterschnittankern und Kopfbolzen bis zu Rißbreiten von ca. 1,5 mm nahezu konstant.
Die niedrigere Bruchlast von in Rissen verankerten Befestigungselementen gegenüber dem für ungerissenen Beton geltenden Wert ist wesentlich auf die Störung des Spannungszustandes im Beton, der bei ungerissenem Ankergrund rotationssymmetrisch ist, durch Risse zurückzuführen. Dadurch ist die Fläche mit hohen Zugspannungen im Beton kleiner als bei Befestigungen im ungerissenen Ankergrund. Weiterhin können benachbarte Risse einen Teil des möglichen Ausbruchkegels abschneiden. Bei Metallspreizdübeln wird zusätzlich die Spreizkraft durch Risse im Beton verringert [9], so daß die Dübel in manchen Fällen keinen Betonausbruch erzeugen, sondern durch Herausziehen versagen. Die letztere Bruchart tritt in der Regel bei Einschlagankern (siehe Bild 1) und bei kraftkontrolliert spreizenden Dübeln mit nicht ausreichendem Nachspreizverhalten auf. In beiden Fällen ist mit einer wesentlich geringeren relativen Tragfähigkeit als nach Bild 13 zu rechnen.

3.2 Einfluß von Spannungen im Bauteil auf die Traglast von Befestigungen

In Stahlbetontragwerken entstehen bei Belastung nicht nur Risse, sondern es werden auch Zugspannungen im Beton geweckt. Hohe Zugspannungen entstehen zum Beispiel im Bereich von Verankerungen und Übergreifungsstößen von Bewehrungsstäben und im Querkraftbereich. Der Einfluß dieser Spannungen auf die Tragkraft von Befestigungen wurde in [10] theoretisch und experimentell untersucht.
In Bild 15 wird angenommen, daß ein Dübel mit einer Setztiefe von 80 mm im Endbereich eines Übergreifungsstoßes dicker Rippenstäbe angeordnet ist. Dargestellt ist die Verteilung der von den Bewehrungsstäben bzw. dem zugbeanspruchten Dübel hervorgerufenen Spannungen im Beton entlang der Oberfläche des Ausbruchkegels. Sie wurden unter Annahme bestimmter Näherungen berechnet. Dabei wurde vorausgesetzt, daß die Bewehrungsstäbe etwa mit der zulässigen Stahlspannung und der Dübel mit der Ausbruchlast beansprucht sind. Die von den gestoßenen Bewehrungsstäben bzw. vom Dübel hervorgerufenen Spannungen überlagern sich teilweise, und es ergeben sich entlang eines Teiles der Oberfläche des Ausbruchkegels höhere Zugspannungen als bei Anordnung des Dübels im ansonsten unbelasteten Beton. Daher ist mit einer Abminderung der Dübelbruchlast zu rechnen. Sie beträgt im vorliegenden Fall theoretisch etwa 25%.
Vergrößert man die Verankerungstiefe oder ordnet man Dübel im Stoßbereich dünner Stäbe an, ergibt sich eine geringere Überlagerung der Spannungen. Beispielsweise ist in dem in Bild 16 dargestellten Fall – 130 mm tiefer Dübel im Bereich eines Stoßes von 14 mm Stäben – die Beeinflussung der Dübeltragkraft vernachlässigbar gering. Ebenso ist mit einer geringe-

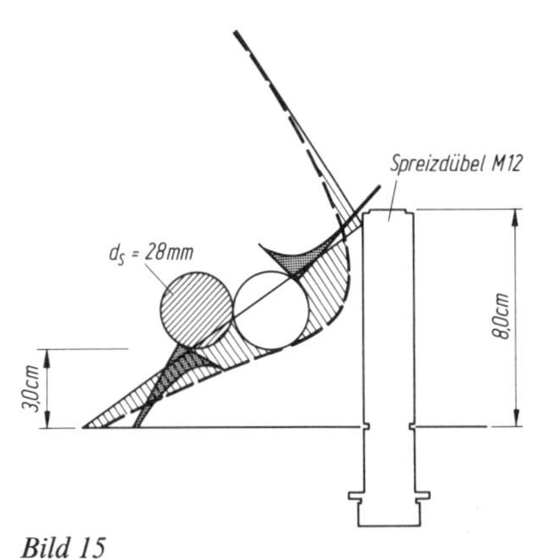

Bild 15
Dübel im Bereich eines Übergreifungsstoßes – Überlagerung der Spannungen (entnommen aus [10])

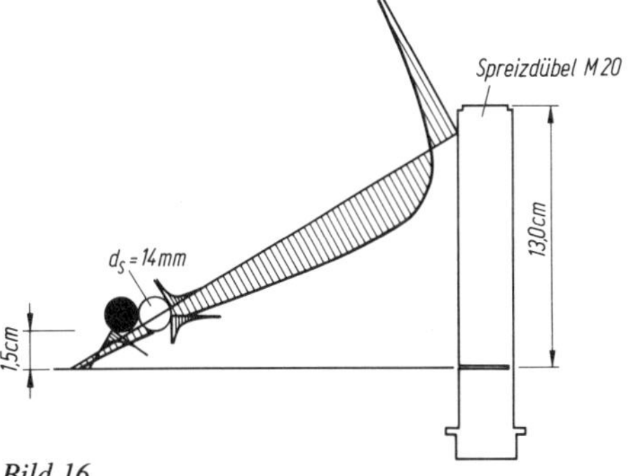

Bild 16
Dübel im Bereich eines Übergreifungsstoßes – Überlagerung der Spannungen (entnommen aus [10])

ren Abminderung der Dübeltragkraft zu rechnen, wenn der Dübel unterhalb der Bewehrung in der Betondeckung verankert ist.
Bei Dübeln im Bereich von Übergreifungsstößen geschweißter Betonstahlmatten sind ähnliche Abminderungen der Bruchlast wie oben angegeben zu erwarten. Die Tragkraft von im Schubbereich angeordneten Dübeln kann bis zu ca. 25% niedriger sein als der Wert, der sich ergibt, wenn der Beton außer den Verankerungslasten keine anderen Lasten zu tragen hat.
Bei anderen Befestigungssystemen (z. B. Kopfbolzen) ist unter sonst gleichen Verhältnissen etwa die gleiche Abminderung der Bruchlast wie bei Dübeln zu erwarten.
Zur Prüfung der Richtigkeit der theoretischen Überlegungen wurden Versuche mit plattenartigen Probekörpern durchgeführt. Die Bewehrung (Platte A: Rippenstäbe mit d_s = 28 mm, Platte B: geschweißte Betonstahlmatten) war im Bereich des konstanten Momentes durch Übergreifung nach DIN 1045 (Vollstoß) gestoßen (Bild 17). Nach Erzeugen von Rissen im

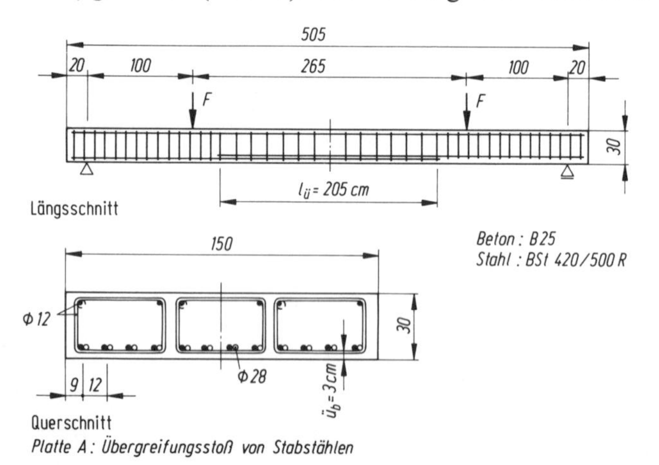

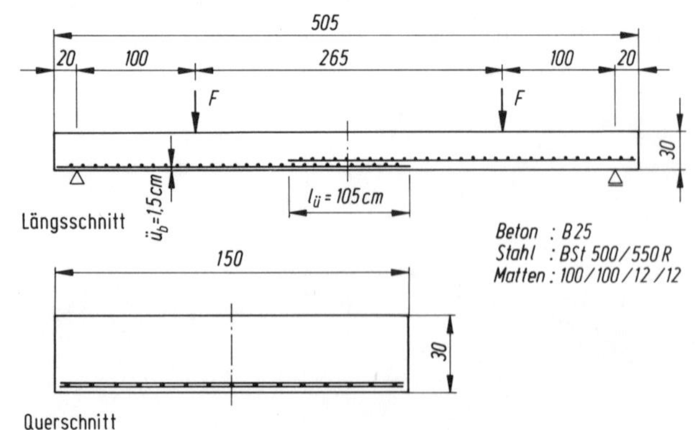

Bild 17
Ausbildung der Probekörper (entnommen aus [10])

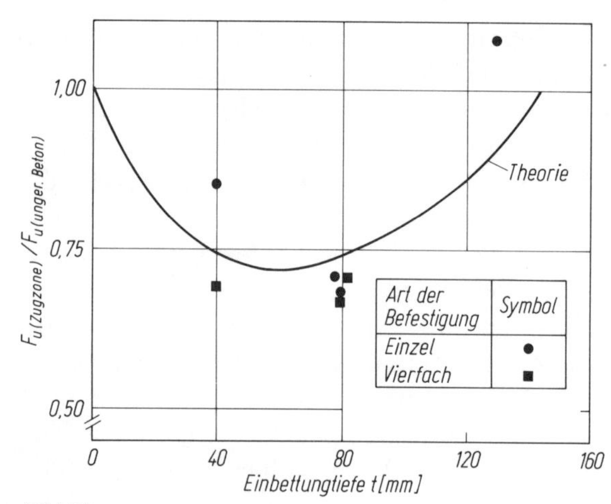

Bild 18
Dübel im Bereich von Übergreifungsstößen – Versuchsergebnisse (nach [10])

Beton wurden die Dübel zwischen diese Risse gesetzt. Dadurch sollte der Einfluß der Risse auf die Dübeltragkraft möglichst gering gehalten werden. Anschließend wurden die Platte durch äußere Lasten bis zur Gebrauchslast beansprucht und die Befestigungen bei belastetem Probekörper ausgezogen. Zum Vergleich wurden auch Befestigungen im ungerissenen und unbelasteten Beton geprüft.
Die wichtigsten Versuchsergebnisse sind aus Bild 18 zu ersehen. Aufgetragen ist das Verhältnis der experimentell gefundenen Bruchlast von im Stoßendbereich angeordneten Dübelbefestigungen zur Bruchlast von identischen Befestigungen im ungerissenen und unbelasteten Beton in Abhängigkeit von der Setztiefe. Zum Vergleich ist die theoretisch zu erwartende relative Tragkraft mit eingetragen. Die Versuchsergebnisse bestätigen im wesentlichen die theoretischen Überlegungen.
Die Versuche zeigen, daß die Bruchlast von Befestigungen i. a. durch Risse im Beton stärker abgemindert wird als durch die Überlagerung der von Befestigungselementen hervorgerufenen Zugspannungen mit denjenigen aus der Tragwerkswirkung. In bestimmten Anwendungsfällen können jedoch beide Einflüsse ungünstig zusammenwirken.

3.3 Einfluß der von Befestigungen hervorgerufenen Spannungen auf die Tragfähigkeit von Stahlbetonbauteilen

Durch die in Abschnitt 3.2 beschriebene Überlagerung der Zugspannungen kann nicht nur die Tragfähigkeit der Verankerung, sondern auch diejenige des als Ankergrund dienenden Stahlbetonbauteils reduziert werden. Allerdings sind die mit der Beeinflussung des Tragverhaltens von Stahlbetonbauteilen durch Befestigungen zusammenhängenden Probleme erst in Ansätzen geklärt.

Man kann davon ausgehen, daß bei Verankerung von relativ geringen Lasten (z. B. 5 bis 6 kN) in großen Abständen das Bauteiltragverhalten nicht wesentlich beeinträchtigt wird, da die örtliche Beanspruchung des Betons gering ist und nur kleine Bereiche des Bauteils betroffen sind. Werden dagegen die von Befestigungselementen maximal übertragbaren (hohen) Lasten in engen Abständen in den Beton eingeleitet, ist eine signifikante Reduzierung der Bauteiltragfähigkeit zu befürchten [11].

Kritisch ist z. B. die Einleitung von hohen Lasten in Bereichen mit Verankerungen und Übergreifungsstößen von Bewehrungen. Eine andere kritische Anwendung ist die Verankerung dieser Lasten im Schubbereich von auf Schub unbewehrten Platten. Dies gilt insbesondere, wenn die Platte – wie im Hochbau vielfach üblich – aus Fertigplatten und einem am Ort gegossenen Aufbeton besteht. Verbindet man Fertig- und Ortbeton nicht durch eine Verbundbewehrung

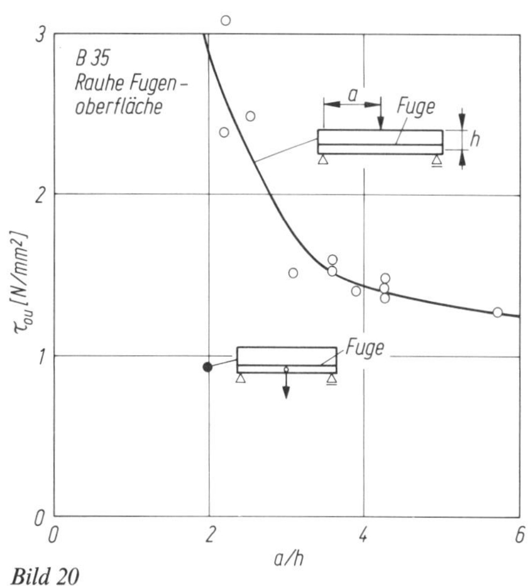

Bild 20
Schubspannung beim Versagen der Fuge zwischen Fertigteil- und Ortbeton in Abhängigkeit von der Schubschlankheit (entnommen aus [11])

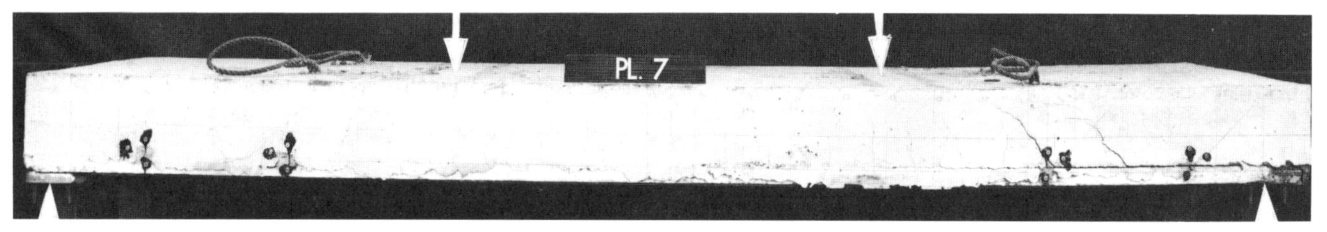

a) Gesamtansicht

b) Detail

Bild 19
Bruch einer aus Fertigteil- und Ortbeton zusammengesetzten Platte ohne Verbundbewehrung (entnommen aus [13])

miteinander, kann der Bruch der Platte durch Überschreiten der Verbundfestigkeit der Fuge erfolgen (Bild 19). Im Regelfall wird die Platte von oben belastet. Dann ist die Fugentragfähigkeit bei rauher Fugenoberfläche so hoch (Bild 20), daß in vielen Fällen auf eine Verbundbewehrung verzichtet werden kann. Entsprechende Vorschläge zur Änderung der DIN 1045 liegen vor. Wird dagegen die Last von unten, z. B. durch Dübel oder Kopfbolzen, in den Fertigbeton eingeleitet, ist die Fugentragfähigkeit wegen der zusätzlichen Beanspruchung der Fuge auf Zug relativ niedrig (Bild 20), so daß immer eine Verbundbewehrung erforderlich ist.

4 Zusammenfassung und Schlußfolgerungen

Zwischen Befestigungstechnik und Stahlbetonbauweise bestehen vielfältige Wechselbeziehungen, die nicht vernachlässigt werden dürfen. Sie können sowohl das Tragverhalten von Befestigungen als auch dasjenige der als Ankergrund dienenden Stahlbetonbauteile ungünstig beeinflussen.

Bisher wurden die Richtlinien für die Bemessung und Ausbildung von Stahlbetonbauwerken ohne Berücksichtigung des möglichen Einflusses der speziellen Lasteinleitung durch Befestigungselemente erarbeitet. Daher war es erforderlich, die Bemessungsregeln für Befestigungen so festzulegen, daß keine wesentliche Beeinflussung des Bauteiltragverhaltens zu erwarten war. Um dies zu gewährleisten, dürfen bisher Verankerungen nur in der aus Lastspannungen erzeugten Druckzone von Stahlbetonbauteilen angeordnet bzw. nur geringe Lasten in die Zugzone eingeleitet werden. Diese Regelung ist praxisfremd und schränkt die Möglichkeiten der modernen Befestigungstechnik wesentlich ein. Daher wird angestrebt, hohe Lasten in die Zugzone einzuleiten. Dies ist bei Verwendung geeigneter Befestigungselemente prinzipiell möglich. Nach den bisherigen Ergebnissen beträgt die Bruchlast von Verankerungen in der Zugzone bei üblichen Rißbreiten unabhängig von der Art des (geeigneten) Befestigungselementes ca. 50% bis 60% des Wertes, der für ungerissenen und ansonsten unbelasteten Beton gilt. Daher kann man z. B. durch eine aus 4 Ankern M 20 (Verankerungstiefe 130 mm) bestehende Ankergruppe bis zu ca. 60 kN in die Betonzugzone einleiten. Bevor man jedoch die Verankerung dieser hohen Lasten auf relativ geringer Tiefe zuläßt, sollte der Einfluß der speziellen Lasteinleitung durch Befestigungselemente auf das Tragverhalten von Stahlbetonbauteilen besser als bisher bekannt sein, um ggf. durch entsprechende konstruktive Maßnahmen eine sonst mögliche Gefährdung der Bauteilsicherheit auszuschließen.

5 Literatur

[1] Schreiben der Landesstelle für Baustatik, Tübingen, an das Institut für Bautechnik, Berlin, vom 25. 11. 81 und 8. 12. 81.

[2] ELIGEHAUSEN, R. und CLAUSNITZER, W.: Befestigungen mit Verbundankern in der aus Lastspannungen erzeugten Druckzone von Beton- und Stahlbetonbauteilen, Bericht Nr. 2/5-83/15 des Instituts für Werkstoffe im Bauwesen, Universität Stuttgart, Dez. 1983.

[3] Zulassungsbescheid Nr. Z-21.3-80 vom 1. 2. 84 des Instituts für Bautechnik, Berlin, für Hilti-Verbundanker HVA.

[4] RÜSCH, H.: Die Ableitung der charakteristischen Werte der Betonzugfestigkeit, beton 2, 1975.

[5] Zulassungsbescheid Nr. Z-21.3-15 vom 22. 12. 1975 des Instituts für Bautechnik, Berlin, für Upat-Verbundanker UKA 3.

[6] LEHMANN, R.: Zum Tragverhalten von Metallspreizdübeln in der aus Lastspannungen erzeugten Zugzone des Betons; Dissertation an der Universität Stuttgart, in Vorbereitung.

[7] ELIGEHAUSEN, R. und LEHMANN, R.: Verankerungen mit Metallspreizdübeln in der aus Lastspannungen erzeugten Zugzone von Stahlbetonbauteilen – Einflüsse auf das Tragverhalten und Vorschlag für Zulassungsversuche, Bericht Nr. 1/4-84/1 des Instituts für Werkstoffe im Bauwesen, Universität Stuttgart, Januar 1984.

[8] REHM, G. und LEHMANN, R.: Untersuchungen mit Metallspreizdübeln in der gerissenen Zugzone von Stahlbetonbauteilen, Bericht der Forschungs- und Materialprüfungsanstalt Baden-Württemberg – Otto-Graf-Institut –, Stuttgart, Juli 1982.

[9] ELIGEHAUSEN, R. und PUSILL-WACHTSMUTH, P.: Stand der Befestigungstechnik im Stahlbetonbau, IVBH-Bericht S 19/82, IVBH-Periodica 1/82, Febr. 1982.

[10] ELIGEHAUSEN, R. und SILVA, J.: Dübel in der aus Lastspannungen erzeugten Zugzone von Stahlbetonbauteilen – Theoretische und experimentelle Untersuchungen, Abschlußbericht in Vorbereitung.

[11] REHM, G. und ELIGEHAUSEN, R.: Auswirkungen der modernen Befestigungstechnik auf die konstruktive Gestaltung im Stahlbetonbau, Betonwerk + Fertigteil-Technik, Heft 6, 1984.

[12] ELIGEHAUSEN, R.: Übergreifungsstöße zugbeanspruchter Rippenstäbe mit geraden Stabenden, Schriftenreihe des DAfStb, Heft 301, Berlin, 1979.

[13] REHM, G., ELIGEHAUSEN, R. und PAUL, F.: Verbundbewehrung in Fugen von Platten ohne Schubbewehrung, Bericht des Instituts für Werkstoffe im Bauwesen, Universität Stuttgart, Januar 1980.

[14] ELIGEHAUSEN, R., MALLÉE, R. und REHM, G.: Verankerungen mit Verbundankern, erscheint demnächst in Betonwerk + Fertigteil-Technik.

Autorenverzeichnis

A. W. Beeby, BSC, PhD
Cement and Concrete Association
Wexham Springs/Großbritannien

Ing. A. Bragard
Centre de Recherches Métallurgiques
Liège/Belgien

Dr.-Ing. D. Briesemann
Hebel Emmering GmbH & Co
Emmering/Fürstenfeldbruck

Prof. Ir. Dr.-Ing. E. h. A. S. G. Bruggeling
Technische Hogeschool Delft
Afdeling der Civiele Techniek
Delft/Niederlande

Ing. J. Defourny
Centre de Recherches Métallurgiques
Liège/Belgien

Dipl.-Ing. W. Dening
Vorstandsmitglied der Badischen Stahlwerke AG
Kehl/Rhein

Prof. Dr.-Ing. J. Eibl
Universität Karlsruhe
Lehrstuhl für Massivbau
Karlsruhe

Prof. Dr.-Ing. R. Eligehausen
Universität Stuttgart
Institut für Werkstoffe im Bauwesen
Stuttgart

Dipl.-Ing. W. Fastenau
Ed. Züblin AG
Stuttgart

Senator E. h. Dr. phil. h. c. A. Fischer
Fischer-Forschung
Tumlingen/Waldachtal

Prof. Dr.-Ing. L. Franke
Technische Universität Hamburg-Harburg
Arbeitsbereich Bauphysik und Werkstoffe
im Bauwesen
Hamburg-Harburg

Dipl.-Ing. H. R. Ganz
Eidgenössische Technische Hochschule Zürich
Institut für Baustatik und Konstruktion
Zürich/Schweiz

Dr.-Ing. E. h. H. Goffin
Ltd. Ministerialrat
Vorsitzender des Deutschen Ausschusses
für Stahlbeton
Berlin

Dipl.-Ing. M. Günter
Universität Karlsruhe
Institut für Massivbau und Baustofftechnologie
Karlsruhe

Dipl.-Ing. F. Herkommer
Direktor der Baustahlgewebe GmbH
Düsseldorf

Prof. Dr.-Ing. H. K. Hilsdorf
Universität Karlsruhe
Institut für Massivbau und Baustofftechnologie
Karlsruhe

Dr.-Ing. D. Jungwirth
Direktor der Entwicklungsabteilung
der Dyckerhoff & Widmann AG
München

Dr.-Ing. H. P. KILLING
Geschäftsführendes Vorstandsmitglied
der Walzstahlvereinigung
Düsseldorf

Dipl.-Ing. J. KOBARG
Universität Karlsruhe
Lehrstuhl für Massivbau
Karlsruhe

Prof. Dr.-Ing. G. KÖNIG
Technische Hochschule Darmstadt
Institut für Massivbau
Darmstadt

o. Prof. Dr.-Ing. Dr.-Ing. E.h. K. KORDINA
Technische Universität Braunschweig
Institut für Baustoffe, Massivbau und Brandschutz
Braunschweig

Dipl.-Ing. A. KRIPS
Technische Hochschule Darmstadt
Institut für Massivbau
Darmstadt

Dr.-Ing. J. KROPP
Universität Karlsruhe
Institut für Massivbau und Baustofftechnologie
Karlsruhe

o. Prof. Dr.-Ing. H. KUPFER
Technische Universität München
Lehrstuhl für Massivbau
München

Dipl.-Ing. H. MALONN
Technische Universität Braunschweig
Lehrstuhl für Baukonstruktion und Vorfertigung
Braunschweig

Dipl.-Ing. R. MANG
Technische Universität München
Lehrstuhl für Massivbau
München

Dr.-Ing. H. MARTIN
Institut für Betonstahl und Stahlbetonbau e.V.
München

Prof. Dr.-Ing. C. MENN
Eidgenössische Hochschule Zürich
Zürich/Schweiz

Dipl.-Ing. W. MENZ
Universität Stuttgart
Institut für Massivbau
Stuttgart

Dipl.-Ing. B. NEUBERT
Forschungs- und Materialprüfungsanstalt Baden-Württemberg (Otto-Graf-Institut)
Stuttgart

Dr.-Ing. U. NÜRNBERGER
Forschungs- und Materialprüfungsanstalt Baden-Württemberg (Otto-Graf-Institut)
Stuttgart

o. Prof. em. Dr.-Ing. H. PASCHEN
Technische Universität Braunschweig
Institut für Baukonstruktion und Vorfertigung
Braunschweig

Dipl.-Ing. L. PREIS
Strabag Bau-AG
Köln-Deutz

Prof. Dr.-Ing. U. QUAST
Technische Universität Braunschweig
Institut für Baustoffe, Massivbau und Brandschutz
Braunschweig

Prof. Dr.-Ing. H.W. REINHARDT
Technische Hogeschool Delft
Afdeling der Civiele Techniek
Delft/Niederlande

Prof. Dr.-Ing. F.S. ROSTÁSY
Technische Universität Braunschweig
Institut für Baustoffe, Massivbau und Brandschutz
Braunschweig

Dr.-Ing. D. RUSSWURM
Prüfstelle für Betonstahl
München

Prof. Dr.-Ing. J. SCHLAICH
Universität Stuttgart
Institut für Massivbau
Stuttgart

Dr. H. D. SEGHEZZI
Vorstandsmitglied Forschung und Entwicklung der HILTI AG
Schaan/Fürstentum Liechtenstein

Dr. M. STOCKER
Karl Bauer Spezialtiefbau GmbH & Co KG
Schrobenhausen

Prof. Dr.-Ing. Dr.-Ing. E. h. B. THÜRLIMANN
Eidgenössische Technische Hochschule Zürich
Institut für Baustatik und Konstruktion
Zürich/Schweiz

Dr.-Ing. O. WAGNER
Ministerialrat
Bayerisches Staatsministerium des Inneren
München

Prof. Dr.-Ing. N. V. WAUBKE
Universität Dortmund
Lehrstuhl für Werkstoffe des Bauwesens
Dortmund

Dipl.-Ing. M. WEISER
Strabag Bau-AG
Köln-Deutz

Dipl.-Ing. H. WEITZMANN
Sprecher des Vorstandes der
Badischen Stahlwerke AG
Kehl/Rhein

Prof. Dr.-Ing. G. WISCHERS
Direktor des Forschungsinstitutes
der Zementindustrie
Düsseldorf

Verzeichnis der Förderer

Die Herausgabe der Festschrift wurde durch
großzügige Spenden folgender Firmen ermöglicht

Arbed, Luxembourg
Artur Fischer Forschung, Tumlingen
Badische Stahlwerke AG, Kehl
Baustahlgewebe GmbH, Düsseldorf
Betonstahlgemeinschaft Deutscher Hüttenwerke, Düsseldorf
Bilfinger & Berger Bau-AG, Mannheim
Centre De Recherches Métallurgiques, Lüttich
Dyckerhoff & Widmann AG, München
Ed. Züblin AG, Stuttgart
Hilti AG, Liechtenstein
Hochtief AG, Essen
Isar-Baustahl GmbH, Dinkelscherben
Karl Bauer, Schrobenhausen
Philipp Holzmann AG, Frankfurt
Strabag Bau-AG, Köln
Tempcore Stahl Deutschland GmbH, Düsseldorf
Wayss & Freytag AG, Frankfurt